Advanced
Engineering
Mathematics

Fourth Edition

Dennis G. Zill   Warren S. Wright

Warren S. Wright
Loyola Marymount University

Carol D. Wright

D1205423

JONES & BARTLETT
LEARNING

World Headquarters

Jones & Bartlett Learning
40 Tall Pine Drive
Sudbury, MA 01776
978-443-5000
info@jblearning.com
www.jblearning.com

Jones & Bartlett Learning
Canada
6339 Ormindale Way
Mississauga, Ontario L5V 1J2
Canada

Jones & Bartlett Learning
International
Barb House, Barb Mews
London W6 7PA
United Kingdom

Jones & Bartlett Learning books and products are available through most bookstores and online booksellers. To contact Jones & Bartlett Learning directly, call 800-832-0034, fax 978-443-8000, or visit our website, www.jblearning.com.

Substantial discounts on bulk quantities of Jones & Bartlett Learning publications are available to corporations, professional associations, and other qualified organizations. For details and specific discount information, contact the special sales department at Jones & Bartlett Learning via the above contact information or send an email to specialsales@jblearning.com.

Copyright © 2011 by Jones & Bartlett Learning, LLC

ISBN-13: 978-0-7637-9075-2
ISBN-10: 0-7637-9075-2

All rights reserved. No part of the material protected by this copyright may be reproduced or utilized in any form, electronic or mechanical, including photocopying, recording, or by any information storage and retrieval system, without written permission from the copyright owner.

Production Credits
Senior Acquisitions Editor: Timothy Anderson
Editorial Assistant: Melissa Potter
Production Director: Amy Rose
Production Assistant: Lindsey Jones
V.P., Manufacturing and Inventory Control: Therese Connell
Senior Marketing Manager: Andrea DeFronzo
Cover Design: Kate Ternullo
Cover Image: © Stephen Strathdee/ShutterStock, Inc.
Printing and Binding: Courier Stoughton
Cover Printing: Courier Stoughton

6048
Printed in the United States of America
14 13 12 11 10   10 9 8 7 6 5 4 3 2 1

Table of Contents

IV Fourier Series and Partial Differential Equations

V Complex Analysis

Appendices

1 Introduction to Differential Equations

EXERCISES 1.1
Definitions and Terminology

3. Fourth order; linear

6. Second order; nonlinear because of R^2

9. Writing the differential equation in the form $x(dy/dx) + y^2 = 1$, we see that it is nonlinear in y because of y^2. However, writing it in the form $(y^2 - 1)(dx/dy) + x = 0$, we see that it is linear in x.

12. From $y = \frac{6}{5} - \frac{6}{5}e^{-20t}$ we obtain $dy/dt = 24e^{-20t}$, so that

$$\frac{dy}{dt} + 20y = 24e^{-20t} + 20\left(\frac{6}{5} - \frac{6}{5}e^{-20t}\right) = 24.$$

15. The domain of the function, found by solving $x + 2 \geq 0$, is $[-2, \infty)$. From $y' = 1 + 2(x+2)^{-1/2}$ we have

$$(y - x)y' = (y - x)[1 + (2(x+2)^{-1/2}]$$
$$= y - x + 2(y - x)(x + 2)^{-1/2}$$
$$= y - x + 2[x + 4(x+2)^{1/2} - x](x + 2)^{-1/2}$$
$$= y - x + 8(x+2)^{1/2}(x+2)^{-1/2} = y - x + 8.$$

An interval of definition for the solution of the differential equation is $(-2, \infty)$ because y' is not defined at $x = -2$.

18. The function is $y = 1/\sqrt{1 - \sin x}$, whose domain is obtained from $1 - \sin x \neq 0$ or $\sin x \neq 1$. Thus, the domain is $\{x \mid x \neq \pi/2 + 2n\pi\}$. From $y' = -\frac{1}{2}(1 - \sin x)^{-3/2}(-\cos x)$ we have

$$2y' = (1 - \sin x)^{-3/2}\cos x = [(1 - \sin x)^{-1/2}]^3 \cos x = y^3 \cos x.$$

An interval of definition for the solution of the differential equation is $(\pi/2, 5\pi/2)$. Another one is $(5\pi/2, 9\pi/2)$, and so on.

21. Differentiating $P = c_1 e^t / (1 + c_1 e^t)$ we obtain

$$\frac{dP}{dt} = \frac{(1 + c_1 e^t)\, c_1 e^t - c_1 e^t \cdot c_1 e^t}{(1 + c_1 e^t)^2} = \frac{c_1 e^t}{1 + c_1 e^t}\,\frac{[(1 + c_1 e^t) - c_1 e^t]}{1 + c_1 e^t}$$

$$= \frac{c_1 e^t}{1 + c_1 e^t}\left[1 - \frac{c_1 e^t}{1 + c_1 e^t}\right] = P(1 - P).$$

1

24. From $y = c_1 x^{-1} + c_2 x + c_3 x \ln x + 4x^2$ we obtain

$$\frac{dy}{dx} = -c_1 x^{-2} + c_2 + c_3 + c_3 \ln x + 8x,$$

$$\frac{d^2 y}{dx^2} = 2c_1 x^{-3} + c_3 x^{-1} + 8,$$

and

$$\frac{d^3 y}{dx^3} = -6c_1 x^{-4} - c_3 x^{-2},$$

so that

$$x^3 \frac{d^3 y}{dx^3} + 2x^2 \frac{d^2 y}{dx^2} - x \frac{dy}{dx} + y = (-6c_1 + 4c_1 + c_1 + c_1)x^{-1} + (-c_3 + 2c_3 - c_2 - c_3 + c_2)x$$

$$+ (-c_3 + c_3)x \ln x + (16 - 8 + 4)x^2$$

$$= 12x^2.$$

27. (a) From $y = e^{mx}$ we obtain $y' = me^{mx}$. Then $y' + 2y = 0$ implies

$$me^{mx} + 2e^{mx} = (m + 2)e^{mx} = 0.$$

Since $e^{mx} > 0$ for all x, $m = -2$. Thus $y = e^{-2x}$ is a solution.

(b) From $y = e^{mx}$ we obtain $y' = me^{mx}$ and $y'' = m^2 e^{mx}$. Then $y'' - 5y' + 6y = 0$ implies

$$m^2 e^{mx} - 5me^{mx} + 6e^{mx} = (m - 2)(m - 3)e^{mx} = 0.$$

Since $e^{mx} > 0$ for all x, $m = 2$ and $m = 3$. Thus $y = e^{2x}$ and $y = e^{3x}$ are solutions.

30. In this problem we substitute $y = c$ into the differential equations and use $y' = 0$ and $y'' = 0$.
Solving $c^2 + 2c - 3 = (c + 3)(c - 1) = 0$ we see that $y = -3$ and $y = 1$ are constant solutions.

33. From $x = e^{-2t} + 3e^{6t}$ and $y = -e^{-2t} + 5e^{6t}$ we obtain

$$\frac{dx}{dt} = -2e^{-2t} + 18e^{6t} \quad \text{and} \quad \frac{dy}{dt} = 2e^{-2t} + 30e^{6t}.$$

Then

$$x + 3y = (e^{-2t} + 3e^{6t}) + 3(-e^{-2t} + 5e^{6t}) = -2e^{-2t} + 18e^{6t} = \frac{dx}{dt}$$

and

$$5x + 3y = 5(e^{-2t} + 3e^{6t}) + 3(-e^{-2t} + 5e^{6t}) = 2e^{-2t} + 30e^{6t} = \frac{dy}{dt}.$$

36. The only solution of $(y')^2 + y^2 = 0$ is $y = 0$, since if $y \neq 0$, $y^2 > 0$ and $(y')^2 + y^2 \geq y^2 > 0$.

39. We first note that $\sqrt{1 - y^2} = \sqrt{1 - \sin^2 x} = \sqrt{\cos^2 x} = |\cos x|$. This prompts us to consider values of x for which $\cos x < 0$, such as $x = \pi$. In this case

$$\frac{dy}{dx}\bigg|_{x=\pi} = \frac{d}{dx}(\sin x)\bigg|_{x=\pi} = \cos x\big|_{x=\pi} = \cos \pi = -1,$$

but

$$\sqrt{1 - y^2}\big|_{x=\pi} = \sqrt{1 - \sin^2 \pi} = \sqrt{1} = 1.$$

Thus, $y = \sin x$ will only be a solution of $y' = \sqrt{1 - y^2}$ when $\cos x > 0$. An interval of definition is then $(-\pi/2, \pi/2)$. Other intervals are $(3\pi/2, 5\pi/2)$, $(7\pi/2, 9\pi/2)$, and so on.

42. One solution, with domain approximately $(-\infty, 1.6)$ is the portion of the graph in the second quadrant together with the lower part of the graph in the first quadrant. A second solution, with domain approximately $(0, 1.6)$

is the upper part of the graph in the first quadrant. The third solution, with domain $(0, \infty)$, is the part of the graph in the fourth quadrant.

45. The derivatives of the functions are $\phi_1'(x) = -x/\sqrt{25 - x^2}$ and $\phi_2'(x) = x/\sqrt{25 - x^2}$, neither of which is defined at $x = \pm 5$.

48. Solving for y' using the quadratic formula we obtain the two differential equations

$$y' = \frac{1}{x}\left(2 + 2\sqrt{1 + 3x^6}\right) \quad \text{and} \quad y' = \frac{1}{x}\left(2 - 2\sqrt{1 + 3x^6}\right),$$

so the differential equation cannot be put in the form $dy/dx = f(x, y)$.

51. (a) Since e^{-x^2} is positive for all values of x, $dy/dx > 0$ for all x, and a solution, $y(x)$, of the differential equation must be increasing on any interval.

(b) $\displaystyle\lim_{x \to -\infty} \frac{dy}{dx} = \lim_{x \to -\infty} e^{-x^2} = 0$ and $\displaystyle\lim_{x \to \infty} \frac{dy}{dx} = \lim_{x \to \infty} e^{-x^2} = 0$. Since dy/dx approaches 0 as x approaches $-\infty$ and ∞, the solution curve has horizontal asymptotes to the left and to the right.

(c) To test concavity we consider the second derivative

$$\frac{d^2 y}{dx^2} = \frac{d}{dx}\left(\frac{dy}{dx}\right) = \frac{d}{dx}\left(e^{-x^2}\right) = -2xe^{-x^2}.$$

Since the second derivative is positive for $x < 0$ and negative for $x > 0$, the solution curve is concave up on $(-\infty, 0)$ and concave down on $(0, \infty)$.

(d)

54. (a) If $y = c$ is a constant solution then $y' = 0$, but $c^2 + 4$ is never 0 for any real value of c.

(b) Since $y' = y^2 + 4 > 0$ for all x where a solution $y = \phi(x)$ is defined, any solution must be increasing on any interval on which it is defined. Thus it cannot have any relative extrema.

(c) Using implicit differentiation we compute $d^2y/dx^2 = 2yy' = 2y(y^2 + 4)$. Setting $d^2y/dx^2 = 0$ we see that $y = 0$ corresponds to the only possible point of inflection. Since $d^2y/dx^2 < 0$ for $y < 0$ and $d^2y/dx^2 > 0$ for $y > 0$, there is a point of inflection where $y = 0$.

(d)

EXERCISES 1.2

Initial-Value Problems

3. Letting $x = 2$ and solving $1/3 = 1/(4 + c)$ we get $c = -1$. The solution is $y = 1/(x^2 - 1)$. This solution is defined on the interval $(1, \infty)$.

6. Letting $x = 1/2$ and solving $-4 = 1/(1/4 + c)$ we get $c = -1/2$. The solution is $y = 1/(x^2 - 1/2) = 2/(2x^2 - 1)$. This solution is defined on the interval $(-1/\sqrt{2}, 1/\sqrt{2})$.

9. In this problem we use $x = c_1 \cos t + c_2 \sin t$ and $x' = -c_1 \sin t + c_2 \cos t$ to obtain a system of two equations in the two unknowns c_1 and c_2.

From the initial conditions we obtain

$$\frac{\sqrt{3}}{2} c_1 + \frac{1}{2} c_2 = \frac{1}{2}$$

$$-\frac{1}{2} c_1 + \frac{\sqrt{3}}{2} c_2 = 0.$$

Solving, we find $c_1 = \sqrt{3}/4$ and $c_2 = 1/4$. The solution of the initial-value problem is

$$x = (\sqrt{3}/4) \cos t + (1/4) \sin t.$$

12. In this problem we use $y = c_1 e^x + c_2 e^{-x}$ and $y' = c_1 e^x - c_2 e^{-x}$ to obtain a system of two equations in the two unknowns c_1 and c_2.

From the initial conditions we obtain

$$e c_1 + e^{-1} c_2 = 0$$

$$e c_1 - e^{-1} c_2 = e.$$

Solving, we find $c_1 = \frac{1}{2}$ and $c_2 = -\frac{1}{2} e^2$. The solution of the initial-value problem is

$$y = \frac{1}{2} e^x - \frac{1}{2} e^2 e^{-x} = \frac{1}{2} e^x - \frac{1}{2} e^{2-x}.$$

15. Two solutions are $y = 0$ and $y = x^3$.

18. For $f(x, y) = \sqrt{xy}$ we have $\partial f/\partial y = \frac{1}{2}\sqrt{x/y}$. Thus, the differential equation will have a unique solution in any region where $x > 0$ and $y > 0$ or where $x < 0$ and $y < 0$.

21. For $f(x, y) = x^2/(4 - y^2)$ we have $\partial f/\partial y = 2x^2 y/(4 - y^2)^2$. Thus the differential equation will have a unique solution in any region where $y < -2$, $-2 < y < 2$, or $y > 2$.

24. For $f(x, y) = (y + x)/(y - x)$ we have $\partial f/\partial y = -2x/(y - x)^2$. Thus the differential equation will have a unique solution in any region where $y < x$ or where $y > x$.

27. In this problem we identify $f(x, y) = \sqrt{y^2 - 9}$ and $\partial f/\partial y = y/\sqrt{y^2 - 9}$. We see that f and $\partial f/\partial y$ are both continuous in the regions of the plane determined by $y < -3$ and $y > 3$ with no restrictions on x.

Since $(2, -3)$ is not in either of the regions defined by $y < -3$ or $y > 3$, there is no guarantee of a unique solution through $(2, -3)$.

30. (a) Since $\dfrac{d}{dx} \tan(x+c) = \sec^2(x+c) = 1 + \tan^2(x+c)$, we see that $y = \tan(x+c)$ satisfies the differential equation.

(b) Solving $y(0) = \tan c = 0$ we obtain $c = 0$ and $y = \tan x$. Since $\tan x$ is discontinuous at $x = \pm\pi/2$, the solution is not defined on $(-2, 2)$ because it contains $\pm\pi/2$.

(c) The largest interval on which the solution can exist is $(-\pi/2, \pi/2)$.

33. (a) Differentiating $3x^2 - y^2 = c$ we get $6x - 2yy' = 0$ or $yy' = 3x$.

(b) Solving $3x^2 - y^2 = 3$ for y we get

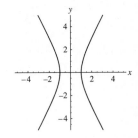

$$y = \phi_1(x) = \sqrt{3(x^2 - 1)}, \qquad 1 < x < \infty,$$
$$y = \phi_2(x) = -\sqrt{3(x^2 - 1)}, \qquad 1 < x < \infty,$$
$$y = \phi_3(x) = \sqrt{3(x^2 - 1)}, \qquad -\infty < x < -1,$$
$$y = \phi_4(x) = -\sqrt{3(x^2 - 1)}, \qquad -\infty < x < -1.$$

(c) Only $y = \phi_3(x)$ satisfies $y(-2) = 3$.

36. In this problem we consider the points on the graphs with x-coordinates $x_0 = -1$, $x_0 = 0$, and $x_0 = 1$. The slopes of the tangent lines at these points are compared with the slopes given by $y'(x_0)$ in (a) through (f).

The graph satisfies the conditions in (e).

39. Integrating $y' = 8e^{2x} + 6x$ we obtain

$$y = \int (8e^{2x} + 6x)\,dx = 4e^{2x} + 3x^2 + c.$$

Setting $x = 0$ and $y = 9$ we have $9 = 4 + c$ so $c = 5$ and $y = 4e^{2x} + 3x^2 + 5$.

42. If the solution is tangent to the x-axis at $(x_0, 0)$, then $y' = 0$ when $x = x_0$ and $y = 0$. Substituting these values into $y' + 2y = 3x - 6$ we get $0 + 0 = 3x_0 - 6$ or $x_0 = 2$.

45. At $t = 0$, $dP/dt = 0.15P(0) + 20 = 0.15(100) + 20 = 35$. Thus, the population is increasing at a rate of 3,500 individuals per year.

If the population is 500 at time $t = T$ then

$$\left. \frac{dP}{dt} \right|_{t=T} = 0.15P(T) + 20 = 0.15(500) + 20 = 95.$$

Thus, at this time, the population is increasing at a rate of 9,500 individuals per year.

EXERCISES 1.3

Differential Equations as Mathematical Models

3. Let b be the rate of births and d the rate of deaths. Then $b = k_1 P$ and $d = k_2 P^2$. Since $dP/dt = b - d$, the differential equation is $dP/dt = k_1 P - k_2 P^2$.

6. By inspecting the graph in the text we take T_m to be $T_m(t) = 80 - 30 \cos \pi t/12$. Then the temperature of the body at time t is determined by the differential equation

$$\frac{dT}{dt} = k\left[T - \left(80 - 30\cos\frac{\pi}{12}t\right)\right], \quad t > 0.$$

9. The rate at which salt is leaving the tank is

$$R_{out} \; (3 \text{ gal/min}) \cdot \left(\frac{A}{300} \text{ lb/gal}\right) = \frac{A}{100} \text{ lb/min}.$$

Thus $dA/dt = A/100$. The initial amount is $A(0) = 50$.

12. The rate at which salt is entering the tank is

$$R_{in} = (c_{in} \text{ lb/gal}) \cdot (r_{in} \text{ gal/min}) = c_{in}r_{in} \text{ lb/min}.$$

Now let $A(t)$ denote the number of pounds of salt and $N(t)$ the number of gallons of brine in the tank at time t. The concentration of salt in the tank as well as in the outflow is $c(t) = x(t)/N(t)$. But the number of gallons of brine in the tank remains steady, is increased, or is decreased depending on whether $r_{in} = r_{out}$, $r_{in} > r_{out}$, or $r_{in} < r_{out}$. In any case, the number of gallons of brine in the tank at time t is $N(t) = N_0 + (r_{in} - r_{out})t$. The output rate of salt is then

$$R_{out} = \left(\frac{A}{N_0 + (r_{in} - r_{out})t} \text{ lb/gal}\right) \cdot (r_{out} \text{ gal/min}) = r_{out}\frac{A}{N_0 + (r_{in} - r_{out})t} \text{ lb/min}.$$

The differential equation for the amount of salt, $dA/dt = R_{in} - R_{out}$, is

$$\frac{dA}{dt} = c_{in}r_{in} - r_{out}\frac{A}{N_0 + (r_{in} - r_{out})t} \quad \text{or} \quad \frac{dA}{dt} + \frac{r_{out}}{N_0 + (r_{in} - r_{out})t}A = c_{in}r_{in}.$$

15. Since $i = dq/dt$ and $L\,d^2q/dt^2 + R\,dq/dt = E(t)$, we obtain $L\,di/dt + Ri = E(t)$.

18. Since the barrel in Figure 1.35(b) in the text is submerged an additional y feet below its equilibrium position the number of cubic feet in the additional submerged portion is the volume of the circular cylinder: $\pi \times (\text{radius})^2 \times \text{height}$ or $\pi(s/2)^2 y$. Then we have from Archimedes' principle

$$\text{upward force of water on barrel} = \text{weight of water displaced}$$

$$= (62.4) \times (\text{volume of water displaced})$$

$$= (62.4)\pi(s/2)^2 y = 15.6\pi s^2 y.$$

It then follows from Newton's second law that

$$\frac{w}{g}\frac{d^2y}{dt^2} = -15.6\pi s^2 y \quad \text{or} \quad \frac{d^2y}{dt^2} + \frac{15.6\pi s^2 g}{w}y = 0,$$

where $g = 32$ and w is the weight of the barrel in pounds.

21. Let $x(t)$ denote the height of the top of the chain at time t with the positive direction upward. The weight of the portion of chain off the ground is $W = (x \text{ ft}) \cdot (1 \text{ lb/ft}) = x$. The mass of the chain is $m = W/g = x/32$. The net force is $F = 5 - W = 5 - x$. By Newton's second law,

$$\frac{d}{dt}\left(\frac{x}{32}v\right) = 5 - x \qquad \text{or} \qquad x\frac{dv}{dt} + v\frac{dx}{dt} = 160 - 32x.$$

Thus, the differential equation is

$$x\frac{d^2x}{dt^2} + \left(\frac{dx}{dt}\right)^2 + 32x = 160.$$

24. The gravitational force on m is $F = -kM_r m/r^2$. Since $M_r = 4\pi\delta r^3/3$ and $M = 4\pi\delta R^3/3$ we have $M_r = r^3 M/R^3$ and

$$F = -k\frac{M_r m}{r^2} = -k\frac{r^3 Mm/R^3}{r^2} = -k\frac{mM}{R^3}r.$$

Now from $F = ma = d^2r/dt^2$ we have

$$m\frac{d^2r}{dt^2} = -k\frac{mM}{R^3}r \qquad \text{or} \qquad \frac{d^2r}{dt^2} = -\frac{kM}{R^3}r.$$

27. The differential equation is $x'(t) = r - kx(t)$ where $k > 0$.

30. The differential equation is $dP/dt = kP$, so from Problem 37 in Exercises 1.1, $P = e^{kt}$, and a one-parameter family of solutions is $P = ce^{kt}$.

33. This differential equation could describe a population that undergoes periodic fluctuations.

36. (a) If ρ is the mass density of the raindrop, then $m = \rho V$ and

$$\frac{dm}{dt} = \rho\frac{dV}{dt} = \rho\frac{d}{dt}\left[\frac{4}{3}\pi r^3\right] = \rho\left(4\pi r^2\frac{dr}{dt}\right) = \rho S\frac{dr}{dt}.$$

If dr/dt is a constant, then $dm/dt = kS$ where $\rho\, dr/dt = k$ or $dr/dt = k/\rho$. Since the radius is decreasing, $k < 0$. Solving $dr/dt = k/\rho$ we get $r = (k/\rho)t + c_0$. Since $r(0) = r_0$, $c_0 = r_0$ and $r = kt/\rho + r_0$.

(b) From Newton's second law, $\dfrac{d}{dt}[mv] = mg$, where v is the velocity of the raindrop. Then

$$m\frac{dv}{dt} + v\frac{dm}{dt} = mg \qquad \text{or} \qquad \rho\left(\frac{4}{3}\pi r^3\right)\frac{dv}{dt} + v(k4\pi r^2) = \rho\left(\frac{4}{3}\pi r^3\right)g.$$

Dividing by $4\rho\pi r^3/3$ we get

$$\frac{dv}{dt} + \frac{3k}{\rho r}v = g \qquad \text{or} \qquad \frac{dv}{dt} + \frac{3k/\rho}{kt/\rho + r_0}v = g, \quad k < 0.$$

39. At time t, when the population is 2 million cells, the differential equation $P'(t) = 0.15P(t)$ gives the rate of increase at time t. Thus, when $P(t) = 2$ (million cells), the rate of increase is $P'(t) = 0.15(2) = 0.3$ million cells per hour or 300,000 cells per hour.

CHAPTER 1 REVIEW EXERCISES

3. $\dfrac{d}{dx}(c_1 \cos kx + c_2 \sin kx) = -kc_1 \sin kx + kc_2 \cos kx$;

$\dfrac{d^2}{dx^2}(c_1 \cos kx + c_2 \sin kx) = -k^2 c_1 \cos kx - k^2 c_2 \sin kx = -k^2(c_1 \cos kx + c_2 \sin kx)$;

$\dfrac{d^2 y}{dx^2} = -k^2 y \quad \text{or} \quad \dfrac{d^2 y}{dx^2} + k^2 y = 0$

6. $y' = -c_1 e^x \sin x + c_1 e^x \cos x + c_2 e^x \cos x + c_2 e^x \sin x$;

$y'' = -c_1 e^x \cos x - c_1 e^x \sin x - c_1 e^x \sin x + c_1 e^x \cos x - c_2 e^x \sin x + c_2 e^x \cos x + c_2 e^x \cos x + c_2 e^x \sin x$

$\quad = -2c_1 e^x \sin x + 2c_2 e^x \cos x$;

$y'' - 2y' = -2c_1 e^x \cos x - 2c_2 e^x \sin x = -2y; \qquad y'' - 2y' + 2y = 0$

9. b

12. a,b,d

15. The slope of the tangent line at (x, y) is y', so the differential equation is $y' = x^2 + y^2$.

18. (a) Differentiating $y^2 - 2y = x^2 - x + c$ we obtain $2yy' - 2y' = 2x - 1$ or $(2y - 2)y' = 2x - 1$.

(b) Setting $x = 0$ and $y = 1$ in the solution we have $1 - 2 = 0 - 0 + c$ or $c = -1$. Thus, a solution of the initial-value problem is $y^2 - 2y = x^2 - x - 1$.

(c) Solving $y^2 - 2y - (x^2 - x - 1) = 0$ by the quadratic formula we get $y = (2 \pm \sqrt{4 + 4(x^2 - x - 1)})/2$ $= 1 \pm \sqrt{x^2 - x} = 1 \pm \sqrt{x(x - 1)}$. Since $x(x-1) \geq 0$ for $x \leq 0$ or $x \geq 1$, we see that neither $y = 1 + \sqrt{x(x - 1)}$ nor $y = 1 - \sqrt{x(x - 1)}$ is differentiable at $x = 0$. Thus, both functions are solutions of the differential equation, but neither is a solution of the initial-value problem.

21. (a)

$y = x^2 + c_1$

$y = -x^2 + c_2$

(b) When $y = x^2 + c_1$, $y' = 2x$ and $(y')^2 = 4x^2$. When $y = -x^2 + c_2$, $y' = -2x$ and $(y')^2 = 4x^2$.

(c) Pasting together x^2, $x \geq 0$, and $-x^2$, $x \leq 0$, we get $y = \begin{cases} -x^2, & x \leq 0 \\ x^2, & x > 0. \end{cases}$

24. Differentiating $y = x \sin x + (\cos x) \ln(\cos x)$ we get

$$y' = x \cos x + \sin x + \cos x \left(\frac{-\sin x}{\cos x} \right) - (\sin x) \ln(\cos x)$$

$$= x \cos x + \sin x - \sin x - (\sin x) \ln(\cos x)$$

$$= x \cos x - (\sin x) \ln(\cos x)$$

and

$$y'' = -x\sin x + \cos x - \sin x\left(\frac{-\sin x}{\cos x}\right) - (\cos x)\ln(\cos x)$$

$$= -x\sin x + \cos x + \frac{\sin^2 x}{\cos x} - (\cos x)\ln(\cos x)$$

$$= -x\sin x + \cos x + \frac{1-\cos^2 x}{\cos x} - (\cos x)\ln(\cos x)$$

$$= -x\sin x + \cos x + \sec x - \cos x - (\cos x)\ln(\cos x)$$

$$= -x\sin x + \sec x - (\cos x)\ln(\cos x).$$

Thus

$$y'' + y = -x\sin x + \sec x - (\cos x)\ln(\cos x) + x\sin x + (\cos x)\ln(\cos x) = \sec x.$$

To obtain an interval of definition we note that the domain of $\ln x$ is $(0,\infty)$, so we must have $\cos x > 0$. Thus, an interval of definition is $(-\pi/2, \pi/2)$.

27. From the graph we see that estimates for y_0 and y_1 are $y_0 = -3$ and $y_1 = 0$.

30. Here we are given that the variable mass of the rocket is $m(t) = m_p + m_\nu + m_f(t)$, where m_p and m_ν are the constant masses of the payload and vehicle, respectively, and $m_f(t)$ is the variable mass of the fuel.

(a) Since

$$\frac{d}{dt}m(t) = \frac{d}{dt}\left(m_p + m_\nu + m_f(t)\right) = \frac{d}{dt}m_f(t),$$

the rates at which the mass of the rocket and the mass of the fuel change are the same.

(b) If the rocket loses fuel at a constant rate λ then we take $dm/dt = -\lambda$. We use $-\lambda$ instead of λ because the fuel is decreasing over time. We next divide the resulting differential equation in Problem 29 by m, obtaining

$$\frac{dv}{dt} + \frac{v}{m}(-\lambda) + \frac{kv}{m} = -g + \frac{R}{m}\frac{dv}{dt} + \frac{k-\lambda}{m}v = -g + \frac{R}{m}.$$

Integrating $dm/dt = -\lambda$ with respect to t we have $m(t) = -\lambda + C$. Since $m(0) = m_0$, $C = m_0$ and $m(t) = -\lambda t + m_0$. The differential equationthen may be written as

$$\frac{dv}{dt} + \frac{k-\lambda}{m_0 - \lambda t}v = -g + \frac{R}{m_0 - \lambda t}.$$

(c) We integrate $dm_f/dt = -\lambda$ to obtain $m_f(t) = -\lambda t + C$. Since $m_f(0) = C$ we have $m_f(t - \lambda t + m_f(0)$. At burnout $m_f(t_b) = -\lambda t_b + m_f(0) = 0$, so $t_b = m_f(0)/\lambda$.

2 First-Order Differential Equations

3.

6.

9.

12.

15. (a) The isoclines have the form $y = -x + c$, which are straight lines with slope -1.

(b) The isoclines have the form $x^2 + y^2 = c$, which are circles centered at the origin.

18. (a) Any horizontal lineal element should be at a point on a nullcline. In Problem 1 the nullclines are $x^2 - y^2 = 0$ or $y = \pm x$. In Problem 3 the nullclines are $1 - xy = 0$ or $y = 1/x$. In Problem 4 the nullclines are $(\sin x)\cos y = 0$ or $x = n\pi$ and $y = \pi/2 + n\pi$, where n is an integer. The graphs on the next page show the nullclines for the differential equation s in Problems 1, 3, and 4 superimposed on the corresponding direction field.

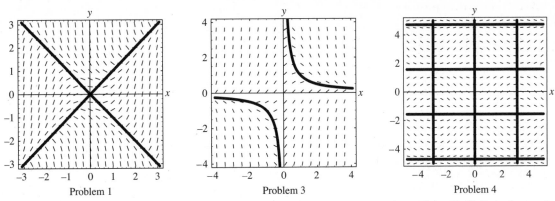

Problem 1   Problem 3   Problem 4

(b) An autonomous first-order differential equation has the form $y' = f(y)$. Nullclines have the form $y = c$ where $f(c) = 0$. These are the graphs of the equilibrium solutions of the differential equation .

21. Solving $y^2 - 3y = y(y - 3) = 0$ we obtain the critical points 0 and 3. From the phase portrait we see that 0 is asymptotically stable (attractor) and 3 is unstable (repeller).

24. Solving $10 + 3y - y^2 = (5 - y)(2 + y) = 0$ we obtain the critical points -2 and 5. From the phase portrait we see that 5 is asymptotically stable (attractor) and -2 is unstable (repeller).

27. Solving $y \ln(y + 2) = 0$ we obtain the critical points -1 and 0. From the phase portrait we see that -1 is asymptotically stable (attractor) and 0 is unstable (repeller).

11

30. The critical points are approximately at $-2, 2, 0.5$, and 1.7. Since $f(y) > 0$ for $y < -2.2$ and $0.5 < y < 1.7$, the graph of the solution is increasing on $(-\infty, -2.2)$ and $(0.5, 1.7)$. Since $f(y) < 0$ for $-2.2 < y < 0.5$ and $y > 1.7$, the graph is decreasing on $(-2.2, 0.5)$ and $(1.7, \infty)$.

33. Recall that for $dy/dx = f(y)$ we are assuming that f and f' are continuous functions of y on some interval I. Now suppose that the graph of a nonconstant solution of the differential equation crosses the line $y = c$. If the point of intersection is taken as an initial condition we have two distinct solutions of the initial-value problem. This violates uniqueness, so the graph of any nonconstant solution must lie entirely on one side of any equilibrium solution. Since f is continuous it can only change signs at a point where it is 0. But this is a critical point. Thus, $f(y)$ is completely positive or completely negative in each region R_i. If $y(x)$ is oscillatory or has a relative extremum, then it must have a horizontal tangent line at some point (x_0, y_0). In this case y_0 would be a critical point of the differential equation, but we saw above that the graph of a nonconstant solution cannot intersect the graph of the equilibrium solution $y = y_0$.

36. Solving $y^2 - y - 6 = (y-3)(y+2) = 0$ we see that 3 and -2 are critical points. Now $d^2y/dx^2 = (2y-1)\,dy/dx = (2y-1)(y-3)(y+2)$, so the only possible point of inflection is at $y = \frac{1}{2}$, although the concavity of solutions can be different on either side of $y = -2$ and $y = 3$. Since $y''(x) < 0$ for $y < -2$ and $\frac{1}{2} < y < 3$, and $y''(x) > 0$ for $-2 < y < \frac{1}{2}$ and $y > 3$, we see that solution curves are concave down for $y < -2$ and $\frac{1}{2} < y < 3$ and concave up for $-2 < y < \frac{1}{2}$ and $y > 3$. Points of inflection of solutions of autonomous differential equations will have the same y-coordinates because between critical points they are horizontal translates of each other.

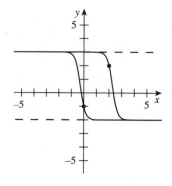

39. (a) Writing the differential equation in the form

$$\frac{dv}{dt} = \frac{k}{m}\left(\frac{mg}{k} - v\right)$$

we see that a critical point is mg/k. From the phase portrait we see that mg/k is an asymptotically stable critical point. Thus, $\lim_{t\to\infty} v = mg/k$.

(b) Writing the differential equation in the form

$$\frac{dv}{dt} = \frac{k}{m}\left(\frac{mg}{k} - v^2\right) = \frac{k}{m}\left(\sqrt{\frac{mg}{k}} - v\right)\left(\sqrt{\frac{mg}{k}} + v\right)$$

we see that the only physically meaningful critical point is $\sqrt{mg/k}$.

From the phase portrait we see that $\sqrt{mg/k}$ is an asymptotically stable critical point. Thus, $\lim_{t\to\infty} v = \sqrt{mg/k}$.

EXERCISES 2.2

Separable Variables

In many of the following problems we will encounter an expression of the form $\ln |g(y)| = f(x) + c$. To solve for $g(y)$ we exponentiate both sides of the equation. This yields $|g(y)| = e^{f(x)+c} = e^c e^{f(x)}$ which implies $g(y) = \pm e^c e^{f(x)}$. Letting $c_1 = \pm e^c$ we obtain $g(y) = c_1 e^{f(x)}$.

3. From $dy = -e^{-3x} \, dx$ we obtain $y = \frac{1}{3} e^{-3x} + c$.

6. From $\frac{1}{y^2} \, dy = -2x \, dx$ we obtain $-\frac{1}{y} = -x^2 + c$ or $y = \frac{1}{x^2 + c_1}$.

9. From $\left(y + 2 + \frac{1}{y} \right) dy = x^2 \ln x \, dx$ we obtain $\frac{y^2}{2} + 2y + \ln |y| = \frac{x^3}{3} \ln |x| - \frac{1}{9} x^3 + c$.

12. From $2y \, dy = -\frac{\sin 3x}{\cos^3 3x} \, dx$ or $2y \, dy = -\tan 3x \sec^2 3x \, dx$ we obtain $y^2 = -\frac{1}{6} \sec^2 3x + c$.

15. From $\frac{1}{S} \, dS = k \, dr$ we obtain $S = c e^{kr}$.

18. From $\frac{1}{N} \, dN = \left(te^{t+2} - 1 \right) dt$ we obtain $\ln |N| = te^{t+2} - e^{t+2} - t + c$ or $N = c_1 e^{te^{t+2} - e^{t+2} - t}$.

21. From $x \, dx = \frac{1}{\sqrt{1 - y^2}} \, dy$ we obtain $\frac{1}{2} x^2 = \sin^{-1} y + c$ or $y = \sin \left(\frac{x^2}{2} + c_1 \right)$.

24. From $\frac{1}{y^2 - 1} \, dy = \frac{1}{x^2 - 1} \, dx$ or $\frac{1}{2} \left(\frac{1}{y-1} - \frac{1}{y+1} \right) dy = \frac{1}{2} \left(\frac{1}{x-1} - \frac{1}{x+1} \right) dx$ we obtain

$\ln |y - 1| - \ln |y + 1| = \ln |x - 1| - \ln |x + 1| + \ln c$ or $\frac{y-1}{y+1} = \frac{c(x-1)}{x+1}$. Using $y(2) = 2$ we find

$c = 1$. A solution of the initial-value problem is $\frac{y-1}{y+1} = \frac{x-1}{x+1}$ or $y = x$.

27. Separating variables and integrating we obtain

$$\frac{dx}{\sqrt{1 - x^2}} - \frac{dy}{\sqrt{1 - y^2}} = 0 \quad \text{and} \quad \sin^{-1} x - \sin^{-1} y = c.$$

Setting $x = 0$ and $y = \sqrt{3}/2$ we obtain $c = -\pi/3$. Thus, an implicit solution of the initial-value problem is $\sin^{-1} x - \sin^{-1} y = \pi/3$. Solving for y and using an addition formula from trigonometry, we get

$$y = \sin \left(\sin^{-1} x + \frac{\pi}{3} \right) = x \cos \frac{\pi}{3} + \sqrt{1 - x^2} \sin \frac{\pi}{3} = \frac{x}{2} + \frac{\sqrt{3}\sqrt{1 - x^2}}{2}.$$

30. Separating variables, we have

$$\frac{dy}{y^2 - y} = \frac{dx}{x} \quad \text{or} \quad \int \frac{dy}{y(y - 1)} = \ln |x| + c.$$

Using partial fractions, we obtain

$$\int \left(\frac{1}{y - 1} - \frac{1}{y} \right) dy = \ln |x| + c$$

$$\ln |y - 1| - \ln |y| = \ln |x| + c$$

$$\ln \left| \frac{y - 1}{xy} \right| = c$$

$$\frac{y - 1}{xy} = e^c = c_1.$$

Solving for y we get $y = 1/(1 - c_1 x)$. We note by inspection that $y = 0$ is a singular solution of the differential equation .

(a) Setting $x = 0$ and $y = 1$ we have $1 = 1/(1 - 0)$, which is true for all values of c_1. Thus, solutions passing through $(0, 1)$ are $y = 1/(1 - c_1 x)$.

(b) Setting $x = 0$ and $y = 0$ in $y = 1/(1 - c_1 x)$ we get $0 = 1$. Thus, the only solution passing through $(0, 0)$ is $y = 0$.

(c) Setting $x = \frac{1}{2}$ and $y = \frac{1}{2}$ we have $\frac{1}{2} = 1/(1 - \frac{1}{2} c_1)$, so $c_1 = -2$ and $y = 1/(1 + 2x)$.

(d) Setting $x = 2$ and $y = \frac{1}{4}$ we have $\frac{1}{4} = 1/(1 - 2c_1)$, so $c_1 = -\frac{3}{2}$ and $y = 1/(1 + \frac{3}{2} x) = 2/(2 + 3x)$.

33. The singular solution $y = 1$ satisfies the initial-value problem.

36. Separating variables we obtain $\dfrac{dy}{(y - 1)^2 - 0.01} = dx$. Then, from (11) in this section of the manual with $u = y - 1$ and $a = \frac{1}{10}$, we get

$$5 \ln \left| \frac{10y - 11}{10y - 9} \right| = x + c.$$

Setting $x = 0$ and $y = 1$ we obtain $c = 5 \ln 1 = 0$. The solution is

$$5 \ln \left| \frac{10y - 11}{10y - 9} \right| = x.$$

Solving for y we obtain

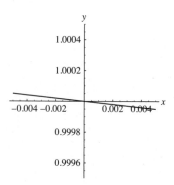

$$y = \frac{11 + 9e^{x/5}}{10 + 10e^{x/5}}.$$

Alternatively, we can use the fact that

$$\int \frac{dy}{(y - 1)^2 - 0.01} = -\frac{1}{0.1} \tanh^{-1} \frac{y - 1}{0.1} = -10 \tanh^{-1} 10(y - 1).$$

(We use the inverse hyperbolic tangent because $|y - 1| < 0.1$ or $0.9 < y < 1.1$. This follows from the initial condition $y(0) = 1$.) Solving the above equation for y we get $y = 1 + 0.1 \tanh(x/10)$.

39. (a) Separating variables we have $2y\,dy = (2x+1)dx$. Integrating gives $y^2 = x^2 + x + c$. When $y(-2) = -1$ we find $c = -1$, so $y^2 = x^2 + x - 1$ and $y = -\sqrt{x^2 + x - 1}$. The negative square root is chosen because of the initial condition.

(b) From the figure, the largest interval of definition appears to be approximately $(-\infty, -1.65)$.

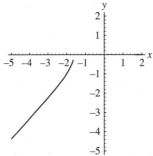

(c) Solving $x^2 + x - 1 = 0$ we get $x = -\frac{1}{2} \pm \frac{1}{2}\sqrt{5}$, so the largest interval of definition is $(-\infty, -\frac{1}{2} - \frac{1}{2}\sqrt{5})$. The right-hand endpoint of the interval is excluded because $y = -\sqrt{x^2 + x - 1}$ is not differentiable at this point.

42. (a) Separating variables and integrating we obtain $x^2 - y^2 = c$. For $c \neq 0$ the graph is a hyperbola centered at the origin. All four initial conditions imply $c = 0$ and $y = \pm x$. Since the differential equation is not defined for $y = 0$, solutions are $y = \pm x$, $x < 0$ and $y = \pm x$, $x > 0$. The solution for $y(a) = a$ is $y = x$, $x > 0$; for $y(a) = -a$ is $y = -x$; for $y(-a) = a$ is $y = -x$, $x < 0$; and for $y(-a) = -a$ is $y = x$, $x < 0$.

(b) Since x/y is not defined when $y = 0$, the initial-value problem has no solution.

(c) Setting $x = 1$ and $y = 2$ in $x^2 - y^2 = c$ we get $c = -3$, so $y^2 = x^2 + 3$ and $y(x) = \sqrt{x^2 + 3}$, where the positive square root is chosen because of the initial condition. The domain is all real numbers since $x^2 + 3 > 0$ for all x.

45. We are looking for a function $y(x)$ such that

$$y^2 + \left(\frac{dy}{dx}\right)^2 = 1.$$

Using the positive square root gives

$$\frac{dy}{dx} = \sqrt{1 - y^2} \implies \frac{dy}{\sqrt{1 - y^2}} = dx \implies \sin^{-1} y = x + c.$$

Thus a solution is $y = \sin(x + c)$. If we use the negative square root we obtain

$$y = \sin(c - x) = -\sin(x - c) = -\sin(x + c_1).$$

Note that when $c = c_1 = 0$ and when $c = c_1 = \pi/2$ we obtain the well known particular solutions $y = \sin x$, $y = -\sin x$, $y = \cos x$, and $y = -\cos x$. Note also that $y = 1$ and $y = -1$ are singular solutions.

48. (a) Separating variables and integrating, we have $(3y^2+1)dy = -(8x+5)dx$ and $y^3 + y = -4x^2 - 5x + c$. Using a CAS we show various contours of $f(x,y) = y^3+y+4x^2+5x$. The plots shown on $[-5,5] \times [-5,5]$ correspond to c-values of $0, \pm 5, \pm 20, \pm 40, \pm 80,$ and ± 125.

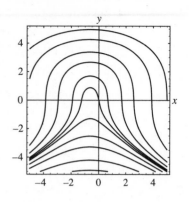

(b) The value of c corresponding to $y(0) = -1$ is $f(0,-1) = -2$; to $y(0) = 2$ is $f(0,2) = 10$; to $y(-1) = 4$ is $f(-1,4) = 67$; and to $y(-1) = -3$ is -31.

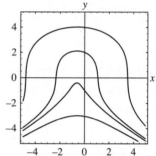

EXERCISES 2.3

Linear Equations

3. For $y'+y = e^{3x}$ an integrating factor is $e^{\int dx} = e^x$ so that $\frac{d}{dx}\left[e^x y\right] = e^{4x}$ and $y = \frac{1}{4}e^{3x}+ce^{-x}$ for $-\infty < x < \infty$. The transient term is ce^{-x}.

6. For $y' + 2xy = x^3$ an integrating factor is $e^{\int 2x\,dx} = e^{x^2}$ so that $\frac{d}{dx}\left[e^{x^2}y\right] = x^3 e^{x^2}$ and $y = \frac{1}{2}x^2 - \frac{1}{2} + ce^{-x^2}$ for $-\infty < x < \infty$. The transient term is ce^{-x^2}.

9. For $y' - \frac{1}{x}y = x\sin x$ an integrating factor is $e^{-\int (1/x)dx} = \frac{1}{x}$ so that $\frac{d}{dx}\left[\frac{1}{x}y\right] = \sin x$ and $y = cx - x\cos x$ for $0 < x < \infty$.

12. For $y' - \frac{x}{(1+x)}y = x$ an integrating factor is $e^{-\int [x/(1+x)]dx} = (x+1)e^{-x}$ so that $\frac{d}{dx}\left[(x+1)e^{-x}y\right] = x(x+1)e^{-x}$ and $y = -x - \frac{2x+3}{x+1} + \frac{ce^x}{x+1}$ for $-1 < x < \infty$.

15. For $\frac{dx}{dy} - \frac{4}{y}x = 4y^5$ an integrating factor is $e^{-\int (4/y)dy} = e^{\ln y^{-4}} = y^{-4}$ so that $\frac{d}{dy}\left[y^{-4}x\right] = 4y$ and $x = 2y^6 + cy^4$ for $0 < y < \infty$.

18. For $y' + (\cot x)y = \sec^2 x \csc x$ an integrating factor is $e^{\int \cot x\,dx} = e^{\ln |\sin x|} = \sin x$ so that $\frac{d}{dx}\left[(\sin x)\,y\right] = \sec^2 x$ and $y = \sec x + c\csc x$ for $0 < x < \pi/2$.

21. For $\frac{dr}{d\theta} + r\sec\theta = \cos\theta$ an integrating factor is $e^{\int \sec\theta\,d\theta} = e^{\ln |\sec x + \tan x|} = \sec\theta + \tan\theta$ so that $\frac{d}{d\theta}\left[(\sec\theta + \tan\theta)r\right] = 1 + \sin\theta$ and $(\sec\theta + \tan\theta)r = \theta - \cos\theta + c$ for $-\pi/2 < \theta < \pi/2$.

24. For $y' + \dfrac{2}{x^2 - 1}\, y = \dfrac{x+1}{x-1}$ an integrating factor is $e^{\int [2/(x^2-1)]\,dx} = \dfrac{x-1}{x+1}$ so that $\dfrac{d}{dx}\left[\dfrac{x-1}{x+1}\,y\right] = 1$ and $(x-1)y = x(x+1) + c(x+1)$ for $-1 < x < 1$.

27. For $\dfrac{di}{dt} + \dfrac{R}{L}\,i = \dfrac{E}{L}$ an integrating factor is $e^{\int (R/L)\,dt} = e^{Rt/L}$ so that $\dfrac{d}{dt}\left[e^{Rt/L}\,i\right] = \dfrac{E}{L}e^{Rt/L}$ and $i = \dfrac{E}{R} + ce^{-Rt/L}$ for $-\infty < t < \infty$. If $i(0) = i_0$ then $c = i_0 - E/R$ and $i = \dfrac{E}{R} + \left(i_0 - \dfrac{E}{R}\right)e^{-Rt/L}$.

30. For $y' + (\tan x)y = \cos^2 x$ an integrating factor is $e^{\int \tan x\,dx} = e^{\ln|\sec x|} = \sec x$ so that $\dfrac{d}{dx}\left[(\sec x)\,y\right] = \cos x$ and $y = \sin x \cos x + c \cos x$ for $-\pi/2 < x < \pi/2$. If $y(0) = -1$ then $c = -1$ and $y = \sin x \cos x - \cos x$.

33. For $y' + 2y = f(x)$ an integrating factor is e^{2x} so that

$$ye^{2x} = \begin{cases} \frac{1}{2}e^{2x} + c_1, & 0 \le x \le 3 \\ c_2, & x > 3. \end{cases}$$

If $y(0) = 0$ then $c_1 = -1/2$ and for continuity we must have $c_2 = \frac{1}{2}e^6 - \frac{1}{2}$ so that

$$y = \begin{cases} \frac{1}{2}(1 - e^{-2x}), & 0 \le x \le 3 \\ \frac{1}{2}(e^6 - 1)e^{-2x}, & x > 3. \end{cases}$$

36. For

$$y' + \dfrac{2x}{1+x^2}\,y = \begin{cases} \dfrac{x}{1+x^2}, & 0 \le x \le 1 \\ \dfrac{-x}{1+x^2}, & x > 1, \end{cases}$$

an integrating factor is $1 + x^2$ so that

$$(1 + x^2)\,y = \begin{cases} \frac{1}{2}x^2 + c_1, & 0 \le x \le 1 \\ -\frac{1}{2}x^2 + c_2, & x > 1. \end{cases}$$

If $y(0) = 0$ then $c_1 = 0$ and for continuity we must have $c_2 = 1$ so that

$$y = \begin{cases} \dfrac{1}{2} - \dfrac{1}{2(1+x^2)}, & 0 \le x \le 1 \\ \dfrac{3}{2(1+x^2)} - \dfrac{1}{2}, & x > 1. \end{cases}$$

39. An integrating factor for $y' - 2xy = 1$ is e^{-x^2}. Thus

$$\dfrac{d}{dx}[e^{-x^2}y] = e^{-x^2}$$

$$e^{-x^2}y = \int_0^x e^{-t^2}\,dt = \dfrac{\sqrt{\pi}}{2}\,\mathrm{erf}(x) + c$$

$$y = \dfrac{\sqrt{\pi}}{2}e^{x^2}\,\mathrm{erf}(x) + ce^{x^2}.$$

From $y(1) = (\sqrt{\pi}/2)e\,\mathrm{erf}(1) + ce = 1$ we get $c = e^{-1} - \dfrac{\sqrt{\pi}}{2}\,\mathrm{erf}(1)$. The solution of the initial-value problem is

$$y = \dfrac{\sqrt{\pi}}{2}e^{x^2}\,\mathrm{erf}(x) + \left(e^{-1} - \dfrac{\sqrt{\pi}}{2}\,\mathrm{erf}(1)\right)e^{x^2}$$

$$= e^{x^2-1} + \dfrac{\sqrt{\pi}}{2}e^{x^2}\,(\mathrm{erf}(x) - \mathrm{erf}(1)).$$

42. On the interval $(-3, 3)$ the integrating factor is

$$e^{\int x \, dx/(x^2-9)} = e^{-\int x \, dx/(9-x^2)} = e^{\frac{1}{2}\ln(9-x^2)} = \sqrt{9-x^2}$$

and so

$$\frac{d}{dx}\left[\sqrt{9-x^2}\, y\right] = 0 \quad \text{and} \quad y = \frac{c}{\sqrt{9-x^2}}.$$

45. (a) Differentiating $y_c = c/x^3$ we get

$$y_c' = -\frac{3c}{x^4} = -\frac{3}{x}\frac{c}{x^3} = -\frac{3}{x}y_c$$

so a differential equation with general solution $y_c = c/x^3$ is $xy' + 3y = 0$. Now

$$xy_p' + 3y_p = x(3x^2) + 3(x^3) = 6x^3$$

so a differential equation with general solution $y = c/x^3 + x^3$ is $xy' + 3y = 6x^3$. This will be a general solution on $(0, \infty)$.

(b) Since $y(1) = 1^3 - 1/1^3 = 0$, an initial condition is $y(1) = 0$. Since $y(1) = 1^3 + 2/1^3 = 3$, an initial condition is $y(1) = 3$. In each case the interval of definition is $(0, \infty)$. The initial-value problem $xy' + 3y = 6x^3$, $y(0) = 0$ has solution $y = x^3$ for $-\infty < x < \infty$. In the figure the lower curve is the graph of $y(x) = x^3 - 1/x^3$, while the upper curve is the graph of $y = x^3 - 2/x^3$.

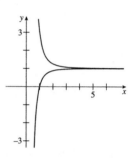

(c) The first two initial-value problems in part (b) are not unique. For example, setting $y(2) = 2^3 - 1/2^3 = 63/8$, we see that $y(2) = 63/8$ is also an initial condition leading to the solution $y = x^3 - 1/x^3$.

48. The solution of the first equation is $x = c_1 e^{-\lambda_1 t}$. From $x(0) = x_0$ we obtain $c_1 = x_0$ and so $x = x_0 e^{-\lambda_1 t}$. The second equation then becomes

$$\frac{dy}{dt} = x_0 \lambda_1 e^{-\lambda_1 t} - \lambda_2 y \quad \text{or} \quad \frac{dy}{dt} + \lambda_2 y = x_0 \lambda_1 e^{-\lambda_1 t}$$

which is linear. An integrating factor is $e^{\lambda_2 t}$. Thus

$$\frac{d}{dt}[e^{\lambda_2 t} y] = x_0 \lambda_1 e^{-\lambda_1 t} e^{\lambda_2 t} = x_0 \lambda_1 e^{(\lambda_2 - \lambda_1)t}$$

$$e^{\lambda_2 t} y = \frac{x_0 \lambda_1}{\lambda_2 - \lambda_1} e^{(\lambda_2 - \lambda_1)t} + c_2$$

$$y = \frac{x_0 \lambda_1}{\lambda_2 - \lambda_1} e^{-\lambda_1 t} + c_2 e^{-\lambda_2 t}.$$

From $y(0) = y_0$ we obtain $c_2 = (y_0 \lambda_2 - y_0 \lambda_1 - x_0 \lambda_1)/(\lambda_2 - \lambda_1)$. The solution is

$$y = \frac{x_0 \lambda_1}{\lambda_2 - \lambda_1} e^{-\lambda_1 t} + \frac{y_0 \lambda_2 - y_0 \lambda_1 - x_0 \lambda_1}{\lambda_2 - \lambda_1} e^{-\lambda_2 t}.$$

51. (a) An integrating factor for

$$y' + \frac{2}{x}y = \frac{10\sin x}{x^3}$$

is x^2. Thus

$$\frac{d}{dx}[x^2 y] = 10\frac{\sin x}{x}$$

$$x^2 y = 10\int_0^x \frac{\sin t}{t}\,dt + c$$

$$y = 10x^{-2}\mathrm{Si}(x) + cx^{-2}.$$

From $y(1) = 0$ we get $c = -10\mathrm{Si}(1)$. Thus

$$y = 10x^{-2}\mathrm{Si}(x) - 10x^{-2}\mathrm{Si}(1) = 10x^{-2}(\mathrm{Si}(x) - \mathrm{Si}(1)).$$

(b)

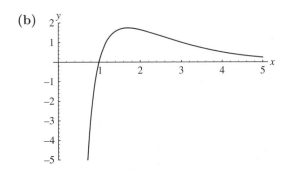

(c) From the graph in part (b) we see that the absolute maximum occurs around $x = 1.7$. Using the root-finding capability of a CAS and solving $y'(x) = 0$ for x we see that the absolute maximum is $(1.688, 1.742)$.

EXERCISES 2.4

Exact Equations

3. Let $M = 5x + 4y$ and $N = 4x - 8y^3$ so that $M_y = 4 = N_x$. From $f_x = 5x + 4y$ we obtain $f = \frac{5}{2}x^2 + 4xy + h(y)$, $h'(y) = -8y^3$, and $h(y) = -2y^4$. A solution is $\frac{5}{2}x^2 + 4xy - 2y^4 = c$.

6. Let $M = 4x^3 - 3y\sin 3x - y/x^2$ and $N = 2y - 1/x + \cos 3x$ so that $M_y = -3\sin 3x - 1/x^2$ and $N_x = 1/x^2 - 3\sin 3x$. The equation is not exact.

9. Let $M = y^3 - y^2\sin x - x$ and $N = 3xy^2 + 2y\cos x$ so that $M_y = 3y^2 - 2y\sin x = N_x$. From $f_x = y^3 - y^2\sin x - x$ we obtain $f = xy^3 + y^2\cos x - \frac{1}{2}x^2 + h(y)$, $h'(y) = 0$, and $h(y) = 0$. A solution is $xy^3 + y^2\cos x - \frac{1}{2}x^2 = c$.

12. Let $M = 3x^2 y + e^y$ and $N = x^3 + xe^y - 2y$ so that $M_y = 3x^2 + e^y = N_x$. From $f_x = 3x^2 y + e^y$ we obtain $f = x^3 y + xe^y + h(y)$, $h'(y) = -2y$, and $h(y) = -y^2$. A solution is $x^3 y + xe^y - y^2 = c$.

15. Let $M = x^2 y^3 - 1/(1 + 9x^2)$ and $N = x^3 y^2$ so that $M_y = 3x^2 y^2 = N_x$. From $f_x = x^2 y^3 - 1/(1 + 9x^2)$ we obtain $f = \frac{1}{3}x^3 y^3 - \frac{1}{3}\arctan(3x) + h(y)$, $h'(y) = 0$, and $h(y) = 0$. A solution is $x^3 y^3 - \arctan(3x) = c$.

18. Let $M = 2y\sin x\cos x - y + 2y^2 e^{xy^2}$ and $N = -x + \sin^2 x + 4xye^{xy^2}$ so that

$$M_y = 2\sin x\cos x - 1 + 4xy^3 e^{xy^2} + 4ye^{xy^2} = N_x.$$

From $f_x = 2y\sin x\cos x - y + 2y^2 e^{xy^2}$ we obtain $f = y\sin^2 x - xy + 2e^{xy^2} + h(y)$, $h'(y) = 0$, and $h(y) = 0$. A solution is $y\sin^2 x - xy + 2e^{xy^2} = c$.

21. Let $M = x^2 + 2xy + y^2$ and $N = 2xy + x^2 - 1$ so that $M_y = 2(x+y) = N_x$. From $f_x = x^2 + 2xy + y^2$ we obtain $f = \frac{1}{3}x^3 + x^2y + xy^2 + h(y)$, $h'(y) = -1$, and $h(y) = -y$. The solution is $\frac{1}{3}x^3 + x^2y + xy^2 - y = c$. If $y(1) = 1$ then $c = 4/3$ and a solution of the initial-value problem is $\frac{1}{3}x^3 + x^2y + xy^2 - y = \frac{4}{3}$.

24. Let $M = t/2y^4$ and $N = \left(3y^2 - t^2\right)/y^5$ so that $M_y = -2t/y^5 = N_t$. From $f_t = t/2y^4$ we obtain $f = \dfrac{t^2}{4y^4} + h(y)$, $h'(y) = \dfrac{3}{y^3}$, and $h(y) = -\dfrac{3}{2y^2}$. The solution is $\dfrac{t^2}{4y^4} - \dfrac{3}{2y^2} = c$. If $y(1) = 1$ then $c = -5/4$ and a solution of the initial-value problem is $\dfrac{t^2}{4y^4} - \dfrac{3}{2y^2} = -\dfrac{5}{4}$.

27. Equating $M_y = 3y^2 + 4kxy^3$ and $N_x = 3y^2 + 40xy^3$ we obtain $k = 10$.

30. Let $M = (x^2 + 2xy - y^2)/(x^2 + 2xy + y^2)$ and $N = (y^2 + 2xy - x^2)/(y^2 + 2xy + x^2)$ so that $M_y = -4xy/(x+y)^3 = N_x$. From $f_x = \left(x^2 + 2xy + y^2 - 2y^2\right)/(x+y)^2$ we obtain $f = x + \dfrac{2y^2}{x+y} + h(y)$, $h'(y) = -1$, and $h(y) = -y$. A solution of the differential equation is $x^2 + y^2 = c(x+y)$.

33. We note that $(N_x - M_y)/M = 2/y$, so an integrating factor is $e^{\int 2dy/y} = y^2$. Let $M = 6xy^3$ and $N = 4y^3 + 9x^2y^2$ so that $M_y = 18xy^2 = N_x$. From $f_x = 6xy^3$ we obtain $f = 3x^2y^3 + h(y)$, $h'(y) = 4y^3$, and $h(y) = y^4$. A solution of the differential equation is $3x^2y^3 + y^4 = c$.

36. We note that $(N_x - M_y)/M = -3/y$, so an integrating factor is $e^{-3\int dy/y} = 1/y^3$. Let $M = (y^2 + xy^3)/y^3 = 1/y + x$ and $N = (5y^2 - xy + y^3 \sin y)/y^3 = 5/y - x/y^2 + \sin y$, so that $M_y = -1/y^2 = N_x$. From $f_x = 1/y + x$ we obtain $f = x/y + \frac{1}{2}x^2 + h(y)$, $h'(y) = 5/y + \sin y$, and $h(y) = 5\ln|y| - \cos y$. A solution of the differential equation is $x/y + \frac{1}{2}x^2 + 5\ln|y| - \cos y = c$.

39. (a) Implicitly differentiating $x^3 + 2x^2y + y^2 = c$ and solving for dy/dx we obtain

$$3x^2 + 2x^2\frac{dy}{dx} + 4xy + 2y\frac{dy}{dx} = 0 \quad \text{and} \quad \frac{dy}{dx} = -\frac{3x^2 + 4xy}{2x^2 + 2y}.$$

By writing the last equation in differential form we get $(4xy + 3x^2)dx + (2y + 2x^2)dy = 0$.

(b) Setting $x = 0$ and $y = -2$ in $x^3 + 2x^2y + y^2 = c$ we find $c = 4$, and setting $x = y = 1$ we also find $c = 4$. Thus, both initial conditions determine the same implicit solution.

(c) Solving $x^3 + 2x^2y + y^2 = 4$ for y we get

$$y_1(x) = -x^2 - \sqrt{4 - x^3 + x^4}$$

and

$$y_2(x) = -x^2 + \sqrt{4 - x^3 + x^4}.$$

Observe in the figure that $y_1(0) = -2$ and $y_2(1) = 1$.

42. (a) Since $f_y = N(x, y) = xe^{xy} + 2xy + 1/x$ we obtain $f = e^{xy} + xy^2 + \dfrac{y}{x} + h(x)$ so that $f_x = ye^{xy} + y^2 - \dfrac{y}{x^2} + h'(x)$. Let $M(x, y) = ye^{xy} + y^2 - \dfrac{y}{x^2}$.

(b) Since $f_x = M(x,y) = y^{1/2}x^{-1/2} + x(x^2+y)^{-1}$ we obtain $f = 2y^{1/2}x^{1/2} + \dfrac{1}{2}\ln|x^2+y| + g(y)$ so that $f_y = y^{-1/2}x^{1/2} + \dfrac{1}{2}(x^2+y)^{-1} + g'(x)$. Let $N(x,y) = y^{-1/2}x^{1/2} + \dfrac{1}{2}(x^2+y)^{-1}$.

45. (a) In differential form we have $(v^2 - 32x)dx + xv\,dv = 0$. This is not an exact form, but $\mu(x) = x$ is an integrating factor. Multiplying by x we get $(xv^2 - 32x^2)dx + x^2 v\,dv = 0$. This form is the total differential of $u = \frac{1}{2}x^2v^2 - \frac{32}{3}x^3$, so an implicit solution is $\frac{1}{2}x^2v^2 - \frac{32}{3}x^3 = c$. Letting $x = 3$ and $v = 0$ we find $c = -288$. Solving for v we get

$$v = 8\sqrt{\frac{x}{3} - \frac{9}{x^2}}\,.$$

(b) The chain leaves the platform when $x = 8$, so the velocity at this time is

$$v(8) = 8\sqrt{\frac{8}{3} - \frac{9}{64}} \approx 12.7 \text{ ft/s}.$$

EXERCISES 2.5

Solutions by Substitutions

3. Letting $x = vy$ we have

$$vy(v\,dy + y\,dv) + (y - 2vy)\,dy = 0$$

$$vy^2\,dv + y\left(v^2 - 2v + 1\right)dy = 0$$

$$\frac{v\,dv}{(v-1)^2} + \frac{dy}{y} = 0$$

$$\ln|v-1| - \frac{1}{v-1} + \ln|y| = c$$

$$\ln\left|\frac{x}{y} - 1\right| - \frac{1}{x/y - 1} + \ln y = c$$

$$(x - y)\ln|x - y| - y = c(x - y).$$

6. Letting $y = ux$ and using partial fractions, we have

$$\left(u^2x^2 + ux^2\right)dx + x^2(u\,dx + x\,du) = 0$$

$$x^2\left(u^2 + 2u\right)dx + x^3\,du = 0$$

$$\frac{dx}{x} + \frac{du}{u(u+2)} = 0$$

$$\ln|x| + \frac{1}{2}\ln|u| - \frac{1}{2}\ln|u+2| = c$$

$$\frac{x^2u}{u+2} = c_1$$

$$x^2\frac{y}{x} = c_1\left(\frac{y}{x} + 2\right)$$

$$x^2y = c_1(y + 2x).$$

9. Letting $y = ux$ we have

$$-ux\,dx + (x + \sqrt{u}\,x)(u\,dx + x\,du) = 0$$

$$(x^2 + x^2\sqrt{u})\,du + xu^{3/2}\,dx = 0$$

$$\left(u^{-3/2} + \frac{1}{u}\right)du + \frac{dx}{x} = 0$$

$$-2u^{-1/2} + \ln|u| + \ln|x| = c$$

$$\ln|y/x| + \ln|x| = 2\sqrt{x/y} + c$$

$$y(\ln|y| - c)^2 = 4x.$$

12. Letting $y = ux$ we have

$$(x^2 + 2u^2x^2)dx - ux^2(u\,dx + x\,du) = 0$$

$$x^2(1 + u^2)dx - ux^3\,du = 0$$

$$\frac{dx}{x} - \frac{u\,du}{1 + u^2} = 0$$

$$\ln|x| - \frac{1}{2}\ln(1 + u^2) = c$$

$$\frac{x^2}{1 + u^2} = c_1$$

$$x^4 = c_1(x^2 + y^2).$$

Using $y(-1) = 1$ we find $c_1 = 1/2$. The solution of the initial-value problem is $2x^4 = y^2 + x^2$.

15. From $y' + \frac{1}{x}y = \frac{1}{x}y^{-2}$ and $w = y^3$ we obtain $\frac{dw}{dx} + \frac{3}{x}w = \frac{3}{x}$. An integrating factor is x^3 so that $x^3w = x^3 + c$ or $y^3 = 1 + cx^{-3}$.

18. From $y' - \left(1 + \frac{1}{x}\right)y = y^2$ and $w = y^{-1}$ we obtain $\frac{dw}{dx} + \left(1 + \frac{1}{x}\right)w = -1$. An integrating factor is xe^x so that $xe^xw = -xe^x + e^x + c$ or $y^{-1} = -1 + \frac{1}{x} + \frac{c}{x}e^{-x}$.

21. From $y' - \frac{2}{x}y = \frac{3}{x^2}y^4$ and $w = y^{-3}$ we obtain $\frac{dw}{dx} + \frac{6}{x}w = -\frac{9}{x^2}$. An integrating factor is x^6 so that $x^6w = -\frac{9}{5}x^5 + c$ or $y^{-3} = -\frac{9}{5}x^{-1} + cx^{-6}$. If $y(1) = \frac{1}{2}$ then $c = \frac{49}{5}$ and $y^{-3} = -\frac{9}{5}x^{-1} + \frac{49}{5}x^{-6}$.

24. Let $u = x + y$ so that $du/dx = 1 + dy/dx$. Then $\frac{du}{dx} - 1 = \frac{1 - u}{u}$ or $u\,du = dx$. Thus $\frac{1}{2}u^2 = x + c$ or $u^2 = 2x + c_1$, and $(x + y)^2 = 2x + c_1$.

27. Let $u = y - 2x + 3$ so that $du/dx = dy/dx - 2$. Then $\frac{du}{dx} + 2 = 2 + \sqrt{u}$ or $\frac{1}{\sqrt{u}}du = dx$. Thus $2\sqrt{u} = x + c$ and $2\sqrt{y - 2x + 3} = x + c$.

30. Let $u = 3x + 2y$ so that $du/dx = 3 + 2\,dy/dx$. Then $\frac{du}{dx} = 3 + \frac{2u}{u + 2} = \frac{5u + 6}{u + 2}$ and $\frac{u + 2}{5u + 6}du = dx$. Now by long division

$$\frac{u + 2}{5u + 6} = \frac{1}{5} + \frac{4}{25u + 30}$$

so we have

$$\int \left(\frac{1}{5} + \frac{4}{25u + 30}\right)du = dx$$

and $\frac{1}{5}u + \frac{4}{25}\ln|25u + 30| = x + c$. Thus

$$\frac{1}{5}(3x + 2y) + \frac{4}{25}\ln|75x + 50y + 30| = x + c.$$

Setting $x = -1$ and $y = -1$ we obtain $c = \frac{4}{25}\ln 95$. The solution is

$$\frac{1}{5}(3x + 2y) + \frac{4}{25}\ln|75x + 50y + 30| = x + \frac{4}{25}\ln 95$$

or

$$5y - 5x + 2\ln|75x + 50y + 30| = 2\ln 95.$$

33. **(a)** By inspection $y = x$ and $y = -x$ are solutions of the differential equation and not members of the family $y = x\sin(\ln x + c_2)$.

(b) Letting $x = 5$ and $y = 0$ in $\sin^{-1}(y/x) = \ln x + c_2$ we get $\sin^{-1} 0 = \ln 5 + c$ or $c = -\ln 5$. Then $\sin^{-1}(y/x) = \ln x - \ln 5 = \ln(x/5)$. Because the range of the arcsine function is $[-\pi/2, \pi/2]$ we must have

$$-\frac{\pi}{2} \leq \ln\frac{x}{5} \leq \frac{\pi}{2}$$

$$e^{-\pi/2} \leq \frac{x}{5} \leq e^{\pi/2}$$

$$5e^{-\pi/2} \leq x \leq 5e^{\pi/2}.$$

The interval of definition of the solution is approximately $[1.04, 24.05]$.

36. Write the differential equation in the form $x(y'/y) = \ln x + \ln y$ and let $u = \ln y$. Then $du/dx = y'/y$ and the differential equation becomes $x(du/dx) = \ln x + u$ or $du/dx - u/x = (\ln x)/x$, which is first-order and linear. An integrating factor is $e^{-\int dx/x} = 1/x$, so that (using integration by parts)

$$\frac{d}{dx}\left[\frac{1}{x}u\right] = \frac{\ln x}{x^2} \quad \text{and} \quad \frac{u}{x} = -\frac{1}{x} - \frac{\ln x}{x} + c.$$

The solution is

$$\ln y = -1 - \ln x + cx \quad \text{or} \quad y = \frac{e^{cx-1}}{x}.$$

EXERCISES 2.6

A Numerical Method

3. Separating variables and integrating, we have

$$\frac{dy}{y} = dx \quad \text{and} \quad \ln|y| = x + c.$$

Thus $y = c_1 e^x$ and, using $y(0) = 1$, we find $c = 1$, so $y = e^x$ is the solution of the initial-value problem.

$h = 0.1$

x_n	y_n	Actual Value	Abs. Error	% Rel. Error
0.00	1.0000	1.0000	0.0000	0.00
1.10	1.1000	1.1052	0.0052	0.47
0.20	1.2100	1.2214	0.0114	0.93
0.30	1.3310	1.3499	0.0189	1.40
0.40	1.4641	1.4918	0.0277	1.86
0.50	1.6105	1.6487	0.0382	2.32
0.60	1.7716	1.8221	0.0506	2.77
0.70	1.9487	2.0138	0.0650	3.23
0.80	2.1436	2.2255	0.0820	3.68
0.90	2.3579	2.4596	0.1017	4.13
1.00	2.5937	2.7183	0.1245	4.58

$h = 0.05$

x_n	y_n	Actual Value	Abs. Error	% Rel. Error
0.00	1.0000	1.0000	0.0000	0.00
0.05	1.0500	1.0513	0.0013	0.12
0.10	1.1025	1.1052	0.0027	0.24
0.15	1.1576	1.1618	0.0042	0.36
0.20	1.2155	1.2214	0.0059	0.48
0.25	1.2763	1.2840	0.0077	0.60
0.30	1.3401	1.3499	0.0098	0.72
0.35	1.4071	1.4191	0.0120	0.84
0.40	1.4775	1.4918	0.0144	0.96
0.45	1.5513	1.5683	0.0170	1.08
0.50	1.6289	1.6487	0.0198	1.20
0.55	1.7103	1.7333	0.0229	1.32
0.60	1.7959	1.8221	0.0263	1.44
0.65	1.8856	1.9155	0.0299	1.56
0.70	1.9799	2.0138	0.0338	1.68
0.75	2.0789	2.1170	0.0381	1.80
0.80	2.1829	2.2255	0.0427	1.92
0.85	2.2920	2.3396	0.0476	2.04
0.90	2.4066	2.4596	0.0530	2.15
0.95	2.5270	2.5857	0.0588	2.27
1.00	2.6533	2.7183	0.0650	2.39

6. $h = 0.1$

x_n	y_n
0.00	1.0000
0.10	1.1000
0.20	1.2220
0.30	1.3753
0.40	1.5735
0.50	1.8371

$h = 0.05$

x_n	y_n
0.00	1.0000
0.05	1.0500
0.10	1.1053
0.15	1.1668
0.20	1.2360
0.25	1.3144
0.30	1.4039
0.35	1.5070
0.40	1.6267
0.45	1.7670
0.50	1.9332

9.

$h = 0.1$	
x_n	y_n
1.00	1.0000
1.10	1.0000
1.20	1.0191
1.30	1.0588
1.40	1.1231
1.50	1.2194

$h = 0.05$	
x_n	y_n
1.00	1.0000
1.05	1.0000
1.10	1.0049
1.15	1.0147
1.20	1.0298
1.25	1.0506
1.30	1.0775
1.35	1.1115
1.40	1.1538
1.45	1.2057
1.50	1.2696

12. See the comments in Problem 11 above.

EXERCISES 2.7

Linear Models

3. Let $P = P(t)$ be the population at time t. Then $dP/dt = kP$ and $P = ce^{kt}$. From $P(0) = c = 500$ we see that $P = 500e^{kt}$. Since 15% of 500 is 75, we have $P(10) = 500e^{10k} = 575$. Solving for k, we get $k = \frac{1}{10}\ln\frac{575}{500} = \frac{1}{10}\ln 1.15$. When $t = 30$,

$$P(30) = 500e^{(1/10)(\ln 1.15)30} = 500e^{3\ln 1.15} = 760 \text{ years}$$

and

$$P'(30) = kP(30) = \frac{1}{10}(\ln 1.15)760 = 10.62 \text{ persons/year.}$$

6. Let $A = A(t)$ be the amount present at time t. From $dA/dt = kA$ and $A(0) = 100$ we obtain $A = 100e^{kt}$. Using $A(6) = 97$ we find $k = \frac{1}{6}\ln 0.97$. Then $A(24) = 100e^{(1/6)(\ln 0.97)24} = 100(0.97)^4 \approx 88.5$ mg.

9. Let $I = I(t)$ be the intensity, t the thickness, and $I(0) = I_0$. If $dI/dt = kI$ and $I(3) = 0.25I_0$, then $I = I_0e^{kt}$, $k = \frac{1}{3}\ln 0.25$, and $I(15) = 0.00098I_0$.

12. From Example 3 in the text, the amount of carbon present at time t is $A(t) = A_0e^{-0.00012378t}$. Letting $t = 660$ and solving for A_0 we have $A(660) = A_0e^{-0.0001237(660)} = 0.921553A_0$. Thus, approximately 92% of the original amount of C-14 remained in the cloth as of 1988.

15. Assume that $dT/dt = k(T - 100)$ so that $T = 100 + ce^{kt}$. If $T(0) = 20°$ and $T(1) = 22°$, then $c = -80$ and $k = \ln(39/40)$ so that $T(t) = 90°$, which implies $t = 82.1$ seconds. If $T(t) = 98°$ then $t = 145.7$ seconds.

18. (a) The initial temperature of the bath is $T_m(0) = 60°$, so in the short term the temperature of the chemical, which starts at $80°$, should decrease or cool. Over time, the temperature of the bath will increase toward $100°$ since $e^{-0.1t}$ decreases from 1 toward 0 as t increases from 0. Thus, in the long term, the temperature of the chemical should increase or warm toward $100°$.

(b) Adapting the model for Newton's law of cooling, we have

$$\frac{dT}{dt} = -0.1(T - 100 + 40e^{-0.1t}), \quad T(0) = 80.$$

Writing the differential equation in the form

$$\frac{dT}{dt} + 0.1T = 10 - 4e^{-0.1t}$$

we see that it is linear with integrating factor $e^{\int 0.1\,dt} = e^{0.1t}$. Thus

$$\frac{d}{dt}[e^{0.1t}T] = 10e^{0.1t} - 4$$

$$e^{0.1t}T = 100e^{0.1t} - 4t + c$$

and

$$T(t) = 100 - 4te^{-0.1t} + ce^{-0.1t}.$$

Now $T(0) = 80$ so $100 + c = 80$, $c = -20$ and

$$T(t) = 100 - 4te^{-0.1t} - 20e^{-0.1t} = 100 - (4t + 20)e^{-0.1t}.$$

The thinner curve verifies the prediction of cooling followed by warming toward $100°$. The wider curve shows the temperature T_m of the liquid bath.

21. From $dA/dt = 4 - A/50$ we obtain $A = 200 + ce^{-t/50}$. If $A(0) = 30$ then $c = -170$ and $A = 200 - 170e^{-t/50}$.

24. From Problem 20 the number of pounds of salt in the tank at time t is $A(t) = 1000 - 1000e^{-t/100}$. The concentration at time t is $c(t) = A(t)/500 = 2 - 2e^{-t/100}$. Therefore $c(5) = 2 - 2e^{-1/20} = 0.0975\,\text{lb/gal}$ and $\lim_{t\to\infty} c(t) = 2$. Solving $c(t) = 1 = 2 - 2e^{-t/100}$ for t we obtain $t = 100\ln 2 \approx 69.3\,\text{min}$.

27. From

$$\frac{dA}{dt} = 3 - \frac{4A}{100 + (6-4)t} = 3 - \frac{2A}{50 + t}$$

we obtain $A = 50 + t + c(50 + t)^{-2}$. If $A(0) = 10$ then $c = -100{,}000$ and $A(30) = 64.38$ pounds.

30. Assume $L\,di/dt + Ri = E(t)$, $E(t) = E_0 \sin\omega t$, and $i(0) = i_0$ so that

$$i = \frac{E_0 R}{L^2\omega^2 + R^2}\sin\omega t - \frac{E_0 L\omega}{L^2\omega^2 + R^2}\cos\omega t + ce^{-Rt/L}.$$

Since $i(0) = i_0$ we obtain $c = i_0 + \dfrac{E_0 L\omega}{L^2\omega^2 + R^2}$.

33. For $0 \le t \le 20$ the differential equation is $20\,di/dt + 2i = 120$. An integrating factor is $e^{t/10}$, so $(d/dt)[e^{t/10}i] = 6e^{t/10}$ and $i = 60 + c_1e^{-t/10}$. If $i(0) = 0$ then $c_1 = -60$ and $i = 60 - 60e^{-t/10}$. For $t > 20$ the differential equation is $20\,di/dt + 2i = 0$ and $i = c_2e^{-t/10}$. At $t = 20$ we want $c_2e^{-2} = 60 - 60e^{-2}$ so that $c_2 = 60\left(e^2 - 1\right)$. Thus

$$i(t) = \begin{cases} 60 - 60e^{-t/10}, & 0 \le t \le 20 \\ 60\left(e^2 - 1\right)e^{-t/10}, & t > 20. \end{cases}$$

36. (a) Integrating $d^2s/dt^2 = -g$ we get $v(t) = ds/dt = -gt + c$. From $v(0) = 300$ we find $c = 300$, and we are given $g = 32$, so the velocity is $v(t) = -32t + 300$.

(b) Integrating again and using $s(0) = 0$ we get $s(t) = -16t^2 + 300t$. The maximum height is attained when $v = 0$, that is, at $t_a = 9.375$. The maximum height will be $s(9.375) = 1406.25\,\text{ft}$.

39. (a) The differential equation is first-order and linear. Letting $b = k/\rho$, the integrating factor is $e^{\int 3b\,dt/(bt+r_0)} = (r_0 + bt)^3$. Then

$$\frac{d}{dt}[(r_0 + bt)^3 v] = g(r_0 + bt)^3 \quad \text{and} \quad (r_0 + bt)^3 v = \frac{g}{4b}(r_0 + bt)^4 + c.$$

The solution of the differential equation is $v(t) = (g/4b)(r_0 + bt) + c(r_0 + bt)^{-3}$. Using $v(0) = 0$ we find $c = -gr_0^4/4b$, so that

$$v(t) = \frac{g}{4b}(r_0 + bt) - \frac{gr_0^4}{4b(r_0 + bt)^3} = \frac{g\rho}{4k}\left(r_0 + \frac{k}{\rho}t\right) - \frac{g\rho r_0^4}{4k(r_0 + kt/\rho)^3}.$$

(b) Integrating $dr/dt = k/\rho$ we get $r = kt/\rho + c$. Using $r(0) = r_0$ we have $c = r_0$, so $r(t) = kt/\rho + r_0$.

(c) If $r = 0.007\,\text{ft}$ when $t = 10\,\text{s}$, then solving $r(10) = 0.007$ for k/ρ, we obtain $k/\rho = -0.0003$ and $r(t) = 0.01 - 0.0003t$. Solving $r(t) = 0$ we get $t = 33.3$, so the raindrop will have evaporated completely at 33.3 seconds.

42. (a) Solving $k_1(M - A) - k_2 A = 0$ for A we find the equilibrium solution $A = k_1 M/(k_1 + k_2)$. From the phase portrait we see that $\lim_{t \to \infty} A(t) = k_1 M/(k_1 + k_2)$. Since $k_2 > 0$, the material will never be completely memorized and the larger k_2 is, the less the amount of material will be memorized over time.

(b) Write the differential equation in the form $dA/dt + (k_1 + k_2)A = k_1 M$. Then an integrating factor is $e^{(k_1 + k_2)t}$, and

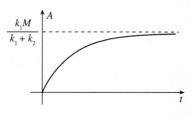

$$\frac{d}{dt}\left[e^{(k_1 + k_2)t} A\right] = k_1 M e^{(k_1 + k_2)t}$$

$$e^{(k_1 + k_2)t} A = \frac{k_1 M}{k_1 + k_2} e^{(k_1 + k_2)t} + c$$

$$A = \frac{k_1 M}{k_1 + k_2} + c e^{-(k_1 + k_2)t}.$$

Using $A(0) = 0$ we find $c = -\dfrac{k_1 M}{k_1 + k_2}$ and $A = \dfrac{k_1 M}{k_1 + k_2}\left(1 - e^{-(k_1 + k_2)t}\right)$. As $t \to \infty$,

$$A \to \frac{k_1 M}{k_1 + k_2}.$$

45. (a) For $0 \leq t < 4$, $6 \leq t < 10$ and $12 \leq t < 16$, no voltage is applied to the heart and $E(t) = 0$. At the other times, the differential equation is $dE/dt = -E/RC$. Separating variables, integrating, and solving for e, we get $E = ke^{-t/RC}$, subject to $E(4) = E(10) = E(16) = 12$. These intitial conditions yield, respectively,

$k = 12e^{4/RC}$, $k = 12e^{10/RC}$, $k = 12e^{16/RC}$, and $k = 12e^{22/RC}$. Thus

$$E(t) = \begin{cases} 0, & 0 \le t < 4, \ 6 \le t < 10, \ 12 \le t < 16 \\ 12e^{(4-t)/RC}, & 4 \le t < 6 \\ 12e^{(10-t)/RC}, & 10 \le t < 12 \\ 12e^{(16-t)/RC}, & 16 \le t < 18 \\ 12e^{(22-t)/RC}, & 22 \le t < 24. \end{cases}$$

(b)

48. (a) (*i*) Using Newton's second law of motion, $F = ma = m\,dv/dt$, the differential equation for the velocity v is

$$m\frac{dv}{dt} = mg\sin\theta \qquad \text{or} \qquad \frac{dv}{dt} = g\sin\theta,$$

where $mg\sin\theta$, $0 < \theta < \pi/2$, is the component of the weight along the plane in the direction of motion.

(*ii*) The model now becomes

$$m\frac{dv}{dt} = mg\sin\theta - \mu mg\cos\theta,$$

where $\mu mg\cos\theta$ is the component of the force of sliding friction (which acts perpendicular to the plane) along the plane. The negative sign indicates that this component of force is a retarding force which acts in the direction opposite to that of motion.

(*iii*) If air resistance is taken to be proportional to the instantaneous velocity of the body, the model becomes

$$m\frac{dv}{dt} = mg\sin\theta - \mu mg\cos\theta - kv,$$

where k is a constant of proportionality.

(b) (*i*) With $m = 3$ slugs, the differential equation is

$$3\frac{dv}{dt} = (96) \cdot \frac{1}{2} \qquad \text{or} \qquad \frac{dv}{dt} = 16.$$

Integrating the last equation gives $v(t) = 16t + c_1$. Since $v(0) = 0$, we have $c_1 = 0$ and so $v(t) = 16t$.

(*ii*) With $m = 3$ slugs, the differential equation is

$$3\frac{dv}{dt} = (96) \cdot \frac{1}{2} - \frac{\sqrt{3}}{4} \cdot (96) \cdot \frac{\sqrt{3}}{2} \qquad \text{or} \qquad \frac{dv}{dt} = 4.$$

In this case $v(t) = 4t$.

(*iii*) When the retarding force due to air resistance is taken into account, the differential equation for velocity v becomes

$$3\frac{dv}{dt} = (96) \cdot \frac{1}{2} - \frac{\sqrt{3}}{4} \cdot (96) \cdot \frac{\sqrt{3}}{2} - \frac{1}{4}v \qquad \text{or} \qquad 3\frac{dv}{dt} = 12 - \frac{1}{4}v.$$

The last differential equation is linear and has solution $v(t) = 48 + c_1 e^{-t/12}$. Since $v(0) = 0$, we find $c_1 = -48$, so $v(t) = 48 - 48e^{-t/12}$.

EXERCISES 2.8

Nonlinear Models

3. From $dP/dt = P\left(10^{-1} - 10^{-7}P\right)$ and $P(0) = 5000$ we obtain $P = 500/(0.0005 + 0.0995e^{-0.1t})$ so that $P \to 1{,}000{,}000$ as $t \to \infty$. If $P(t) = 500{,}000$ then $t = 52.9$ months.

6. Solving $P(5 - P) - \frac{25}{4} = 0$ for P we obtain the equilibrium solution $P = \frac{5}{2}$. For $P \neq \frac{5}{2}$, $dP/dt < 0$. Thus, if $P_0 < \frac{5}{2}$, the population becomes extinct (otherwise there would be another equilibrium solution.) Using separation of variables to solve the initial-value problem, we get

$$P(t) = [4P_0 + (10P_0 - 25)t]/[4 + (4P_0 - 10)t].$$

To find when the population becomes extinct for $P_0 < \frac{5}{2}$ we solve $P(t) = 0$ for t. We see that the time of extinction is $t = 4P_0/5(5 - 2P_0)$.

9. Let $X = X(t)$ be the amount of C at time t and $dX/dt = k(120 - 2X)(150 - X)$. If $X(0) = 0$ and $X(5) = 10$, then

$$X(t) = \frac{150 - 150e^{180kt}}{1 - 2.5e^{180kt}},$$

where $k = .0001259$ and $X(20) = 29.3$ grams. Now by L'Hôpital's rule, $X \to 60$ as $t \to \infty$, so that the amount of $A \to 0$ and the amount of $B \to 30$ as $t \to \infty$.

12. To obtain the solution of this differential equation we use $h(t)$ from Problem 13 in Exercises 1.3. Then $h(t) = (A_w\sqrt{H} - 4cA_h t)^2/A_w^2$. Solving $h(t) = 0$ with $c = 0.6$ and the values from Problem 11 we see that the tank empties in 3035.79 seconds or 50.6 minutes.

15. (a) After separating variables we obtain

$$\frac{m\,dv}{mg - kv^2} = dt$$

$$\frac{1}{g}\frac{dv}{1 - (\sqrt{k}\,v/\sqrt{mg})^2} = dt$$

$$\frac{\sqrt{mg}}{\sqrt{k}\,g}\frac{\sqrt{k/mg}\,dv}{1 - (\sqrt{k}\,v/\sqrt{mg})^2} = dt$$

$$\sqrt{\frac{m}{kg}}\,\tanh^{-1}\frac{\sqrt{k}\,v}{\sqrt{mg}} = t + c$$

$$\tanh^{-1}\frac{\sqrt{k}\,v}{\sqrt{mg}} = \sqrt{\frac{kg}{m}}\,t + c_1.$$

Thus the velocity at time t is

$$v(t) = \sqrt{\frac{mg}{k}}\,\tanh\left(\sqrt{\frac{kg}{m}}t + c_1\right).$$

Setting $t = 0$ and $v = v_0$ we find $c_1 = \tanh^{-1}(\sqrt{k}\,v_0/\sqrt{mg})$.

(b) Since $\tanh t \to 1$ as $t \to \infty$, we have $v \to \sqrt{mg/k}$ as $t \to \infty$.

(c) Integrating the expression for $v(t)$ in part (a) we obtain an integral of the form $\int du/u$:

$$s(t) = \sqrt{\frac{mg}{k}} \int \tanh\left(\sqrt{\frac{kg}{m}}t + c_1\right) dt = \frac{m}{k} \ln\left[\cosh\left(\sqrt{\frac{kg}{m}}t + c_1\right)\right] + c_2.$$

Setting $t = 0$ and $s = 0$ we find $c_2 = -(m/k)\ln(\cosh c_1)$, where c_1 is given in part (a).

18. (a) Writing the equation in the form $(x - \sqrt{x^2 + y^2})dx + y\, dy = 0$ we identify $M = x - \sqrt{x^2 + y^2}$ and $N = y$. Since M and N are both homogeneous functions of degree 1 we use the substitution $y = ux$. It follows that

$$\left(x - \sqrt{x^2 + u^2 x^2}\right) dx + ux(u\, dx + x\, du) = 0$$

$$x\left[1 - \sqrt{1 + u^2} + u^2\right] dx + x^2 u\, du = 0$$

$$-\frac{u\, du}{1 + u^2 - \sqrt{1 + u^2}} = \frac{dx}{x}$$

$$\frac{u\, du}{\sqrt{1 + u^2}\,(1 - \sqrt{1 + u^2})} = \frac{dx}{x}.$$

Letting $w = 1 - \sqrt{1 + u^2}$ we have $dw = -u\, du/\sqrt{1 + u^2}$ so that

$$-\ln\left|1 - \sqrt{1 + u^2}\right| = \ln|x| + c$$

$$\frac{1}{1 - \sqrt{1 + u^2}} = c_1 x$$

$$1 - \sqrt{1 + u^2} = -\frac{c_2}{x} \qquad (-c_2 = 1/c_1)$$

$$1 + \frac{c_2}{x} = \sqrt{1 + \frac{y^2}{x^2}}$$

$$1 + \frac{2c_2}{x} + \frac{c_2^2}{x^2} = 1 + \frac{y^2}{x^2}.$$

Solving for y^2 we have

$$y^2 = 2c_2 x + c_2^2 = 4\left(\frac{c_2}{2}\right)\left(x + \frac{c_2}{2}\right)$$

which is a family of parabolas symmetric with respect to the x-axis with vertex at $(-c_2/2, 0)$ and focus at the origin.

(b) Let $u = x^2 + y^2$ so that

$$\frac{du}{dx} = 2x + 2y\frac{dy}{dx}.$$

Then

$$y\frac{dy}{dx} = \frac{1}{2}\frac{du}{dx} - x.$$

and the differential equation can be written in the form

$$\frac{1}{2}\frac{du}{dx} - x = -x + \sqrt{u} \quad \text{or} \quad \frac{1}{2}\frac{du}{dx} = \sqrt{u}.$$

Separating variables and integrating gives

$$\frac{du}{2\sqrt{u}} = dx$$

$$\sqrt{u} = x + c$$

$$u = x^2 + 2cx + c^2$$

$$x^2 + y^2 = x^2 + 2cx + c^2$$

$$y^2 = 2cx + c^2.$$

21. (a)

t	P(t)	Q(t)
0	3.929	0.035
10	5.308	0.036
20	7.240	0.033
30	9.638	0.033
40	12.866	0.033
50	17.069	0.036
60	23.192	0.036
70	31.433	0.023
80	38.558	0.030
90	50.156	0.026
100	62.948	0.021
110	75.996	0.021
120	91.972	0.015
130	105.711	0.016
140	122.775	0.007
150	131.669	0.014
160	150.697	0.019
170	179.300	

(b) The regression line is $Q = 0.0348391 - 0.000168222P$.

(c) The solution of the logistic equation is given in equation (5) in the text. Identifying $a = 0.0348391$ and $b = 0.000168222$ we have

$$P(t) = \frac{aP_0}{bP_0 + (a - bP_0)e^{-at}}.$$

(d) With $P_0 = 3.929$ the solution becomes

$$P(t) = \frac{0.136883}{0.000660944 + 0.0341781e^{-0.0348391t}}.$$

(e)

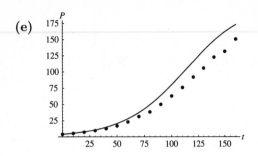

(f) We identify $t = 180$ with 1970, $t = 190$ with 1980, and $t = 200$ with 1990. The model predicts $P(180) = 188.661$, $P(190) = 193.735$, and $P(200) = 197.485$. The actual population figures for these years are 203.303, 226.542, and 248.765 millions. As $t \to \infty$, $P(t) \to a/b = 207.102$.

24. (a) Solving $v_t = \sqrt{mg/k}$ for k we obtain $k = mg/v_t^2$. The differential equation then becomes

$$m\frac{dv}{dt} = mg - \frac{mg}{v_t^2}v^2 \quad \text{or} \quad \frac{dv}{dt} = g\left(1 - \frac{1}{v_t^2}v^2\right).$$

Separating variables and integrating gives

$$v_t \tanh^{-1}\frac{v}{v_t} = gt + c_1.$$

The initial condition $v(0) = 0$ implies $c_1 = 0$, so

$$v(t) = v_t \tanh\frac{gt}{v_t}.$$

We find the distance by integrating:

$$s(t) = \int v_t \tanh\frac{gt}{v_t}\, dt = \frac{v_t^2}{g}\ln\left(\cosh\frac{gt}{v_t}\right) + c_2.$$

The initial condition $s(0) = 0$ implies $c_2 = 0$, so

$$s(t) = \frac{v_t^2}{g}\ln\left(\cosh\frac{gt}{v_t}\right).$$

In 25 seconds she has fallen $20{,}000 - 14{,}800 = 5{,}200$ feet. Using a CAS to solve

$$5200 = (v_t^2/32)\ln\left(\cosh\frac{32(25)}{v_t}\right)$$

for v_t gives $v_t \approx 271.711$ ft/s. Then

$$s(t) = \frac{v_t^2}{g}\ln\left(\cosh\frac{gt}{v_t}\right) = 2307.08\ln(\cosh 0.117772t).$$

(b) At $t = 15$, $s(15) = 2{,}542.94$ ft and $v(15) = s'(15) = 256.287$ ft/sec.

EXERCISES 2.9

Modeling with Systems of First-Order DEs

3. The amounts x and y are the same at about $t = 5$ days. The amounts x and z are the same at about $t = 20$ days. The amounts y and z are the same at about $t = 147$ days. The time when y and z are the same makes sense because most of A and half of B are gone, so half of C should have been formed.

6. Let x_1, x_2, and x_3 be the amounts of salt in tanks A, B, and C, respectively, so that

$$x_1' = \frac{1}{100}x_2 \cdot 2 - \frac{1}{100}x_1 \cdot 6 = \frac{1}{50}x_2 - \frac{3}{50}x_1$$

$$x_2' = \frac{1}{100}x_1 \cdot 6 + \frac{1}{100}x_3 - \frac{1}{100}x_2 \cdot 2 - \frac{1}{100}x_2 \cdot 5 = \frac{3}{50}x_1 - \frac{7}{100}x_2 + \frac{1}{100}x_3$$

$$x_3' = \frac{1}{100}x_2 \cdot 5 - \frac{1}{100}x_3 - \frac{1}{100}x_3 \cdot 4 = \frac{1}{20}x_2 - \frac{1}{20}x_3.$$

9. Zooming in on the graph it can be seen that the populations are first equal at about $t = 5.6$. The approximate periods of x and y are both 45.

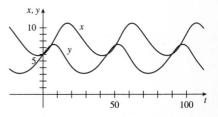

12. By Kirchhoff's first law we have $i_1 = i_2 + i_3$. By Kirchhoff's second law, on each loop we have $E(t) = Li_1' + R_1i_2$ and $E(t) = Li_1' + R_2i_3 + q/C$ so that $q = CR_1i_2 - CR_2i_3$. Then $i_3 = q' = CR_1i_2' - CR_2i_3'$ so that the system is

$$Li_2' + Li_3' + R_1i_2 = E(t)$$

$$-R_1i_2' + R_2i_3' + \frac{1}{C}i_3 = 0.$$

15. We first note that $s(t) + i(t) + r(t) = n$. Now the rate of change of the number of susceptible persons, $s(t)$, is proportional to the number of contacts between the number of people infected and the number who are susceptible; that is, $ds/dt = -k_1si$. We use $-k_1 < 0$ because $s(t)$ is decreasing. Next, the rate of change of the number of persons who have recovered is proportional to the number infected; that is, $dr/dt = k_2i$ where $k_2 > 0$ since r is increasing. Finally, to obtain di/dt we use

$$\frac{d}{dt}(s + i + r) = \frac{d}{dt}n = 0.$$

This gives

$$\frac{di}{dt} = -\frac{dr}{dt} - \frac{ds}{dt} = -k_2i + k_1si.$$

The system of differential equations is then

$$\frac{ds}{dt} = -k_1si$$

$$\frac{di}{dt} = -k_2i + k_1si$$

$$\frac{dr}{dt} = k_2i.$$

A reasonable set of initial conditions is $i(0) = i_0$, the number of infected people at time 0, $s(0) = n - i_0$, and $r(0) = 0$.

3. $\dfrac{dy}{dx} = (y-1)^2(y-3)^2$

6. Using a CAS we find that the zero of f occurs at approximately $P = 1.3214$. From the graph we observe that $dP/dt > 0$ for $P < 1.3214$ and $dP/dt < 0$ for $P > 1.3214$, so $P = 1.3214$ is an asymptotically stable critical point. Thus, $\lim_{t\to\infty} P(t) = 1.3214$.

9. Separating variables and using the identity $\cos^2 x = \frac{1}{2}(1 + \cos 2x)$, we have

$$\cos^2 x \, dx = \frac{y}{y^2 + 1}\, dy,$$

$$\frac{1}{2}x + \frac{1}{4}\sin 2x = \frac{1}{2}\ln\left(y^2 + 1\right) + c,$$

and

$$2x + \sin 2x = 2\ln\left(y^2 + 1\right) + c.$$

12. Write the differential equation in the form $(3y^2 + 2x)dx + (4y^2 + 6xy)dy = 0$. Letting $M = 3y^2 + 2x$ and $N = 4y^2 + 6xy$ we see that $M_y = 6y = N_x$, so the differential equation is exact. From $f_x = 3y^2 + 2x$ we obtain $f = 3xy^2 + x^2 + h(y)$. Then $f_y = 6xy + h'(y) = 4y^2 + 6xy$ and $h'(y) = 4y^2$ so $h(y) = \frac{4}{3}y^3$. A one-parameter family of solutions is

$$3xy^2 + x^2 + \frac{4}{3}y^3 = c.$$

15. Write the equation in the form

$$\frac{dy}{dx} + \frac{8x}{x^2 + 4}y = \frac{2x}{x^2 + 4}.$$

An integrating factor is $\left(x^2 + 4\right)^4$, so

$$\frac{d}{dx}\left[\left(x^2 + 4\right)^4 y\right] = 2x\left(x^2 + 4\right)^3$$

$$\left(x^2 + 4\right)^4 y = \frac{1}{4}\left(x^2 + 4\right)^4 + c$$

and

$$y = \frac{1}{4} + c\left(x^2 + 4\right)^{-4}.$$

18. Separating variables and integrating we have

$$\frac{dy}{y^2} = -2(t+1)\, dt$$

$$-\frac{1}{y} = -(t+1)^2 + c$$

$$y = \frac{1}{(t+1)^2 + c_1}, \qquad \text{where } -c = c_1.$$

The initial condition $y(0) = -\frac{1}{8}$ implies $c_1 = -9$, so a solution of the initial-value problem is

$$y = \frac{1}{(t+1)^2 - 9} \qquad \text{or} \qquad y = \frac{1}{t^2 + 2t - 8},$$

where $-4 < t < 2$.

21. The graph of $y_1(x)$ is the portion of the closed black curve lying in the fourth quadrant. Its interval of definition is approximately $(0.7, 4.3)$. The graph of $y_2(x)$ is the portion of the left-hand black curve lying in the third quadrant. Its interval of definition is $(-\infty, 0)$.

24. (a) We assume that the rate of disintegration of Iodine-131 is proportional to the amount remaining. If $A(t)$ is the amount of Iodine-131 remaining at time t then

$$\frac{dA}{dt} = kA, \quad A(0) = A_0,$$

where k is the constant of proportionality and A_0 is the initial amount. The solution of the initial-value problemis $A(t) = A_0 e^{kt}$. After one day the amount of Iodine-131 is

$$(1 - 0.083)A_0 = 0.917A_0 \quad \text{so} \quad A(1) = A_0 e^k = 0.917A_0$$

and $e^k = 0.917$. After eight days

$$A(8) = A_0 e^{8k} = A_0 (e^k)^8 = A_0 (0.917)^8 \approx 0.49998A_0.$$

(b) Since $A(8) \approx \frac{1}{2}A_0$, the half-life of Iodine-131 is approximately 8 days.

27. Separating variables, we have

$$\frac{\sqrt{s^2 - y^2}}{y} \, dy = -dx.$$

Substituting $y = s \sin \theta$, this becomes

$$\frac{\sqrt{s^2 - s^2 \sin^2 \theta}}{s \sin \theta} (s \cos \theta) d\theta = -dx$$

$$s \int \frac{\cos^2}{\sin \theta} \, d\theta = -\int dx$$

$$s \int \frac{1 - \sin^2 \theta}{\sin \theta} \, d\theta = -\dot{x} + c$$

$$s \int (\csc \theta - \sin \theta) d\theta = -x + c$$

$$s \ln |\csc \theta - \cot \theta| + s \cos \theta = -x + c$$

$$s \ln \left| \frac{s}{y} - \frac{\sqrt{s^2 - y^2}}{y} \right| + s \frac{\sqrt{s^2 - y^2}}{s} = -x + c.$$

Letting $s = 10$, this is

$$10 \ln \left| \frac{10}{y} - \frac{\sqrt{100 - y^2}}{y} \right| + \sqrt{100 - y^2} = -x + c.$$

Letting $x = 0$ and $y = 10$ we determine that $c = 0$, so the solution is

$$10 \ln \left| \frac{10}{y} - \frac{\sqrt{100 - y^2}}{y} \right| + \sqrt{100 - y^2} = -x.$$

30. (a) By separation of variables and partial fractions,

$$\ln \left| \frac{T - T_m}{T + T_m} \right| - 2 \tan^{-1} \left(\frac{T}{T_m} \right) = 4T_m^3 kt + c.$$

Then rewrite the right-hand side of the differential equation as

$$\frac{dT}{dt} = k(T^4 - T_m^4) = [(T_m + (T - T_m))^4 - T_m^4]$$

$$= kT_m^4 \left[\left(1 + \frac{T - T_m}{T_m} \right)^4 - 1 \right]$$

$$= kT_m^4 \left[\left(1 + 4\frac{T - T_m}{T_m} + 6 \left(\frac{T - T_m}{T_m} \right)^2 \cdots \right) - 1 \right] \leftarrow \text{binomial expansion}$$

(b) When $T - T_m$ is small compared to T_m, every term in the expansion after the first two can be ignored, giving

$$\frac{dT}{dt} \approx k_1(T - T_m), \quad \text{where} \quad k_1 = 4kT_m^3.$$

33. Letting $c = 0.6$, $A_h = \pi(\frac{1}{32} \cdot \frac{1}{12})^2$, $A_w = \pi \cdot 1^2 = \pi$, and $g = 32$, the differential equation becomes $dh/dt = -0.00003255\sqrt{h}$. Separating variables and integrating, we get $2\sqrt{h} = -0.00003255t + c$, so $h = (c_1 - 0.00001628t)^2$. Setting $h(0) = 2$, we find $c = \sqrt{2}$, so $h(t) = (\sqrt{2} - 0.00001628t)^2$, where h is measured in feet and t in seconds.

36. If we let r_h denote the radius of the hole and $A_w = \pi[f(h)]^2$, then the differential equation $dh/dt = -k\sqrt{h}$, where $k = cA_h\sqrt{2g}/A_w$, becomes

$$\frac{dh}{dt} = -\frac{c\pi r_h^2\sqrt{2g}}{\pi[f(h)]^2}\sqrt{h} = -\frac{8cr_h^2\sqrt{h}}{[f(h)]^2}.$$

For the time marks to be equally spaced, the rate of change of the height must be a constant; that is, $dh/dt = -a$. (The constant is negative because the height is decreasing.) Thus

$$-a = -\frac{8cr_h^2\sqrt{h}}{[f(h)]^2}, \qquad [f(h)]^2 = \frac{8cr_h^2\sqrt{h}}{a}, \qquad \text{and} \qquad r = f(h) = 2r_h\sqrt{\frac{2c}{a}}\,h^{1/4}.$$

Solving for h, we have

$$h = \frac{a^2}{64c^2r_h^4}\,r^4.$$

The shape of the tank with $c = 0.6$, $a = 2$ ft/12 hr = 1 ft/21,600 s, and $r_h = 1/32(12) = 1/384$ is shown in the above figure.

39. From $y = -x - 1 + c_1e^x$ we obtain $y' = y + x$ so that the differential equation of the orthogonal family is

$$\frac{dy}{dx} = -\frac{1}{y + x} \qquad \text{or} \qquad \frac{dx}{dy} + x = -y.$$

This is a linear differential equation and has integrating factor $e^{\int dy} = e^y$, so

$$\frac{d}{dy}[e^y x] = -ye^y$$

$$e^y x = -ye^y + e^y + c_2$$

$$x = -y + 1 + c_2e^{-y}.$$

3 Higher-Order Differential Equations

EXERCISES 3.1

Preliminary Theory: Linear Equations

3. From $y = c_1 x + c_2 x \ln x$ we find $y' = c_1 + c_2(1 + \ln x)$. Then $y(1) = c_1 = 3$, $y'(1) = c_1 + c_2 = -1$ so that $c_1 = 3$ and $c_2 = -4$. The solution is $y = 3x - 4x \ln x$.

6. In this case we have $y(0) = c_1 = 0$, $y'(0) = 2c_2 \cdot 0 = 0$ so $c_1 = 0$ and c_2 is arbitrary. Two solutions are $y = x^2$ and $y = 2x^2$.

9. Since $a_2(x) = x - 2$ and $x_0 = 0$ the problem has a unique solution for $-\infty < x < 2$.

12. In this case we have $y(0) = c_1 = 1$, $y'(1) = 2c_2 = 6$ so that $c_1 = 1$ and $c_2 = 3$. The solution is $y = 1 + 3x^2$.

15. Since $(-4)x + (3)x^2 + (1)(4x - 3x^2) = 0$ the set of functions is linearly dependent.

18. Since $(1)\cos 2x + (1)1 + (-2)\cos^2 x = 0$ the set of functions is linearly dependent.

21. Suppose $c_1(1 + x) + c_2 x + c_3 x^2 = 0$. Then $c_1 + (c_1 + c_2)x + c_3 x^2 = 0$ and so $c_1 = 0$, $c_1 + c_2 = 0$, and $c_3 = 0$. Since $c_1 = 0$ we also have $c_2 = 0$. Thus, the set of functions is linearly independent.

24. The functions satisfy the differential equation and are linearly independent since

$$W(\cosh 2x, \sinh 2x) = 2$$

for $-\infty < x < \infty$. The general solution is

$$y = c_1 \cosh 2x + c_2 \sinh 2x.$$

27. The functions satisfy the differential equation and are linearly independent since

$$W\left(x^3, x^4\right) = x^6 \neq 0$$

for $0 < x < \infty$. The general solution on this interval is

$$y = c_1 x^3 + c_2 x^4.$$

30. The functions satisfy the differential equation and are linearly independent since

$$W(1, x, \cos x, \sin x) = 1$$

for $-\infty < x < \infty$. The general solution on this interval is

$$y = c_1 + c_2 x + c_3 \cos x + c_4 \sin x.$$

33. The functions $y_1 = e^{2x}$ and $y_2 = xe^{2x}$ form a fundamental set of solutions of the associated homogeneous equation, and $y_p = x^2e^{2x} + x - 2$ is a particular solution of the nonhomogeneous equation.

36. (a) $y_{p_1} = 5$

(b) $y_{p_2} = -2x$

(c) $y_p = y_{p_1} + y_{p_2} = 5 - 2x$

(d) $y_p = \frac{1}{2}y_{p_1} - 2y_{p_2} = \frac{5}{2} + 4x$

39. (a) From the graphs of $y_1 = x^3$ and $y_2 = |x|^3$ we see that the functions are linearly independent since they cannot be multiples of each other. It is easily shown that $y_1 = x^3$ is a solution of $x^2y'' - 4xy' + 6y = 0$. To show that $y_2 = |x|^3$ is a solution let $y_2 = x^3$ for $x \geq 0$ and let $y_2 = -x^3$ for $x < 0$.

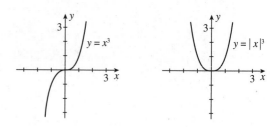

(b) If $x \geq 0$ then $y_2 = x^3$ and

$$W(y_1, y_2) = \begin{vmatrix} x^3 & x^3 \\ 3x^2 & 3x^2 \end{vmatrix} = 0.$$

If $x < 0$ then $y_2 = -x^3$ and

$$W(y_1, y_2) = \begin{vmatrix} x^3 & -x^3 \\ 3x^2 & -3x^2 \end{vmatrix} = 0.$$

This does not violate Theorem 3.3 since $a_2(x) = x^2$ is zero at $x = 0$.

(c) The functions $Y_1 = x^3$ and $Y_2 = x^2$ are solutions of $x^2y'' - 4xy' + 6y = 0$. They are linearly independent since $W\left(x^3, x^2\right) = x^4 \neq 0$ for $-\infty < x < \infty$.

(d) The function $y = x^3$ satisfies $y(0) = 0$ and $y'(0) = 0$.

(e) Neither is the general solution on $(-\infty, \infty)$ since we form a general solution on an interval for which $a_2(x) \neq 0$ for every x in the interval.

42. The set of solutions is linearly dependent. Suppose n of the solutions are linearly independent (if not, then the set of $n + 1$ solutions is linearly dependent). Without loss of generality, let this set be $y_1, y_2, \ldots, y_n$. Then $y = c_1y_1 + c_2y_2 + \cdots + c_ny_n$ is the general solution of the nth-order differential equation and for some choice, $c_1^*, c_2^*, \ldots, c_n^*$, of the coefficients $y_{n+1} = c_1^*y_1 + c_2^*y_2 + \cdots + c_n^*y_n$. But then the set $y_1, y_2, \ldots, y_n, y_{n+1}$ is linearly dependent.

EXERCISES 3.2

Reduction of Order

3. In this problem we use reduction of order to find a second solution.

Define $y = u(x)\cos 4x$ so

$$y' = -4u\sin 4x + u'\cos 4x, \quad y'' = u''\cos 4x - 8u'\sin 4x - 16u\cos 4x$$

and

$$y'' + 16y = (\cos 4x)u'' - 8(\sin 4x)u' = 0 \quad \text{or} \quad u'' - 8(\tan 4x)u' = 0.$$

If $w = u'$ we obtain the linear first-order equation $w' - 8(\tan 4x)w = 0$ which has the integrating factor $e^{-8\int \tan 4x\, dx} = \cos^2 4x$. Now

$$\frac{d}{dx}[(\cos^2 4x)w] = 0 \quad \text{gives} \quad (\cos^2 4x)w = c.$$

Therefore $w = u' = c\sec^2 4x$ and $u = c_1\tan 4x$. A second solution is $y_2 = \tan 4x\cos 4x = \sin 4x$.

6. In this problem we use reduction of order to find a second solution.

Define $y = u(x)e^{5x}$ so

$$y' = 5e^{5x}u + e^{5x}u', \quad y'' = e^{5x}u'' + 10e^{5x}u' + 25e^{5x}u$$

and

$$y'' - 25y = e^{5x}(u'' + 10u') = 0 \quad \text{or} \quad u'' + 10u' = 0.$$

If $w = u'$ we obtain the linear first-order equation $w' + 10w = 0$ which has the integrating factor $e^{10\int dx} = e^{10x}$. Now

$$\frac{d}{dx}[e^{10x}w] = 0 \quad \text{gives} \quad e^{10x}w = c.$$

Therefore $w = u' = ce^{-10x}$ and $u = c_1 e^{-10x}$. A second solution is $y_2 = e^{-10x}e^{5x} = e^{-5x}$.

9. In this problem we use formula (5) from the text.

Identifying $P(x) = -7/x$ we have

$$y_2 = x^4 \int \frac{e^{-\int (-7/x)\, dx}}{x^8}\, dx = x^4 \int \frac{1}{x}\, dx = x^4 \ln|x|.$$

A second solution is $y_2 = x^4 \ln|x|$.

12. In this problem we use formula (5) from the text.

Identifying $P(x) = 0$ we have

$$y_2 = x^{1/2}\ln x \int \frac{e^{-\int 0\, dx}}{x(\ln x)^2}\, dx = x^{1/2}\ln x\left(-\frac{1}{\ln x}\right) = -x^{1/2}.$$

A second solution is $y_2 = x^{1/2}$.

15. In this problem we use formula (5) from the text.

Identifying $P(x) = 2(1+x)/\left(1 - 2x - x^2\right)$ we have

$$y_2 = (x+1) \int \frac{e^{-\int 2(1+x)dx/\left(1-2x-x^2\right)}}{(x+1)^2} \, dx = (x+1) \int \frac{e^{\ln\left(1-2x-x^2\right)}}{(x+1)^2} \, dx$$

$$= (x+1) \int \frac{1 - 2x - x^2}{(x+1)^2} \, dx = (x+1) \int \left[\frac{2}{(x+1)^2} - 1\right] dx$$

$$= (x+1) \left[-\frac{2}{x+1} - x\right] = -2 - x^2 - x.$$

A second solution is $y_2 = x^2 + x + 2$.

18. Define $y = u(x) \cdot 1$ so

$$y' = u', \quad y'' = u'' \quad \text{and} \quad y'' + y' = u'' + u' = 1.$$

If $w = u'$ we obtain the linear first-order equation $w' + w = 1$ which has the integrating factor $e^{\int dx} = e^x$. Now

$$\frac{d}{dx}[e^x w] = e^x \quad \text{gives} \quad e^x w = e^x + c.$$

Therefore $w = u' = 1 + ce^{-x}$ and $u = x + c_1 e^{-x} + c_2$. The general solution is

$$y = u = x + c_1 e^{-x} + c_2.$$

21. (a) For m_1 constant, let $y_1 = e^{m_1 x}$. Then $y_1' = m_1 e^{m_1 x}$ and $y_1'' = m_1^2 e^{m_1 x}$. Substituting into the differential equation we obtain

$$ay_1'' + by_1' + cy_1 = am_1^2 e^{m_1 x} + bm_1 e^{m_1 x} + ce^{m_1 x}$$

$$= e^{m_1 x}(am_1^2 + bm_1 + c) = 0.$$

Thus, $y_1 = e^{m_1 x}$ will be a solution of the differential equation whenever $am_1^2 + bm_1 + c = 0$. Since a quadratic equation always has at least one real or complex root, the differential equation must have a solution of the form $y_1 = e^{m_1 x}$.

(b) Write the differential equation in the form

$$y'' + \frac{b}{a} y' + \frac{c}{a} y = 0,$$

and let $y_1 = e^{m_1 x}$ be a solution. Then a second solution is given by

$$y_2 = e^{m_1 x} \int \frac{e^{-bx/a}}{e^{2m_1 x}} \, dx$$

$$= e^{m_1 x} \int e^{-(b/a + 2m_1)x} dx$$

$$= -\frac{1}{b/a + 2m_1} e^{m_1 x} e^{-(b/a + 2m_1)x} \qquad (m_1 \neq -b/2a)$$

$$= -\frac{1}{b/a + 2m_1} e^{-(b/a + m_1)x}.$$

Thus, when $m_1 \neq -b/2a$, a second solution is given by $y_2 = e^{m_2 x}$ where $m_2 = -b/a - m_1$. When $m_1 = -b/2a$ a second solution is given by

$$y_2 = e^{m_1 x} \int dx = xe^{m_1 x}.$$

(c) The functions

$$\sin x = \frac{1}{2i}(e^{ix} - e^{-ix}) \qquad \cos x = \frac{1}{2}(e^{ix} + e^{-ix})$$

$$\sinh x = \frac{1}{2}(e^x - e^{-x}) \qquad \cosh x = \frac{1}{2}(e^x + e^{-x})$$

are all expressible in terms of exponential functions.

EXERCISES 3.3

Homogeneous Linear Equations with Constant Coefficients

3. From $m^2 - m - 6 = 0$ we obtain $m = 3$ and $m = -2$ so that $y = c_1 e^{3x} + c_2 e^{-2x}$.

6. From $m^2 - 10m + 25 = 0$ we obtain $m = 5$ and $m = 5$ so that $y = c_1 e^{5x} + c_2 x e^{5x}$.

9. From $m^2 + 9 = 0$ we obtain $m = 3i$ and $m = -3i$ so that $y = c_1 \cos 3x + c_2 \sin 3x$.

12. From $2m^2 + 2m + 1 = 0$ we obtain $m = -1/2 \pm i/2$ so that

$$y = e^{-x/2}[c_1 \cos(x/2) + c_2 \sin(x/2)].$$

15. From $m^3 - 4m^2 - 5m = 0$ we obtain $m = 0$, $m = 5$, and $m = -1$ so that

$$y = c_1 + c_2 e^{5x} + c_3 e^{-x}.$$

18. From $m^3 + 3m^2 - 4m - 12 = 0$ we obtain $m = -2$, $m = 2$, and $m = -3$ so that

$$y = c_1 e^{-2x} + c_2 e^{2x} + c_3 e^{-3x}.$$

21. From $m^3 + 3m^2 + 3m + 1 = 0$ we obtain $m = -1$, $m = -1$, and $m = -1$ so that

$$y = c_1 e^{-x} + c_2 x e^{-x} + c_3 x^2 e^{-x}.$$

24. From $m^4 - 2m^2 + 1 = 0$ we obtain $m = 1$, $m = 1$, $m = -1$, and $m = -1$ so that

$$y = c_1 e^x + c_2 x e^x + c_3 e^{-x} + c_4 x e^{-x}.$$

27. From $m^5 + 5m^4 - 2m^3 - 10m^2 + m + 5 = 0$ we obtain $m = -1$, $m = -1$, $m = 1$, and $m = 1$, and $m = -5$ so that

$$u = c_1 e^{-r} + c_2 r e^{-r} + c_3 e^r + c_4 r e^r + c_5 e^{-5r}.$$

30. From $m^2 + 1 = 0$ we obtain $m = \pm i$ so that $y = c_1 \cos\theta + c_2 \sin\theta$. If $y(\pi/3) = 0$ and $y'(\pi/3) = 2$ then

$$\frac{1}{2}c_1 + \frac{\sqrt{3}}{2}c_2 = 0$$

$$-\frac{\sqrt{3}}{2}c_1 + \frac{1}{2}c_2 = 2,$$

so $c_1 = -\sqrt{3}$, $c_2 = 1$, and $y = -\sqrt{3}\cos\theta + \sin\theta$.

33. From $m^2 + m + 2 = 0$ we obtain $m = -1/2 \pm \sqrt{7}\,i/2$ so that $y = e^{-x/2}[c_1 \cos(\sqrt{7}\,x/2) + c_2 \sin(\sqrt{7}\,x/2)]$. If $y(0) = 0$ and $y'(0) = 0$ then $c_1 = 0$ and $c_2 = 0$ so that $y = 0$.

36. From $m^3 + 2m^2 - 5m - 6 = 0$ we obtain $m = -1$, $m = 2$, and $m = -3$ so that

$$y = c_1 e^{-x} + c_2 e^{2x} + c_3 e^{-3x}.$$

If $y(0) = 0$, $y'(0) = 0$, and $y''(0) = 1$ then

$$c_1 + c_2 + c_3 = 0, \quad -c_1 + 2c_2 - 3c_3 = 0, \quad c_1 + 4c_2 + 9c_3 = 1,$$

so $c_1 = -1/6$, $c_2 = 1/15$, $c_3 = 1/10$, and

$$y = -\frac{1}{6} e^{-x} + \frac{1}{15} e^{2x} + \frac{1}{10} e^{-3x}.$$

39. From $m^2 + 1 = 0$ we obtain $m = \pm i$ so that $y = c_1 \cos x + c_2 \sin x$ and $y' = -c_1 \sin x + c_2 \cos x$. From $y'(0) = c_1(0) + c_2(1) = c_2 = 0$ and $y'(\pi/2) = -c_1(1) = 0$ we find $c_1 = c_2 = 0$. A solution of the boundary-value problem is $y = 0$.

42. The auxiliary equation is $m^2 - 1 = 0$ which has roots -1 and 1. By (10) the general solution is $y = c_1 e^x + c_2 e^{-x}$. By (11) the general solution is $y = c_1 \cosh x + c_2 \sinh x$. For $y = c_1 e^x + c_2 e^{-x}$ the boundary conditions imply $c_1 + c_2 = 1$, $c_1 e - c_2 e^{-1} = 0$. Solving for c_1 and c_2 we find $c_1 = 1/(1 + e^2)$ and $c_2 = e^2/(1 + e^2)$ so $y = e^x/(1 + e^2) + e^2 e^{-x}/(1 + e^2)$. For $y = c_1 \cosh x + c_2 \sinh x$ the boundary conditions imply $c_1 = 1$, $c_2 = -\tanh 1$, so $y = \cosh x - (\tanh 1) \sinh x$.

45. The auxiliary equation should have a pair of complex roots $\alpha \pm \beta i$ where $\alpha < 0$, so that the solution has the form $e^{\alpha x}(c_1 \cos \beta x + c_2 \sin \beta x)$. Thus, the differential equation is (e).

48. The differential equation should have the form $y'' + k^2 y = 0$ where $k = 2$ so that the period of the solution is π. Thus, the differential equation is (b).

51. From the solution $y_1 = e^{-4x} \cos x$ we conclude that $m_1 = -4 + i$ and $m_2 = -4 - i$ are roots of the auxiliary equation. Hence another solution must be $y_2 = e^{-4x} \sin x$. Now dividing the polynomial $m^3 + 6m^2 + m - 34$ by $\left[m - (-4 + i) \right]\left[m - (-4 - i) \right] = m^2 + 8m + 17$ gives $m - 2$. Therefore $m_3 = 2$ is the third root of the auxiliary equation, and the general solution of the differential equation is

$$y = c_1 e^{-4x} \cos x + c_2 e^{-4x} \sin x + c_3 e^{2x}.$$

54. The auxiliary equation is $m^2 + \alpha = 0$ and we consider three cases where $\lambda = 0$, $\lambda = \alpha^2 > 0$, and $\lambda = -\alpha^2 < 0$:

Case I When $\alpha = 0$ the general solution of the differential equation is $y = c_1 + c_2 x$. The boundary conditions imply $0 = y(0) = c_1$ and $0 = y(\pi/2) = c_2 \pi/2$, so that $c_1 = c_2 = 0$ and the problem possesses only the trivial solution.

Case II When $\lambda = -\alpha^2 < 0$ the general solution of the differential equation is $y = c_1 e^{\alpha x} + c_2 e^{-\alpha x}$, or alternatively, $y = c_1 \cosh \alpha x + c_2 \sinh \alpha x$. Again, $y(0) = 0$ implies $c_1 = 0$ so $y = c_2 \sinh \alpha x$. The second boundary condition implies $0 = y(\pi/2) = c_2 \sinh \alpha \pi/2$ or $c_2 = 0$. In this case also, the problem possesses only the trivial solution.

Case III When $\lambda = \alpha^2 > 0$ the general solution of the differential equation is $y = c_1 \cos \alpha x + c_2 \sin \alpha x$. In this case also, $y(0) = 0$ yields $c_1 = 0$, so that $y = c_2 \sin \alpha x$. The second boundary condition implies $0 = c_2 \sin \alpha \pi/2$. When $\alpha \pi/2$ is an integer multiple of π, that is, when $\alpha = 2k$ for k a nonzero integer, the problem will have nontrivial solutions. Thus, for $\lambda = \alpha^2 = 4k^2$ the boundary-value problem will have nontrivial solutions $y = c_2 \sin 2kx$, where k is a nonzero integer. On the other hand, when α is not an even integer, the boundary-value problem will have only the trivial solution.

57. Using a CAS to solve the auxiliary equation $m^3 - 6m^2 + 2m + 1$ we find $m_1 = -0.270534$,

$m_2 = 0.658675$, and $m_3 = 5.61186$. The general solution is

$$y = c_1 e^{-0.270534x} + c_2 e^{0.658675x} + c_3 e^{5.61186x}.$$

60. Using a CAS to solve the auxiliary equation $m^4 + 2m^2 - m + 2 = 0$ we find $m_1 = 1/2 + \sqrt{3}\,i/2$, $m_2 = 1/2 - \sqrt{3}\,i/2$, $m_3 = -1/2 + \sqrt{7}\,i/2$, and $m_4 = -1/2 - \sqrt{7}\,i/2$. The general solution is

$$y = e^{x/2}\left(c_1 \cos \frac{\sqrt{3}}{2}x + c_2 \sin \frac{\sqrt{3}}{2}x\right) + e^{-x/2}\left(c_3 \cos \frac{\sqrt{7}}{2}x + c_4 \sin \frac{\sqrt{7}}{2}x\right).$$

EXERCISES 3.4

Undetermined Coefficients

3. From $m^2 - 10m + 25 = 0$ we find $m_1 = m_2 = 5$. Then $y_c = c_1 e^{5x} + c_2 x e^{5x}$ and we assume $y_p = Ax + B$. Substituting into the differential equation we obtain $25A = 30$ and $-10A + 25B = 3$. Then $A = \frac{6}{5}$, $B = \frac{6}{5}$, $y_p = \frac{6}{5}x + \frac{6}{5}$, and

$$y = c_1 e^{5x} + c_2 x e^{5x} + \frac{6}{5}x + \frac{6}{5}.$$

6. From $m^2 - 8m + 20 = 0$ we find $m_1 = 4 + 2i$ and $m_2 = 4 - 2i$. Then $y_c = e^{4x}(c_1 \cos 2x + c_2 \sin 2x)$ and we assume $y_p = Ax^2 + Bx + C + (Dx + E)e^x$. Substituting into the differential equation we obtain

$$2A - 8B + 20C = 0$$

$$-6D + 13E = 0$$

$$-16A + 20B = 0$$

$$13D = -26$$

$$20A = 100.$$

Then $A = 5$, $B = 4$, $C = \frac{11}{10}$, $D = -2$, $E = -\frac{12}{13}$, $y_p = 5x^2 + 4x + \frac{11}{10} + \left(-2x - \frac{12}{13}\right)e^x$ and

$$y = e^{4x}(c_1 \cos 2x + c_2 \sin 2x) + 5x^2 + 4x + \frac{11}{10} + \left(-2x - \frac{12}{13}\right)e^x.$$

9. From $m^2 - m = 0$ we find $m_1 = 1$ and $m_2 = 0$. Then $y_c = c_1 e^x + c_2$ and we assume $y_p = Ax$. Substituting into the differential equation we obtain $-A = -3$. Then $A = 3$, $y_p = 3x$ and $y = c_1 e^x + c_2 + 3x$.

12. From $m^2 - 16 = 0$ we find $m_1 = 4$ and $m_2 = -4$. Then $y_c = c_1 e^{4x} + c_2 e^{-4x}$ and we assume $y_p = Axe^{4x}$. Substituting into the differential equation we obtain $8A = 2$. Then $A = \frac{1}{4}$, $y_p = \frac{1}{4}xe^{4x}$ and

$$y = c_1 e^{4x} + c_2 e^{-4x} + \frac{1}{4}xe^{4x}.$$

15. From $m^2 + 1 = 0$ we find $m_1 = i$ and $m_2 = -i$. Then $y_c = c_1 \cos x + c_2 \sin x$ and we assume $y_p = (Ax^2 + Bx)\cos x + (Cx^2 + Dx)\sin x$. Substituting into the differential equation we obtain $4C = 0$, $2A + 2D = 0$, $-4A = 2$, and $-2B + 2C = 0$. Then $A = -\frac{1}{2}$, $B = 0$, $C = 0$, $D = \frac{1}{2}$, $y_p = -\frac{1}{2}x^2 \cos x + \frac{1}{2}x \sin x$, and

$$y = c_1 \cos x + c_2 \sin x - \frac{1}{2}x^2 \cos x + \frac{1}{2}x \sin x.$$

18. From $m^2 - 2m + 2 = 0$ we find $m_1 = 1 + i$ and $m_2 = 1 - i$. Then $y_c = e^x(c_1 \cos x + c_2 \sin x)$ and we assume $y_p = Ae^{2x} \cos x + Be^{2x} \sin x$. Substituting into the differential equation we obtain $A + 2B = 1$ and $-2A + B = -3$. Then $A = \frac{7}{5}$, $B = -\frac{1}{5}$, $y_p = \frac{7}{5}e^{2x} \cos x - \frac{1}{5}e^{2x} \sin x$ and

$$y = e^x(c_1 \cos x + c_2 \sin x) + \frac{7}{5}e^{2x} \cos x - \frac{1}{5}e^{2x} \sin x.$$

21. From $m^3 - 6m^2 = 0$ we find $m_1 = m_2 = 0$ and $m_3 = 6$. Then $y_c = c_1 + c_2 x + c_3 e^{6x}$ and we assume $y_p = Ax^2 + B \cos x + C \sin x$. Substituting into the differential equation we obtain $-12A = 3$, $6B - C = -1$, and $B + 6C = 0$. Then $A = -\frac{1}{4}$, $B = -\frac{6}{37}$, $C = \frac{1}{37}$, $y_p = -\frac{1}{4}x^2 - \frac{6}{37} \cos x + \frac{1}{37} \sin x$, and

$$y = c_1 + c_2 x + c_3 e^{6x} - \frac{1}{4}x^2 - \frac{6}{37} \cos x + \frac{1}{37} \sin x.$$

24. From $m^3 - m^2 - 4m + 4 = 0$ we find $m_1 = 1$, $m_2 = 2$, and $m_3 = -2$. Then $y_c = c_1 e^x + c_2 e^{2x} + c_3 e^{-2x}$ and we assume $y_p = A + Bxe^x + Cxe^{2x}$. Substituting into the differential equation we obtain $4A = 5$, $-3B = -1$, and $4C = 1$. Then $A = \frac{5}{4}$, $B = \frac{1}{3}$, $C = \frac{1}{4}$, $y_p = \frac{5}{4} + \frac{1}{3}xe^x + \frac{1}{4}xe^{2x}$, and

$$y = c_1 e^x + c_2 e^{2x} + c_3 e^{-2x} + \frac{5}{4} + \frac{1}{3}xe^x + \frac{1}{4}xe^{2x}.$$

27. We have $y_c = c_1 \cos 2x + c_2 \sin 2x$ and we assume $y_p = A$. Substituting into the differential equation we find $A = -\frac{1}{2}$. Thus $y = c_1 \cos 2x + c_2 \sin 2x - \frac{1}{2}$. From the initial conditions we obtain $c_1 = 0$ and $c_2 = \sqrt{2}$, so

$$y = \sqrt{2} \sin 2x - \frac{1}{2}.$$

30. We have $y_c = c_1 e^{-2x} + c_2 xe^{-2x}$ and we assume $y_p = (Ax^3 + Bx^2)e^{-2x}$. Substituting into the differential equation we find $A = \frac{1}{6}$ and $B = \frac{3}{2}$. Thus $y = c_1 e^{-2x} + c_2 xe^{-2x} + \left(\frac{1}{6}x^3 + \frac{3}{2}x^2\right)e^{-2x}$. From the initial conditions we obtain $c_1 = 2$ and $c_2 = 9$, so

$$y = 2e^{-2x} + 9xe^{-2x} + \left(\frac{1}{6}x^3 + \frac{3}{2}x^2\right)e^{-2x}.$$

33. We have $x_c = c_1 \cos \omega t + c_2 \sin \omega t$ and we assume $x_p = At \cos \omega t + Bt \sin \omega t$. Substituting into the differential equation we find $A = -F_0/2\omega$ and $B = 0$. Thus $x = c_1 \cos \omega t + c_2 \sin \omega t - (F_0/2\omega)t \cos \omega t$. From the initial conditions we obtain $c_1 = 0$ and $c_2 = F_0/2\omega^2$, so

$$x = (F_0/2\omega^2) \sin \omega t - (F_0/2\omega)t \cos \omega t.$$

36. We have $y_c = c_1 e^{-2x} + e^x(c_2 \cos \sqrt{3}\,x + c_3 \sin \sqrt{3}\,x)$ and we assume $y_p = Ax + B + Cxe^{-2x}$. Substituting into the differential equation we find $A = \frac{1}{4}$, $B = -\frac{5}{8}$, and $C = \frac{2}{3}$. Thus

$$y = c_1 e^{-2x} + e^x(c_2 \cos \sqrt{3}\,x + c_3 \sin \sqrt{3}\,x) + \frac{1}{4}x - \frac{5}{8} + \frac{2}{3}xe^{-2x}.$$

From the initial conditions we obtain $c_1 = -\frac{23}{12}$, $c_2 = -\frac{59}{24}$, and $c_3 = \frac{17}{72}\sqrt{3}$, so

$$y = -\frac{23}{12}e^{-2x} + e^x\left(-\frac{59}{24} \cos \sqrt{3}\,x + \frac{17}{72}\sqrt{3} \sin \sqrt{3}\,x\right) + \frac{1}{4}x - \frac{5}{8} + \frac{2}{3}xe^{-2x}.$$

39. The general solution of the differential equation $y'' + 3y = 6x$ is $y = c_1 \cos\sqrt{3}x + c_2 \sin\sqrt{3}x + 2x$. The condition $y(0) = 0$ implies $c_1 = 0$ and so $y = c_2 \sin\sqrt{3}x + 2x$. The condition $y(1) + y'(1) = 0$ implies $c_2 \sin\sqrt{3} + 2 + c_2\sqrt{3}\cos\sqrt{3} + 2 = 0$ so $c_2 = -4/(\sin\sqrt{3} + \sqrt{3}\cos\sqrt{3})$. The solution is

$$y = \frac{-4\sin\sqrt{3}x}{\sin\sqrt{3} + \sqrt{3}\cos\sqrt{3}} + 2x.$$

42. We have $y_c = e^x(c_1\cos 3x + c_2 \sin 3x)$ and we assume $y_p = A$ on $[0, \pi]$. Substituting into the differential equation we find $A = 2$. Thus, $y = e^x(c_1\cos 3x + c_2\sin 3x) + 2$ on $[0, \pi]$. On (π, ∞) we have $y = e^x(c_3\cos 3x + c_4\sin 3x)$. From $y(0) = 0$ and $y'(0) = 0$ we obtain

$$c_1 = -2, \qquad c_1 + 3c_2 = 0.$$

Solving this system, we find $c_1 = -2$ and $c_2 = \frac{2}{3}$. Thus $y = e^x(-2\cos 3x + \frac{2}{3}\sin 3x) + 2$ on $[0, \pi]$. Now, continuity of y at $x = \pi$ implies

$$e^\pi\left(-2\cos 3\pi + \frac{2}{3}\sin 3\pi\right) + 2 = e^\pi(c_3\cos 3\pi + c_4\sin 3\pi)$$

or $2 + 2e^\pi = -c_3 e^\pi$ or $c_3 = -2e^{-\pi}(1 + e^\pi)$. Continuity of y' at π implies

$$\frac{20}{3}e^\pi \sin 3\pi = e^\pi[(c_3 + 3c_4)\cos 3\pi + (-3c_3 + c_4)\sin 3\pi]$$

or $-c_3 e^\pi - 3c_4 e^\pi = 0$. Since $c_3 = -2e^{-\pi}(1 + e^\pi)$ we have $c_4 = \frac{2}{3}e^{-\pi}(1 + e^\pi)$. The solution of the initial-value problem is

$$y(x) = \begin{cases} e^x(-2\cos 3x + \frac{2}{3}\sin 3x) + 2, & 0 \le x \le \pi \\ (1 + e^\pi)e^{x-\pi}(-2\cos 3x + \frac{2}{3}\sin 3x), & x > \pi. \end{cases}$$

45. (a) $f(x) = e^x\sin x$. We see that $y_p \to \infty$ as $x \to \infty$ and $y_p \to 0$ as $x \to -\infty$.

(b) $f(x) = e^{-x}$. We see that $y_p \to \infty$ as $x \to \infty$ and $y_p \to \infty$ as $x \to -\infty$.

(c) $f(x) = \sin 2x$. We see that y_p is sinusoidal.

(d) $f(x) = 1$. We see that y_p is constant and simply translates y_c vertically.

EXERCISES 3.5

Variation of Parameters

The particular solution, $y_p = u_1 y_1 + u_2 y_2$, in the following problems can take on a variety of forms, especially where trigonometric functions are involved. The validity of a particular form can best be checked by substituting it back into the differential equation.

3. The auxiliary equation is $m^2 + 1 = 0$, so $y_c = c_1 \cos x + c_2 \sin x$ and

$$W = \begin{vmatrix} \cos x & \sin x \\ -\sin x & \cos x \end{vmatrix} = 1.$$

Identifying $f(x) = \sin x$ we obtain

$$u_1' = -\sin^2 x$$
$$u_2' = \cos x \sin x.$$

Then

$$u_1 = \frac{1}{4}\sin 2x - \frac{1}{2}x = \frac{1}{2}\sin x \cos x - \frac{1}{2}x$$

$$u_2 = -\frac{1}{2}\cos^2 x.$$

and

$$y = c_1 \cos x + c_2 \sin x + \frac{1}{2}\sin x \cos^2 x - \frac{1}{2}x \cos x - \frac{1}{2}\cos^2 x \sin x$$

$$= c_1 \cos x + c_2 \sin x - \frac{1}{2}x \cos x.$$

6. The auxiliary equation is $m^2 + 1 = 0$, so $y_c = c_1 \cos x + c_2 \sin x$ and

$$W = \begin{vmatrix} \cos x & \sin x \\ -\sin x & \cos x \end{vmatrix} = 1.$$

Identifying $f(x) = \sec^2 x$ we obtain

$$u_1' = -\frac{\sin x}{\cos^2 x}$$

$$u_2' = \sec x.$$

Then

$$u_1 = -\frac{1}{\cos x} = -\sec x$$

$$u_2 = \ln|\sec x + \tan x|$$

and

$$y = c_1 \cos x + c_2 \sin x - \cos x \sec x + \sin x \ln|\sec x + \tan x|$$

$$= c_1 \cos x + c_2 \sin x - 1 + \sin x \ln|\sec x + \tan x|.$$

9. The auxiliary equation is $m^2 - 4 = 0$, so $y_c = c_1 e^{2x} + c_2 e^{-2x}$ and

$$W = \begin{vmatrix} e^{2x} & e^{-2x} \\ 2e^{2x} & -2e^{-2x} \end{vmatrix} = -4.$$

Identifying $f(x) = e^{2x}/x$ we obtain $u_1' = 1/4x$ and $u_2' = -e^{4x}/4x$. Then

$$u_1 = \frac{1}{4}\ln|x|,$$

$$u_2 = -\frac{1}{4}\int_{x_0}^{x}\frac{e^{4t}}{t}\,dt$$

and

$$y = c_1 e^{2x} + c_2 e^{-2x} + \frac{1}{4}\left(e^{2x}\ln|x| - e^{-2x}\int_{x_0}^{x}\frac{e^{4t}}{t}\,dt\right), \qquad x_0 > 0.$$

12. The auxiliary equation is $m^2 - 2m + 1 = (m-1)^2 = 0$, so $y_c = c_1 e^x + c_2 x e^x$ and

$$W = \begin{vmatrix} e^x & xe^x \\ e^x & xe^x + e^x \end{vmatrix} = e^{2x}.$$

Identifying $f(x) = e^x/\left(1 + x^2\right)$ we obtain

$$u_1' = -\frac{xe^x e^x}{e^{2x}\left(1 + x^2\right)} = -\frac{x}{1 + x^2}$$

$$u_2' = \frac{e^x e^x}{e^{2x}\left(1 + x^2\right)} = \frac{1}{1 + x^2}.$$

Then $u_1 = -\frac{1}{2}\ln\left(1 + x^2\right)$, $u_2 = \tan^{-1}x$, and

$$y = c_1 e^x + c_2 x e^x - \frac{1}{2}e^x \ln\left(1 + x^2\right) + xe^x \tan^{-1}x.$$

15. The auxiliary equation is $m^2 + 2m + 1 = (m+1)^2 = 0$, so $y_c = c_1 e^{-t} + c_2 t e^{-t}$ and

$$W = \begin{vmatrix} e^{-t} & te^{-t} \\ -e^{-t} & -te^{-t} + e^{-t} \end{vmatrix} = e^{-2t}.$$

Identifying $f(t) = e^{-t}\ln t$ we obtain

$$u_1' = -\frac{te^{-t}e^{-t}\ln t}{e^{-2t}} = -t\ln t$$

$$u_2' = \frac{e^{-t}e^{-t}\ln t}{e^{-2t}} = \ln t.$$

Then

$$u_1 = -\frac{1}{2}t^2 \ln t + \frac{1}{4}t^2$$

$$u_2 = t\ln t - t$$

and

$$y = c_1 e^{-t} + c_2 t e^{-t} - \frac{1}{2}t^2 e^{-t}\ln t + \frac{1}{4}t^2 e^{-t} + t^2 e^{-t}\ln t - t^2 e^{-t}$$

$$= c_1 e^{-t} + c_2 t e^{-t} + \frac{1}{2}t^2 e^{-t}\ln t - \frac{3}{4}t^2 e^{-t}.$$

18. The auxiliary equation is $4m^2 - 4m + 1 = (2m-1)^2 = 0$, so $y_c = c_1 e^{x/2} + c_2 x e^{x/2}$ and

$$W = \begin{vmatrix} e^{x/2} & xe^{x/2} \\ \frac{1}{2}e^{x/2} & \frac{1}{2}xe^{x/2} + e^{x/2} \end{vmatrix} = e^x.$$

Identifying $f(x) = \frac{1}{4}e^{x/2}\sqrt{1-x^2}$ we obtain

$$u_1' = -\frac{xe^{x/2}e^{x/2}\sqrt{1-x^2}}{4e^x} = -\frac{1}{4}x\sqrt{1-x^2}$$

$$u_2' = \frac{e^{x/2}e^{x/2}\sqrt{1-x^2}}{4e^x} = \frac{1}{4}\sqrt{1-x^2}.$$

To find u_1 and u_2 we use the substitution $v = 1 - x^2$ and the trig substitution $x = \sin\theta$, respectively:

$$u_1 = \frac{1}{12}\left(1 - x^2\right)^{3/2}$$

$$u_2 = \frac{x}{8}\sqrt{1 - x^2} + \frac{1}{8}\sin^{-1}x.$$

Thus

$$y = c_1 e^{x/2} + c_2 x e^{x/2} + \frac{1}{12}e^{x/2}\left(1 - x^2\right)^{3/2} + \frac{1}{8}x^2 e^{x/2}\sqrt{1 - x^2} + \frac{1}{8}x e^{x/2}\sin^{-1}x.$$

21. The auxiliary equation is $m^2 + 2m - 8 = (m - 2)(m + 4) = 0$, so $y_c = c_1 e^{2x} + c_2 e^{-4x}$ and

$$W = \begin{vmatrix} e^{2x} & e^{-4x} \\ 2e^{2x} & -4e^{-4x} \end{vmatrix} = -6e^{-2x}.$$

Identifying $f(x) = 2e^{-2x} - e^{-x}$ we obtain

$$u_1' = \frac{1}{3}e^{-4x} - \frac{1}{6}e^{-3x}$$

$$u_2' = \frac{1}{6}e^{3x} - \frac{1}{3}e^{2x}.$$

Then

$$u_1 = -\frac{1}{12}e^{-4x} + \frac{1}{18}e^{-3x}$$

$$u_2 = \frac{1}{18}e^{3x} - \frac{1}{6}e^{2x}.$$

Thus

$$y = c_1 e^{2x} + c_2 e^{-4x} - \frac{1}{12}e^{-2x} + \frac{1}{18}e^{-x} + \frac{1}{18}e^{-x} - \frac{1}{6}e^{-2x}$$

$$= c_1 e^{2x} + c_2 e^{-4x} - \frac{1}{4}e^{-2x} + \frac{1}{9}e^{-x}$$

and

$$y' = 2c_1 e^{2x} - 4c_2 e^{-4x} + \frac{1}{2}e^{-2x} - \frac{1}{9}e^{-x}.$$

The initial conditions imply

$$c_1 + \ c_2 - \frac{5}{36} = 1$$

$$2c_1 - 4c_2 + \frac{7}{18} = 0.$$

Thus $c_1 = 25/36$ and $c_2 = 4/9$, and

$$y = \frac{25}{36}e^{2x} + \frac{4}{9}e^{-4x} - \frac{1}{4}e^{-2x} + \frac{1}{9}e^{-x}.$$

24. Write the equation in the form

$$y'' + \frac{1}{x}y' + \frac{1}{x^2}y = \frac{\sec(\ln x)}{x^2}$$

and identify $f(x) = \sec(\ln x)/x^2$. From $y_1 = \cos(\ln x)$ and $y_2 = \sin(\ln x)$ we compute

$$W = \begin{vmatrix} \cos(\ln x) & \sin(\ln x) \\ -\dfrac{\sin(\ln x)}{x} & \dfrac{\cos(\ln x)}{x} \end{vmatrix} = \frac{1}{x}.$$

Now

$$u_1' = -\frac{\tan(\ln x)}{x} \quad \text{so} \quad u_1 = \ln|\cos(\ln x)|,$$

and

$$u_2' = \frac{1}{x} \quad \text{so} \quad u_2 = \ln x.$$

Thus, a particular solution is

$$y_p = \cos(\ln x) \ln|\cos(\ln x)| + (\ln x)\sin(\ln x),$$

and the general solution is

$$y = c_1 \cos(\ln x) + c_2 \sin(\ln x) + \cos(\ln x)\ln|\cos(\ln x)| + (\ln x)\sin(\ln x).$$

27. The auxiliary equation is $3m^2 - 6m + 30 = 0$, which has roots $1 \pm 3i$, so $y_c = e^x(c_1 \cos 3x + c_2 \sin 3x)$. We consider first the differential equation $3y'' - 6y' + 30y = 15\sin x$, which can be solved using undetermined coefficients. Letting $y_{p_1} = A\cos x + B\sin x$ and substituting into the differential equation we get

$$(27A - 6B)\cos x + (6A + 27B)\sin x = 15\sin x.$$

Then

$$27A - 6B = 0 \quad \text{and} \quad 6A + 27B = 15,$$

so $A = \frac{2}{17}$ and $B = \frac{9}{17}$. Thus, $y_{p_1} = \frac{2}{17}\cos x + \frac{9}{17}\sin x$. Next, we consider the differential equation $3y'' - 6y' + 30y$, for which a particular solution y_{p_2} can be found using variation of parameters. The Wronskian is

$$W = \begin{vmatrix} e^x \cos 3x & e^x \sin 3x \\ e^x \cos 3x - 3e^x \sin 3x & 3e^x \cos 3x + e^x \sin 3x \end{vmatrix} = 3e^{2x}.$$

Identifying $f(x) = \frac{1}{3}e^x \tan x$ we obtain

$$u_1' = -\frac{1}{9}\sin 3x \tan 3x = -\frac{1}{9}\left(\frac{\sin^2 3x}{\cos 3x}\right) = -\frac{1}{9}\left(\frac{1 - \cos^2 3x}{\cos 3x}\right) = -\frac{1}{9}(\sec 3x - \cos 3x)$$

so

$$u_1 = -\frac{1}{27}\ln|\sec 3x + \tan 3x| + \frac{1}{27}\sin 3x.$$

Next

$$u_2' = \frac{1}{9}\sin 3x \quad \text{so} \quad u_2 = -\frac{1}{27}\cos 3x.$$

Thus

$$y_{p_2} = -\frac{1}{27}e^x \cos 3x(\ln|\sec 3x + \tan 3x| - \sin 3x) - \frac{1}{27}e^x \sin 3x \cos 3x$$

$$= -\frac{1}{27}e^x(\cos 3x)\ln|\sec 3x + \tan 3x|$$

and the general solution of the original differential equation is

$$y = e^x(c_1 \cos 3x + c_2 \sin 3x) + y_{p_1}(x) + y_{p_2}(x).$$

30. We are given that $y_1 = x^2$ is a solution of $x^4 y'' + x^3 y' - 4x^2 y = 0$. To find a second solution we use reduction of order. Let $y = x^2 u(x)$. Then the product rule gives

$$y' = x^2 u' + 2xu \quad \text{and} \quad y'' = x^2 u'' + 4xu' + 2u,$$

so

$$x^4 y'' + x^3 y' - 4x^2 y = x^5(xu'' + 5u') = 0.$$

Letting $w = u'$, this becomes $xw' + 5w = 0$. Separating variables and integrating we have

$$\frac{dw}{w} = -\frac{5}{x}\,dx \quad \text{and} \quad \ln|w| = -5\ln x + c.$$

49

Thus, $w = x^{-5}$ and $u = -\frac{1}{4}x^{-4}$. A second solution is then $y_2 = x^2 x^{-4} = 1/x^2$, and the general solution of the homogeneous differential equation is $y_c = c_1 x^2 + c_2/x^2$. To find a particular solution, y_p, we use variation of parameters. The Wronskian is

$$W = \begin{vmatrix} x^2 & 1/x^2 \\ 2x & -2/x^3 \end{vmatrix} = -\frac{4}{x}.$$

Identifying $f(x) = 1/x^4$ we obtain $u'_1 = \frac{1}{4}x^{-5}$ and $u'_2 = -\frac{1}{4}x^{-1}$. Then $u_1 = -\frac{1}{16}x^{-4}$ and $u_2 = -\frac{1}{4}\ln x$, so

$$y_p = -\frac{1}{16}x^{-4}x^2 - \frac{1}{4}(\ln x)x^{-2} = -\frac{1}{16}x^{-2} - \frac{1}{4}x^{-2}\ln x.$$

The general solution is

$$y = c_1 x^2 + \frac{c_2}{x^2} - \frac{1}{16x^2} - \frac{1}{4x^2}\ln x.$$

EXERCISES 3.6

Cauchy-Euler Equation

3. The auxiliary equation is $m^2 = 0$ so that $y = c_1 + c_2 \ln x$.

6. The auxiliary equation is $m^2 + 4m + 3 = (m+1)(m+3) = 0$ so that $y = c_1 x^{-1} + c_2 x^{-3}$.

9. The auxiliary equation is $25m^2 + 1 = 0$ so that $y = c_1 \cos\left(\frac{1}{5}\ln x\right) + c_2 \sin\left(\frac{1}{5}\ln x\right)$.

12. The auxiliary equation is $m^2 + 7m + 6 = (m+1)(m+6) = 0$ so that $y = c_1 x^{-1} + c_2 x^{-6}$.

15. Assuming that $y = x^m$ and substituting into the differential equation we obtain

$$m(m-1)(m-2) - 6 = m^3 - 3m^2 + 2m - 6 = (m-3)(m^2+2) = 0.$$

Thus

$$y = c_1 x^3 + c_2 \cos\left(\sqrt{2}\ln x\right) + c_3 \sin\left(\sqrt{2}\ln x\right).$$

18. Assuming that $y = x^m$ and substituting into the differential equation we obtain

$$m(m-1)(m-2)(m-3) + 6m(m-1)(m-2) + 9m(m-1) + 3m + 1 = m^4 + 2m^2 + 1 = (m^2+1)^2 = 0.$$

Thus

$$y = c_1 \cos(\ln x) + c_2 \sin(\ln x) + c_3(\ln x)\cos(\ln x) + c_4(\ln x)\sin(\ln x).$$

21. The auxiliary equation is $m^2 - 2m + 1 = (m-1)^2 = 0$ so that $y_c = c_1 x + c_2 x \ln x$ and

$$W(x, x\ln x) = \begin{vmatrix} x & x\ln x \\ 1 & 1 + \ln x \end{vmatrix} = x.$$

Identifying $f(x) = 2/x$ we obtain $u'_1 = -2\ln x/x$ and $u'_2 = 2/x$. Then $u_1 = -(\ln x)^2$, $u_2 = 2\ln x$, and

$$y = c_1 x + c_2 x \ln x - x(\ln x)^2 + 2x(\ln x)^2$$

$$= c_1 x + c_2 x \ln x + x(\ln x)^2, \qquad x > 0.$$

24. The auxiliary equation $m(m-1) + m - 1 = m^2 - 1 = 0$ has roots $m_1 = -1$, $m_2 = 1$, so $y_c = c_1 x^{-1} + c_2 x$. With $y_1 = x^{-1}$, $y_2 = x$, and the identification $f(x) = 1/x^2(x+1)$, we get

$$W = 2x^{-1}, \qquad W_1 = -1/x(x+1), \qquad \text{and} \qquad W_2 = 1/x^3(x+1).$$

Then $u_1' = W_1/W = -1/2(x+1)$, $u_2' = W_2/W = 1/2x^2(x+1)$, and integration (by partial fractions for u_2') gives

$$u_1 = -\frac{1}{2}\ln(x+1)$$

$$u_2 = -\frac{1}{2}x^{-1} - \frac{1}{2}\ln x + \frac{1}{2}\ln(x+1),$$

so

$$y_p = u_1 y_1 + u_2 y_2 = \left[-\frac{1}{2}\ln(x+1)\right]x^{-1} + \left[-\frac{1}{2}x^{-1} - \frac{1}{2}\ln x + \frac{1}{2}\ln(x+1)\right]x$$

$$= -\frac{1}{2} - \frac{1}{2}x\ln x + \frac{1}{2}x\ln(x+1) - \frac{\ln(x+1)}{2x} = -\frac{1}{2} + \frac{1}{2}x\ln\left(1+\frac{1}{x}\right) - \frac{\ln(x+1)}{2x}$$

and

$$y = y_c + y_p = c_1 x^{-1} + c_2 x - \frac{1}{2} + \frac{1}{2}x\ln\left(1+\frac{1}{x}\right) - \frac{\ln(x+1)}{2x}, \qquad x>0.$$

27. The auxiliary equation is $m^2+1=0$, so that

$$y = c_1\cos(\ln x) + c_2\sin(\ln x)$$

and

$$y' = -c_1\frac{1}{x}\sin(\ln x) + c_2\frac{1}{x}\cos(\ln x).$$

The initial conditions imply $c_1=1$ and $c_2=2$. Thus $y = \cos(\ln x) + 2\sin(\ln x)$. The graph is given to the right.

30. The auxiliary equation is $m^2 - 6m + 8 = (m-2)(m-4) = 0$, so that $y_c = c_1 x^2 + c_2 x^4$ and

$$W = \begin{vmatrix} x^2 & x^4 \\ 2x & 4x^3 \end{vmatrix} = 2x^5.$$

Identifying $f(x) = 8x^4$ we obtain $u_1' = -4x^3$ and $u_2' = 4x$. Then $u_1 = -x^4$, $u_2 = 2x^2$, and $y = c_1 x^2 + c_2 x^4 + x^6$. The initial conditions imply

$$\frac{1}{4}c_1 + \frac{1}{16}c_2 = -\frac{1}{64}$$

$$c_1 + \frac{1}{2}c_2 = -\frac{3}{16}.$$

Thus $c_1 = \frac{1}{16}$, $c_2 = -\frac{1}{2}$, and $y = \frac{1}{16}x^2 - \frac{1}{2}x^4 + x^6$. The graph is given above.

33. Substituting $x = e^t$ into the differential equation we obtain

$$\frac{d^2 y}{dt^2} + 9\frac{dy}{dt} + 8y = e^{2t}.$$

The auxiliary equation is $m^2 + 9m + 8 = (m+1)(m+8) = 0$ so that $y_c = c_1 e^{-t} + c_2 e^{-8t}$. Using undetermined coefficients we try $y_p = Ae^{2t}$. This leads to $30Ae^{2t} = e^{2t}$, so that $A = 1/30$ and

$$y = c_1 e^{-t} + c_2 e^{-8t} + \frac{1}{30}e^{2t} = c_1 x^{-1} + c_2 x^{-8} + \frac{1}{30}x^2.$$

36. From

$$\frac{d^2 y}{dx^2} = \frac{1}{x^2}\left(\frac{d^2 y}{dt^2} - \frac{dy}{dt}\right)$$

it follows that

$$\frac{d^3y}{dx^3} = \frac{1}{x^2}\frac{d}{dx}\left(\frac{d^2y}{dt^2} - \frac{dy}{dt}\right) - \frac{2}{x^3}\left(\frac{d^2y}{dt^2} - \frac{dy}{dt}\right)$$

$$= \frac{1}{x^2}\frac{d}{dx}\left(\frac{d^2y}{dt^2}\right) - \frac{1}{x^2}\frac{d}{dx}\left(\frac{dy}{dt}\right) - \frac{2}{x^3}\frac{d^2y}{dt^2} + \frac{2}{x^3}\frac{dy}{dt}$$

$$= \frac{1}{x^2}\frac{d^3y}{dt^3}\left(\frac{1}{x}\right) - \frac{1}{x^2}\frac{d^2y}{dt^2}\left(\frac{1}{x}\right) - \frac{2}{x^3}\frac{d^2y}{dt^2} + \frac{2}{x^3}\frac{dy}{dt}$$

$$= \frac{1}{x^3}\left(\frac{d^3y}{dt^3} - 3\frac{d^2y}{dt^2} + 2\frac{dy}{dt}\right).$$

Substituting into the differential equation we obtain

$$\frac{d^3y}{dt^3} - 3\frac{d^2y}{dt^2} + 2\frac{dy}{dt} - 3\left(\frac{d^2y}{dt^2} - \frac{dy}{dt}\right) + 6\frac{dy}{dt} - 6y = 3 + 3t$$

or

$$\frac{d^3y}{dt^3} - 6\frac{d^2y}{dt^2} + 11\frac{dy}{dt} - 6y = 3 + 3t.$$

The auxiliary equation is $m^3 - 6m^2 + 11m - 6 = (m-1)(m-2)(m-3) = 0$ so that $y_c = c_1e^t + c_2e^{2t} + c_3e^{3t}$. Using undetermined coefficients we try $y_p = A + Bt$. This leads to $(11B - 6A) - 6Bt = 3 + 3t$, so that $A = -17/12$, $B = -1/2$, and

$$y = c_1e^t + c_2e^{2t} + c_3e^{3t} - \frac{17}{12} - \frac{1}{2}t = c_1x + c_2x^2 + c_3x^3 - \frac{17}{12} - \frac{1}{2}\ln x.$$

39. Letting $u = x + 2$ we obtain $dy/dx = dy/du$ and, using the Chain Rule,

$$\frac{d^2y}{dx^2} = \frac{d}{dx}\left(\frac{dy}{du}\right) = \frac{d^2y}{du^2}\frac{du}{dx} = \frac{d^2y}{du^2}(1) = \frac{d^2y}{du^2}.$$

Substituting into the differential equation we obtain

$$u^2\frac{d^2y}{du^2} + u\frac{dy}{du} + y = 0.$$

The auxiliary equation is $m^2 + 1 = 0$ so that

$$y = c_1\cos(\ln u) + c_2\sin(\ln u) = c_1\cos[\ln(x+2)] + c_2\sin[\ln(x+2)].$$

42. The function $y(x) = -\sqrt{x}\cos(\ln x)$ is defined for $x > 0$ and has x-intercepts where $\ln x = \pi/2 + k\pi$ for k an integer or where $x = e^{\pi/2+k\pi}$. Solving $\pi/2 + k\pi = 0.5$ we get $k \approx -0.34$, so $e^{\pi/2+k\pi} < 0.5$ for all negative integers and the graph has infinitely many x-intercepts in the interval $(0, 0.5)$.

45. The auxiliary equation is $m(m-1)(m-2)(m-3) + 6m(m-1)(m-2) + 3m(m-1) - 3m + 4 = 0$, so that $m_1 = m_2 = \sqrt{2}$ and $m_3 = m_4 = -\sqrt{2}$. The general solution of the differential equation is

$$y = c_1x^{\sqrt{2}} + c_2x^{\sqrt{2}}\ln x + c_3x^{-\sqrt{2}} + c_4x^{-\sqrt{2}}\ln x.$$

EXERCISES 3.7

Nonlinear Equations

3. Let $u = y'$ so that $u' = y''$. The equation becomes $u' = -u - 1$ which is separable. Thus

$$\frac{du}{u^2 + 1} = -dx \implies \tan^{-1} u = -x + c_1 \implies y' = \tan(c_1 - x) \implies y = \ln|\cos(c_1 - x)| + c_2.$$

6. Let $u = y'$ so that $y'' = u\, du/dy$. The equation becomes $(y + 1)u\, du/dy = u^2$. Separating variables we obtain

$$\frac{du}{u} = \frac{dy}{y + 1} \implies \ln|u| = \ln|y + 1| + \ln c_1 \implies u = c_1(y + 1)$$

$$\implies \frac{dy}{dx} = c_1(y + 1) \implies \frac{dy}{y + 1} = c_1\, dx$$

$$\implies \ln|y + 1| = c_1 x + c_2 \implies y + 1 = c_3 e^{c_1 x}.$$

9. (a)

(b) Let $u = y'$ so that $y'' = u\, du/dy$. The equation becomes $u\, du/dy + yu = 0$. Separating variables we obtain

$$du = -y\, dy \implies u = -\frac{1}{2}y^2 + c_1 \implies y' = -\frac{1}{2}y^2 + c_1.$$

When $x = 0$, $y = 1$ and $y' = -1$ so $-1 = -1/2 + c_1$ and $c_1 = -1/2$. Then

$$\frac{dy}{dx} = -\frac{1}{2}y^2 - \frac{1}{2} \implies \frac{dy}{y^2 + 1} = -\frac{1}{2}\, dx \implies \tan^{-1} y = -\frac{1}{2}x + c_2$$

$$\implies y = \tan\left(-\frac{1}{2}x + c_2\right).$$

When $x = 0$, $y = 1$ so $1 = \tan c_2$ and $c_2 = \pi/4$. The solution of the initial-value problem is

$$y = \tan\left(\frac{\pi}{4} - \frac{1}{2}x\right).$$

The graph is shown in part (a).

(c) The interval of definition is $-\pi/2 < \pi/4 - x/2 < \pi/2$ or $-\pi/2 < x < 3\pi/2$.

12. Let $u = y'$ so that $u' = y''$. The equation becomes $u' - (1/x)u = u^2$, which is a Bernoulli differential equation. Using the substitution $w = u^{-1}$ we obtain $dw/dx + (1/x)w = -1$. An integrating factor is x, so

$$\frac{d}{dx}[xw] = -x \implies w = -\frac{1}{2}x + \frac{1}{x}c \implies \frac{1}{u} = \frac{c_1 - x^2}{2x} \implies u = \frac{2x}{c_1 - x^2} \implies y = -\ln\left|c_1 - x^2\right| + c_2.$$

15. In this problem the thinner curve is obtained using a numerical solver, while the thicker curve is the graph of the Taylor polynomial.

We look for a solution of the form

$$y(x) = y(0) + y'(0)x + \frac{1}{2!}y''(0)x^2 + \frac{1}{3!}y'''(0)x^3 + \frac{1}{4!}y^{(4)}(0)x^4 + \frac{1}{5!}y^{(5)}(0)x^5.$$

From $y''(x) = x^2 + y^2 - 2y'$ we compute

$$y'''(x) = 2x + 2yy' - 2y''$$
$$y^{(4)}(x) = 2 + 2(y')^2 + 2yy'' - 2y'''$$
$$y^{(5)}(x) = 6y'y'' + 2yy''' - 2y^{(4)}.$$

Using $y(0) = 1$ and $y'(0) = 1$ we find

$$y''(0) = -1, \quad y'''(0) = 4, \quad y^{(4)}(0) = -6, \quad y^{(5)}(0) = 14.$$

An approximate solution is

$$y(x) = 1 + x - \frac{1}{2}x^2 + \frac{2}{3}x^3 - \frac{1}{4}x^4 + \frac{7}{60}x^5.$$

18. When $y = \sin x$, $y' = \cos x$, $y'' = -\sin x$, and

$$(y'')^2 - y^2 = \sin^2 x - \sin^2 x = 0.$$

When $y = e^{-x}$, $y' = -e^{-x}$, $y'' = e^{-x}$, and

$$(y'')^2 - y^2 = e^{-2x} - e^{-2x} = 0.$$

From $(y'')^2 - y^2 = 0$ we have $y'' = \pm y$, which can be treated as two linear equations. Since linear combinations of solutions of linear homogeneous differential equations are also solutions, we see that $y = c_1 e^x + c_2 e^{-x}$ and $y = c_3 \cos x + c_4 \sin x$ must satisfy the differential equation. However, linear combinations that involve both exponential and trigonometric functions will not be solutions since the differential equation is not linear and each type of function satisfies a different linear differential equation that is part of the original differential equation.

21. Let $u = dx/dt$ so that $d^2x/dt^2 = u\,du/dx$. The equation becomes $u\,du/dx = -k^2/x^2$. Separating variables we obtain

$$u\,du = -\frac{k^2}{x^2}\,dx \implies \frac{1}{2}u^2 = \frac{k^2}{x} + c \implies \frac{1}{2}v^2 = \frac{k^2}{x} + c.$$

When $t = 0$, $x = x_0$ and $v = 0$ so $0 = (k^2/x_0) + c$ and $c = -k^2/x_0$. Then

$$\frac{1}{2}v^2 = k^2\left(\frac{1}{x} - \frac{1}{x_0}\right) \quad \text{and} \quad \frac{dx}{dt} = -k\sqrt{2}\sqrt{\frac{x_0 - x}{xx_0}}.$$

Separating variables we have

$$-\sqrt{\frac{xx_0}{x_0 - x}}\,dx = k\sqrt{2}\,dt \implies t = -\frac{1}{k}\sqrt{\frac{x_0}{2}}\int\sqrt{\frac{x}{x_0 - x}}\,dx.$$

Using *Mathematica* to integrate we obtain

$$t = -\frac{1}{k}\sqrt{\frac{x_0}{2}}\left[-\sqrt{x(x_0 - x)} - \frac{x_0}{2}\tan^{-1}\frac{(x_0 - 2x)}{2x}\sqrt{\frac{x}{x_0 - x}}\right]$$

$$= \frac{1}{k}\sqrt{\frac{x_0}{2}}\left[\sqrt{x(x_0 - x)} + \frac{x_0}{2}\tan^{-1}\frac{x_0 - 2x}{2\sqrt{x(x_0 - x)}}\right].$$

EXERCISES 3.8

Linear Models: Initial-Value Problems

3. From $\frac{3}{4}x'' + 72x = 0$, $x(0) = -1/4$, and $x'(0) = 0$ we obtain $x = -\frac{1}{4}\cos 4\sqrt{6}\,t$.

6. From $50x'' + 200x = 0$, $x(0) = 0$, and $x'(0) = -10$ we obtain $x = -5\sin 2t$ and $x' = -10\cos 2t$.

9. From $\frac{1}{4}x'' + x = 0$, $x(0) = 1/2$, and $x'(0) = 3/2$ we obtain

$$x = \frac{1}{2}\cos 2t + \frac{3}{4}\sin 2t = \frac{\sqrt{13}}{4}\sin(2t + 0.588).$$

12. From $x'' + 9x = 0$, $x(0) = -1$, and $x'(0) = -\sqrt{3}$ we obtain

$$x = -\cos 3t - \frac{\sqrt{3}}{3}\sin 3t = \frac{2}{\sqrt{3}}\sin\left(3t + \frac{4\pi}{3}\right)$$

and $x' = 2\sqrt{3}\cos(3t + 4\pi/3)$. If $x' = 3$ then $t = -7\pi/18 + 2n\pi/3$ and $t = -\pi/2 + 2n\pi/3$ for $n = 1, 2, 3, \ldots$.

15. For large values of t the differential equation is approximated by $x'' = 0$. The solution of this equation is the linear function $x = c_1 t + c_2$. Thus, for large time, the restoring force will have decayed to the point where the spring is incapable of returning the mass, and the spring will simply keep on stretching.

18. (a) below   (b) from rest

21. From $\frac{1}{8}x'' + x' + 2x = 0$, $x(0) = -1$, and $x'(0) = 8$ we obtain $x = 4te^{-4t} - e^{-4t}$ and $x' = 8e^{-4t} - 16te^{-4t}$. If $x = 0$ then $t = 1/4$ second. If $x' = 0$ then $t = 1/2$ second and the extreme displacement is $x = e^{-2}$ feet.

24. (a) $x = \frac{1}{3}e^{-8t}\left(4e^{6t} - 1\right)$ is not zero for $t \geq 0$; the extreme displacement is $x(0) = 1$ meter.

   (b) $x = \frac{1}{3}e^{-8t}\left(5 - 2e^{6t}\right) = 0$ when $t = \frac{1}{6}\ln\frac{5}{2} \approx 0.153$ second; if $x' = \frac{4}{3}e^{-8t}\left(e^{6t} - 10\right) = 0$ then $t = \frac{1}{6}\ln 10 \approx 0.384$ second and the extreme displacement is $x = -0.232$ meter.

27. From $\frac{5}{16}x'' + \beta x' + 5x = 0$ we find that the roots of the auxiliary equation are $m = -\frac{8}{5}\beta \pm \frac{4}{5}\sqrt{4\beta^2 - 25}$.

   (a) If $4\beta^2 - 25 > 0$ then $\beta > 5/2$.
   (b) If $4\beta^2 - 25 = 0$ then $\beta = 5/2$.
   (c) If $4\beta^2 - 25 < 0$ then $0 < \beta < 5/2$.

30. (a) If $x'' + 2x' + 5x = 12\cos 2t + 3\sin 2t$, $x(0) = 1$, and $x'(0) = 5$ then $x_c = e^{-t}(c_1\cos 2t + c_2\sin 2t)$ and $x_p = 3\sin 2t$ so that the equation of motion is

$$x = e^{-t}\cos 2t + 3\sin 2t.$$

(b)

(c)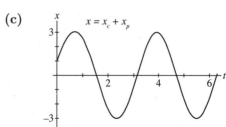

33. From $2x'' + 32x = 68e^{-2t}\cos 4t$, $x(0) = 0$, and $x'(0) = 0$ we obtain $x_c = c_1\cos 4t + c_2\sin 4t$ and $x_p = \frac{1}{2}e^{-2t}\cos 4t - 2e^{-2t}\sin 4t$ so that

$$x = -\frac{1}{2}\cos 4t + \frac{9}{4}\sin 4t + \frac{1}{2}e^{-2t}\cos 4t - 2e^{-2t}\sin 4t.$$

36. (a) From $100x'' + 1600x = 1600\sin 8t$, $x(0) = 0$, and $x'(0) = 0$ we obtain $x_c = c_1\cos 4t + c_2\sin 4t$ and $x_p = -\frac{1}{3}\sin 8t$ so that by a trig identity

$$x = \frac{2}{3}\sin 4t - \frac{1}{3}\sin 8t = \frac{2}{3}\sin 4t - \frac{2}{3}\sin 4t\cos 4t.$$

(b) If $x = \frac{1}{3}\sin 4t(2 - 2\cos 4t) = 0$ then $t = n\pi/4$ for $n = 0, 1, 2, \ldots$.

(c) If $x' = \frac{8}{3}\cos 4t - \frac{8}{3}\cos 8t = \frac{8}{3}(1 - \cos 4t)(1 + 2\cos 4t) = 0$ then $t = \pi/3 + n\pi/2$ and $t = \pi/6 + n\pi/2$ for $n = 0, 1, 2, \ldots$ at the extreme values. *Note*: There are many other values of t for which $x' = 0$.

(d) $x(\pi/6 + n\pi/2) = \sqrt{3}/2$ cm and $x(\pi/3 + n\pi/2) = -\sqrt{3}/2$ cm.

(e)

39. (a) From $x'' + \omega^2 x = F_0\cos\gamma t$, $x(0) = 0$, and $x'(0) = 0$ we obtain $x_c = c_1\cos\omega t + c_2\sin\omega t$ and $x_p = (F_0\cos\gamma t)/(\omega^2 - \gamma^2)$ so that

$$x = -\frac{F_0}{\omega^2 - \gamma^2}\cos\omega t + \frac{F_0}{\omega^2 - \gamma^2}\cos\gamma t.$$

(b) $\displaystyle\lim_{\gamma\to\omega}\frac{F_0}{\omega^2 - \gamma^2}(\cos\gamma t - \cos\omega t) = \lim_{\gamma\to\omega}\frac{-F_0 t\sin\gamma t}{-2\gamma} = \frac{F_0}{2\omega}t\sin\omega t.$

42. See the article "Distinguished Oscillations of a Forced Harmonic Oscillator" by T.G. Procter in *The College Mathematics Journal*, March, 1995. In this article the author illustrates that for $F_0 = 1$, $\lambda = 0.01$, $\gamma = 22/9$, and $\omega = 2$ the system exhibits beats oscillations on the interval $[0, 9\pi]$, but that this phenomenon is transient as $t \to \infty$.

45. Solving $\frac{1}{20}q'' + 2q' + 100q = 0$ we obtain $q(t) = e^{-20t}(c_1\cos 40t + c_2\sin 40t)$. The initial conditions $q(0) = 5$ and $q'(0) = 0$ imply $c_1 = 5$ and $c_2 = 5/2$. Thus

$$q(t) = e^{-20t}\left(5\cos 40t + \frac{5}{2}\sin 40t\right) = \sqrt{25 + 25/4}\,e^{-20t}\sin(40t + 1.1071)$$

and $q(0.01) \approx 4.5676$ coulombs. The charge is zero for the first time when $40t + 1.1071 = \pi$ or $t \approx 0.0509$ second.

48. Solving $q'' + 100q' + 2500q = 30$ we obtain $q(t) = c_1 e^{-50t} + c_2 t e^{-50t} + 0.012$. The initial conditions $q(0) = 0$ and $q'(0) = 2$ imply $c_1 = -0.012$ and $c_2 = 1.4$. Thus, using $i(t) = q'(t)$ we get

$$q(t) = -0.012e^{-50t} + 1.4te^{-50t} + 0.012 \quad\text{and}\quad i(t) = 2e^{-50t} - 70te^{-50t}.$$

Solving $i(t) = 0$ we see that the maximum charge occurs when $t = 1/35$ second and $q(1/35) \approx 0.01871$ coulomb.

51. The differential equation is $\frac{1}{2}q'' + 20q' + 1000q = 100\sin 60t$. To use Example 10 in the text we identify $E_0 = 100$ and $\gamma = 60$. Then

$$X = L\gamma - \frac{1}{c\gamma} = \frac{1}{2}(60) - \frac{1}{0.001(60)} \approx 13.3333,$$

$$Z = \sqrt{X^2 + R^2} = \sqrt{X^2 + 400} \approx 24.0370,$$

and

$$\frac{E_0}{Z} = \frac{100}{Z} \approx 4.1603.$$

From Problem 50, then

$$i_p(t) \approx 4.1603\sin(60t + \phi)$$

where $\sin\phi = -X/Z$ and $\cos\phi = R/Z$. Thus $\tan\phi = -X/R \approx -0.6667$ and ϕ is a fourth quadrant angle. Now $\phi \approx -0.5880$ and

$$i_p(t) = 4.1603\sin(60t - 0.5880).$$

54. In Problem 50 it is shown that the amplitude of the steady-state current is E_0/Z, where $Z = \sqrt{X^2 + R^2}$ and $X = L\gamma - 1/C\gamma$. Since E_0 is constant the amplitude will be a maximum when Z is a minimum. Since R is constant, Z will be a minimum when $X = 0$. Solving $L\gamma - 1/C\gamma = 0$ for γ we obtain $\gamma = 1/\sqrt{LC}$. The maximum amplitude will be E_0/R.

57. In an LC-series circuit there is no resistor, so the differential equation is

$$L\frac{d^2q}{dt^2} + \frac{1}{C}q = E(t).$$

Then $q(t) = c_1\cos\left(t/\sqrt{LC}\right) + c_2\sin\left(t/\sqrt{LC}\right) + q_p(t)$ where $q_p(t) = A\sin\gamma t + B\cos\gamma t$. Substituting $q_p(t)$ into the differential equation we find

$$\left(\frac{1}{C} - L\gamma^2\right)A\sin\gamma t + \left(\frac{1}{C} - L\gamma^2\right)B\cos\gamma t = E_0\cos\gamma t.$$

Equating coefficients we obtain $A = 0$ and $B = E_0 C/(1 - LC\gamma^2)$. Thus, the charge is

$$q(t) = c_1\cos\frac{1}{\sqrt{LC}}t + c_2\sin\frac{1}{\sqrt{LC}}t + \frac{E_0 C}{1 - LC\gamma^2}\cos\gamma t.$$

The initial conditions $q(0) = q_0$ and $q'(0) = i_0$ imply $c_1 = q_0 - E_0 C/(1 - LC\gamma^2)$ and $c_2 = i_0\sqrt{LC}$. The current is $i(t) = q'(t)$ or

$$i(t) = -\frac{c_1}{\sqrt{LC}}\sin\frac{1}{\sqrt{LC}}t + \frac{c_2}{\sqrt{LC}}\cos\frac{1}{\sqrt{LC}}t - \frac{E_0 C\gamma}{1 - LC\gamma^2}\sin\gamma t$$

$$= i_0\cos\frac{1}{\sqrt{LC}}t - \frac{1}{\sqrt{LC}}\left(q_0 - \frac{E_0 C}{1 - LC\gamma^2}\right)\sin\frac{1}{\sqrt{LC}}t - \frac{E_0 C\gamma}{1 - LC\gamma^2}\sin\gamma t.$$

EXERCISES 3.9

Linear Models: Boundary-Value Problems

3. (a) The general solution is

$$y(x) = c_1 + c_2 x + c_3 x^2 + c_4 x^3 + \frac{w_0}{24EI} x^4.$$

The boundary conditions are $y(0) = 0$, $y'(0) = 0$, $y(L) = 0$, $y''(L) = 0$. The first two conditions give $c_1 = 0$ and $c_2 = 0$. The conditions at $x = L$ give the system

$$c_3 L^2 + c_4 L^3 + \frac{w_0}{24EI} L^4 = 0$$

$$2c_3 + 6c_4 L + \frac{w_0}{2EI} L^2 = 0.$$

Solving, we obtain $c_3 = w_0 L^2/16EI$ and $c_4 = -5w_0 L/48EI$. The deflection is

$$y(x) = \frac{w_0}{48EI}(3L^2 x^2 - 5Lx^3 + 2x^4).$$

(b)

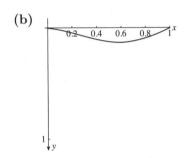

6. (a) $y_{\max} = y(L) = w_0 L^4/8EI$

(b) Replacing both L and x by $L/2$ in $y(x)$ we obtain $w_0 L^4/128EI$, which is $1/16$ of the maximum deflection when the length of the beam is L.

(c) $y_{\max} = y(L/2) = 5w_0 L^4/384EI$

(d) The maximum deflection in Example 1 is $y(L/2) = (w_0/24EI)L^4/16 = w_0 L^4/384EI$, which is $1/5$ of the maximum displacement of the beam in part c.

9. This is Example 2 in the text with $L = \pi$. The eigenvalues are $\lambda_n = n^2\pi^2/\pi^2 = n^2$, $n = 1, 2, 3, \ldots$ and the corresponding eigenfunctions are $y_n = \sin(n\pi x/\pi) = \sin nx$, $n = 1, 2, 3, \ldots$.

12. For $\lambda \le 0$ the only solution of the boundary-value problem is $y = 0$. For $\lambda = \alpha^2 > 0$ we have

$$y = c_1 \cos \alpha x + c_2 \sin \alpha x.$$

Since $y(0) = 0$ implies $c_1 = 0$, $y = c_2 \sin x\, dx$. Now

$$y'\left(\frac{\pi}{2}\right) = c_2 \alpha \cos \alpha \frac{\pi}{2} = 0$$

gives

$$\alpha\frac{\pi}{2} = \frac{(2n-1)\pi}{2} \quad \text{or} \quad \lambda = \alpha^2 = (2n-1)^2, \ n = 1, 2, 3, \ldots.$$

The eigenvalues $\lambda_n = (2n-1)^2$ correspond to the eigenfunctions $y_n = \sin(2n-1)x$.

15. The auxiliary equation has solutions

$$m = \frac{1}{2}\left(-2 \pm \sqrt{4 - 4(\lambda + 1)}\right) = -1 \pm \alpha.$$

For $\lambda = -\alpha^2 < 0$ we have

$$y = e^{-x}\left(c_1 \cosh \alpha x + c_2 \sinh \alpha x\right).$$

The boundary conditions imply

$$y(0) = c_1 = 0$$

$$y(5) = c_2 e^{-5} \sinh 5\alpha = 0$$

so $c_1 = c_2 = 0$ and the only solution of the boundary-value problem is $y = 0$.

For $\lambda = 0$ we have

$$y = c_1 e^{-x} + c_2 x e^{-x}$$

and the only solution of the boundary-value problem is $y = 0$.

For $\lambda = \alpha^2 > 0$ we have

$$y = e^{-x}\left(c_1 \cos \alpha x + c_2 \sin \alpha x\right).$$

Now $y(0) = 0$ implies $c_1 = 0$, so

$$y(5) = c_2 e^{-5} \sin 5\alpha = 0$$

gives

$$5\alpha = n\pi \quad \text{or} \quad \lambda = \alpha^2 = \frac{n^2 \pi^2}{25}, \quad n = 1, 2, 3, \ldots.$$

The eigenvalues $\lambda_n = \dfrac{n^2 \pi^2}{25}$ correspond to the eigenfunctions $y_n = e^{-x} \sin \dfrac{n\pi}{5} x$ for $n = 1, 2, 3, \ldots.$

18. For $\lambda = 0$ the general solution is $y = c_1 + c_2 \ln x$. Now $y' = c_2/x$, so $y'(e^{-1}) = c_2 e = 0$ implies $c_2 = 0$. Then $y = c_1$ and $y(1) = 0$ gives $c_1 = 0$. Thus $y(x) = 0$.

For $\lambda = -\alpha^2 < 0$, $y = c_1 x^{-\alpha} + c_2 x^{\alpha}$. The boundary conditions give $c_2 = c_1 e^{2\alpha}$ and $c_1 = 0$, so that $c_2 = 0$ and $y(x) = 0$.

For $\lambda = \alpha^2 > 0$, $y = c_1 \cos(\alpha \ln x) + c_2 \sin(\alpha \ln x)$. From $y(1) = 0$ we obtain $c_1 = 0$ and $y = c_2 \sin(\alpha \ln x)$. Now $y' = c_2(\alpha/x) \cos(\alpha \ln x)$, so $y'(e^{-1}) = c_2 e \alpha \cos \alpha = 0$ implies $\cos \alpha = 0$ or $\alpha = (2n-1)\pi/2$ and $\lambda = \alpha^2 = (2n-1)^2 \pi^2/4$ for $n = 1, 2, 3, \ldots.$ The corresponding eigenfunctions are

$$y_n = \sin\left(\frac{2n-1}{2}\pi \ln x\right).$$

21. If restraints are put on the column at $x = L/4$, $x = L/2$, and $x = 3L/4$, then the critical load will be P_4.

24. (a) The boundary-value problem is

$$\frac{d^4 y}{dx^4} + \lambda \frac{d^2 y}{dx^2} = 0, \quad y(0) = 0, y''(0) = 0, \ y(L) = 0, y'(L) = 0,$$

where $\lambda = \alpha^2 = P/EI$. The solution of the differential equation is $y = c_1 \cos \alpha x + c_2 \sin \alpha x + c_3 x + c_4$ and the conditions $y(0) = 0$, $y''(0) = 0$ yield $c_1 = 0$ and $c_4 = 0$. Next, by applying $y(L) = 0$, $y'(L) = 0$ to $y = c_2 \sin \alpha x + c_3 x$ we get the system of equations

$$c_2 \sin \alpha L + c_3 L = 0$$

$$\alpha c_2 \cos \alpha L + c_3 = 0.$$

To obtain nontrivial solutions c_2, c_3, we must have the determinant of the coefficients equal to zero:

$$\begin{vmatrix} \sin \alpha L & L \\ \alpha \cos \alpha L & 1 \end{vmatrix} = 0 \quad \text{or} \quad \tan \beta = \beta,$$

where $\beta = \alpha L$. If β_n denotes the positive roots of the last equation, then the eigenvalues are found from $\beta_n = \alpha_n L = \sqrt{\lambda_n} L$ or $\lambda_n = (\beta_n/L)^2$. From $\lambda = P/EI$ we see that the critical loads are $P_n = \beta_n^2 EI/L^2$. With the aid of a CAS we find that the first positive root of $\tan \beta = \beta$ is (approximately) $\beta_1 = 4.4934$, and so the Euler load is (approximately) $P_1 = 20.1907 EI/L^2$. Finally, if we use $c_3 = -c_2 \alpha \cos \alpha L$, then the deflection curves are

$$y_n(x) = c_2 \sin \alpha_n x + c_3 x = c_2 \left[\sin \left(\frac{\beta_n}{L} x \right) - \left(\frac{\beta_n}{L} \cos \beta_n \right) x \right].$$

(b) With $L = 1$ and c_2 appropriately chosen, the general shape of the first buckling mode,

$$y_1(x) = c_2 \left[\sin \left(\frac{4.4934}{L} x \right) - \left(\frac{4.4934}{L} \cos(4.4934) \right) x \right],$$

is shown below.

27. The auxiliary equation is $m^2 + m = m(m+1) = 0$ so that $u(r) = c_1 r^{-1} + c_2$. The boundary conditions $u(a) = u_0$ and $u(b) = u_1$ yield the system $c_1 a^{-1} + c_2 = u_0$, $c_1 b^{-1} + c_2 = u_1$. Solving gives

$$c_1 = \left(\frac{u_0 - u_1}{b - a} \right) ab \quad \text{and} \quad c_2 = \frac{u_1 b - u_0 a}{b - a}.$$

Thus

$$u(r) = \left(\frac{u_0 - u_1}{b - a} \right) \frac{ab}{r} + \frac{u_1 b - u_0 a}{b - a}.$$

30. The initial-value problem is

$$x'' + \frac{2}{m} x' + \frac{k}{m} x = 0, \quad x(0) = 0, \ x'(0) = v_0.$$

With $k = 10$, the auxiliary equation has roots $\gamma = -1/m \pm \sqrt{1 - 10m}/m$. Consider the three cases:

(*i*) $m = \frac{1}{10}$. The roots are $\gamma_1 = \gamma_2 = 10$ and the solution of the differential equation is $x(t) = c_1 e^{-10t} + c_2 t e^{-10t}$. The initial conditions imply $c_1 = 0$ and $c_2 = v_0$ and so $x(t) = v_0 t e^{-10t}$. The condition $x(1) = 0$ implies $v_0 e^{-10} = 0$ which is impossible because $v_0 \neq 0$.

(*ii*) $1 - 10m > 0$ or $0 < m < \frac{1}{10}$. The roots are

$$\gamma_1 = -\frac{1}{m} - \frac{1}{m} \sqrt{1 - 10m} \quad \text{and} \quad \gamma_2 = -\frac{1}{m} + \frac{1}{m} \sqrt{1 - 10m}$$

and the solution of the differential equation is $x(t) = c_1 e^{\gamma_1 t} + c_2 e^{\gamma_2 t}$. The initial conditions imply

$$c_1 + c_2 = 0$$

$$\gamma_1 c_1 + \gamma_2 c_2 = v_0$$

so $c_1 = v_0/(\gamma_1 - \gamma_2)$, $c_2 = -v_0/(\gamma_1 - \gamma_2)$, and

$$x(t) = \frac{v_0}{\gamma_1 - \gamma_2}(e^{\gamma_1 t} - e^{\gamma_2 t}).$$

Again, $x(1) = 0$ is impossible because $v_0 \neq 0$.

(*iii*) $1 - 10m < 0$ or $m > \frac{1}{10}$. The roots of the auxiliary equation are

$$\gamma_1 = -\frac{1}{m} - \frac{1}{m}\sqrt{10m - 1}\,i \qquad \text{and} \qquad \gamma_2 = -\frac{1}{m} + \frac{1}{m}\sqrt{10m - 1}\,i$$

and the solution of the differential equation is

$$x(t) = c_1 e^{-t/m}\cos\frac{1}{m}\sqrt{10m - 1}\,t + c_2 e^{-t/m}\sin\frac{1}{m}\sqrt{10m - 1}\,t.$$

The initial conditions imply $c_1 = 0$ and $c_2 = mv_0/\sqrt{10m - 1}$, so that

$$x(t) = \frac{mv_0}{\sqrt{10m - 1}}\,e^{-t/m}\sin\left(\frac{1}{m}\sqrt{10m - 1}\,t\right),$$

The condition $x(1) = 0$ implies

$$\frac{mv_0}{\sqrt{10m - 1}}e^{-1/m}\sin\frac{1}{m}\sqrt{10m - 1} = 0$$

$$\sin\frac{1}{m}\sqrt{10m - 1} = 0$$

$$\frac{1}{m}\sqrt{10m - 1} = n\pi$$

$$\frac{10m - 1}{m^2} = n^2\pi^2, \;\; n = 1, 2, 3, \ldots$$

$$(n^2\pi^2)m^2 - 10m + 1 = 0$$

$$m = \frac{10\sqrt{100 - 4n^2\pi^2}}{2n^2\pi^2} = \frac{5 \pm \sqrt{25 - n^2\pi^2}}{n^2\pi^2}.$$

Since m is real, $25 - n^2\pi^2 \geq 0$. If $25 - n^2\pi^2 = 0$, then $n^2 = 25/\pi^2$, and n is not an integer. Thus, $25 - n^2\pi^2 = (5 - n\pi)(5 + n\pi) > 0$ and since $n > 0$, $5 + n\pi > 0$, so $5 - n\pi > 0$ also. Then $n < 5/\pi$, and so $n = 1$. Therefore, the mass m will pass through the equilibrium position when $t = 1$ for

$$m_1 = \frac{5 + \sqrt{25 - \pi^2}}{\pi^2} \qquad \text{and} \qquad m_2 = \frac{5 - \sqrt{25 - \pi^2}}{\pi^2}.$$

33. (a) A solution curve has the same y-coordinate at both ends of the interval $[-\pi, \pi]$ and the tangent lines at the endpoints of the interval are parallel.

(b) For $\lambda = 0$ the solution of $y'' = 0$ is $y = c_1 x + c_2$. From the first boundary condition we have

$$y(-\pi) = -c_1\pi + c_2 = y(\pi) = c_1\pi + c_2$$

or $2c_1\pi = 0$. Thus, $c_1 = 0$ and $y = c_2$. This constant solution is seen to satisfy the boundary-value problem.

For $\lambda = -\alpha^2 < 0$ we have $y = c_1\cosh\alpha x + c_2\sinh\alpha x$. In this case the first boundary condition gives

$$y(-\pi) = c_1\cosh(-\alpha\pi) + c_2\sinh(-\alpha\pi)$$
$$= c_1\cosh\alpha\pi - c_2\sinh\alpha\pi$$
$$= y(\pi) = c_1\cosh\alpha\pi + c_2\sinh\alpha\pi$$

or $2c_2\sinh\alpha\pi = 0$. Thus $c_2 = 0$ and $y = c_1\cosh\alpha x$. The second boundary condition implies in a similar fashion that $c_1 = 0$. Thus, for $\lambda < 0$, the only solution of the boundary-value problem is $y = 0$.

For $\lambda = \alpha^2 > 0$ we have $y = c_1 \cos \alpha x + c_2 \sin \alpha x$. The first boundary condition implies

$$y(-\pi) = c_1 \cos(-\alpha \pi) + c_2 \sin(-\alpha \pi)$$

$$= c_1 \cos \alpha \pi - c_2 \sin \alpha \pi$$

$$= y(\pi) = c_1 \cos \alpha \pi + c_2 \sin \alpha \pi$$

or $2c_2 \sin \alpha \pi = 0$. Similarly, the second boundary condition implies $2c_1 \alpha \sin \alpha \pi = 0$. If $c_1 = c_2 = 0$ the solution is $y = 0$. However, if $c_1 \neq 0$ or $c_2 \neq 0$, then $\sin \alpha \pi = 0$, which implies that α must be an integer, n. Therefore, for c_1 and c_2 not both 0, $y = c_1 \cos nx + c_2 \sin nx$ is a nontrivial solution of the boundary-value problem. Since $\cos(-nx) = \cos nx$ and $\sin(-nx) = -\sin nx$, we may assume without loss of generality that the eigenvalues are $\lambda_n = \alpha^2 = n^2$, for n a positive integer. The corresponding eigenfunctions are $y_n = \cos nx$ and $y_n = \sin nx$.

(c)

$y = 2 \sin 3x$

$y = \sin 4x - 2 \cos 3x$

36. Using a CAS we find that the first four nonnegative roots of $\tan x = -x$ are approximately $2.02876, 4.91318,$ $7.97867,$ and 11.0855. The corresponding eigenvalues are $4.11586, 24.1393, 63.6591,$ and 122.889, with eigenfunctions $\sin(2.02876x), \sin(4.91318x), \sin(7.97867x),$ and $\sin(11.0855x)$.

EXERCISES 3.10
Greens's Functions

3. $y'' + 2y' + y = f(x) \Rightarrow y'' + 2y' + y = 0 \Rightarrow y_1 = e^{-x}, \; y_2 = xe^{-x} \Rightarrow$

$$W(e^{-x}, xe^{-x}) = \begin{vmatrix} e^{-x} & xe^{-x} \\ -e^{-x} & -xe^{-x} + e^{-x} \end{vmatrix} = e^{-2x}$$

$$G(x,t) = \frac{e^{-t}xe^{-x} - e^{-x}te^{-t}}{e^{-2t}} = (x-t)e^{-(x-t)} \Rightarrow y_p(x) = \int_{x_0}^{x} (x-t)e^{-(x-t)} f(t) dt$$

6. $y'' - 2y' + 2y = f(x) \Rightarrow y'' - 2y' + 2y = 0 \Rightarrow y_1 = e^x \cos x, \; y_2 = e^x \sin x \Rightarrow$

$$W(e^x \cos x, e^x \sin x) = \begin{vmatrix} e^x \cos x & e^x \sin x \\ -e^x \sin x + e^x \cos x & e^x \cos x + e^x \sin x \end{vmatrix} = e^{2x}$$

$$G(x,t) = \frac{e^t \cos t e^x \sin x - e^x \cos x e^t \sin t}{e^{2t}} = e^{x-t} \sin(x-t) \Rightarrow y_p(x) = \int_{x_0}^{x} e^{x-t} \sin(x-t) f(t) dt$$

9. $y'' + 2y' + y = e^{-x} \Rightarrow y = y_c + y_p \Rightarrow y = c_1 e^{-x} + c_2 x e^{-x} + y_p \Rightarrow$

$$y = c_1 e^{-x} + c_2 x e^{-x} + \int_{x_0}^{x} (x-t)e^{-(x-t)} e^{-t} dt$$

12. $y'' - 2y' + 2y = \cos^2 x \Rightarrow y = y_c + y_p \Rightarrow y = c_1 e^x \cos x + c_2 e^x \sin x + y_p \Rightarrow$

$y = c_1 e^x \cos x + c_2 e^x \sin x = \int_{x_0}^{x} e^{(x-t)} \sin(x - t) \cos^2 t\, dt$

15. The initial-value problemis $y'' - 10y' + 25y = e^{5x}$, $y(0) = 0$, $y'(0) = 0$. Then we find that

$$y_1 = e^{5x}, \; y_2 = xe^{5x} \quad \text{so} \quad W(e^{5x}, xe^{5x}) = \begin{vmatrix} e^{5x} & xe^{5x} \\ 5e^{5x} & 5xe^{5x} + e^{5x} \end{vmatrix} = e^{10x}.$$

Then $G(x, t) = \dfrac{e^{5t} xe^{5x} - e^{5x} te^{5t}}{e^{10t}} = (x - t)e^{5(x-t)}$ and the solution of the initial-value problemis

$$y_p(x) = \int_0^x (x - t)e^{5(x-t)} e^{5t}\, dt = xe^{5x} \int_0^x dt - e^{5x} \int_0^x t\, dt = x^2 e^{5x} - \frac{1}{2}x^2 e^{5x} = \frac{1}{2}x^2 e^{5x}.$$

18. The initial-value problemis $y'' + y = \sec^2 x$, $y(\pi) = 0$, $y'(\pi) = 0$. Then we find that

$$y_1 = \cos x, \; y_2 = \sin x \quad \text{so} \quad W(\cos x, \sin x) = \begin{vmatrix} \cos x & \sin x \\ -\sin x & \cos x \end{vmatrix} = 1.$$

Then $G(x, t) = \cos t \sin x - \cos x \sin t = \sin(x - t)$ and the solution of the initial-value problemis

$$y_p(x) = \int_\pi^x (\cos t \sin x - \cos x \sin t) \sec^2 t\, dt$$

$$= \sin x \int_\pi^x \sec t\, dt - \cos x \int_\pi^x \sec t \tan t\, dt$$

$$= -\cos x - 1 + \sin x \ln|\sec x + \tan x|.$$

21. The initial-value problemis $y'' - 10y' + 25y = e^{5x}$, $y(0) = -1$, $y'(0) = 1$, so $y(x) = c_1 e^{5x} + c_2 xe^{5x}$. The initial conditions give

$$c_1 = -1$$
$$5c_1 + c_2 = 1,$$

so $c_1 = -1$ and $c_2 = 6$, which implies that $y_h = -e^{5x} + 6xe^{5x}$. Now, y_p found in the solution of Problem 15 in this section gives

$$y = y_h + y_p = -e^{5x} + 6xe^{5x} + \frac{1}{2}x^2 e^{5x}.$$

24. The initial-value problemis $y'' + y = \sec^2 x$, $y(\pi) = \frac{1}{2}$, $y'(\pi) = -1$, so $y(x) = c_1 \cos x + c_2 \sin x$. The initial conditions give

$$-c_1 = \frac{1}{2}$$
$$-c_2 = -1,$$

so $c_1 = -\frac{1}{2}$ and $c_2 = 1$, which implies that $y_h = -\frac{1}{2}\cos x + \sin x$. Now, y_p found in the solution of Problem 18 in this section gives

$$y = y_h + y_p = -\frac{1}{2}\cos x + \sin x + (-\cos x - 1 + \sin x \ln|\sec x + \tan x|)$$

$$= -\frac{3}{2}\cos x + \sin x - 1 + \sin x \ln|\sec x + \tan x|.$$

27. The Cauchy-Euler initial-value problem $x^2 y'' - 2xy' + 2y = x$, $y(1) = 2$, $y'(1) = -1$, has auxiliary equation $m(m - 1) - 2m + 2 = m^2 - 3m + 2 = (m - 1)(m - 2) = 0$ so $m_1 = 1$, $m_2 = 2$, $y(x) = c_1 x + c_2 x^2$, and $y' = c_1 + 2c_2 x$. The initial conditions give

$$c_1 + c_2 = 2$$
$$c_1 + 2c_2 = -1,$$

so $c_1 = 5$ and $c_2 = -3$, which implies that $y_h = 5x - 3x^2$. The Wronskian is

$$W(x, x^2) = \begin{vmatrix} x & x^2 \\ 1 & 2x \end{vmatrix} = x^2.$$

Then $G(x,t) = \dfrac{tx^2 - xt^2}{t^2} = \dfrac{x(x-t)}{t}$. From the standard form of the differential equationwe identify the forcing function $f(t) = \dfrac{1}{t}$. Then, for $x > 1$,

$$y_p(x) = \int_1^x \frac{x(x-t)}{t} \frac{1}{t} dt = x^2 \int_1^x \frac{1}{t^2} dt - x \int_1^x \frac{1}{t} dt$$

$$= x^2 \left(-\frac{1}{x} + 1 \right) - x(\ln x - \ln 1) = x^2 - x - x \ln x.$$

The solution of the initial-value problemis

$$y = y_h + y_p = (5x - 3x^2) + (x^2 - x - x \ln x) = 4x - 2x^2 - x \ln x.$$

30. The Cauchy-Euler initial-value problem$x^2 y'' - xy' + y = x^2$, $y(1) = 4$, $y'(1) = 3$, has auxiliary equation $m(m-1) - m + 1 = m^2 - 2m + 1 = (m-1)^2 = 0$ so $m_1 = m_2 = 1$, $y(x) = c_1 x + c_2 x \ln x$, and $y' = c_1 + c_2(1 + \ln x)$. The initial conditions give

$$c_1 = 4$$

$$c_1 + c_2 = 3,$$

so $c_1 = 4$ and $c_2 = -1$, which implies that $y_h = 4x - x \ln x$. The Wronskian is

$$W(x, x \ln x) = \begin{vmatrix} x & x \ln x \\ 1 & 1 + \ln x \end{vmatrix} = x.$$

Then $G(x,t) = \dfrac{tx \ln x - xt \ln t}{t} = x \ln x - x \ln t$. From the standard form of the differential equationwe identify the forcing function $f(t) = 1$. Then, for $x > 1$,

$$y_p(x) = \int_1^x (x \ln x - x \ln t) dt = x \ln x \int_1^x dt - x \int_1^x \ln t \, dt$$

$$= x \ln x (x - 1) - x(x \ln x - x + 1) = x^2 - x - x \ln x.$$

The solution of the initial-value problemis

$$y = y_h + y_p = 4x - x \ln x + (x^2 - x - x \ln x) = x^2 + 3x - 2x \ln x.$$

33. The initial-value problemis $y'' + y = f(x)$, $y(0) = 1$, $y'(0) = -1$, where

$$f(x) = \begin{cases} 0, & x < 0 \\ 10, & 0 \leq x \leq 3\pi \\ 0, & x > 3\pi. \end{cases}$$

We first find

$$y_h(x) = \cos x - \sin x \quad \text{and} \quad y_p(x) = \int_0^x \sin(x - t) f(t) dt.$$

Then for $x < 0$

$$y_p(x) = \int_0^x \sin(x - t) \, 0 \, dt = 0,$$

64

for $0 \leq x \leq 3\pi$

$$y_p(x) = 10 \int_0^x \sin(x - t)dt = 10 - 10\cos x,$$

and for $x > 3\pi$

$$y_p(x) = 10 \int_0^{3\pi} \sin(x - t)dt + \int_{3\pi}^x \sin(x - t)0dt = -20\cos x.$$

The solution is

$$y(x) = y_h(x) + y_p(x) = \cos x - \sin x + y_p(x),$$

where

$$y_p(x) = \begin{cases} 0, & x < 0 \\ 10 - 10\cos x, & 0 \leq x \leq 3\pi \\ -20\cos x, & x > 3\pi. \end{cases}$$

36. The boundary-value problemis $y'' = f(x)$, $y(0) = 0$, $y(1) + y'(1) = 0$. The solution of the associated homogeneous equation is $y = c_1 + c_2 x$.

(a) To satisfy $y(0) = 0$ we take $y_1(x) = x$ and to satisfy $y(1) + y'(1) = 0$ we note that $y(1) + y'(1) = (c_1 + c_2) + c_2 = 0$ which implies that $c_1 = -2c_2$. Taking $c_2 = 1$, we find that $c_1 = -2$, so we have $y_2(x) = -2 + x$. The Wronskian of y_1 and y_2 iss

$$W(y_1, y_2) = \begin{vmatrix} x & -2 + x \\ 1 & 1 \end{vmatrix} = 2,$$

so

$$G(x, t) = \begin{cases} \frac{1}{2}t(x - 2), & 0 \leq t \leq x \\ \frac{1}{2}x(t - 2), & x \leq t \leq 1. \end{cases}$$

Therefore

$$y_p(x) = \int_0^1 G(x, t)f(t)dt = \frac{1}{2}(x - 2)\int_0^x tf(t)dt + \frac{1}{2}x\int_x^1 (t - 2)f(t)dt.$$

(b) By the Product Rule and the Fundamental Theorem of Calculus, the first two derivative of $y_p(x)$ are

$$y_p'(x) = \frac{1}{2}(x - 2)xf(x) + \frac{1}{2}\int_0^x tf(t)dt + \frac{1}{2}x[-(x - 2)f(x)] + \frac{1}{2}\int_x^1 (t - 2)f(t)dt$$

$$y_p''(x) = \frac{1}{2}(x - 2)[xf'(x) + f(x)] + \frac{1}{2}xf(x) + \frac{1}{2}xf(x)$$

$$- \frac{1}{2}(x^2 - 2x)f'(x) - \frac{1}{2}(2x - 2)f(x) - \frac{1}{2}(x - 2)f(x)$$

$$= \frac{1}{2}\left[x^2 f'(x) + xf(x) - 2xf'(x) - 2f(x) + xf(x) + xf(x) - x^2 f'(x)\right.$$

$$\left. + 2xf'(x) - 2xf(x) + 2f(x) - xf(x) + 2f(x)\right] = f(x)$$

Thus, $y_p(x)$ satisfies the differential equation. To see that the boundary conditions are satisfied we first compute

$$y_p(0) = \frac{1}{2}(0 - 2)\int_0^0 tf(t)dt + \frac{1}{2}(0)\int_0^1 (t - 2)f(t)dt = 0.$$

Next we use $y_p'(x)$ found at the beginning of this part of the solution to compute

$$y_p(1) + y_p'(1) = \frac{1}{2}(1 - 2)\int_0^1 tf(t)dt + \frac{1}{2}(1)\int_1^1 (t - 2)f(t)dt + \frac{1}{2}(1 - 2)f(x)$$

$$+ \frac{1}{2}\int_0^1 tf(t)dt + \frac{1}{2}(1)[-(1 - 2)f(x)] + \frac{1}{2}(1)\int_1^1 (1 - 2)f(t)dt$$

65

$$= \frac{1}{2}(-1)\int_0^1 tf(t)dt + \frac{1}{2}(-1)f(1) + \frac{1}{2}\int_0^1 tf(t)dt + \frac{1}{2}f(1) = 0.$$

39. The boundary-value problemis $y'' + y = 1$, $y(0) = 0$, $y(1) = 0$. The solution of the associated homogeneous equation is $y = c_1\cos x + c_2\sin x$. Since $y(0) = c_1\cos 0 + c_2\sin 0 = c_1 = 0$, we take $y_1(x) = \sin x$. To satisfy $y(1) = 0$ we note that $y(1) = c_1\cos 1 + c_2\sin 1 = 0$ which implies that $c_1 = -c_2\sin 1/\cos 1$ so

$$y(x) = -c_2\frac{\sin 1}{\cos 1}\cos x + c_2\sin x = -\frac{c_2}{\cos 1}(\sin x\cos 1 - \cos x\sin 1).$$

Taking $c_2 = -\cos 1$, we have

$$y_2(x) = \sin x\cos 1 - \cos x\sin 1 = \sin(x-1).$$

The Wronskian of y_1 and y_2 is

$$W(y_1, y_2) = \begin{vmatrix} \sin x & \sin(x-1) \\ \cos x & \cos(x-1) \end{vmatrix} = \sin x\cos(x-1) - \cos x\sin(x-1) = \sin[x - (x-1)] = \sin 1,$$

so

$$G(x,t) = \begin{cases} \dfrac{\sin t\sin(x-1)}{\sin 1}, & 0 \le t \le x \\ \dfrac{\sin x\sin(t-1)}{\sin 1}, & x \le t \le 1. \end{cases}$$

Therefore, taking $f(t) = 1$,

$$y_p(x) = \int_0^1 G(x,t)dt = \frac{\sin(x-1)}{\sin 1}\int_0^x \sin t\,dt + \frac{\sin x}{\sin 1}\int_x^1 \sin(t-1)dt$$

$$= \frac{\sin(x-1)}{\sin 1}(-\cos x + 1) + \frac{\sin x}{\sin 1}[-1 + \cos(x-1)] = \frac{\sin(x-1)}{\sin 1} - \frac{\sin x}{\sin 1} + 1.$$

42. The boundary-value problemis $y'' - y' = e^{2x}$, $y(0) = 0$, $y(1) = 0$. The solution of the associated homogeneous equation is $y = c_1 + c_2 e^x$. Since $y(0) = c_1 + c_2 = 0$, we see that $c_1 = -c_2$ and we take $y_1(x) = 1 - e^x$. To satisfy $y(1) = 0$ we note that $y(1) = c_1 + c_2 e = 0$ which implies that $c_1 = -c_2 e$ so

$$y(x) = -c_2 e + c_2 e^x = -c_2 e(1 - e^{x-1}).$$

Taking $c_2 = -1/e$, we have $y_2(x) = 1 - e^{x-1}$. The Wronskian of y_1 and y_2 is

$$W(y_1, y_2) = \begin{vmatrix} 1 - e^x & 1 - e^{x-1} \\ -e^x & -e^{x-1} \end{vmatrix} = -e^{x-1} + e^{2x-1} + e^x - e^{2x-1} = e^x - e^{x-1} = e^x(1 - e^{-1}),$$

so

$$G(x,t) = \begin{cases} \dfrac{(1 - e^t)(1 - e^{x-1})}{e^t(1 - e^{-1})}, & 0 \le t \le x \\ \dfrac{(1 - e^x)(1 - e^{t-1})}{e^t(1 - e^{-1})}, & x \le t \le 1. \end{cases}$$

Therefore, taking $f(t) = e^{2t}$,

$$y_p(x) = \int_0^1 G(x,t)e^{2t}dt = \frac{1 - e^{x-1}}{1 - e^{-1}}\int_0^x (e^t - e^{2t})dt + \frac{1 - e^x}{1 - e^{-1}}\int_x^1 (e^t - e^{2t-1})dt$$

$$= \frac{1 - e^{x-1}}{1 - e^{-1}}\left(e^x - \frac{1}{2}e^{2x} - \frac{1}{2}\right) + \frac{1 - e^x}{1 - e^{-1}}\left(\frac{1}{2}e - e^x + \frac{1}{2}e^{2x-1}\right)$$

$$= \frac{1}{2}e^{2x} - \frac{1}{2}e^x = \frac{1}{2}e^{x+1} + \frac{1}{2}e.$$

45. Let $L = \dfrac{d^2}{dx^2} + P\dfrac{d}{dx} + Q$ be a differential operator. If

$$y(x) = y_p(x) + \frac{B}{y_1(b)}y_1(x) + \frac{A}{y_2(a)}y_2(x)$$

then

$$L(y) = \overbrace{L(y_p)}^{f(x)} + \frac{B}{y_1(b)}\overbrace{L(y_1)}^{0} + \frac{A}{y_2(a)}\overbrace{L(y_2)}^{0} = f(x).$$

Moreover,

$$y(a) = \overbrace{y_p(a)}^{0} + \frac{B}{y_1(b)}\overbrace{y_1(a)}^{0} + \frac{A}{y_2(a)}y_2(a) = A$$

and

$$y(b) = \overbrace{y_p(b)}^{0} + \frac{B}{y_1(b)}y_1(b) + \frac{A}{y_2(a)}\overbrace{y_2(b)}^{0} = B.$$

Now it was assumed at the start of the discussions on boundary-value problems that

$$y'' + Py' + Qy = 0 \quad y(a) = 0, \ y(b) = 0$$

possessed only the trivial solution $y = 0$. But if the general solution of the differential equationis $y(x) = c_1 y_1(x) + c_2 y_2(x)$, then

$$y(a) = c_1 y_1(a) + c_2 y_2(a) = 0$$
$$y(b) = c_1 y_1(b) + c_2 y_2(b) = 0.$$

This system will have a unique *trivial* solution for c_1 and c_2 if and only if

$$W(y_1, y_2) = \begin{vmatrix} y_1(a) & y_2(a) \\ y_1(b) & y_2(b) \end{vmatrix} = \overbrace{y_1(a)}^{0} + \overbrace{y_2(b)}^{0} - y_2(a)y_1(b) \neq 0.$$

This requires $y_1(b) \neq 0$ and $y_2(a) \neq 0$.

EXERCISES 3.11

Nonlinear Models

3. The period corresponding to $x(0) = 1$, $x'(0) = 1$ is approximately 5.8. The second initial-value problem does not have a periodic solution.

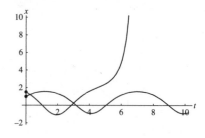

6. From the graphs we see that the interval is approximately $(-0.8, 1.1)$.

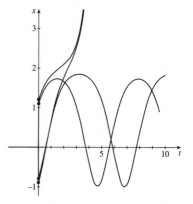

9. This is a damped hard spring, so x will approach 0 as t approaches ∞.

12. (a)

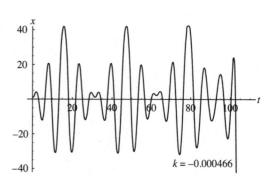

The system appears to be oscillatory for $-0.000465 \leq k_1 < 0$ and nonoscillatory for $k_1 \leq -0.000466$.

(b)

The system appears to be oscillatory for $-0.3493 \le k_1 < 0$ and nonoscillatory for $k_1 \le -0.3494$.

15. (a) Intuitively, one might expect that only half of a 10-pound chain could be lifted by a 5-pound vertical force.

(b) Since $x = 0$ when $t = 0$, and $v = dx/dt = \sqrt{160 - 64x/3}$, we have $v(0) = \sqrt{160} \approx 12.65$ ft/s.

(c) Since x should always be positive, we solve $x(t) = 0$, getting $t = 0$ and $t = \frac{3}{2}\sqrt{5/2} \approx 2.3717$. Since the graph of $x(t)$ is a parabola, the maximum value occurs at $t_m = \frac{3}{4}\sqrt{5/2}$. (This can also be obtained by solving $x'(t) = 0$.) At this time the height of the chain is $x(t_m) \approx 7.5$ ft. This is higher than predicted because of the momentum generated by the force. When the chain is 5 feet high it still has a positive velocity of about 7.3 ft/s, which keeps it going higher for a while.

(d) Once the length of the chain passes 5 feet in the air, the weight of the chain is a force greater than the constant 5 pounds upward force and so the weight of the chain will pull it back down and it will continue to go down due to momentum. But then at some point when the length of the chain is less than 5 feet the upward constant force is greater than the downward force and so the chain is again pulled upward.

18. (a) There are two forces acting on the chain as it falls from the platform. One is the force due to gravity on the portion of the chain hanging over the edge of the platform. This is $F_1 = 2x$. The second is due to the motion of the portion of the chain stretched out on the platform. By Newton's second law this is

$$F_2 = \frac{d}{dt}[mv] = \frac{d}{dt}\left[\frac{(8-x)2}{32}v\right] = \frac{d}{dt}\left[\frac{8-x}{16}v\right]$$

$$= \frac{8-x}{16}\frac{dv}{dt} - \frac{1}{16}v\frac{dx}{dt} = \frac{1}{16}\left[(8-x)\frac{dv}{dt} - v^2\right].$$

From $\dfrac{d}{dt}[mv] = F_1 - F_2$ we have

$$\frac{d}{dt}\left[\frac{2x}{32}v\right] = 2x - \frac{1}{16}\left[(8-x)\frac{dv}{dt} - v^2\right]$$

$$\frac{x}{16}\frac{dv}{dt} + \frac{1}{16}v\frac{dx}{dt} = 2x - \frac{1}{16}\left[(8-x)\frac{dv}{dt} - v^2\right]$$

$$x\frac{dv}{dt} + v^2 = 32x - (8-x)\frac{dv}{dt} + v^2$$

$$x\frac{dv}{dt} = 32x - 8\frac{dv}{dt} + x\frac{dv}{dt}$$

$$8\frac{dv}{dt} = 32x.$$

By the Chain Rule, $dv/dt = (dv/dx)(dx/dt) = v\,dv/dx$, so

$$8\frac{dv}{dt} = 8v\frac{dv}{dx} = 32x \quad \text{and} \quad v\frac{dv}{dx} = 4x.$$

(b) Integrating $v\,dv = 4x\,dx$ we get $\frac{1}{2}v^2 = 2x^2 + c$. Since $v = 0$ when $x = 3$, we have $c = -18$. Then $v^2 = 4x^2 - 36$ and $v = \sqrt{4x^2 - 36}$. Using $v = dx/dt$, separating variables, and integrating we obtain

$$\frac{dx}{\sqrt{x^2 - 9}} = 2\,dt \quad \text{and} \quad \cosh^{-1}\frac{x}{3} = 2t + c_1.$$

Solving for x we get $x(t) = 3\cosh(2t + c_1)$. Since $x = 3$ when $t = 0$, we have $\cosh c_1 = 1$ and $c_1 = 0$. Thus, $x(t) = 3\cosh 2t$. Differentiating, we find $v(t) = dx/dt = 6\sinh 2t$.

(c) To find the time when the back end of the chain leaves the platform we solve $x(t) = 3\cosh 2t = 8$. This gives $t_1 = \frac{1}{2}\cosh^{-1}\frac{8}{3} \approx 0.8184$ seconds. The velocity at this instant is

$$v(t_1) = 6\sinh\left(\cosh^{-1}\frac{8}{3}\right) = 2\sqrt{55} \approx 14.83 \text{ ft/s}.$$

(d) Replacing 8 with L and 32 with g in part (a) we have $L\,dv/dt = gx$. Then

$$L\frac{dv}{dt} = Lv\frac{dv}{dx} = gx \quad \text{and} \quad v\frac{dv}{dx} = \frac{g}{L}x.$$

Integrating we get $\frac{1}{2}v^2 = (g/2L)x^2 + c$. Setting $x = x_0$ and $v = 0$, we find $c = -(g/2L)x_0^2$. Solving for v we find

$$v(x) = \sqrt{\frac{g}{L}x^2 - \frac{g}{L}x_0^2}\,.$$

Then the velocity at which the end of the chain leaves the edge of the platform is

$$v(L) = \sqrt{\frac{g}{L}(L^2 - x_0^2)}\,.$$

21. Since $(dx/dt)^2$ is always positive, it is necessary to use $|dx/dt|(dx/dt)$ in order to account for the fact that the motion is oscillatory and the velocity (or its square) should be negative when the spring is contracting.

24. (a) The general solution of

$$\frac{d^2\theta}{dt^2} + \theta = 0$$

is $\theta(t) = c_1\cos t + c_2\sin t$. From $\theta(0) = \pi/12$ and $\theta'(0) = -1/3$ we find

$$\theta(t) = (\pi/12)\cos t - (1/3)\sin t.$$

Setting $\theta(t) = 0$ we have $\tan t = \pi/4$ which implies $t_1 = \tan^{-1}(\pi/4) \approx 0.66577$.

(b) We set $\theta(t) = \theta(0) + \theta'(0)t + \frac{1}{2}\theta''(0)t^2 + \frac{1}{6}\theta'''(0)t^3 + \cdots$ and use $\theta''(t) = -\sin\theta(t)$ together with $\theta(0) = \pi/12$ and $\theta'(0) = -1/3$. Then

$$\theta''(0) = -\sin(\pi/12) = -\sqrt{2}\,(\sqrt{3} - 1)/4$$

and

$$\theta'''(0) = -\cos\theta(0)\cdot\theta'(0) = -\cos(\pi/12)(-1/3) = \sqrt{2}\,(\sqrt{3} + 1)/12.$$

Thus

$$\theta(t) = \frac{\pi}{12} - \frac{1}{3}t - \frac{\sqrt{2}\,(\sqrt{3} - 1)}{8}t^2 + \frac{\sqrt{2}\,(\sqrt{3} + 1)}{72}t^3 + \cdots.$$

(c) Setting $\pi/12 - t/3 = 0$ we obtain $t_1 = \pi/4 \approx 0.785398$.

(d) Setting

$$\frac{\pi}{12} - \frac{1}{3}t - \frac{\sqrt{2}\,(\sqrt{3} - 1)}{8}t^2 = 0$$

and using the positive root we obtain $t_1 \approx 0.63088$.

(e) Setting

$$\frac{\pi}{12} - \frac{1}{3}t - \frac{\sqrt{2}\left(\sqrt{3}-1\right)}{8}t^2 + \frac{\sqrt{2}\left(\sqrt{3}+1\right)}{72}t^3 = 0$$

we find with the help of a CAS that $t_1 \approx 0.661973$ is the first positive root.

(f) From the output we see that $y(t)$ is an interpolating function on the interval $0 \le t \le 5$, whose graph is shown. The positive root of $y(t) = 0$ near $t = 1$ is $t_1 = 0.666404$.

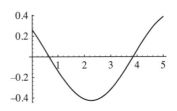

(g) To find the next two positive roots we change the interval used in **NDSolve** and **Plot** from $\{t,0,5\}$ to $\{t,0,10\}$. We see from the graph that the second and third positive roots are near 4 and 7, respectively. Replacing $\{t,1\}$ in **FindRoot** with $\{t,4\}$ and then $\{t,7\}$ we obtain $t_2 = 3.84411$ and $t_3 = 7.0218$.

EXERCISES 3.12
Solving Systems of Linear Equations

3. From $Dx = -y + t$ and $Dy = x - t$ we obtain $y = t - Dx$, $Dy = 1 - D^2x$, and $(D^2 + 1)x = 1 + t$. The solution is

$$x = c_1 \cos t + c_2 \sin t + 1 + t$$
$$y = c_1 \sin t - c_2 \cos t + t - 1.$$

6. From $(D+1)x + (D-1)y = 2$ and $3x + (D+2)y = -1$ we obtain $x = -\frac{1}{3} - \frac{1}{3}(D+2)y$, $Dx = -\frac{1}{3}(D^2 + 2D)y$, and $(D^2 + 5)y = -7$. The solution is

$$y = c_1 \cos \sqrt{5}\,t + c_2 \sin \sqrt{5}\,t - \frac{7}{5}$$

$$x = \left(-\frac{2}{3}c_1 - \frac{\sqrt{5}}{3}c_2\right) \cos \sqrt{5}\,t + \left(\frac{\sqrt{5}}{3}c_1 - \frac{2}{3}c_2\right) \sin \sqrt{5}\,t + \frac{3}{5}.$$

9. From $Dx + D^2y = e^{3t}$ and $(D+1)x + (D-1)y = 4e^{3t}$ we obtain $D(D^2+1)x = 34e^{3t}$ and $D(D^2+1)y = -8e^{3t}$. The solution is

$$y = c_1 + c_2 \sin t + c_3 \cos t - \frac{4}{15}e^{3t}$$

$$x = c_4 + c_5 \sin t + c_6 \cos t + \frac{17}{15}e^{3t}.$$

Substituting into $(D+1)x + (D-1)y = 4e^{3t}$ gives

$$(c_4 - c_1) + (c_5 - c_6 - c_3 - c_2)\sin t + (c_6 + c_5 + c_2 - c_3)\cos t = 0$$

so that $c_4 = c_1$, $c_5 = c_3$, $c_6 = -c_2$, and

$$x = c_1 - c_2 \cos t + c_3 \sin t + \frac{17}{15} e^{3t}.$$

12. From $(2D^2 - D - 1)x - (2D + 1)y = 1$ and $(D - 1)x + Dy = -1$ we obtain $(2D + 1)(D - 1)(D + 1)x = -1$ and $(2D + 1)(D + 1)y = -2$. The solution is

$$x = c_1 e^{-t/2} + c_2 e^{-t} + c_3 e^t + 1$$
$$y = c_4 e^{-t/2} + c_5 e^{-t} - 2.$$

Substituting into $(D - 1)x + Dy = -1$ gives

$$\left(-\frac{3}{2}c_1 - \frac{1}{2}c_4\right) e^{-t/2} + (-2c_2 - c_5)e^{-t} = 0$$

so that $c_4 = -3c_1$, $c_5 = -2c_2$, and

$$y = -3c_1 e^{-t/2} - 2c_2 e^{-t} - 2.$$

15. Multiplying the first equation by $D + 1$ and the second equation by $D^2 + 1$ and subtracting we obtain $(D^4 - D^2)x = 1$. Then

$$x = c_1 + c_2 t + c_3 e^t + c_4 e^{-t} - \frac{1}{2}t^2.$$

Multiplying the first equation by $D + 1$ and subtracting we obtain $D^2(D + 1)y = 1$. Then

$$y = c_5 + c_6 t + c_7 e^{-t} - \frac{1}{2}t^2.$$

Substituting into $(D - 1)x + (D^2 + 1)y = 1$ gives

$$(-c_1 + c_2 + c_5 - 1) + (-2c_4 + 2c_7)e^{-t} + (-1 - c_2 + c_6)t = 1$$

so that $c_5 = c_1 - c_2 + 2$, $c_6 = c_2 + 1$, and $c_7 = c_4$. The solution of the system is

$$x = c_1 + c_2 t + c_3 e^t + c_4 e^{-t} - \frac{1}{2}t^2$$
$$y = (c_1 - c_2 + 2) + (c_2 + 1)t + c_4 e^{-t} - \frac{1}{2}t^2.$$

18. From $Dx + z = e^t$, $(D-1)x + Dy + Dz = 0$, and $x + 2y + Dz = e^t$ we obtain $z = -Dx + e^t$, $Dz = -D^2x + e^t$, and the system $(-D^2 + D - 1)x + Dy = -e^t$ and $(-D^2 + 1)x + 2y = 0$. Then $y = \frac{1}{2}(D^2 - 1)x$, $Dy = \frac{1}{2}D(D^2 - 1)x$, and $(D - 2)(D^2 + 1)x = -2e^t$ so that the solution is

$$x = c_1 e^{2t} + c_2 \cos t + c_3 \sin t + e^t$$
$$y = \frac{3}{2}c_1 e^{2t} - c_2 \cos t - c_3 \sin t$$
$$z = -2c_1 e^{2t} - c_3 \cos t + c_2 \sin t.$$

21. From $(D + 5)x + y = 0$ and $4x - (D + 1)y = 0$ we obtain $y = -(D + 5)x$ so that $Dy = -(D^2 + 5D)x$. Then $4x + (D^2 + 5D)x + (D + 5)x = 0$ and $(D + 3)^2 x = 0$. Thus

$$x = c_1 e^{-3t} + c_2 t e^{-3t}$$
$$y = -(2c_1 + c_2)e^{-3t} - 2c_2 t e^{-3t}.$$

Using $x(1) = 0$ and $y(1) = 1$ we obtain

$$c_1 e^{-3} + c_2 e^{-3} = 0$$
$$-(2c_1 + c_2)e^{-3} - 2c_2 e^{-3} = 1$$

or

$$c_1 + c_2 = 0$$

$$2c_1 + 3c_2 = -e^3.$$

Thus $c_1 = e^3$ and $c_2 = -e^3$. The solution of the initial value problem is

$$x = e^{-3t+3} - te^{-3t+3}$$

$$y = -e^{-3t+3} + 2te^{-3t+3}.$$

24. From Newton's second law in the x-direction we have

$$m\frac{d^2x}{dt^2} = -k\cos\theta = -k\frac{1}{v}\frac{dx}{dt} = -|c|\frac{dx}{dt}.$$

In the y-direction we have

$$m\frac{d^2y}{dt^2} = -mg - k\sin\theta = -mg - k\frac{1}{v}\frac{dy}{dt} = -mg - |c|\frac{dy}{dt}.$$

From $mD^2x + |c|Dx = 0$ we have $D(mD + |c|)x = 0$ so that $(mD + |c|)x = c_1$ or $(D + |c|/m)x = c_2$. This is a linear first-order differential equation. An integrating factor is $e^{\int |c|dt/m} = e^{|c|t/m}$ so that

$$\frac{d}{dt}[e^{|c|t/m}x] = c_2e^{|c|t/m}$$

and $e^{|c|t/m}x = (c_2m/|c|)e^{|c|t/m} + c_3$. The general solution of this equation is $x(t) = c_4 + c_3e^{-|c|t/m}$. From $(mD^2+|c|D)y = -mg$ we have $D(mD+|c|)y = -mg$ so that $(mD+|c|)y = -mgt+c_1$ or $(D+|c|/m)y = -gt+c_2$. This is a linear first-order differential equation with integrating factor $e^{\int |c|dt/m} = e^{|c|t/m}$. Thus

$$\frac{d}{dt}[e^{|c|t/m}y] = (-gt + c_2)e^{|c|t/m}$$

$$e^{|c|t/m}y = -\frac{mg}{|c|}te^{|c|t/m} + \frac{m^2g}{c^2}e^{|c|t/m} + c_3e^{|c|t/m} + c_4$$

and

$$y(t) = -\frac{mg}{|c|}t + \frac{m^2g}{c^2} + c_3 + c_4e^{-|c|t/m}.$$

27. (a) Write the system as

$$(D^2 + 3)x_1 - \qquad\quad x_2 = 0$$
$$-2x_1 + (D^2 + 2)x_2 = 0.$$

Then

$$(D^2 + 2)(D^2 + 3)x_1 - 2x_1 = (D^2 + 1)(D^2 + 4)x_1 = 0,$$

and

$$x_1(t) = c_1\cos t + c_2\sin t + c_3\cos 2t + c_4\sin 2t.$$

Since $x_2 = (D^2 + 3)x_1$, we have

$$x_2(t) = 2c_1\cos t + 2c_2\sin t - c_3\cos 2t - c_4\sin 2t.$$

The initial conditions imply

$$x_1(0) = c_1 + c_3 = 2$$
$$x_1'(0) = c_1 + 2c_4 = 1$$
$$x_2(0) = 2c_1 - c_3 = -1$$
$$x_2'(0) = 2c_2 - 2c_4 = 1,$$

so $c_1 = \frac{1}{3}$, $c_2 = \frac{2}{3}$, $c_3 = \frac{5}{3}$, and $c_4 = \frac{1}{6}$. Thus

$$x_1(t) = \frac{1}{3}\cos t + \frac{2}{3}\sin t + \frac{5}{3}\cos 2t + \frac{1}{6}\sin 2t$$

$$x_2(t) = \frac{2}{3}\cos t + \frac{4}{3}\sin t - \frac{5}{3}\cos 2t - \frac{1}{6}\sin 2t.$$

(b)

In this problem the motion appears to be periodic with period 2π. In Figure 3.59 of Example 4 in the text the motion does not appear to be periodic.

(c)

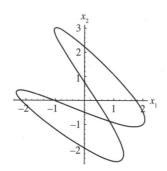

CHAPTER 3 REVIEW EXERCISES

3. It is not true unless the differential equation is homogeneous. For example, $y_1 = x$ is a solution of $y'' + y = x$, but $y_2 = 5x$ is not.

6. $2\pi/5$, since $\frac{1}{4}x'' + 6.25x = 0$

9. The set is linearly independent over $(-\infty, \infty)$ and linearly dependent over $(0, \infty)$.

12. **(a)** The auxiliary equation is $am(m-1) + bm + c = am^2 + (b-a)m + c = 0$. If the roots are 3 and -1, then we want $(m-3)(m+1) = m^2 - 2m - 3 = 0$. Thus, let $a = 1$, $b = -1$, and $c = -3$, so that the differential equation is $x^2y'' - xy' - 3y = 0$.

(b) In this case we want the auxiliary equation to be $m^2 + 1 = 0$, so let $a = 1$, $b = 1$, and $c = 1$. Then the differential equation is $x^2y'' + xy' + y = 0$.

15. From $m^3 + 10m^2 + 25m = 0$ we obtain $m = 0$, $m = -5$, and $m = -5$ so that

$$y = c_1 + c_2 e^{-5x} + c_3 x e^{-5x}.$$

18. From $2m^4 + 3m^3 + 2m^2 + 6m - 4 = 0$ we obtain $m = 1/2$, $m = -2$, and $m = \pm\sqrt{2}\,i$ so that

$$y = c_1 e^{x/2} + c_2 e^{-2x} + c_3 \cos\sqrt{2}\,x + c_4 \sin\sqrt{2}\,x.$$

21. Applying $D(D^2 + 1)$ to the differential equation we obtain

$$D(D^2 + 1)(D^3 - 5D^2 + 6D) = D^2(D^2 + 1)(D - 2)(D - 3) = 0.$$

Then

$$y = \underbrace{c_1 + c_2 e^{2x} + c_3 e^{3x}}_{y_c} + c_4 x + c_5 \cos x + c_6 \sin x$$

and $y_p = Ax + B\cos x + C\sin x$. Substituting y_p into the differential equation yields

$$6A + (5B + 5C)\cos x + (-5B + 5C)\sin x = 8 + 2\sin x.$$

Equating coefficients gives $A = 4/3$, $B = -1/5$, and $C = 1/5$. The general solution is

$$y = c_1 + c_2 e^{2x} + c_3 e^{3x} + \frac{4}{3}x - \frac{1}{5}\cos x + \frac{1}{5}\sin x.$$

24. The auxiliary equation is $m^2 - 1 = 0$, so $y_c = c_1 e^x + c_2 e^{-x}$ and

$$W = \begin{vmatrix} e^x & e^{-x} \\ e^x & -e^{-x} \end{vmatrix} = -2.$$

Identifying $f(x) = 2e^x/(e^x + e^{-x})$ we obtain

$$u_1' = \frac{1}{e^x + e^{-x}} = \frac{e^x}{1 + e^{2x}}$$

$$u_2' = -\frac{e^{2x}}{e^x + e^{-x}} = -\frac{e^{3x}}{1 + e^{2x}} = -e^x + \frac{e^x}{1 + e^{2x}}.$$

Then $u_1 = \tan^{-1} e^x$, $u_2 = -e^x + \tan^{-1} e^x$, and

$$y = c_1 e^x + c_2 e^{-x} + e^x \tan^{-1} e^x - 1 + e^{-x} \tan^{-1} e^x.$$

27. The auxiliary equation is $m^2 - 5m + 6 = (m - 2)(m - 3) = 0$ and a particular solution is $y_p = x^4 - x^2 \ln x$ so that

$$y = c_1 x^2 + c_2 x^3 + x^4 - x^2 \ln x.$$

30. (a) If $y = \sin x$ is a solution then so is $y = \cos x$ and $m^2 + 1$ is a factor of the auxiliary equation $m^4 + 2m^3 + 11m^2 + 2m + 10 = 0$. Dividing by $m^2 + 1$ we get $m^2 + 2m + 10$, which has roots $-1 \pm 3i$. The general solution of the differential equation is

$$y = c_1 \cos x + c_2 \sin x + e^{-x}(c_3 \cos 3x + c_4 \sin 3x).$$

(b) The auxiliary equation is $m(m + 1) = m^2 + m = 0$, so the associated homogeneous differential equation is $y'' + y' = 0$. Letting $y = c_1 + c_2 e^{-x} + \frac{1}{2}x^2 - x$ and computing $y'' + y'$ we get x. Thus, the differential equation is $y'' + y' = x$.

33. The auxiliary equation is $m^2 - 2m + 2 = 0$ so that $m = 1 \pm i$ and $y = e^x(c_1 \cos x + c_2 \sin x)$. Setting $y(\pi/2) = 0$ and $y(\pi) = -1$ we obtain $c_1 = e^{-\pi}$ and $c_2 = 0$. Thus, $y = e^{x-\pi} \cos x$.

36. The auxiliary equation is $m^2 + 1 = 0$, so $y_c = c_1 \cos x + c_2 \sin x$ and

$$W = \begin{vmatrix} \cos x & \sin x \\ -\sin x & \cos x \end{vmatrix} = 1.$$

Identifying $f(x) = \sec^3 x$ we obtain

$$u_1' = -\sin x \sec^3 x = -\frac{\sin x}{\cos^3 x}$$

$$u_2' = \cos x \sec^3 x = \sec^2 x.$$

Then

$$u_1 = -\frac{1}{2}\frac{1}{\cos^2 x} = -\frac{1}{2}\sec^2 x$$

$$u_2 = \tan x.$$

Thus

$$y = c_1 \cos x + c_2 \sin x - \frac{1}{2}\cos x \sec^2 x + \sin x \tan x$$

$$= c_1 \cos x + c_2 \sin x - \frac{1}{2}\sec x + \frac{1-\cos^2 x}{\cos x}$$

$$= c_3 \cos x + c_2 \sin x + \frac{1}{2}\sec x.$$

and

$$y' = -c_3 \sin x + c_2 \cos x + \frac{1}{2}\sec x \tan x.$$

The initial conditions imply

$$c_3 + \frac{1}{2} = 1$$

$$c_2 = \frac{1}{2}.$$

Thus $c_3 = c_2 = 1/2$ and

$$y = \frac{1}{2}\cos x + \frac{1}{2}\sin x + \frac{1}{2}\sec x.$$

39. (a) The auxiliary equation is $12m^4 + 64m^3 + 59m^2 - 23m - 12 = 0$ and has roots -4, $-\frac{3}{2}$, $-\frac{1}{3}$, and $\frac{1}{2}$. The general solution is

$$y = c_1 e^{-4x} + c_2 e^{-3x/2} + c_3 e^{-x/3} + c_4 e^{x/2}.$$

(b) The system of equations is

$$c_1 + c_2 + c_3 + c_4 = -1$$

$$-4c_1 - \frac{3}{2}c_2 - \frac{1}{3}c_3 + \frac{1}{2}c_4 = 2$$

$$16c_1 + \frac{9}{4}c_2 + \frac{1}{9}c_3 + \frac{1}{4}c_4 = 5$$

$$-64c_1 - \frac{27}{8}c_2 - \frac{1}{27}c_3 + \frac{1}{8}c_4 = 0.$$

Using a CAS we find $c_1 = -\frac{73}{495}$, $c_2 = \frac{109}{35}$, $c_3 = -\frac{3726}{385}$, and $c_4 = \frac{257}{45}$. The solution of the initial-value problem is

$$y = -\frac{73}{495}e^{-4x} + \frac{109}{35}e^{-3x/2} - \frac{3726}{385}e^{-x/3} + \frac{257}{45}e^{x/2}.$$

42. From $(D-2)x - y = t - 2$ and $-3x + (D-4)y = -4t$ we obtain $(D-1)(D-5)x = 9 - 8t$. Then

$$x = c_1 e^t + c_2 e^{5t} - \frac{8}{5}t - \frac{3}{25}$$

and

$$y = (D-2)x - t + 2 = -c_1 e^t + 3c_2 e^{5t} + \frac{16}{25} + \frac{11}{25}t.$$

45. The period of a spring/mass system is given by $T = 2\pi/\omega$ where $\omega^2 = k/m = kg/W$, where k is the spring constant, W is the weight of the mass attached to the spring, and g is the acceleration due to gravity. Thus, the period of oscillation is $T = (2\pi/\sqrt{kg})\sqrt{W}$. If the weight of the original mass is W, then $(2\pi/\sqrt{kg})\sqrt{W} = 3$ and $(2\pi/\sqrt{kg})\sqrt{W-8} = 2$. Dividing, we get $\sqrt{W}/\sqrt{W-8} = 3/2$ or $W = \frac{9}{4}(W-8)$. Solving for W we find that the weight of the original mass was 14.4 pounds.

48. From $x'' + \beta x' + 64x = 0$ we see that oscillatory motion results if $\beta^2 - 256 < 0$ or $0 \le \beta < 16$.

51. For $\lambda = \alpha^2 > 0$ the general solution is $y = c_1 \cos \alpha x + c_2 \sin \alpha x$. Now

$$y(0) = c_1 \qquad \text{and} \qquad y(2\pi) = c_1 \cos 2\pi\alpha + c_2 \sin 2\pi\alpha,$$

so the condition $y(0) = y(2\pi)$ implies

$$c_1 = c_1 \cos 2\pi\alpha + c_2 \sin 2\pi\alpha$$

which is true when $\alpha = \sqrt{\lambda} = n$ or $\lambda = n^2$ for $n = 1, 2, 3, \ldots$. Since

$$y' = -\alpha c_1 \sin \alpha x + \alpha c_2 \cos \alpha x = -n c_1 \sin nx + n c_2 \cos nx,$$

we see that $y'(0) = nc_2 = y'(2\pi)$ for $n = 1, 2, 3, \ldots$. Thus, the eigenvalues are n^2 for $n = 1, 2, 3, \ldots$, with corresponding eigenfunctions $\cos nx$ and $\sin nx$. When $\lambda = 0$, the general solution is $y = c_1 x + c_2$ and the corresponding eigenfunction is $y = 1$.

For $\lambda = -\alpha^2 < 0$ the general solution is $y = c_1 \cosh \alpha x + c_2 \sinh \alpha x$. In this case $y(0) = c_1$ and $y(2\pi) = c_1 \cosh 2\pi\alpha + c_2 \sinh 2\pi\alpha$, so $y(0) = y(2\pi)$ can only be valid for $\alpha = 0$. Thus, there are no eigenvalues corresponding to $\lambda < 0$.

54. The force of kinetic friction opposing the motion of the mass in μN, where μ is the coefficient of sliding friction and N is the normal component of the weight. Since friction is a force opposite to the direction of motion and since N is pointed directly downward (it is simply the weight of the mass), Newton's second law gives, for motion to the right $(x' > 0)$,

$$m\frac{d^2 x}{dt^2} = -kx - \mu mg,$$

and for motion to the left $(x' < 0)$,

$$m\frac{d^2 x}{dt^2} = -kx + \mu mg.$$

Traditionally, these two equations are written as one expression

$$m\frac{d^2 x}{dt^2} + f_x \operatorname{sgn}(x') + kx = 0,$$

where $f_k = \mu mg$ and

$$\operatorname{sgn}(x') = \begin{cases} 1, & x' > 0 \\ -1, & x' < 0. \end{cases}$$

57. (a) The auxiliary equation is $m^2 + g/l = 0$, so the general solution of the differential equationis

$$\theta(t) = c_1 \cos \sqrt{\frac{g}{l}}\, t + c_2 \sin \sqrt{\frac{g}{l}}\, t.$$

The initial condtion $\theta(0) = 0$ implies $c_1 = 0$ and $\theta'(0) = \omega_0$ implies $c_2 = \omega_0 \sqrt{l/g}$. Thus,

$$\theta(t) = \omega_0 \sqrt{\frac{l}{g}} \, \sin \sqrt{\frac{g}{l}} \, t.$$

(b) At θ_{max}, $\sin \sqrt{g/l} \, t = 1$, so

$$\theta_{max} = \omega_0 \sqrt{\frac{l}{g}} = \frac{m_b}{m_w + m_b} \frac{v_b}{l} \sqrt{\frac{l}{g}} = \frac{m_b}{m_w + m_b} \frac{v_b}{\sqrt{lg}}$$

and

$$v_b = \frac{m_w + m_b}{m_b} \sqrt{lg} \, \theta_{max}.$$

(c) We have $\cos \theta_{max} = (l - h)/l = 1 - h/l$. Then

$$\cos \theta_{max} \approx 1 - \frac{1}{2} \theta_{max}^2 = 1 - \frac{h}{l}$$

and

$$\theta_{max}^2 = \frac{2h}{l} \qquad \text{or} \qquad \theta_{max} = \sqrt{\frac{2h}{l}}.$$

Thus

$$v_b = \frac{m_w + m_b}{m_b} \sqrt{lg} \sqrt{\frac{2h}{l}} = \frac{m_w + m_b}{m_b} \sqrt{2gh}.$$

4 The Laplace Transform

EXERCISES 4.1

Definition of the Laplace Transform

3. $\mathscr{L}\{f(t)\} = \displaystyle\int_0^1 te^{-st}dt + \int_1^\infty e^{-st}dt = \left(-\dfrac{1}{s}te^{-st} - \dfrac{1}{s^2}e^{-st}\right)\Big|_0^1 - \dfrac{1}{s}e^{-st}\Big|_1^\infty$

$\qquad = \left(-\dfrac{1}{s}e^{-s} - \dfrac{1}{s^2}e^{-s}\right) - \left(0 - \dfrac{1}{s^2}\right) - \dfrac{1}{s}(0 - e^{-s}) = \dfrac{1}{s^2}(1 - e^{-s}), \quad s > 0$

6. $\mathscr{L}\{f(t)\} = \displaystyle\int_{\pi/2}^\infty (\cos t)e^{-st}dt = \left(-\dfrac{s}{s^2+1}e^{-st}\cos t + \dfrac{1}{s^2+1}e^{-st}\sin t\right)\Big|_{\pi/2}^\infty$

$\qquad = 0 - \left(0 + \dfrac{1}{s^2+1}e^{-\pi s/2}\right) = -\dfrac{1}{s^2+1}e^{-\pi s/2}, \quad s > 0$

9. The function is $f(t) = \begin{cases} 1 - t, & 0 < t < 1 \\ 0, & t > 1 \end{cases}$ so

$\qquad \mathscr{L}\{f(t)\} = \displaystyle\int_0^1 (1-t)e^{-st}\,dt + \int_1^\infty 0e^{-st}\,dt = \int_0^1 (1-t)e^{-st}\,dt = \left(-\dfrac{1}{s}(1-t)e^{-st} + \dfrac{1}{s^2}e^{-st}\right)\Big|_0^1$

$\qquad = \dfrac{1}{s^2}e^{-s} + \dfrac{1}{s} - \dfrac{1}{s^2}, \quad s > 0$

12. $\mathscr{L}\{f(t)\} = \displaystyle\int_0^\infty e^{-2t-5}e^{-st}dt = e^{-5}\int_0^\infty e^{-(s+2)t}dt = -\dfrac{e^{-5}}{s+2}e^{-(s+2)t}\Big|_0^\infty = \dfrac{e^{-5}}{s+2}, \quad s > -2$

15. $\mathscr{L}\{f(t)\} = \displaystyle\int_0^\infty e^{-t}(\sin t)e^{-st}dt = \int_0^\infty (\sin t)e^{-(s+1)t}dt$

$\qquad = \left(\dfrac{-(s+1)}{(s+1)^2+1}e^{-(s+1)t}\sin t - \dfrac{1}{(s+1)^2+1}e^{-(s+1)t}\cos t\right)\Big|_0^\infty$

$\qquad = \dfrac{1}{(s+1)^2+1} = \dfrac{1}{s^2+2s+2}, \quad s > -1$

18. $\mathscr{L}\{f(t)\} = \displaystyle\int_0^\infty t(\sin t)e^{-st}dt$

$\qquad = \left[\left(-\dfrac{t}{s^2+1} - \dfrac{2s}{(s^2+1)^2}\right)(\cos t)e^{-st} - \left(\dfrac{st}{s^2+1} + \dfrac{s^2-1}{(s^2+1)^2}\right)(\sin t)e^{-st}\right]_0^\infty$

$\qquad = \dfrac{2s}{(s^2+1)^2}, \quad s > 0$

21. $\mathscr{L}\{4t - 10\} = \dfrac{4}{s^2} - \dfrac{10}{s}$

24. $\mathscr{L}\{-4t^2 + 16t + 9\} = -4\dfrac{2}{s^3} + \dfrac{16}{s^2} + \dfrac{9}{s}$

27. $\mathscr{L}\{1 + e^{4t}\} = \dfrac{1}{s} + \dfrac{1}{s-4}$

30. $\mathscr{L}\{e^{2t} - 2 + e^{-2t}\} = \dfrac{1}{s-2} - \dfrac{2}{s} + \dfrac{1}{s+2}$

33. $\mathscr{L}\{\sinh kt\} = \dfrac{1}{2}\mathscr{L}\{e^{kt} - e^{-kt}\} = \dfrac{1}{2}\left[\dfrac{1}{s-k} - \dfrac{1}{s+k}\right] = \dfrac{k}{s^2 - k^2}$

36. $\mathscr{L}\{e^{-t}\cosh t\} = \mathscr{L}\left\{e^{-t}\dfrac{e^t + e^{-t}}{2}\right\} = \mathscr{L}\left\{\dfrac{1}{2} + \dfrac{1}{2}e^{-2t}\right\} = \dfrac{1}{2s} + \dfrac{1}{2(s+2)}$

39. From the addition formula for the sine function, $\sin(4t + 5) = \sin 4t \cos 5 + \cos 4t \sin 5$ so

$$\mathscr{L}\{\sin(4t+5)\} = (\cos 5)\,\mathscr{L}\{\sin 4t\} + (\sin 5)\,\mathscr{L}\{\cos 4t\} = (\cos 5)\dfrac{4}{s^2+16} + (\sin 5)\dfrac{s}{s^2+16} = \dfrac{4\cos 5 + (\sin 5)s}{s^2+16}.$$

42. **(a)** $\mathscr{L}\{t^{-1/2}\} = \dfrac{\Gamma(1/2)}{s^{1/2}} = \sqrt{\dfrac{\pi}{s}}$ **(b)** $\mathscr{L}\{t^{1/2}\} = \dfrac{\Gamma(3/2)}{s^{3/2}} = \dfrac{\sqrt{\pi}}{2s^{3/2}}$ **(c)** $\mathscr{L}\{t^{3/2}\} = \dfrac{\Gamma(5/2)}{s^{5/2}} = \dfrac{3\sqrt{\pi}}{4s^{5/2}}$

45. Since e^t is an increasing function and $t^2 > \ln M + ct$ for $M > 0$ we have $e^{t^2} > e^{\ln M + ct} = Me^{ct}$ for t sufficiently large and for any c. Thus, e^{t^2} is not of exponential order.

48. Assume that $\mathscr{L}\{t^{n-1}\} = (n-1)!/s^n$. Then, using the definition of the Laplace transform and integration by parts, we have

$$\mathscr{L}\{t^n\} = \int_0^\infty e^{-st}t^n\,dt = -\dfrac{1}{s}e^{-st}t^n\,\Big|_0^\infty + \dfrac{n}{s}\int_0^\infty e^{-st}t^{n-1}\,dt$$

$$= 0 + \dfrac{n}{s}\,\mathscr{L}\{t^{n-1}\} = \dfrac{n}{s}\dfrac{(n-1)!}{s^n} = \dfrac{n!}{s^{n+1}}.$$

EXERCISES 4.2

The Inverse Transform and Transforms of Derivatives

3. $\mathscr{L}^{-1}\left\{\dfrac{1}{s^2} - \dfrac{48}{s^5}\right\} = \mathscr{L}^{-1}\left\{\dfrac{1}{s^2} - \dfrac{48}{24}\cdot\dfrac{4!}{s^5}\right\} = t - 2t^4$

6. $\mathscr{L}^{-1}\left\{\dfrac{(s+2)^2}{s^3}\right\} = \mathscr{L}^{-1}\left\{\dfrac{1}{s} + 4\cdot\dfrac{1}{s^2} + 2\cdot\dfrac{2}{s^3}\right\} = 1 + 4t + 2t^2$

9. $\mathscr{L}^{-1}\left\{\dfrac{1}{4s+1}\right\} = \dfrac{1}{4}\mathscr{L}^{-1}\left\{\dfrac{1}{s+1/4}\right\} = \dfrac{1}{4}e^{-t/4}$

12. $\mathscr{L}^{-1}\left\{\dfrac{10s}{s^2+16}\right\} = 10\cos 4t$

15. $\mathscr{L}^{-1}\left\{\dfrac{2s-6}{s^2+9}\right\} = \mathscr{L}^{-1}\left\{2\cdot\dfrac{s}{s^2+9} - 2\cdot\dfrac{3}{s^2+9}\right\} = 2\cos 3t - 2\sin 3t$

18. $\mathscr{L}^{-1}\left\{\dfrac{s+1}{s^2-4s}\right\} = \mathscr{L}^{-1}\left\{-\dfrac{1}{4}\cdot\dfrac{1}{s} + \dfrac{5}{4}\cdot\dfrac{1}{s-4}\right\} = -\dfrac{1}{4} + \dfrac{5}{4}e^{4t}$

21. $\mathscr{L}^{-1}\left\{\dfrac{0.9s}{(s-0.1)(s+0.2)}\right\} = \mathscr{L}^{-1}\left\{(0.3)\cdot\dfrac{1}{s-0.1} + (0.6)\cdot\dfrac{1}{s+0.2}\right\} = 0.3e^{0.1t} + 0.6e^{-0.2t}$

24. $\mathscr{L}^{-1}\left\{\dfrac{s^2+1}{s(s-1)(s+1)(s-2)}\right\} = \mathscr{L}^{-1}\left\{\dfrac{1}{2}\cdot\dfrac{1}{s} - \dfrac{1}{s-1} - \dfrac{1}{3}\cdot\dfrac{1}{s+1} + \dfrac{5}{6}\cdot\dfrac{1}{s-2}\right\}$

$$= \dfrac{1}{2} - e^t - \dfrac{1}{3}e^{-t} + \dfrac{5}{6}e^{2t}$$

80

27. $\mathscr{L}^{-1}\left\{\dfrac{2s-4}{(s^2+s)(s^2+1)}\right\} = \mathscr{L}^{-1}\left\{\dfrac{2s-4}{s(s+1)(s^2+1)}\right\} = \mathscr{L}^{-1}\left\{-\dfrac{4}{s}+\dfrac{3}{s+1}+\dfrac{s}{s^2+1}+\dfrac{3}{s^2+1}\right\}$

$$= -4 + 3e^{-t} + \cos t + 3\sin t$$

30. $\mathscr{L}^{-1}\left\{\dfrac{6s+3}{(s^2+1)(s^2+4)}\right\} = \mathscr{L}^{-1}\left\{2\cdot\dfrac{s}{s^2+1}+\dfrac{1}{s^2+1}-2\cdot\dfrac{s}{s^2+4}-\dfrac{1}{2}\cdot\dfrac{2}{s^2+4}\right\}$

$$= 2\cos t + \sin t - 2\cos 2t - \dfrac{1}{2}\sin 2t$$

33. The Laplace transform of the initial-value problem is

$$s\mathscr{L}\{y\} - y(0) + 6\mathscr{L}\{y\} = \dfrac{1}{s-4}.$$

Solving for $\mathscr{L}\{y\}$ we obtain

$$\mathscr{L}\{y\} = \dfrac{1}{(s-4)(s+6)} + \dfrac{2}{s+6} = \dfrac{1}{10}\cdot\dfrac{1}{s-4} + \dfrac{19}{10}\cdot\dfrac{1}{s+6}.$$

Thus

$$y = \dfrac{1}{10}e^{4t} + \dfrac{19}{10}e^{-6t}.$$

36. The Laplace transform of the initial-value problem is

$$s^2\mathscr{L}\{y\} - sy(0) - y'(0) - 4[s\mathscr{L}\{y\} - y(0)] = \dfrac{6}{s-3} - \dfrac{3}{s+1}.$$

Solving for $\mathscr{L}\{y\}$ we obtain

$$\mathscr{L}\{y\} = \dfrac{6}{(s-3)(s^2-4s)} - \dfrac{3}{(s+1)(s^2-4s)} + \dfrac{s-5}{s^2-4s}$$

$$= \dfrac{5}{2}\cdot\dfrac{1}{s} - \dfrac{2}{s-3} - \dfrac{3}{5}\cdot\dfrac{1}{s+1} + \dfrac{11}{10}\cdot\dfrac{1}{s-4}.$$

Thus

$$y = \dfrac{5}{2} - 2e^{3t} - \dfrac{3}{5}e^{-t} + \dfrac{11}{10}e^{4t}.$$

39. The Laplace transform of the initial-value problem is

$$2\left[s^3\mathscr{L}\{y\} - s^2(0) - sy'(0) - y''(0)\right] + 3\left[s^2\mathscr{L}\{y\} - sy(0) - y'(0)\right] - 3[s\mathscr{L}\{y\} - y(0)] - 2\mathscr{L}\{y\} = \dfrac{1}{s+1}.$$

Solving for $\mathscr{L}\{y\}$ we obtain

$$\mathscr{L}\{y\} = \dfrac{2s+3}{(s+1)(s-1)(2s+1)(s+2)} = \dfrac{1}{2}\dfrac{1}{s+1} + \dfrac{5}{18}\dfrac{1}{s-1} - \dfrac{8}{9}\dfrac{1}{s+1/2} + \dfrac{1}{9}\dfrac{1}{s+2}.$$

Thus

$$y = \dfrac{1}{2}e^{-t} + \dfrac{5}{18}e^{t} - \dfrac{8}{9}e^{-t/2} + \dfrac{1}{9}e^{-2t}.$$

42. The Laplace transform of the initial-value problem is

$$s^2\mathscr{L}\{y\} - s\cdot 1 - 3 - 2[s\mathscr{L}\{y\} - 1] + 5\mathscr{L}\{y\} = (s^2 - 2s + 5)\mathscr{L}\{y\} - s - 1 = 0.$$

Solving for $\mathscr{L}\{y\}$ we obtain

$$\mathscr{L}\{y\} = \dfrac{s+1}{s^2 - 2s + 5} = \dfrac{s-1+2}{(s-1)^2 + 2^2} = \dfrac{s-1}{(s-1)^2 + 2^2} + \dfrac{2}{(s-1)^2 + 2^2}.$$

Thus

$$y = e^t\cos 2t + e^t\sin 2t.$$

45. For $y'' - 4y' = 6e^{3t} - 3e^{-t}$ the transfer function is $W(s) = 1/(s^2 - 4s)$. The zero-input response is

$$y_0(t) = \mathscr{L}^{-1}\left\{\frac{s-5}{s^2-4s}\right\} = \mathscr{L}^{-1}\left\{\frac{5}{4}\cdot\frac{1}{s} - \frac{1}{4}\cdot\frac{1}{s-4}\right\} = \frac{5}{4} - \frac{1}{4}e^{4t},$$

and the zero-state response is

$$\begin{aligned} y_1(t) &= \mathscr{L}^{-1}\left\{\frac{6}{(s-3)(s^2-4s)} - \frac{3}{(s+1)(s^2-4s)}\right\} \\ &= \mathscr{L}^{-1}\left\{\frac{27}{20}\cdot\frac{1}{s-4} - \frac{2}{s-3} + \frac{5}{4}\cdot\frac{1}{s} - \frac{3}{5}\cdot\frac{1}{s+1}\right\} \\ &= \frac{27}{20}e^{4t} - 2e^{3t} + \frac{5}{4} - \frac{3}{5}e^{-t}. \end{aligned}$$

EXERCISES 4.3

Translation Theorems

3. $\mathscr{L}\{t^3 e^{-2t}\} = \dfrac{3!}{(s+2)^4}$

6. $\mathscr{L}\{e^{2t}(t-1)^2\} = \mathscr{L}\{t^2 e^{2t} - 2te^{2t} + e^{2t}\} = \dfrac{2}{(s-2)^3} - \dfrac{2}{(s-2)^2} + \dfrac{1}{s-2}$

9. $\mathscr{L}\{(1 - e^t + 3e^{-4t})\cos 5t\} = \mathscr{L}\{\cos 5t - e^t \cos 5t + 3e^{-4t}\cos 5t\} = \dfrac{s}{s^2+25} - \dfrac{s-1}{(s-1)^2+25} + \dfrac{3(s+4)}{(s+4)^2+25}$

12. $\mathscr{L}^{-1}\left\{\dfrac{1}{(s-1)^4}\right\} = \dfrac{1}{6}\mathscr{L}^{-1}\left\{\dfrac{3!}{(s-1)^4}\right\} = \dfrac{1}{6}t^3 e^t$

15. $\mathscr{L}^{-1}\left\{\dfrac{s}{s^2+4s+5}\right\} = \mathscr{L}^{-1}\left\{\dfrac{s+2}{(s+2)^2+1^2} - 2\dfrac{1}{(s+2)^2+1^2}\right\} = e^{-2t}\cos t - 2e^{-2t}\sin t$

18. $\mathscr{L}^{-1}\left\{\dfrac{5s}{(s-2)^2}\right\} = \mathscr{L}^{-1}\left\{\dfrac{5(s-2)+10}{(s-2)^2}\right\} = \mathscr{L}^{-1}\left\{\dfrac{5}{s-2} + \dfrac{10}{(s-2)^2}\right\} = 5e^{2t} + 10te^{2t}$

21. The Laplace transform of the differential equation is

$$s\mathscr{L}\{y\} - y(0) + 4\mathscr{L}\{y\} = \frac{1}{s+4}.$$

Solving for $\mathscr{L}\{y\}$ we obtain

$$\mathscr{L}\{y\} = \frac{1}{(s+4)^2} + \frac{2}{s+4}.$$

Thus

$$y = te^{-4t} + 2e^{-4t}.$$

24. The Laplace transform of the differential equation is

$$s^2\mathscr{L}\{y\} - sy(0) - y'(0) - 4[s\mathscr{L}\{y\} - y(0)] + 4\mathscr{L}\{y\} = \frac{6}{(s-2)^4}.$$

Solving for $\mathscr{L}\{y\}$ we obtain $\mathscr{L}\{y\} = \dfrac{1}{20}\dfrac{5!}{(s-2)^6}$. Thus, $y = \dfrac{1}{20}t^5 e^{2t}$.

27. The Laplace transform of the differential equation is

$$s^2 \mathscr{L}\{y\} - sy(0) - y'(0) - 6\left[s\mathscr{L}\{y\} - y(0)\right] + 13\mathscr{L}\{y\} = 0.$$

Solving for $\mathscr{L}\{y\}$ we obtain

$$\mathscr{L}\{y\} = -\frac{3}{s^2 - 6s + 13} = -\frac{3}{2}\frac{2}{(s-3)^2 + 2^2}.$$

Thus

$$y = -\frac{3}{2}e^{3t}\sin 2t.$$

30. The Laplace transform of the differential equation is

$$s^2 \mathscr{L}\{y\} - sy(0) - y'(0) - 2\left[s\mathscr{L}\{y\} - y(0)\right] + 5\mathscr{L}\{y\} = \frac{1}{s} + \frac{1}{s^2}.$$

Solving for $\mathscr{L}\{y\}$ we obtain

$$\mathscr{L}\{y\} = \frac{4s^2 + s + 1}{s^2(s^2 - 2s + 5)} = \frac{7}{25}\frac{1}{s} + \frac{1}{5}\frac{1}{s^2} + \frac{-7s/25 + 109/25}{s^2 - 2s + 5}$$

$$= \frac{7}{25}\frac{1}{s} + \frac{1}{5}\frac{1}{s^2} - \frac{7}{25}\frac{s-1}{(s-1)^2 + 2^2} + \frac{51}{25}\frac{2}{(s-1)^2 + 2^2}.$$

Thus

$$y = \frac{7}{25} + \frac{1}{5}t - \frac{7}{25}e^t\cos 2t + \frac{51}{25}e^t\sin 2t.$$

33. Recall from Section 3.8 that $mx'' = -kx - \beta x'$. Now $m = W/g = 4/32 = \frac{1}{8}$ slug, and $4 = 2k$ so that $k = 2$ lb/ft. Thus, the differential equation is $x'' + 7x' + 16x = 0$. The initial conditions are $x(0) = -3/2$ and $x'(0) = 0$. The Laplace transform of the differential equation is

$$s^2\mathscr{L}\{x\} + \frac{3}{2}s + 7s\mathscr{L}\{x\} + \frac{21}{2} + 16\mathscr{L}\{x\} = 0.$$

Solving for $\mathscr{L}\{x\}$ we obtain

$$\mathscr{L}\{x\} = \frac{-3s/2 - 21/2}{s^2 + 7s + 16} = -\frac{3}{2}\frac{s + 7/2}{(s+7/2)^2 + (\sqrt{15}/2)^2} - \frac{7\sqrt{15}}{10}\frac{\sqrt{15}/2}{(s+7/2)^2 + (\sqrt{15}/2)^2}.$$

Thus

$$x = -\frac{3}{2}e^{-7t/2}\cos\frac{\sqrt{15}}{2}t - \frac{7\sqrt{15}}{10}e^{-7t/2}\sin\frac{\sqrt{15}}{2}t.$$

36. The differential equation is

$$R\frac{dq}{dt} + \frac{1}{C}q = E_0 e^{-kt}, \quad q(0) = 0.$$

The Laplace transform of this equation is

$$Rs\mathscr{L}\{q\} + \frac{1}{C}\mathscr{L}\{q\} = E_0\frac{1}{s + k}.$$

Solving for $\mathscr{L}\{q\}$ we obtain

$$\mathscr{L}\{q\} = \frac{E_0 C}{(s+k)(RCs+1)} = \frac{E_0/R}{(s+k)(s+1/RC)}.$$

When $1/RC \neq k$ we have by partial fractions

$$\mathscr{L}\{q\} = \frac{E_0}{R}\left(\frac{1/(1/RC - k)}{s + k} - \frac{1/(1/RC - k)}{s + 1/RC}\right) = \frac{E_0}{R}\frac{1}{1/RC - k}\left(\frac{1}{s+k} - \frac{1}{s + 1/RC}\right).$$

83

Thus

$$q(t) = \frac{E_0 C}{1 - kRC} \left(e^{-kt} - e^{-t/RC} \right).$$

When $1/RC = k$ we have

$$\mathscr{L}\{q\} = \frac{E_0}{R} \frac{1}{(s+k)^2}.$$

Thus

$$q(t) = \frac{E_0}{R} t e^{-kt} = \frac{E_0}{R} t e^{-t/RC}.$$

39. $\mathscr{L}\{t\,\mathscr{U}(t-2)\} = \mathscr{L}\{(t-2)\,\mathscr{U}(t-2) + 2\,\mathscr{U}(t-2)\} = \dfrac{e^{-2s}}{s^2} + \dfrac{2e^{-2s}}{s}$

Alternatively, (16) of this section could be used:

$$\mathscr{L}\{t\,\mathscr{U}(t-2)\} = e^{-2s}\,\mathscr{L}\{t+2\} = e^{-2s}\left(\frac{1}{s^2} + \frac{2}{s} \right).$$

42. $\mathscr{L}\left\{ \sin t\,\mathscr{U}\left(t - \dfrac{\pi}{2}\right) \right\} = \mathscr{L}\left\{ \cos\left(t - \dfrac{\pi}{2}\right)\mathscr{U}\left(t - \dfrac{\pi}{2}\right) \right\} = \dfrac{se^{-\pi s/2}}{s^2 + 1}$

Alternatively, (16) of this section could be used:

$$\mathscr{L}\left\{ \sin t\,\mathscr{U}\left(t - \frac{\pi}{2}\right) \right\} = e^{-\pi s/2}\,\mathscr{L}\left\{ \sin\left(t + \frac{\pi}{2}\right) \right\} = e^{-\pi s/2}\,\mathscr{L}\{\cos t\} = e^{-\pi s/2}\,\frac{s}{s^2 + 1}.$$

45. $\mathscr{L}^{-1}\left\{ \dfrac{e^{-\pi s}}{s^2 + 1} \right\} = \sin(t-\pi)\,\mathscr{U}(t-\pi) = -\sin t\,\mathscr{U}(t-\pi)$

48. $\mathscr{L}^{-1}\left\{ \dfrac{e^{-2s}}{s^2(s-1)} \right\} = \mathscr{L}^{-1}\left\{ -\dfrac{e^{-2s}}{s} - \dfrac{e^{-2s}}{s^2} + \dfrac{e^{-2s}}{s-1} \right\} = -\mathscr{U}(t-2) - (t-2)\,\mathscr{U}(t-2) + e^{t-2}\,\mathscr{U}(t-2)$

51. (f)

54. (d)

57. $\mathscr{L}\{t^2\,\mathscr{U}(t-1)\} = \mathscr{L}\{[(t-1)^2 + 2t - 1]\,\mathscr{U}(t-1)\} = \mathscr{L}\{[(t-1)^2 + 2(t-1) - 1]\,\mathscr{U}(t-1)\}$

$$= \left(\frac{2}{s^3} + \frac{2}{s^2} + \frac{1}{s} \right)e^{-s}$$

Alternatively, by (16) of this section,

$$\mathscr{L}\{t^2\,\mathscr{U}(t-1)\} = e^{-s}\,\mathscr{L}\{t^2 + 2t + 1\} = e^{-s}\left(\frac{2}{s^3} + \frac{2}{s^2} + \frac{1}{s} \right).$$

60. $\mathscr{L}\{\sin t - \sin t\,\mathscr{U}(t-2\pi)\} = \mathscr{L}\{\sin t - \sin(t-2\pi)\,\mathscr{U}(t-2\pi)\} = \dfrac{1}{s^2+1} - \dfrac{e^{-2\pi s}}{s^2+1}$

63. The Laplace transform of the differential equation is

$$s\mathscr{L}\{y\} - y(0) + \mathscr{L}\{y\} = \frac{5}{s}e^{-s}.$$

Solving for $\mathscr{L}\{y\}$ we obtain

$$\mathscr{L}\{y\} = \frac{5e^{-s}}{s(s+1)} = 5e^{-s}\left[\frac{1}{s} - \frac{1}{s+1} \right].$$

Thus

$$y = 5\,\mathscr{U}(t-1) - 5e^{-(t-1)}\,\mathscr{U}(t-1).$$

66. The Laplace transform of the differential equation is

$$s^2\mathscr{L}\{y\} - sy(0) - y'(0) + 4\mathscr{L}\{y\} = \frac{1}{s} - \frac{e^{-s}}{s}.$$

Solving for $\mathcal{L}\{y\}$ we obtain

$$\mathcal{L}\{y\} = \frac{1-s}{s(s^2+4)} - e^{-s}\frac{1}{s(s^2+4)} = \frac{1}{4}\frac{1}{s} - \frac{1}{4}\frac{s}{s^2+4} - \frac{1}{2}\frac{2}{s^2+4} - e^{-s}\left[\frac{1}{4}\frac{1}{s} - \frac{1}{4}\frac{s}{s^2+4}\right].$$

Thus

$$y = \frac{1}{4} - \frac{1}{4}\cos 2t - \frac{1}{2}\sin 2t - \left[\frac{1}{4} - \frac{1}{4}\cos 2(t-1)\right]\mathcal{U}(t-1).$$

69. The Laplace transform of the differential equation is

$$s^2\,\mathcal{L}\{y\} - sy(0) - y'(0) + \mathcal{L}\{y\} = \frac{e^{-\pi s}}{s} - \frac{e^{-2\pi s}}{s}.$$

Solving for $\mathcal{L}\{y\}$ we obtain

$$\mathcal{L}\{y\} = e^{-\pi s}\left[\frac{1}{s} - \frac{s}{s^2+1}\right] - e^{-2\pi s}\left[\frac{1}{s} - \frac{s}{s^2+1}\right] + \frac{1}{s^2+1}.$$

Thus

$$y = [1 - \cos(t-\pi)]\,\mathcal{U}(t-\pi) - [1 - \cos(t-2\pi)]\,\mathcal{U}(t-2\pi) + \sin t.$$

72. Recall from Section 3.8 that $mx'' = -kx + f(t)$. Now $m = W/g = 32/32 = 1$ slug, and $32 = 2k$ so that $k = 16$ lb/ft. Thus, the differential equation is $x'' + 16x = f(t)$. The initial conditions are $x(0) = 0$, $x'(0) = 0$. Also, since

$$f(t) = \begin{cases} \sin t, & 0 \le t < 2\pi \\ 0, & t \ge 2\pi \end{cases}$$

and $\sin t = \sin(t - 2\pi)$ we can write

$$f(t) = \sin t - \sin(t - 2\pi)\,\mathcal{U}(t - 2\pi).$$

The Laplace transform of the differential equation is

$$s^2\mathcal{L}\{x\} + 16\,\mathcal{L}\{x\} = \frac{1}{s^2+1} - \frac{1}{s^2+1}e^{-2\pi s}.$$

Solving for $\mathcal{L}\{x\}$ we obtain

$$\mathcal{L}\{x\} = \frac{1}{(s^2+16)(s^2+1)} - \frac{1}{(s^2+16)(s^2+1)}e^{-2\pi s}$$

$$= \frac{-1/15}{s^2+16} + \frac{1/15}{s^2+1} - \left[\frac{-1/15}{s^2+16} + \frac{1/15}{s^2+1}\right]e^{-2\pi s}.$$

Thus

$$x(t) = -\frac{1}{60}\sin 4t + \frac{1}{15}\sin t + \frac{1}{60}\sin 4(t-2\pi)\,\mathcal{U}(t-2\pi) - \frac{1}{15}\sin(t-2\pi)\,\mathcal{U}(t-2\pi)$$

$$= \begin{cases} -\frac{1}{60}\sin 4t + \frac{1}{15}\sin t, & 0 \le t < 2\pi \\ 0, & t \ge 2\pi. \end{cases}$$

75. **(a)** The differential equation is

$$\frac{di}{dt} + 10i = \sin t + \cos\left(t - \frac{3\pi}{2}\right)\mathcal{U}\left(t - \frac{3\pi}{2}\right), \quad i(0) = 0.$$

The Laplace transform of this equation is

$$s\mathcal{L}\{i\} + 10\mathcal{L}\{i\} = \frac{1}{s^2+1} + \frac{se^{-3\pi s/2}}{s^2+1}.$$

Solving for $\mathcal{L}\{i\}$ we obtain

$$\mathcal{L}\{i\} = \frac{1}{(s^2+1)(s+10)} + \frac{s}{(s^2+1)(s+10)} e^{-3\pi s/2}$$

$$= \frac{1}{101}\left(\frac{1}{s+10} - \frac{s}{s^2+1} + \frac{10}{s^2+1}\right) + \frac{1}{101}\left(\frac{-10}{s+10} + \frac{10s}{s^2+1} + \frac{1}{s^2+1}\right)e^{-3\pi s/2}.$$

Thus

$$i(t) = \frac{1}{101}\left(e^{-10t} - \cos t + 10\sin t\right)$$

$$+ \frac{1}{101}\left[-10e^{-10(t-3\pi/2)} + 10\cos\left(t - \frac{3\pi}{2}\right) + \sin\left(t - \frac{3\pi}{2}\right)\right]\mathcal{U}\left(t - \frac{3\pi}{2}\right).$$

(b)

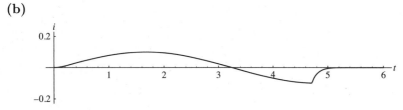

The maximum value of $i(t)$ is approximately 0.1 at $t = 1.7$, the minimum is approximately -0.1 at 4.7.

78. The differential equation is

$$EI\frac{d^4y}{dx^4} = w_0[\mathcal{U}(x - L/3) - \mathcal{U}(x - 2L/3)].$$

Taking the Laplace transform of both sides and using $y(0) = y'(0) = 0$ we obtain

$$s^4\mathcal{L}\{y\} - sy''(0) - y'''(0) = \frac{w_0}{EI}\frac{1}{s}\left(e^{-Ls/3} - e^{-2Ls/3}\right).$$

Letting $y''(0) = c_1$ and $y'''(0) = c_2$ we have

$$\mathcal{L}\{y\} = \frac{c_1}{s^3} + \frac{c_2}{s^4} + \frac{w_0}{EI}\frac{1}{s^5}\left(e^{-Ls/3} - e^{-2Ls/3}\right)$$

so that

$$y(x) = \frac{1}{2}c_1x^2 + \frac{1}{6}c_2x^3 + \frac{1}{24}\frac{w_0}{EI}\left[\left(x - \frac{L}{3}\right)^4\mathcal{U}\left(x - \frac{L}{3}\right) - \left(x - \frac{2L}{3}\right)^4\mathcal{U}\left(x - \frac{2L}{3}\right)\right].$$

To find c_1 and c_2 we compute

$$y''(x) = c_1 + c_2x + \frac{1}{2}\frac{w_0}{EI}\left[\left(x - \frac{L}{3}\right)^2\mathcal{U}\left(x - \frac{L}{3}\right) - \left(x - \frac{2L}{3}\right)^2\mathcal{U}\left(x - \frac{2L}{3}\right)\right]$$

and

$$y'''(x) = c_2 + \frac{w_0}{EI}\left[\left(x - \frac{L}{3}\right)\mathcal{U}\left(x - \frac{L}{3}\right) - \left(x - \frac{2L}{3}\right)\mathcal{U}\left(x - \frac{2L}{3}\right)\right].$$

Then $y''(L) = y'''(L) = 0$ yields the system

$$c_1 + c_2L + \frac{1}{2}\frac{w_0}{EI}\left[\left(\frac{2L}{3}\right)^2 - \left(\frac{L}{3}\right)^2\right] = c_1 + c_2L + \frac{1}{6}\frac{w_0L^2}{EI} = 0$$

$$c_2 + \frac{w_0}{EI}\left[\frac{2L}{3} - \frac{L}{3}\right] = c_2 + \frac{1}{3}\frac{w_0L}{EI} = 0.$$

Solving for c_1 and c_2 we obtain $c_1 = \frac{1}{6}w_0L^2/EI$ and $c_2 = -\frac{1}{3}w_0L/EI$. Thus

$$y(x) = \frac{w_0}{EI}\left(\frac{1}{12}L^2x^2 - \frac{1}{18}Lx^3 + \frac{1}{24}\left[\left(x - \frac{L}{3}\right)^4\mathcal{U}\left(x - \frac{L}{3}\right) - \left(x - \frac{2L}{3}\right)^4\mathcal{U}\left(x - \frac{2L}{3}\right)\right]\right).$$

81. (a) The temperature T of the cake inside the oven is modeled by

$$\frac{dT}{dt} = k(T - T_m)$$

where T_m is the ambient temperature of the oven. For $0 \leq t \leq 4$, we have

$$T_m = 70 + \frac{300 - 70}{4 - 0}\, t = 70 + 57.5t.$$

Hence for $t \geq 0$,

$$T_m = \begin{cases} 70 + 57.5t, & 0 \leq t < 4 \\ 300, & t \geq 4. \end{cases}$$

In terms of the unit step function,

$$T_m = (70 + 57.5t)[1 - \mathscr{U}(t - 4)] + 300\,\mathscr{U}(t - 4) = 70 + 57.5t + (230 - 57.5t)\,\mathscr{U}(t - 4).$$

The initial-value problem is then

$$\frac{dT}{dt} = k[T - 70 - 57.5t - (230 - 57.5t)\,\mathscr{U}(t - 4)], \qquad T(0) = 70.$$

(b) Let $t(s) = \mathscr{L}\{T(t)\}$. Transforming the equation, using $230 - 57.5t = -57.5(t - 4)$ and Theorem 4.7, gives

$$st(s) - 70 = k\left(t(s) - \frac{70}{s} - \frac{57.5}{s^2} + \frac{57.5}{s^2}\, e^{-4s}\right)$$

or

$$t(s) = \frac{70}{s - k} - \frac{70k}{s(s - k)} - \frac{57.5k}{s^2(s - k)} + \frac{57.5k}{s^2(s - k)}\, e^{-4s}.$$

After using partial functions, the inverse transform is then

$$T(t) = 70 + 57.5\left(\frac{1}{k} + t - \frac{1}{k}\, e^{kt}\right) - 57.5\left(\frac{1}{k} + t - 4 - \frac{1}{k}\, e^{k(t-4)}\right)\mathscr{U}(t - 4).$$

Of course, the obvious question is: What is k? If the cake is supposed to bake for, say, 20 minutes, then $T(20) = 300$. That is,

$$300 = 70 + 57.5\left(\frac{1}{k} + 20 - \frac{1}{k}\, e^{20k}\right) - 57.5\left(\frac{1}{k} + 16 - \frac{1}{k}\, e^{16k}\right).$$

But this equation has no physically meaningful solution. This should be no surprise since the model predicts the asymptotic behavior $T(t) \to 300$ as t increases. Using $T(20) = 299$ instead, we find, with the help of a CAS, that $k \approx -0.3$.

EXERCISES 4.4

Additional Operational Properties

3. $\mathcal{L}\{t\cos 2t\} = -\dfrac{d}{ds}\left(\dfrac{s}{s^2+4}\right) = \dfrac{s^2-4}{(s^2+4)^2}$

6. $\mathcal{L}\{t^2\cos t\} = \dfrac{d^2}{ds^2}\left(\dfrac{s}{s^2+1}\right) = \dfrac{d}{ds}\left(\dfrac{1-s^2}{(s^2+1)^2}\right) = \dfrac{2s\left(s^2-3\right)}{(s^2+1)^3}$

9. The Laplace transform of the differential equation is

$$s\,\mathcal{L}\{y\} + \mathcal{L}\{y\} = \dfrac{2s}{(s^2+1)^2}\,.$$

Solving for $\mathcal{L}\{y\}$ we obtain

$$\mathcal{L}\{y\} = \dfrac{2s}{(s+1)(s^2+1)^2} = -\dfrac{1}{2}\dfrac{1}{s+1} - \dfrac{1}{2}\dfrac{1}{s^2+1} + \dfrac{1}{2}\dfrac{s}{s^2+1} + \dfrac{1}{(s^2+1)^2} + \dfrac{s}{(s^2+1)^2}\,.$$

Thus

$$y(t) = -\dfrac{1}{2}e^{-t} - \dfrac{1}{2}\sin t + \dfrac{1}{2}\cos t + \dfrac{1}{2}(\sin t - t\cos t) + \dfrac{1}{2}t\sin t$$

$$= -\dfrac{1}{2}e^{-t} + \dfrac{1}{2}\cos t - \dfrac{1}{2}t\cos t + \dfrac{1}{2}t\sin t.$$

12. The Laplace transform of the differential equation is

$$s^2\,\mathcal{L}\{y\} - sy(0) - y'(0) + \mathcal{L}\{y\} = \dfrac{1}{s^2+1}\,.$$

Solving for $\mathcal{L}\{y\}$ we obtain

$$\mathcal{L}\{y\} = \dfrac{s^3-s^2+s}{(s^2+1)^2} = \dfrac{s}{s^2+1} - \dfrac{1}{s^2+1} + \dfrac{1}{(s^2+1)^2}\,.$$

Thus

$$y = \cos t - \sin t + \left(\dfrac{1}{2}\sin t - \dfrac{1}{2}t\cos t\right) = \cos t - \dfrac{1}{2}\sin t - \dfrac{1}{2}t\cos t.$$

15.

18. From Theorem 4.8 in the text

$$\mathcal{L}\{ty'\} = -\dfrac{d}{ds}\mathcal{L}\{y'\} = -\dfrac{d}{ds}[sY(s) - y(0)] = -s\dfrac{dY}{ds} - Y$$

so that the transform of the given second-order differential equation is the linear first-order differential equation in $Y(s)$:

$$Y' + \left(\dfrac{3}{s} - 2s\right)Y = -\dfrac{10}{s}\,.$$

Using the integrating factor $s^3 e^{-s^2}$, the last equation yields

$$Y(s) = \frac{5}{s^3} + \frac{c}{s^3} e^{s^2}.$$

But if $Y(s)$ is the Laplace transform of a piecewise-continuous function of exponential order, we must have, in view of Theorem 4.5, $\lim_{s\to\infty} Y(s) = 0$. In order to obtain this condition we require $c = 0$. Hence

$$y(t) = \mathscr{L}^{-1}\left\{\frac{5}{s^3}\right\} = \frac{5}{2}t^2.$$

21. $\mathscr{L}\{e^{-t} * e^t \cos t\} = \dfrac{s-1}{(s+1)\left[(s-1)^2 + 1\right]}$

24. $\mathscr{L}\left\{\displaystyle\int_0^t \cos\tau\, d\tau\right\} = \dfrac{1}{s}\mathscr{L}\{\cos t\} = \dfrac{s}{s(s^2+1)} = \dfrac{1}{s^2+1}$

27. $\mathscr{L}\left\{\displaystyle\int_0^t \tau e^{t-\tau}\, d\tau\right\} = \mathscr{L}\{t\}\mathscr{L}\{e^t\} = \dfrac{1}{s^2(s-1)}$

30. $\mathscr{L}\left\{t\displaystyle\int_0^t \tau e^{-\tau}\, d\tau\right\} = -\dfrac{d}{ds}\mathscr{L}\left\{\displaystyle\int_0^t \tau e^{-\tau}\, d\tau\right\} = -\dfrac{d}{ds}\left(\dfrac{1}{s}\dfrac{1}{(s+1)^2}\right) = \dfrac{3s+1}{s^2(s+1)^3}$

33. $\mathscr{L}^{-1}\left\{\dfrac{1}{s^3(s-1)}\right\} = \mathscr{L}^{-1}\left\{\dfrac{1/s^2(s-1)}{s}\right\} = \displaystyle\int_0^t (e^\tau - \tau - 1)\,d\tau = e^t - \dfrac{1}{2}t^2 - t - 1$

36. The Laplace transform of the differential equation is

$$s^2\mathscr{L}\{y\} + \mathscr{L}\{y\} = \frac{1}{(s^2+1)} + \frac{2s}{(s^2+1)^2}.$$

Thus

$$\mathscr{L}\{y\} = \frac{1}{(s^2+1)^2} + \frac{2s}{(s^2+1)^3}$$

and, using Problem 35 with $k = 1$,

$$y = \frac{1}{2}(\sin t - t\cos t) + \frac{1}{4}(t\sin t - t^2\cos t).$$

39. The Laplace transform of the given equation is

$$\mathscr{L}\{f\} = \mathscr{L}\{te^t\} + \mathscr{L}\{t\}\mathscr{L}\{f\}.$$

Solving for $\mathscr{L}\{f\}$ we obtain

$$\mathscr{L}\{f\} = \frac{s^2}{(s-1)^3(s+1)} = \frac{1}{8}\frac{1}{s-1} + \frac{3}{4}\frac{1}{(s-1)^2} + \frac{1}{4}\frac{2}{(s-1)^3} - \frac{1}{8}\frac{1}{s+1}.$$

Thus

$$f(t) = \frac{1}{8}e^t + \frac{3}{4}te^t + \frac{1}{4}t^2e^t - \frac{1}{8}e^{-t}.$$

42. The Laplace transform of the given equation is

$$\mathscr{L}\{f\} = \mathscr{L}\{\cos t\} + \mathscr{L}\{e^{-t}\}\mathscr{L}\{f\}.$$

Solving for $\mathscr{L}\{f\}$ we obtain

$$\mathscr{L}\{f\} = \frac{s}{s^2+1} + \frac{1}{s^2+1}.$$

Thus

$$f(t) = \cos t + \sin t.$$

45. The Laplace transform of the given equation is

$$s\mathcal{L}\{y\} - y(0) = \mathcal{L}\{1\} - \mathcal{L}\{\sin t\} - \mathcal{L}\{1\}\mathcal{L}\{y\}.$$

Solving for $\mathcal{L}\{f\}$ we obtain

$$\mathcal{L}\{y\} = \frac{s^2 - s + 1}{(s^2+1)^2} = \frac{1}{s^2+1} - \frac{1}{2}\frac{2s}{(s^2+1)^2}.$$

Thus

$$y = \sin t - \frac{1}{2}t\sin t.$$

48. The differential equation is

$$0.005\frac{di}{dt} + i + \frac{1}{0.02}\int_0^t i(\tau)d\tau = 100\big[t - (t-1)\,\mathcal{U}(t-1)\big]$$

or

$$\frac{di}{dt} + 200i + 10{,}000\int_0^t i(\tau)d\tau = 20{,}000\big[t - (t-1)\,\mathcal{U}(t-1)\big],$$

where $i(0) = 0$. The Laplace transform of the differential equation is

$$s\mathcal{L}\{i\} + 200\,\mathcal{L}\{i\} + \frac{10{,}000}{s}\mathcal{L}\{i\} = 20{,}000\left(\frac{1}{s^2} - \frac{1}{s^2}e^{-s}\right).$$

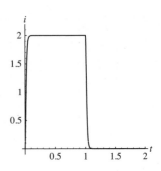

Solving for $\mathcal{L}\{i\}$ we obtain

$$\mathcal{L}\{i\} = \frac{20{,}000}{s(s+100)^2}(1-e^{-s}) = \left[\frac{2}{s} - \frac{2}{s+100} - \frac{200}{(s+100)^2}\right](1-e^{-s}).$$

Thus

$$i(t) = 2 - 2e^{-100t} - 200te^{-100t} - 2\,\mathcal{U}(t-1) + 2e^{-100(t-1)}\,\mathcal{U}(t-1) + 200(t-1)e^{-100(t-1)}\,\mathcal{U}(t-1).$$

51. Using integration by parts,

$$\mathcal{L}\{f(t)\} = \frac{1}{1-e^{-bs}}\int_0^b \frac{a}{b}te^{-st}dt = \frac{a}{s}\left(\frac{1}{bs} - \frac{1}{e^{bs}-1}\right).$$

54. $\mathcal{L}\{f(t)\} = \dfrac{1}{1-e^{-2\pi s}}\displaystyle\int_0^\pi e^{-st}\sin t\, dt = \dfrac{1}{s^2+1}\cdot\dfrac{1}{1-e^{-\pi s}}$

57. The differential equation is $x'' + 2x' + 10x = 20f(t)$, where $f(t)$ is the meander function in Problem 49 with $a = \pi$. Using the initial conditions $x(0) = x'(0) = 0$ and taking the Laplace transform we obtain

$$(s^2 + 2s + 10)\mathcal{L}\{x(t)\} = \frac{20}{s}(1-e^{-\pi s})\frac{1}{1+e^{-\pi s}}$$

$$= \frac{20}{s}(1-e^{-\pi s})(1 - e^{-\pi s} + e^{-2\pi s} - e^{-3\pi s} + \cdots)$$

$$= \frac{20}{s}(1 - 2e^{-\pi s} + 2e^{-2\pi s} - 2e^{-3\pi s} + \cdots)$$

$$= \frac{20}{s} + \frac{40}{s}\sum_{n=1}^{\infty}(-1)^n e^{-n\pi s}.$$

90

Then

$$\mathscr{L}\{x(t)\} = \frac{20}{s(s^2+2s+10)} + \frac{40}{s(s^2+2s+10)} \sum_{n=1}^{\infty} (-1)^n e^{-n\pi s}$$

$$= \frac{2}{s} - \frac{2s+4}{s^2+2s+10} + \sum_{n=1}^{\infty} (-1)^n \left[\frac{4}{s} - \frac{4s+8}{s^2+2s+10}\right] e^{-n\pi s}$$

$$= \frac{2}{s} - \frac{2(s+1)+2}{(s+1)^2+9} + 4\sum_{n=1}^{\infty} (-1)^n \left[\frac{1}{s} - \frac{(s+1)+1}{(s+1)^2+9}\right] e^{-n\pi s}$$

and

$$x(t) = 2\left(1 - e^{-t}\cos 3t - \frac{1}{3}e^{-t}\sin 3t\right) + 4\sum_{n=1}^{\infty} (-1)^n \left[1 - e^{-(t-n\pi)}\cos 3(t-n\pi)\right.$$

$$\left. - \frac{1}{3}e^{-(t-n\pi)}\sin 3(t-n\pi)\right] \mathscr{U}(t-n\pi).$$

The graph of $x(t)$ on the interval $[0, 2\pi)$ is shown below.

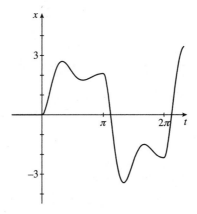

60. The transform of Bessel's equation is

$$-\frac{d}{ds}[s^2 Y(s) - sy(0) - y'(0)] + sY(s) - y(0) - \frac{d}{ds}Y(s) = 0$$

or, after simplifying and using the initial condition, $(s^2+1)Y' + sY = 0$. This equation is both separable and linear. Solving gives $Y(s) = c/\sqrt{s^2+1}$. Now $Y(s) = \mathscr{L}\{J_0(t)\}$, where J_0 has a derivative that is continuous and of exponential order, implies by Problem 46 of Exercises 4.2 that

$$1 = J_0(0) = \lim_{s\to\infty} sY(s) = c \lim_{s\to\infty} \frac{s}{\sqrt{s^2+k^2}} = c$$

so $c = 1$ and

$$Y(s) = \frac{1}{\sqrt{s^2+1}} \qquad \text{or} \qquad \mathscr{L}\{J_0(t)\} = \frac{1}{\sqrt{s^2+1}}.$$

63. The solution is

$$y(t) = \frac{1}{6}e^t - \frac{1}{6}e^{-t/2}\cos\sqrt{15}\,t - \frac{\sqrt{3/5}}{6}e^{-t/2}\sin\sqrt{15}\,t.$$

EXERCISES 4.5

The Dirac Delta Function

3. The Laplace transform of the differential equation yields

$$\mathcal{L}\{y\} = \frac{1}{s^2 + 1}\left(1 + e^{-2\pi s}\right)$$

so that

$$y = \sin t + \sin t\,\mathcal{U}(t - 2\pi).$$

6. The Laplace transform of the differential equation yields

$$\mathcal{L}\{y\} = \frac{s}{s^2 + 1} + \frac{1}{s^2 + 1}\left(e^{-2\pi s} + e^{-4\pi s}\right)$$

so that

$$y = \cos t + \sin t[\mathcal{U}(t - 2\pi) + \mathcal{U}(t - 4\pi)].$$

9. The Laplace transform of the differential equation yields

$$\mathcal{L}\{y\} = \frac{1}{(s + 2)^2 + 1}e^{-2\pi s}$$

so that

$$y = e^{-2(t - 2\pi)}\sin t\,\mathcal{U}(t - 2\pi).$$

12. The Laplace transform of the differential equation yields

$$\mathcal{L}\{y\} = \frac{1}{(s - 1)^2(s - 6)} + \frac{e^{-2s} + e^{-4s}}{(s - 1)(s - 6)}$$

$$= -\frac{1}{25}\frac{1}{s - 1} - \frac{1}{5}\frac{1}{(s - 1)^2} + \frac{1}{25}\frac{1}{s - 6} + \left[-\frac{1}{5}\frac{1}{s - 1} + \frac{1}{5}\frac{1}{s - 6}\right]\left(e^{-2s} + e^{-4s}\right)$$

so that

$$y = -\frac{1}{25}e^t - \frac{1}{5}te^t + \frac{1}{25}e^{6t} + \left[-\frac{1}{5}e^{t-2} + \frac{1}{5}e^{6(t-2)}\right]\mathcal{U}(t - 2) + \left[-\frac{1}{5}e^{t-4} + \frac{1}{5}e^{6(t-4)}\right]\mathcal{U}(t - 4).$$

15. You should disagree. Although formal manipulations of the Laplace transform lead to $y(t) = \frac{1}{3}e^{-t}\sin 3t$ in both cases, this function does not satisfy the initial condition $y'(0) = 0$ of the second initial-value problem.

EXERCISES 4.6

Systems of Linear Differential Equations

3. Taking the Laplace transform of the system gives

$$s\mathcal{L}\{x\} + 1 = \mathcal{L}\{x\} - 2\mathcal{L}\{y\}$$
$$s\mathcal{L}\{y\} - 2 = 5\mathcal{L}\{x\} - \mathcal{L}\{y\}$$

so that

$$\mathcal{L}\{x\} = \frac{-s-5}{s^2+9} = -\frac{s}{s^2+9} - \frac{5}{3}\frac{3}{s^2+9}$$

and

$$x = -\cos 3t - \frac{5}{3}\sin 3t.$$

Then

$$y = \frac{1}{2}x - \frac{1}{2}x' = 2\cos 3t - \frac{7}{3}\sin 3t.$$

6. Taking the Laplace transform of the system gives

$$(s+1)\mathcal{L}\{x\} - (s-1)\mathcal{L}\{y\} = -1$$
$$s\mathcal{L}\{x\} + (s+2)\mathcal{L}\{y\} = 1$$

so that

$$\mathcal{L}\{y\} = \frac{s+1/2}{s^2+s+1} = \frac{s+1/2}{(s+1/2)^2 + (\sqrt{3}/2)^2}$$

and

$$\mathcal{L}\{x\} = \frac{-3/2}{s^2+s+1} = -\sqrt{3}\,\frac{\sqrt{3}/2}{(s+1/2)^2 + (\sqrt{3}/2)^2}.$$

Then

$$y = e^{-t/2}\cos\frac{\sqrt{3}}{2}t \qquad \text{and} \qquad x = -\sqrt{3}\,e^{-t/2}\sin\frac{\sqrt{3}}{2}t.$$

9. Adding the equations and then subtracting them gives

$$\frac{d^2x}{dt^2} = \frac{1}{2}t^2 + 2t$$

$$\frac{d^2y}{dt^2} = \frac{1}{2}t^2 - 2t.$$

Taking the Laplace transform of the system gives

$$\mathcal{L}\{x\} = 8\frac{1}{s} + \frac{1}{24}\frac{4!}{s^5} + \frac{1}{3}\frac{3!}{s^4}$$

and

$$\mathcal{L}\{y\} = \frac{1}{24}\frac{4!}{s^5} - \frac{1}{3}\frac{3!}{s^4}$$

so that

$$x = 8 + \frac{1}{24}t^4 + \frac{1}{3}t^3 \qquad \text{and} \qquad y = \frac{1}{24}t^4 - \frac{1}{3}t^3.$$

12. Taking the Laplace transform of the system gives

$$(s-4)\mathcal{L}\{x\} + 2\mathcal{L}\{y\} = \frac{2e^{-s}}{s}$$

$$-3\mathcal{L}\{x\} + (s+1)\mathcal{L}\{y\} = \frac{1}{2} + \frac{e^{-s}}{s}$$

93

so that

$$\mathscr{L}\{x\} = \frac{-1/2}{(s-1)(s-2)} + e^{-s}\frac{1}{(s-1)(s-2)}$$

$$= \frac{1}{2}\frac{1}{s-1} - \frac{1}{2}\frac{1}{s-2} + e^{-s}\left[-\frac{1}{s-1} + \frac{1}{s-2}\right]$$

and

$$\mathscr{L}\{y\} = \frac{e^{-s}}{s} + \frac{s/4-1}{(s-1)(s-2)} + e^{-s}\frac{-s/2+2}{(s-1)(s-2)}$$

$$= \frac{3}{4}\frac{1}{s-1} - \frac{1}{2}\frac{1}{s-2} + e^{-s}\left[\frac{1}{s} - \frac{3}{2}\frac{1}{s-1} + \frac{1}{s-2}\right].$$

Then

$$x = \frac{1}{2}e^t - \frac{1}{2}e^{2t} + \left[-e^{t-1} + e^{2(t-1)}\right]\mathscr{U}(t-1)$$

and

$$y = \frac{3}{4}e^t - \frac{1}{2}e^{2t} + \left[1 - \frac{3}{2}e^{t-1} + e^{2(t-1)}\right]\mathscr{U}(t-1).$$

15. **(a)** By Kirchhoff's first law we have $i_1 = i_2 + i_3$. By Kirchhoff's second law, on each loop we have $E(t) = Ri_1 + L_1i_2'$ and $E(t) = Ri_1 + L_2i_3'$ or $L_1i_2' + Ri_2 + Ri_3 = E(t)$ and $L_2i_3' + Ri_2 + Ri_3 = E(t)$.

(b) Taking the Laplace transform of the system

$$0.01i_2' + 5i_2 + 5i_3 = 100$$

$$0.0125i_3' + 5i_2 + 5i_3 = 100$$

gives

$$(s+500)\mathscr{L}\{i_2\} + 500\mathscr{L}\{i_3\} = \frac{10,000}{s}$$

$$400\mathscr{L}\{i_2\} + (s+400)\mathscr{L}\{i_3\} = \frac{8,000}{s}$$

so that

$$\mathscr{L}\{i_3\} = \frac{8,000}{s^2 + 900s} = \frac{80}{9}\frac{1}{s} - \frac{80}{9}\frac{1}{s+900}.$$

Then

$$i_3 = \frac{80}{9} - \frac{80}{9}e^{-900t} \qquad \text{and} \qquad i_2 = 20 - 0.0025i_3' - i_3 = \frac{100}{9} - \frac{100}{9}e^{-900t}.$$

(c) $i_1 = i_2 + i_3 = 20 - 20e^{-900t}$

18. Taking the Laplace transform of the system

$$0.5i_1' + 50i_2 = 60$$

$$0.005i_2' + i_2 - i_1 = 0$$

gives

$$s\mathscr{L}\{i_1\} + 100\mathscr{L}\{i_2\} = \frac{120}{s}$$

$$-200\mathscr{L}\{i_1\} + (s+200)\mathscr{L}\{i_2\} = 0$$

so that

$$\mathscr{L}\{i_2\} = \frac{24,000}{s(s^2 + 200s + 20,000)} = \frac{6}{5}\frac{1}{s} - \frac{6}{5}\frac{s+100}{(s+100)^2 + 100^2} - \frac{6}{5}\frac{100}{(s+100)^2 + 100^2}.$$

Then

$$i_2 = \frac{6}{5} - \frac{6}{5}e^{-100t}\cos 100t - \frac{6}{5}e^{-100t}\sin 100t$$

and

$$i_1 = 0.005i_2' + i_2 = \frac{6}{5} - \frac{6}{5}e^{-100t}\cos 100t.$$

21. (a) Taking the Laplace transform of the system

$$4\theta_1'' + \theta_2'' + 8\theta_1 = 0$$
$$\theta_1'' + \theta_2'' + 2\theta_2 = 0$$

gives

$$4\left(s^2 + 2\right)\mathscr{L}\{\theta_1\} + s^2\mathscr{L}\{\theta_2\} = 3s$$
$$s^2\mathscr{L}\{\theta_1\} + \left(s^2 + 2\right)\mathscr{L}\{\theta_2\} = 0$$

so that

$$\left(3s^2 + 4\right)\left(s^2 + 4\right)\mathscr{L}\{\theta_2\} = -3s^3$$

or

$$\mathscr{L}\{\theta_2\} = \frac{1}{2}\frac{s}{s^2 + 4/3} - \frac{3}{2}\frac{s}{s^2 + 4}.$$

Then

$$\theta_2 = \frac{1}{2}\cos\frac{2}{\sqrt{3}}t - \frac{3}{2}\cos 2t \qquad \text{and} \qquad \theta_1'' = -\theta_2'' - 2\theta_2$$

so that

$$\theta_1 = \frac{1}{4}\cos\frac{2}{\sqrt{3}}t + \frac{3}{4}\cos 2t.$$

(b)

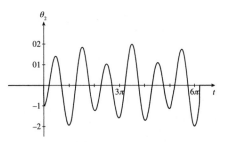

Mass m_2 has extreme displacements of greater magnitude. Mass m_1 first passes through its equilibrium position at about $t = 0.87$, and mass m_2 first passes through its equilibrium position at about $t = 0.66$. The motion of the pendulums is not periodic since $\cos(2t/\sqrt{3})$ has period $\sqrt{3}\,\pi$, $\cos 2t$ has period π, and the ratio of these periods is $\sqrt{3}$, which is not a rational number.

95

(c) The Lissajous curve is plotted for $0 \leq t \leq 30$.

(d)

t	θ_1	θ_2
1	−0.2111	0.8263
2	−0.6585	0.6438
3	0.4830	−1.9145
4	−0.1325	0.1715
5	−0.4111	1.6951
6	0.8327	−0.8662
7	0.0458	−0.3186
8	−0.9639	0.9452
9	0.3534	−1.2741
10	0.4370	−0.3502

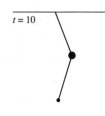

(e) Using a CAS to solve $\theta_1(t) = \theta_2(t)$ we see that $\theta_1 = \theta_2$ (so that the double pendulum is straight out) when t is about 0.75 seconds.

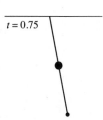

(f) To make a movie of the pendulum it is necessary to locate the mass in the plane as a function of time. Suppose that the upper arm is attached to the origin and that the equilibrium position lies along the negative y-axis. Then mass m_1 is at $(x, (t), y_1(t))$ and mass m_2 is at $(x_2(t), y_2(t))$, where

$$x_1(t) = 16 \sin \theta_1(t) \qquad \text{and} \qquad y_1(t) = -16 \cos \theta_1(t)$$

and

$$x_2(t) = x_1(t) + 16 \sin \theta_2(t) \qquad \text{and} \qquad y_2(t) = y_1(t) - 16 \cos \theta_2(t).$$

A reasonable movie can be constructed by letting t range from 0 to 10 in increments of 0.1 seconds.

3. False; consider $f(t) = t^{-1/2}$.

6. False; consider $f(t) = 1$ and $g(t) = 1$.

9. $\mathscr{L}\{\sin 2t\} = \dfrac{2}{s^2 + 4}$

12. $\mathscr{L}\{\sin 2t\, \mathscr{U}(t - \pi)\} = \mathscr{L}\{\sin 2(t - \pi)\, \mathscr{U}(t - \pi)\} = \dfrac{2}{s^2 + 4}e^{-\pi s}$

15. $\mathscr{L}^{-1}\left\{\dfrac{1}{(s - 5)^3}\right\} = \dfrac{1}{2}\mathscr{L}^{-1}\left\{\dfrac{2}{(s - 5)^3}\right\} = \dfrac{1}{2}t^2 e^{5t}$

18. $\mathscr{L}^{-1}\left\{\dfrac{1}{s^2}e^{-5s}\right\} = (t - 5)\mathscr{U}(t - 5)$

21. $\mathscr{L}\{e^{-5t}\}$ exists for $s > -5$.

24. $\mathscr{L}\left\{\displaystyle\int_0^t e^{a\tau}f(\tau)\,d\tau\right\} = \dfrac{1}{s}\mathscr{L}\{e^{at}f(t)\} = \dfrac{F(s - a)}{s}$, whereas

$\mathscr{L}\left\{e^{at}\displaystyle\int_0^t f(\tau)\,d\tau\right\} = \mathscr{L}\left\{\displaystyle\int_0^t f(\tau)\,d\tau\right\}\bigg|_{s \to s - a} = \dfrac{F(s)}{s}\bigg|_{s \to s - a} = \dfrac{F(s - a)}{s - a}$.

27. $f(t - t_0)\,\mathscr{U}(t - t_0)$

30. $f(t) = \sin t\,\mathscr{U}(t - \pi) - \sin t\,\mathscr{U}(t - 3\pi) = -\sin(t - \pi)\,\mathscr{U}(t - \pi) + \sin(t - 3\pi)\,\mathscr{U}(t - 3\pi)$

$\mathscr{L}\{f(t)\} = -\dfrac{1}{s^2 + 1}e^{-\pi s} + \dfrac{1}{s^2 + 1}e^{-3\pi s}$

$\mathscr{L}\{e^t f(t)\} = -\dfrac{1}{(s - 1)^2 + 1}e^{-\pi(s - 1)} + \dfrac{1}{(s - 1)^2 + 1}e^{-3\pi(s - 1)}$

33. Taking the Laplace transform of the differential equation we obtain

$$\mathscr{L}\{y\} = \dfrac{5}{(s - 1)^2} + \dfrac{1}{2}\dfrac{2}{(s - 1)^3}$$

so that

$$y = 5te^t + \dfrac{1}{2}t^2 e^t.$$

36. Taking the Laplace transform of the differential equation we obtain

$$\mathscr{L}\{y\} = \dfrac{s^3 + 2}{s^3(s - 5)} - \dfrac{2 + 2s + s^2}{s^3(s - 5)}e^{-s}$$

$$= -\dfrac{2}{125}\dfrac{1}{s} - \dfrac{2}{25}\dfrac{1}{s^2} - \dfrac{1}{5}\dfrac{2}{s^3} + \dfrac{127}{125}\dfrac{1}{s - 5} - \left[-\dfrac{37}{125}\dfrac{1}{s} - \dfrac{12}{25}\dfrac{1}{s^2} - \dfrac{1}{5}\dfrac{2}{s^3} + \dfrac{37}{125}\dfrac{1}{s - 5}\right]e^{-s}$$

so that

$$y = -\dfrac{2}{125} - \dfrac{2}{25}t - \dfrac{1}{5}t^2 + \dfrac{127}{125}e^{5t} - \left[-\dfrac{37}{125} - \dfrac{12}{25}(t - 1) - \dfrac{1}{5}(t - 1)^2 + \dfrac{37}{125}e^{5(t - 1)}\right]\mathscr{U}(t - 1).$$

39. Taking the Laplace transform of the system gives

$$s\mathscr{L}\{x\} + \mathscr{L}\{y\} = \frac{1}{s^2} + 1$$

$$4\mathscr{L}\{x\} + s\mathscr{L}\{y\} = 2$$

so that

$$\mathscr{L}\{x\} = \frac{s^2 - 2s + 1}{s(s-2)(s+2)} = -\frac{1}{4}\frac{1}{s} + \frac{1}{8}\frac{1}{s-2} + \frac{9}{8}\frac{1}{s+2}.$$

Then

$$x = -\frac{1}{4} + \frac{1}{8}e^{2t} + \frac{9}{8}e^{-2t} \quad \text{and} \quad y = -x' + t = \frac{9}{4}e^{-2t} - \frac{1}{4}e^{2t} + t.$$

42. The differential equation is

$$\frac{1}{2}\frac{d^2q}{dt^2} + 10\frac{dq}{dt} + 100q = 10 - 10\,\mathscr{U}(t-5).$$

Taking the Laplace transform we obtain

$$\mathscr{L}\{q\} = \frac{20}{s(s^2 + 20s + 200)}\left(1 - e^{-5s}\right)$$

$$= \left[\frac{1}{10}\frac{1}{s} - \frac{1}{10}\frac{s+10}{(s+10)^2 + 10^2} - \frac{1}{10}\frac{10}{(s+10)^2 + 10^2}\right]\left(1 - e^{-5s}\right)$$

so that

$$q(t) = \frac{1}{10} - \frac{1}{10}e^{-10t}\cos 10t - \frac{1}{10}e^{-10t}\sin 10t$$

$$- \left[\frac{1}{10} - \frac{1}{10}e^{-10(t-5)}\cos 10(t-5) - \frac{1}{10}e^{-10(t-5)}\sin 10(t-5)\right]\mathscr{U}(t-5).$$

45. (a) With $\omega^2 = g/l$ and $K = k/m$ the system of differential equations is

$$\theta_1'' + \omega^2\theta_1 = -K(\theta_1 - \theta_2)$$

$$\theta_2'' + \omega^2\theta_2 = K(\theta_1 - \theta_2).$$

Denoting the Laplace transform of $\theta(t)$ by $\Theta(s)$ we have that the Laplace transform of the system is

$$(s^2 + \omega^2)\Theta_1(s) = -K\Theta_1(s) + K\Theta_2(s) + s\theta_0$$

$$(s^2 + \omega^2)\Theta_2(s) = K\Theta_1(s) - K\Theta_2(s) + s\psi_0.$$

If we add the two equations, we get

$$\Theta_1(s) + \Theta_2(s) = (\theta_0 + \psi_0)\frac{s}{s^2 + \omega^2}$$

which implies

$$\theta_1(t) + \theta_2(t) = (\theta_0 + \psi_0)\cos\omega t.$$

This enables us to solve for first, say, $\theta_1(t)$ and then find $\theta_2(t)$ from

$$\theta_2(t) = -\theta_1(t) + (\theta_0 + \psi_0)\cos\omega t.$$

Now solving

$$(s^2 + \omega^2 + K)\Theta_1(s) - K\Theta_2(s) = s\theta_0$$

$$-k\Theta_1(s) + (s^2 + \omega^2 + K)\Theta_2(s) = s\psi_0$$

gives

$$[(s^2 + \omega^2 + K)^2 - K^2]\Theta_1(s) = s(s^2 + \omega^2 + K)\theta_0 + Ks\psi_0.$$

Factoring the difference of two squares and using partial fractions we get

$$\Theta_1(s) = \frac{s(s^2 + \omega^2 + K)\theta_0 + Ks\psi_0}{(s^2 + \omega^2)(s^2 + \omega^2 + 2K)} = \frac{\theta_0 + \psi_0}{2}\frac{s}{s^2 + \omega^2} + \frac{\theta_0 - \psi_0}{2}\frac{s}{s^2 + \omega^2 + 2K},$$

so

$$\theta_1(t) = \frac{\theta_0 + \psi_0}{2}\cos\omega t + \frac{\theta_0 - \psi_0}{2}\cos\sqrt{\omega^2 + 2K}\,t.$$

Then from $\theta_2(t) = -\theta_1(t) + (\theta_0 + \psi_0)\cos\omega t$ we get

$$\theta_2(t) = \frac{\theta_0 + \psi_0}{2}\cos\omega t - \frac{\theta_0 - \psi_0}{2}\cos\sqrt{\omega^2 + 2K}\,t.$$

(b) With the initial conditions $\theta_1(0) = \theta_0$, $\theta_1'(0) = 0$, $\theta_2(0) = \theta_0$, $\theta_2'(0) = 0$ we have

$$\theta_1(t) = \theta_0\cos\omega t, \qquad \theta_2(t) = \theta_0\cos\omega t.$$

Physically this means that both pendulums swing in the same direction as if they were free since the spring exerts no influence on the motion ($\theta_1(t)$ and $\theta_2(t)$ are free of K).

With the initial conditions $\theta_1(0) = \theta_0$, $\theta_1'(0) = 0$, $\theta_2(0) = -\theta_0$, $\theta_2'(0) = 0$ we have

$$\theta_1(t) = \theta_0\cos\sqrt{\omega^2 + 2K}\,t, \qquad \theta_2(t) = -\theta_0\cos\sqrt{\omega^2 + 2K}\,t.$$

Physically this means that both pendulums swing in the opposite directions, stretching and compressing the spring. The amplitude of both displacements is $|\theta_0|$. Moreover, $\theta_1(t) = \theta_0$ and $\theta_2(t) = -\theta_0$ at precisely the same times. At these times the spring is stretched to its maximum.

5 Series Solutions of Linear Differential Equations

EXERCISES 5.1

Solutions About Ordinary Points

3. By the ratio test,

$$\lim_{n\to\infty}\left|\frac{(x-5)^{n+1}/10^{n+1}}{(x-5)^n/10^n}\right| = \lim_{n\to\infty}\frac{1}{10}|x-5| = \frac{1}{10}|x-5|.$$

The series is absolutely convergent for $\frac{1}{10}|x-5| < 1$, $|x-5| < 10$, or on $(-5, 15)$. The radius of convergence is $R = 10$. At $x = -5$, the series $\sum_{n=1}^{\infty}(-1)^n(-10)^n/10^n = \sum_{n=1}^{\infty}1$ diverges by the nth term test. At $x = 15$, the series $\sum_{n=1}^{\infty}(-1)^n10^n/10^n = \sum_{n=1}^{\infty}(-1)^n$ diverges by the nth term test. Thus, the series converges on $(-5, 15)$.

6. $e^{-x}\cos x = \left(1 - x + \dfrac{x^2}{2} - \dfrac{x^3}{6} + \dfrac{x^4}{24} - \cdots\right)\left(1 - \dfrac{x^2}{2} + \dfrac{x^4}{24} - \cdots\right) = 1 - x + \dfrac{x^3}{3} - \dfrac{x^4}{6} + \cdots$

9. Let $k = n+2$ so that $n = k-2$ and

$$\sum_{n=1}^{\infty} nc_n x^{n+2} = \sum_{k=3}^{\infty}(k-2)c_{k-2}x^k.$$

12. $\displaystyle\sum_{n=2}^{\infty}n(n-1)c_n x^n + 2\sum_{n=2}^{\infty}n(n-1)c_n x^{n-2} + 3\sum_{n=1}^{\infty}nc_n x^n$

$= 2\cdot 2\cdot 1c_2 x^0 + 2\cdot 3\cdot 2c_3 x^1 + 3\cdot 1\cdot c_1 x^1 + \underbrace{\sum_{n=2}^{\infty}n(n-1)c_n x^n}_{k=n} + 2\underbrace{\sum_{n=4}^{\infty}n(n-1)c_n x^{n-2}}_{k=n-2} + 3\underbrace{\sum_{n=2}^{\infty}nc_n x^n}_{k=n}$

$= 4c_2 + (3c_1 + 12c_3)x + \displaystyle\sum_{k=2}^{\infty}k(k-1)c_k x^k + 2\sum_{k=2}^{\infty}(k+2)(k+1)c_{k+2}x^k + 3\sum_{k=2}^{\infty}kc_k x^k$

$= 4c_2 + (3c_1 + 12c_3)x + \displaystyle\sum_{k=2}^{\infty}\left[(k(k-1) + 3k)c_k + 2(k+2)(k+1)c_{k+2}\right]x^k$

$= 4c_2 + (3c_1 + 12c_3)x + \displaystyle\sum_{k=2}^{\infty}\left[k(k+2)c_k + 2(k+1)(k+2)c_{k+2}\right]x^k$

15. The singular points of $(x^2 - 25)y'' + 2xy' + y = 0$ are -5 and 5. The distance from 0 to either of these points is 5. The distance from 1 to the closest of these points is 4.

18. Substituting $y = \sum_{n=0}^{\infty} c_n x^n$ into the differential equation we have

$$y'' + x^2 y = \underbrace{\sum_{n=2}^{\infty} n(n-1)c_n x^{n-2}}_{k=n-2} + \underbrace{\sum_{n=0}^{\infty} c_n x^{n+2}}_{k=n+2} = \sum_{k=0}^{\infty} (k+2)(k+1)c_{k+2} x^k + \sum_{k=2}^{\infty} c_{k-2} x^k$$

$$= 2c_2 + 6c_3 x + \sum_{k=2}^{\infty} [(k+2)(k+1)c_{k+2} + c_{k-2}]x^k = 0.$$

Thus

$$c_2 = c_3 = 0$$

$$(k+2)(k+1)c_{k+2} + c_{k-2} = 0$$

and

$$c_{k+2} = -\frac{1}{(k+2)(k+1)} c_{k-2}, \quad k = 2, 3, 4, \ldots.$$

Choosing $c_0 = 1$ and $c_1 = 0$ we find

$$c_4 = -\frac{1}{12}$$

$$c_5 = c_6 = c_7 = 0$$

$$c_8 = \frac{1}{672}$$

and so on. For $c_0 = 0$ and $c_1 = 1$ we obtain

$$c_4 = 0$$

$$c_5 = -\frac{1}{20}$$

$$c_6 = c_7 = c_8 = 0$$

$$c_9 = \frac{1}{1440}$$

and so on. Thus, two solutions are

$$y_1 = 1 - \frac{1}{12}x^4 + \frac{1}{672}x^8 - \cdots \quad \text{and} \quad y_2 = x - \frac{1}{20}x^5 + \frac{1}{1440}x^9 - \cdots.$$

21. Substituting $y = \sum_{n=0}^{\infty} c_n x^n$ into the differential equation we have

$$y'' + x^2 y' + xy = \underbrace{\sum_{n=2}^{\infty} n(n-1)c_n x^{n-2}}_{k=n-2} + \underbrace{\sum_{n=1}^{\infty} nc_n x^{n+1}}_{k=n+1} + \underbrace{\sum_{n=0}^{\infty} c_n x^{n+1}}_{k=n+1}$$

$$= \sum_{k=0}^{\infty} (k+2)(k+1)c_{k+2} x^k + \sum_{k=2}^{\infty} (k-1)c_{k-1} x^k + \sum_{k=1}^{\infty} c_{k-1} x^k$$

$$= 2c_2 + (6c_3 + c_0)x + \sum_{k=2}^{\infty} [(k+2)(k+1)c_{k+2} + kc_{k-1}]x^k = 0.$$

Thus

$$c_2 = 0$$

$$6c_3 + c_0 = 0$$

$$(k+2)(k+1)c_{k+2} + kc_{k-1} = 0$$

and

$$c_2 = 0$$

$$c_3 = -\frac{1}{6}c_0$$

$$c_{k+2} = -\frac{k}{(k+2)(k+1)}c_{k-1}, \quad k = 2, 3, 4, \ldots .$$

Choosing $c_0 = 1$ and $c_1 = 0$ we find

$$c_3 = -\frac{1}{6}$$

$$c_4 = c_5 = 0$$

$$c_6 = \frac{1}{45}$$

and so on. For $c_0 = 0$ and $c_1 = 1$ we obtain

$$c_3 = 0$$

$$c_4 = -\frac{1}{6}$$

$$c_5 = c_6 = 0$$

$$c_7 = \frac{5}{252}$$

and so on. Thus, two solutions are

$$y_1 = 1 - \frac{1}{6}x^3 + \frac{1}{45}x^6 - \cdots \quad \text{and} \quad y_2 = x - \frac{1}{6}x^4 + \frac{5}{252}x^7 - \cdots .$$

24. Substituting $y = \sum_{n=0}^{\infty} c_n x^n$ into the differential equation we have

$$(x+2)y'' + xy' - y = \underbrace{\sum_{n=2}^{\infty} n(n-1)c_n x^{n-1}}_{k=n-1} + \underbrace{\sum_{n=2}^{\infty} 2n(n-1)c_n x^{n-2}}_{k=n-2} + \underbrace{\sum_{n=1}^{\infty} nc_n x^n}_{k=n} - \underbrace{\sum_{n=0}^{\infty} c_n x^n}_{k=n}$$

$$= \sum_{k=1}^{\infty}(k+1)kc_{k+1}x^k + \sum_{k=0}^{\infty} 2(k+2)(k+1)c_{k+2}x^k + \sum_{k=1}^{\infty} kc_k x^k - \sum_{k=0}^{\infty} c_k x^k$$

$$= 4c_2 - c_0 + \sum_{k=1}^{\infty}\Big[(k+1)kc_{k+1} + 2(k+2)(k+1)c_{k+2} + (k-1)c_k\Big]x^k = 0.$$

Thus

$$4c_2 - c_0 = 0$$

$$(k+1)kc_{k+1} + 2(k+2)(k+1)c_{k+2} + (k-1)c_k = 0, \quad k = 1, 2, 3, \ldots$$

and

$$c_2 = \frac{1}{4}c_0$$

$$c_{k+2} = -\frac{(k+1)kc_{k+1} + (k-1)c_k}{2(k+2)(k+1)}, \quad k = 1, 2, 3, \ldots .$$

Choosing $c_0 = 1$ and $c_1 = 0$ we find

$$c_1 = 0, \quad c_2 = \frac{1}{4}, \quad c_3 = -\frac{1}{24}, \quad c_4 = 0, \quad c_5 = \frac{1}{480}$$

and so on. For $c_0 = 0$ and $c_1 = 1$ we obtain

$$c_2 = 0$$

$$c_3 = 0$$

$$c_4 = c_5 = c_6 = \cdots = 0.$$

Thus, two solutions are

$$y_1 = c_0 \left[1 + \frac{1}{4}x^2 - \frac{1}{24}x^3 + \frac{1}{480}x^5 + \cdots \right] \quad \text{and} \quad y_2 = c_1 x.$$

27. Substituting $y = \sum_{n=0}^{\infty} c_n x^n$ into the differential equation we have

$$\left(x^2 + 2\right)y'' + 3xy' - y = \underbrace{\sum_{n=2}^{\infty} n(n-1)c_n x^n}_{k=n} + 2\underbrace{\sum_{n=2}^{\infty} n(n-1)c_n x^{n-2}}_{k=n-2} + 3\underbrace{\sum_{n=1}^{\infty} nc_n x^n}_{k=n} - \underbrace{\sum_{n=0}^{\infty} c_n x^n}_{k=n}$$

$$= \sum_{k=2}^{\infty} k(k-1)c_k x^k + 2\sum_{k=0}^{\infty} (k+2)(k+1)c_{k+2} x^k + 3\sum_{k=1}^{\infty} kc_k x^k - \sum_{k=0}^{\infty} c_k x^k$$

$$= (4c_2 - c_0) + (12c_3 + 2c_1)x + \sum_{k=2}^{\infty} \left[2(k+2)(k+1)c_{k+2} + \left(k^2 + 2k - 1\right)c_k \right] x^k = 0.$$

Thus

$$4c_2 - c_0 = 0$$

$$12c_3 + 2c_1 = 0$$

$$2(k+2)(k+1)c_{k+2} + \left(k^2 + 2k - 1\right)c_k = 0$$

and

$$c_2 = \frac{1}{4}c_0$$

$$c_3 = -\frac{1}{6}c_1$$

$$c_{k+2} = -\frac{k^2 + 2k - 1}{2(k+2)(k+1)}c_k, \quad k = 2, 3, 4, \ldots.$$

Choosing $c_0 = 1$ and $c_1 = 0$ we find

$$c_2 = \frac{1}{4}$$

$$c_3 = c_5 = c_7 = \cdots = 0$$

$$c_4 = -\frac{7}{96}$$

and so on. For $c_0 = 0$ and $c_1 = 1$ we obtain

$$c_2 = c_4 = c_6 = \cdots = 0$$

$$c_3 = -\frac{1}{6}$$

$$c_5 = \frac{7}{120}$$

and so on. Thus, two solutions are

$$y_1 = 1 + \frac{1}{4}x^2 - \frac{7}{96}x^4 + \cdots \quad \text{and} \quad y_2 = x - \frac{1}{6}x^3 + \frac{7}{120}x^5 - \cdots .$$

30. Substituting $y = \sum_{n=0}^{\infty} c_n x^n$ into the differential equation we have

$$(x+1)y'' - (2-x)y' + y$$

$$= \sum_{n=2}^{\infty} n(n-1)c_n x^{n-1} + \sum_{n=2}^{\infty} n(n-1)c_n x^{n-2} - 2\sum_{n=1}^{\infty} nc_n x^{n-1} + \sum_{n=1}^{\infty} nc_n x^n + \sum_{n=0}^{\infty} c_n x^n$$
$$\underbrace{\qquad}_{k=n-1} \qquad \underbrace{\qquad}_{k=n-2} \qquad \underbrace{\qquad}_{k=n-1} \qquad \underbrace{\qquad}_{k=n} \qquad \underbrace{\qquad}_{k=n}$$

$$= \sum_{k=1}^{\infty}(k+1)kc_{k+1}x^k + \sum_{k=0}^{\infty}(k+2)(k+1)c_{k+2}x^k - 2\sum_{k=0}^{\infty}(k+1)c_{k+1}x^k + \sum_{k=1}^{\infty}kc_k x^k + \sum_{k=0}^{\infty}c_k x^k$$

$$= 2c_2 - 2c_1 + c_0 + \sum_{k=1}^{\infty}[(k+2)(k+1)c_{k+2} - (k+1)c_{k+1} + (k+1)c_k]x^k = 0.$$

Thus

$$2c_2 - 2c_1 + c_0 = 0$$

$$(k+2)(k+1)c_{k+2} - (k+1)c_{k+1} + (k+1)c_k = 0$$

and

$$c_2 = c_1 - \frac{1}{2}c_0$$

$$c_{k+2} = \frac{1}{k+2}c_{k+1} - \frac{1}{k+2}c_k, \quad k = 1, 2, 3, \dots.$$

Choosing $c_0 = 1$ and $c_1 = 0$ we find

$$c_2 = -\frac{1}{2}, \qquad c_3 = -\frac{1}{6}, \qquad c_4 = \frac{1}{12},$$

and so on. For $c_0 = 0$ and $c_1 = 1$ we obtain

$$c_2 = 1, \qquad c_3 = 0, \qquad c_4 = -\frac{1}{4},$$

and so on. Thus,

$$y = C_1\left(1 - \frac{1}{2}x^2 - \frac{1}{6}x^3 + \frac{1}{12}x^4 + \cdots\right) + C_2\left(x + x^2 - \frac{1}{4}x^4 + \cdots\right)$$

and

$$y' = C_1\left(-x - \frac{1}{2}x^2 + \frac{1}{3}x^3 + \cdots\right) + C_2\left(1 + 2x - x^3 + \cdots\right).$$

The initial conditions imply $C_1 = 2$ and $C_2 = -1$, so

$$y = 2\left(1 - \frac{1}{2}x^2 - \frac{1}{6}x^3 + \frac{1}{12}x^4 + \cdots\right) - \left(x + x^2 - \frac{1}{4}x^4 + \cdots\right)$$

$$= 2 - x - 2x^2 - \frac{1}{3}x^3 + \frac{5}{12}x^4 + \cdots.$$

33. Substituting $y = \sum_{n=0}^{\infty} c_n x^n$ into the differential equation we have

$$y'' + (\sin x)y = \sum_{n=2}^{\infty} n(n-1)c_n x^{n-2} + \left(x - \frac{1}{6}x^3 + \frac{1}{120}x^5 - \cdots\right)(c_0 + c_1 x + c_2 x^2 + \cdots)$$

$$= [2c_2 + 6c_3 x + 12c_4 x^2 + 20c_5 x^3 + \cdots] + \left[c_0 x + c_1 x^2 + \left(c_2 - \frac{1}{6}c_0\right)x^3 + \cdots\right]$$

$$= 2c_2 + (6c_3 + c_0)x + (12c_4 + c_1)x^2 + \left(20c_5 + c_2 - \frac{1}{6}c_0\right)x^3 + \cdots = 0.$$

Thus

$$2c_2 = 0$$

$$6c_3 + c_0 = 0$$

$$12c_4 + c_1 = 0$$

$$20c_5 + c_2 - \frac{1}{6}c_0 = 0$$

and

$$c_2 = 0$$

$$c_3 = -\frac{1}{6}c_0$$

$$c_4 = -\frac{1}{12}c_1$$

$$c_5 = -\frac{1}{20}c_2 + \frac{1}{120}c_0.$$

Choosing $c_0 = 1$ and $c_1 = 0$ we find

$$c_2 = 0, \qquad c_3 = -\frac{1}{6}, \qquad c_4 = 0, \qquad c_5 = \frac{1}{120}$$

and so on. For $c_0 = 0$ and $c_1 = 1$ we obtain

$$c_2 = 0, \qquad c_3 = 0, \qquad c_4 = -\frac{1}{12}, \qquad c_5 = 0$$

and so on. Thus, two solutions are

$$y_1 = 1 - \frac{1}{6}x^3 + \frac{1}{120}x^5 + \cdots \qquad \text{and} \qquad y_2 = x - \frac{1}{12}x^4 + \cdots .$$

36. Substituting $y = \sum_{n=0}^{\infty} c_n x^n$ into the first differential equation leads to

$$y'' - xy = \underbrace{\sum_{n=2}^{\infty} n(n-1)c_n x^{n-2}}_{k=n-2} - \underbrace{\sum_{n=0}^{\infty} c_n x^{n+1}}_{k=n+1} = \sum_{k=0}^{\infty} (k+2)(k+1)c_{k+2}x^k - \sum_{k=1}^{\infty} c_{k-1}x^k$$

$$= 2c_2 + \sum_{k=1}^{\infty} [(k+2)(k+1)c_{k+2} - c_{k-1}]x^k = 1.$$

Thus

$$2c_2 = 1$$

$$(k+2)(k+1)c_{k+2} - c_{k-1} = 0$$

and

$$c_2 = \frac{1}{2}$$

$$c_{k+2} = \frac{c_{k-1}}{(k+2)(k+1)}, \qquad k = 1, 2, 3, \dots .$$

c_0 and c_1 be arbitrary and iterate to find

$$c_2 = \frac{1}{2}$$

$$c_3 = \frac{1}{6}c_0$$

$$c_4 = \frac{1}{12}c_1$$

$$c_5 = \frac{1}{20}c_2 = \frac{1}{40}$$

105

and so on. The solution is

$$y = c_0 + c_1 x + \frac{1}{2}x^2 + \frac{1}{6}c_0 x^3 + \frac{1}{12}c_1 x^4 + \frac{1}{40}c_5 + \cdots$$

$$= c_0 \left(1 + \frac{1}{6}x^3 + \cdots \right) + c_1 \left(x + \frac{1}{12}x^4 + \cdots \right) + \frac{1}{2}x^2 + \frac{1}{40}x^5 + \cdots .$$

Substituting $y = \sum_{n=0}^{\infty} c_n x^n$ into the second differential equation leads to

$$y'' - 4xy' - 4y = \underbrace{\sum_{n=2}^{\infty} n(n-1)c_n x^{n-2}}_{k=n-2} - \underbrace{\sum_{n=1}^{\infty} 4n c_n x^n}_{k=n} - \underbrace{\sum_{n=0}^{\infty} 4 c_n x^n}_{k=n}$$

$$= \sum_{k=0}^{\infty} (k+2)(k+1)c_{k+2} x^k - \sum_{k=1}^{\infty} 4k c_k x^k - \sum_{k=0}^{\infty} 4 c_k x^k$$

$$= 2c_2 - 4c_0 + \sum_{k=1}^{\infty} \left[(k+2)(k+1)c_{k+2} - 4(k+1)c_k \right] x^k$$

$$= e^x = 1 + \sum_{k=1}^{\infty} \frac{1}{k!} x^k .$$

Thus

$$2c_2 - 4c_0 = 1$$

$$(k+2)(k+1)c_{k+2} - 4(k+1)c_k = \frac{1}{k!}$$

and

$$c_2 = \frac{1}{2} + 2c_0$$

$$c_{k+2} = \frac{1}{(k+2)!} + \frac{4}{k+2}c_k, \qquad k = 1, 2, 3, \ldots .$$

Let c_0 and c_1 be arbitrary and iterate to find

$$c_2 = \frac{1}{2} + 2c_0$$

$$c_3 = \frac{1}{3!} + \frac{4}{3}c_1 = \frac{1}{3!} + \frac{4}{3}c_1$$

$$c_4 = \frac{1}{4!} + \frac{4}{4}c_2 = \frac{1}{4!} + \frac{1}{2} + 2c_0 = \frac{13}{4!} + 2c_0$$

$$c_5 = \frac{1}{5!} + \frac{4}{5}c_3 = \frac{1}{5!} + \frac{4}{5 \cdot 3!} + \frac{16}{15}c_1 = \frac{17}{5!} + \frac{16}{15}c_1$$

$$c_6 = \frac{1}{6!} + \frac{4}{6}c_4 = \frac{1}{6!} + \frac{4 \cdot 13}{6 \cdot 4!} + \frac{8}{6}c_0 = \frac{261}{6!} + \frac{4}{3}c_0$$

$$c_7 = \frac{1}{7!} + \frac{4}{7}c_5 = \frac{1}{7!} + \frac{4 \cdot 17}{7 \cdot 5!} + \frac{64}{105}c_1 = \frac{409}{7!} + \frac{64}{105}c_1$$

and so on. The solution is

$$y = c_0 + c_1 x + \left(\frac{1}{2} + 2c_0\right)x^2 + \left(\frac{1}{3!} + \frac{4}{3}c_1\right)x^3 + \left(\frac{13}{4!} + 2c_0\right)x^4 + \left(\frac{17}{5!} + \frac{16}{15}c_1\right)x^5$$

$$+ \left(\frac{261}{6!} + \frac{4}{3}c_0\right)x^6 + \left(\frac{409}{7!} + \frac{64}{105}c_1\right)x^7 + \cdots$$

$$= c_0\left[1 + 2x^2 + 2x^4 + \frac{4}{3}x^6 + \cdots\right] + c_1\left[x + \frac{4}{3}x^3 + \frac{16}{15}x^5 + \frac{64}{105}x^7 + \cdots\right]$$

$$+ \frac{1}{2}x^2 + \frac{1}{3!}x^3 + \frac{13}{4!}x^4 + \frac{17}{5!}x^5 + \frac{261}{6!}x^6 + \frac{409}{7!}x^7 + \cdots.$$

39. (a) Substituting $y = \sum_{n=0}^{\infty} c_n x^n$ into the differential equation we have

$$y'' + xy' + y = \underbrace{\sum_{n=2}^{\infty} n(n-1)c_n x^{n-2}}_{k=n-2} + \underbrace{\sum_{n=1}^{\infty} nc_n x^n}_{k=n} + \underbrace{\sum_{n=0}^{\infty} c_n x^n}_{k=n}$$

$$= \sum_{k=0}^{\infty}(k+2)(k+1)c_{k+2}x^k + \sum_{k=1}^{\infty} kc_k x^k + \sum_{k=0}^{\infty} c_k x^k$$

$$= (2c_2 + c_0) + \sum_{k=1}^{\infty}\left[(k+2)(k+1)c_{k+2} + (k+1)c_k\right]x^k = 0.$$

Thus

$$2c_2 + c_0 = 0$$

$$(k+2)(k+1)c_{k+2} + (k+1)c_k = 0$$

and

$$c_2 = -\frac{1}{2}c_0$$

$$c_{k+2} = -\frac{1}{k+2}c_k, \quad k = 1, 2, 3, \ldots.$$

Choosing $c_0 = 1$ and $c_1 = 0$ we find

$$c_2 = -\frac{1}{2}$$

$$c_3 = c_5 = c_7 = \cdots = 0$$

$$c_4 = -\frac{1}{4}\left(-\frac{1}{2}\right) = \frac{1}{2^2 \cdot 2}$$

$$c_6 = -\frac{1}{6}\left(\frac{1}{2^2 \cdot 2}\right) = -\frac{1}{2^3 \cdot 3!}$$

and so on. For $c_0 = 0$ and $c_1 = 1$ we obtain

$$c_2 = c_4 = c_6 = \cdots = 0$$

$$c_3 = -\frac{1}{3} = -\frac{2}{3!}$$

$$c_5 = -\frac{1}{5}\left(-\frac{1}{3}\right) = \frac{1}{5 \cdot 3} = \frac{4 \cdot 2}{5!}$$

$$c_7 = -\frac{1}{7}\left(\frac{4 \cdot 2}{5!}\right) = -\frac{6 \cdot 4 \cdot 2}{7!}$$

and so on. Thus, two solutions are

$$y_1 = \sum_{k=0}^{\infty} \frac{(-1)^k}{2^k \cdot k!} x^{2k} \quad \text{and} \quad y_2 = \sum_{k=0}^{\infty} \frac{(-1)^k 2^k k!}{(2k+1)!} x^{2k+1}.$$

(b) For y_1, $S_3 = S_2$ and $S_5 = S_4$, so we plot S_2, S_4, S_6, S_8, and S_{10}.

For y_2, $S_3 = S_4$ and $S_5 = S_6$, so we plot S_2, S_4, S_6, S_8, and S_{10}.

(c)

The graphs of y_1 and y_2 obtained from a numerical solver are shown. We see that the partial sum representations indicate the even and odd natures of the solution, but don't really give a very accurate representation of the true solution. Increasing N to about 20 gives a much more accurate representation on $[-4, 4]$.

(d) From $e^x = \sum_{k=0}^{\infty} x^k/k!$ we see that $e^{-x^2/2} = \sum_{k=0}^{\infty} (-x^2/2)^k/k! = \sum_{k=0}^{\infty} (-1)^k x^{2k}/2^k k!$. From (5) of Section 3.2 we have

$$y_2 = y_1 \int \frac{e^{-\int x\,dx}}{y_1^2}\,dx = e^{-x^2/2} \int \frac{e^{-x^2/2}}{(e^{-x^2/2})^2}\,dx = e^{-x^2/2} \int \frac{e^{-x^2/2}}{e^{-x^2}}\,dx = e^{-x^2/2} \int e^{x^2/2}\,dx$$

$$= \sum_{k=0}^{\infty} \frac{(-1)^k}{2^k k!} x^{2k} \int \sum_{k=0}^{\infty} \frac{1}{2^k k!} x^{2k}\,dx = \left(\sum_{k=0}^{\infty} \frac{(-1)^k}{2^k k!} x^{2k}\right)\left(\sum_{k=0}^{\infty} \int \frac{1}{2^k k!} x^{2k}\,dx\right)$$

$$= \left(\sum_{k=0}^{\infty} \frac{(-1)^k}{2^k k!} x^{2k}\right)\left(\sum_{k=0}^{\infty} \frac{1}{(2k+1)2^k k!} x^{2k+1}\right)$$

$$= \left(1 - \frac{1}{2}x^2 + \frac{1}{2^2 \cdot 2}x^4 - \frac{1}{2^3 \cdot 3!}x^6 + \cdots\right)\left(x + \frac{1}{3 \cdot 2}x^3 + \frac{1}{5 \cdot 2^2 \cdot 2}x^5 + \frac{1}{7 \cdot 2^3 \cdot 3!}x^7 + \cdots\right)$$

$$= x - \frac{2}{3!}x^3 + \frac{4 \cdot 2}{5!}x^5 - \frac{6 \cdot 4 \cdot 2}{7!}x^7 + \cdots = \sum_{k=0}^{\infty} \frac{(-1)^k 2^k k!}{(2k+1)!} x^{2k+1}.$$

EXERCISES 5.2

Solutions About Singular Points

3. Irregular singular point: $x = 3$; regular singular point: $x = -3$

6. Irregular singular point: $x = 5$; regular singular point: $x = 0$

9. Irregular singular point: $x = 0$; regular singular points: $x = 2, \pm 5$

12. Writing the differential equation in the form

$$y'' + \frac{x+3}{x}y' + 7xy = 0$$

we see that $x_0 = 0$ is a regular singular point. Multiplying by x^2, the differential equation can be put in the form

$$x^2 y'' + x(x+3)y' + 7x^3 y = 0.$$

We identify $p(x) = x + 3$ and $q(x) = 7x^3$.

15. Substituting $y = \sum_{n=0}^{\infty} c_n x^{n+r}$ into the differential equation and collecting terms, we obtain

$$2xy'' - y' + 2y = \left(2r^2 - 3r\right)c_0 x^{r-1} + \sum_{k=1}^{\infty}[2(k+r-1)(k+r)c_k - (k+r)c_k + 2c_{k-1}]x^{k+r-1} = 0,$$

which implies

$$2r^2 - 3r = r(2r - 3) = 0$$

and

$$(k+r)(2k+2r-3)c_k + 2c_{k-1} = 0.$$

The indicial roots are $r = 0$ and $r = 3/2$. For $r = 0$ the recurrence relation is

$$c_k = -\frac{2c_{k-1}}{k(2k-3)}, \quad k = 1, 2, 3, \ldots,$$

and

$$c_1 = 2c_0, \qquad c_2 = -2c_0, \qquad c_3 = \frac{4}{9}c_0,$$

and so on. For $r = 3/2$ the recurrence relation is

$$c_k = -\frac{2c_{k-1}}{(2k+3)k}, \qquad k = 1, 2, 3, \ldots,$$

and

$$c_1 = -\frac{2}{5}c_0, \qquad c_2 = \frac{2}{35}c_0, \qquad c_3 = -\frac{4}{945}c_0,$$

and so on. The general solution on $(0, \infty)$ is

$$y = C_1 \left(1 + 2x - 2x^2 + \frac{4}{9}x^3 + \cdots\right) + C_2 x^{3/2} \left(1 - \frac{2}{5}x + \frac{2}{35}x^2 - \frac{4}{945}x^3 + \cdots\right).$$

18. Substituting $y = \sum_{n=0}^{\infty} c_n x^{n+r}$ into the differential equation and collecting terms, we obtain

$$2x^2 y'' - xy' + (x^2 + 1)\, y = (2r^2 - 3r + 1)\, c_0 x^r + (2r^2 + r)\, c_1 x^{r+1}$$

$$+ \sum_{k=2}^{\infty} [2(k+r)(k+r-1)c_k - (k+r)c_k + c_k + c_{k-2}]x^{k+r}$$

$$= 0,$$

which implies

$$2r^2 - 3r + 1 = (2r - 1)(r - 1) = 0,$$

$$(2r^2 + r)\, c_1 = 0,$$

and

$$[(k+r)(2k + 2r - 3) + 1]c_k + c_{k-2} = 0.$$

The indicial roots are $r = 1/2$ and $r = 1$, so $c_1 = 0$. For $r = 1/2$ the recurrence relation is

$$c_k = -\frac{c_{k-2}}{k(2k-1)}, \qquad k = 2, 3, 4, \ldots,$$

and

$$c_2 = -\frac{1}{6}c_0, \qquad c_3 = 0, \qquad c_4 = \frac{1}{168}c_0,$$

and so on. For $r = 1$ the recurrence relation is

$$c_k = -\frac{c_{k-2}}{k(2k+1)}, \qquad k = 2, 3, 4, \ldots,$$

and

$$c_2 = -\frac{1}{10}c_0, \qquad c_3 = 0, \qquad c_4 = \frac{1}{360}c_0,$$

and so on. The general solution on $(0, \infty)$ is

$$y = C_1 x^{1/2} \left(1 - \frac{1}{6}x^2 + \frac{1}{168}x^4 + \cdots\right) + C_2 x \left(1 - \frac{1}{10}x^2 + \frac{1}{360}x^4 + \cdots\right).$$

21. Substituting $y = \sum_{n=0}^{\infty} c_n x^{n+r}$ into the differential equation and collecting terms, we obtain

$$2xy'' - (3 + 2x)y' + y = (2r^2 - 5r)\, c_0 x^{r-1} + \sum_{k=1}^{\infty} [2(k+r)(k+r-1)c_k$$

$$- 3(k+r)c_k - 2(k+r-1)c_{k-1} + c_{k-1}]x^{k+r-1}$$

$$= 0,$$

which implies

$$2r^2 - 5r = r(2r - 5) = 0$$

and

$$(k + r)(2k + 2r - 5)c_k - (2k + 2r - 3)c_{k-1} = 0.$$

The indicial roots are $r = 0$ and $r = 5/2$. For $r = 0$ the recurrence relation is

$$c_k = \frac{(2k - 3)c_{k-1}}{k(2k - 5)}, \quad k = 1, 2, 3, \ldots,$$

and

$$c_1 = \frac{1}{3}c_0, \qquad c_2 = -\frac{1}{6}c_0, \qquad c_3 = -\frac{1}{6}c_0,$$

and so on. For $r = 5/2$ the recurrence relation is

$$c_k = \frac{2(k + 1)c_{k-1}}{k(2k + 5)}, \quad k = 1, 2, 3, \ldots,$$

and

$$c_1 = \frac{4}{7}c_0, \qquad c_2 = \frac{4}{21}c_0, \qquad c_3 = \frac{32}{693}c_0,$$

and so on. The general solution on $(0, \infty)$ is

$$y = C_1\left(1 + \frac{1}{3}x - \frac{1}{6}x^2 - \frac{1}{6}x^3 + \cdots\right) + C_2 x^{5/2}\left(1 + \frac{4}{7}x + \frac{4}{21}x^2 + \frac{32}{693}x^3 + \cdots\right).$$

24. Substituting $y = \sum_{n=0}^{\infty} c_n x^{n+r}$ into the differential equation and collecting terms, we obtain

$$2x^2 y'' + 3xy' + (2x - 1)y = \left(2r^2 + r - 1\right)c_0 x^r + \sum_{k=1}^{\infty}[2(k + r)(k + r - 1)c_k + 3(k + r)c_k - c_k + 2c_{k-1}]x^{k+r} = 0,$$

which implies

$$2r^2 + r - 1 = (2r - 1)(r + 1) = 0$$

and

$$[(k + r)(2k + 2r + 1) - 1]c_k + 2c_{k-1} = 0.$$

The indicial roots are $r = -1$ and $r = 1/2$. For $r = -1$ the recurrence relation is

$$c_k = -\frac{2c_{k-1}}{k(2k - 3)}, \quad k = 1, 2, 3, \ldots,$$

and

$$c_1 = 2c_0, \qquad c_2 = -2c_0, \qquad c_3 = \frac{4}{9}c_0,$$

and so on. For $r = 1/2$ the recurrence relation is

$$c_k = -\frac{2c_{k-1}}{k(2k + 3)}, \quad k = 1, 2, 3, \ldots,$$

and

$$c_1 = -\frac{2}{5}c_0, \qquad c_2 = \frac{2}{35}c_0, \qquad c_3 = -\frac{4}{945}c_0,$$

and so on. The general solution on $(0, \infty)$ is

$$y = C_1 x^{-1}\left(1 + 2x - 2x^2 + \frac{4}{9}x^3 + \cdots\right) + C_2 x^{1/2}\left(1 - \frac{2}{5}x + \frac{2}{35}x^2 - \frac{4}{945}x^3 + \cdots\right).$$

5.2 Solutions About Singular Points

27. Substituting $y = \sum_{n=0}^{\infty} c_n x^{n+r}$ into the differential equation and collecting terms, we obtain

$$xy'' - xy' + y = \left(r^2 - r\right) c_0 x^{r-1} + \sum_{k=0}^{\infty}[(k+r+1)(k+r)c_{k+1} - (k+r)c_k + c_k]x^{k+r} = 0$$

which implies

$$r^2 - r = r(r-1) = 0$$

and

$$(k+r+1)(k+r)c_{k+1} - (k+r-1)c_k = 0.$$

The indicial roots are $r_1 = 1$ and $r_2 = 0$. For $r_1 = 1$ the recurrence relation is

$$c_{k+1} = \frac{kc_k}{(k+2)(k+1)}, \quad k = 0, 1, 2, \ldots,$$

and one solution is $y_1 = c_0 x$. A second solution is

$$y_2 = x \int \frac{e^{-\int -1\,dx}}{x^2}\,dx = x \int \frac{e^x}{x^2}\,dx = x \int \frac{1}{x^2}\left(1 + x + \frac{1}{2}x^2 + \frac{1}{3!}x^3 + \cdots\right)dx$$

$$= x \int \left(\frac{1}{x^2} + \frac{1}{x} + \frac{1}{2} + \frac{1}{3!}x + \frac{1}{4!}x^2 + \cdots\right)dx = x\left[-\frac{1}{x} + \ln x + \frac{1}{2}x + \frac{1}{12}x^2 + \frac{1}{72}x^3 + \cdots\right]$$

$$= x\ln x - 1 + \frac{1}{2}x^2 + \frac{1}{12}x^3 + \frac{1}{72}x^4 + \cdots.$$

The general solution on $(0, \infty)$ is

$$y = C_1 x + C_2 y_2(x).$$

30. Substituting $y = \sum_{n=0}^{\infty} c_n x^{n+r}$ into the differential equation and collecting terms, we obtain

$$xy'' + y' + y = r^2 c_0 x^{r-1} + \sum_{k=1}^{\infty}[(k+r)(k+r-1)c_k + (k+r)c_k + c_{k-1}]x^{k+r-1} = 0$$

which implies $r^2 = 0$ and

$$(k+r)^2 c_k + c_{k-1} = 0.$$

The indicial roots are $r_1 = r_2 = 0$ and the recurrence relation is

$$c_k = -\frac{c_{k-1}}{k^2}, \quad k = 1, 2, 3, \ldots.$$

One solution is

$$y_1 = c_0\left(1 - x + \frac{1}{2^2}x^2 - \frac{1}{(3!)^2}x^3 + \frac{1}{(4!)^2}x^4 - \cdots\right) = c_0 \sum_{n=0}^{\infty} \frac{(-1)^n}{(n!)^2}x^n.$$

112

A second solution is

$$y_2 = y_1 \int \frac{e^{-\int (1/x)\,dx}}{y_1^2}\,dx = y_1 \int \frac{dx}{x\left(1 - x + \frac{1}{4}x^2 - \frac{1}{36}x^3 + \cdots\right)^2}$$

$$= y_1 \int \frac{dx}{x\left(1 - 2x + \frac{3}{2}x^2 - \frac{5}{9}x^3 + \frac{35}{288}x^4 - \cdots\right)}$$

$$= y_1 \int \frac{1}{x}\left(1 + 2x + \frac{5}{2}x^2 + \frac{23}{9}x^3 + \frac{677}{288}x^4 + \cdots\right)dx$$

$$= y_1 \int \left(\frac{1}{x} + 2 + \frac{5}{2}x + \frac{23}{9}x^2 + \frac{677}{288}x^3 + \cdots\right)dx$$

$$= y_1 \left[\ln x + 2x + \frac{5}{4}x^2 + \frac{23}{27}x^3 + \frac{677}{1,152}x^4 + \cdots\right]$$

$$= y_1 \ln x + y_1\left(2x + \frac{5}{4}x^2 + \frac{23}{27}x^3 + \frac{677}{1,152}x^4 + \cdots\right).$$

The general solution on $(0, \infty)$ is

$$y = C_1 y_1(x) + C_2 y_2(x).$$

33. (a) From $t = 1/x$ we have $dt/dx = -1/x^2 = -t^2$. Then

$$\frac{dy}{dx} = \frac{dy}{dt}\frac{dt}{dx} = -t^2\frac{dy}{dt}$$

and

$$\frac{d^2y}{dx^2} = \frac{d}{dx}\left(\frac{dy}{dx}\right) = \frac{d}{dx}\left(-t^2\frac{dy}{dt}\right) = -t^2\frac{d^2y}{dt^2}\frac{dt}{dx} - \frac{dy}{dt}\left(2t\frac{dt}{dx}\right) = t^4\frac{d^2y}{dt^2} + 2t^3\frac{dy}{dt}.$$

Now

$$x^4\frac{d^2y}{dx^2} + \lambda y = \frac{1}{t^4}\left(t^4\frac{d^2y}{dt^2} + 2t^3\frac{dy}{dt}\right) + \lambda y = \frac{d^2y}{dt^2} + \frac{2}{t}\frac{dy}{dt} + \lambda y = 0$$

becomes

$$t\frac{d^2y}{dt^2} + 2\frac{dy}{dt} + \lambda t y = 0.$$

(b) Substituting $y = \sum_{n=0}^{\infty} c_n t^{n+r}$ into the differential equation and collecting terms, we obtain

$$t\frac{d^2y}{dt^2} + 2\frac{dy}{dt} + \lambda t y = (r^2 + r)c_0 t^{r-1} + (r^2 + 3r + 2)c_1 t^r$$

$$+ \sum_{k=2}^{\infty}[(k+r)(k+r-1)c_k + 2(k+r)c_k + \lambda c_{k-2}]t^{k+r-1}$$

$$= 0,$$

which implies

$$r^2 + r = r(r+1) = 0,$$
$$\left(r^2 + 3r + 2\right)c_1 = 0,$$

and

$$(k+r)(k+r+1)c_k + \lambda c_{k-2} = 0.$$

113

The indicial roots are $r_1 = 0$ and $r_2 = -1$, so $c_1 = 0$. For $r_1 = 0$ the recurrence relation is

$$c_k = -\frac{\lambda c_{k-2}}{k(k+1)}, \quad k = 2, 3, 4, \ldots,$$

and

$$c_2 = -\frac{\lambda}{3!}c_0$$

$$c_3 = c_5 = c_7 = \cdots = 0$$

$$c_4 = \frac{\lambda^2}{5!}c_0$$

$$\vdots$$

$$c_{2n} = (-1)^n \frac{\lambda^n}{(2n+1)!}c_0.$$

For $r_2 = -1$ the recurrence relation is

$$c_k = -\frac{\lambda c_{k-2}}{k(k-1)}, \quad k = 2, 3, 4, \ldots,$$

and

$$c_2 = -\frac{\lambda}{2!}c_0$$

$$c_3 = c_5 = c_7 = \cdots = 0$$

$$c_4 = \frac{\lambda^2}{4!}c_0$$

$$\vdots$$

$$c_{2n} = (-1)^n \frac{\lambda^n}{(2n)!}c_0.$$

The general solution on $(0, \infty)$ is

$$y(t) = c_1 \sum_{n=0}^{\infty} \frac{(-1)^n}{(2n+1)!}(\sqrt{\lambda}\,t)^{2n} + c_2 t^{-1} \sum_{n=0}^{\infty} \frac{(-1)^n}{(2n)!}(\sqrt{\lambda}\,t)^{2n}$$

$$= \frac{1}{t}\left[C_1 \sum_{n=0}^{\infty} \frac{(-1)^n}{(2n+1)!}(\sqrt{\lambda}\,t)^{2n+1} + C_2 \sum_{n=0}^{\infty} \frac{(-1)^n}{(2n)!}(\sqrt{\lambda}\,t)^{2n} \right]$$

$$= \frac{1}{t}[C_1 \sin\sqrt{\lambda}\,t + C_2 \cos\sqrt{\lambda}\,t].$$

(c) Using $t = 1/x$, the solution of the original equation is

$$y(x) = C_1 x \sin\frac{\sqrt{\lambda}}{x} + C_2 x \cos\frac{\sqrt{\lambda}}{x}.$$

36. Substituting $y = \sum_{n=0}^{\infty} c_n x^{n+r}$ into the first differential equation and collecting terms, we obtain

$$x^3 y'' + y = c_0 x^r + \sum_{k=1}^{\infty}[c_k + (k+r-1)(k+r-2)c_{k-1}]x^{k+r} = 0.$$

It follows that $c_0 = 0$ and

$$c_k = -(k+r-1)(k+r-2)c_{k-1}.$$

The only solution we obtain is $y(x) = 0$.

Substituting $y = \sum_{n=0}^{\infty} c_n x^{n+r}$ into the second differential equation and collecting terms, we obtain

$$x^2 y'' + (3x - 1)y' + y = -rc_0 + \sum_{k=0}^{\infty} [(k + r + 1)^2 c_k - (k + r + 1)c_{k+1}]x^{k+r} = 0,$$

which implies

$$-rc_0 = 0$$

$$(k + r + 1)^2 c_k - (k + r + 1)c_{k+1} = 0.$$

If $c_0 = 0$, then the solution of the differential equation is $y = 0$. Thus, we take $r = 0$, from which we obtain

$$c_{k+1} = (k + 1)c_k, \quad k = 0, 1, 2, \ldots.$$

Letting $c_0 = 1$ we get $c_1 = 2$, $c_2 = 3!$, $c_3 = 4!$, and so on. The solution of the differential equation is then $y = \sum_{n=0}^{\infty} (n + 1)! x^n$, which converges only at $x = 0$.

EXERCISES 5.3
Special Functions

3. Since $\nu^2 = 25/4$ the general solution is $y = c_1 J_{5/2}(x) + c_2 J_{-5/2}(x)$.

6. Since $\nu^2 = 4$ the general solution is $y = c_1 J_2(x) + c_2 Y_2(x)$.

9. We identify $\alpha = 5$ and $\nu = \frac{2}{3}$. Then the general solution is $y = c_1 J_{2/3}(5x) + c_2 J_{-2/3}(5x)$.

12. If $y = \sqrt{x}\, v(x)$ then

$$y' = x^{1/2} v'(x) + \frac{1}{2} x^{-1/2} v(x)$$

$$y'' = x^{1/2} v''(x) + x^{-1/2} v'(x) - \frac{1}{4} x^{-3/2} v(x)$$

and

$$x^2 y'' + \left(\alpha^2 x^2 - \nu^2 + \frac{1}{4}\right) y = x^{5/2} v''(x) + x^{3/2} v'(x) - \frac{1}{4} x^{1/2} v(x) + \left(\alpha^2 x^2 - \nu^2 + \frac{1}{4}\right) x^{1/2} v(x)$$

$$= x^{5/2} v''(x) + x^{3/2} v'(x) + (\alpha^2 x^{5/2} - \nu^2 x^{1/2}) v(x) = 0.$$

Multiplying by $x^{-1/2}$ we obtain

$$x^2 v''(x) + x v'(x) + (\alpha^2 x^2 - \nu^2) v(x) = 0,$$

whose solution is $v(x) = c_1 J_\nu(\alpha x) + c_2 Y_\nu(\alpha x)$. Then $y = c_1 \sqrt{x}\, J_\nu(\alpha x) + c_2 \sqrt{x}\, Y_\nu(\alpha x)$.

15. Write the differential equation in the form $y'' - (1/x)y' + y = 0$. This is the form of (18) in the text with $a = 1$, $c = 1$, $b = 1$, and $p = 1$, so, by (19) in the text, the general solution is

$$y = x[c_1 J_1(x) + c_2 Y_1(x)].$$

18. Write the differential equation in the form $y'' + (4 + 1/4x^2)y = 0$. This is the form of (18) in the text with $a = \frac{1}{2}$, $c = 1$, $b = 2$, and $p = 0$, so, by (19) in the text, the general solution is

$$y = x^{1/2}[c_1 J_0(2x) + c_2 Y_0(2x)].$$

21. Using the fact that $i^2 = -1$, along with the definition of $J_\nu(x)$ in (7) in the text, we have

$$I_\nu(x) = i^{-\nu} J_\nu(ix) = i^{-\nu} \sum_{n=0}^\infty \frac{(-1)^n}{n!\Gamma(1+\nu+n)} \left(\frac{ix}{2}\right)^{2n+\nu}$$

$$= \sum_{n=0}^\infty \frac{(-1)^n}{n!\Gamma(1+\nu+n)} i^{2n+\nu-\nu} \left(\frac{x}{2}\right)^{2n+\nu}$$

$$= \sum_{n=0}^\infty \frac{(-1)^n}{n!\Gamma(1+\nu+n)} (i^2)^n \left(\frac{x}{2}\right)^{2n+\nu}$$

$$= \sum_{n=0}^\infty \frac{(-1)^{2n}}{n!\Gamma(1+\nu+n)} \left(\frac{x}{2}\right)^{2n+\nu}$$

$$= \sum_{n=0}^\infty \frac{1}{n!\Gamma(1+\nu+n)} \left(\frac{x}{2}\right)^{2n+\nu},$$

which is a real function.

24. Write the differential equation in the form $y'' + (4/x)y' + (1+2/x^2)y = 0$. This is the form of (18) in the text with

$$1 - 2a = 4 \implies a = -\frac{3}{2}$$

$$2c - 2 = 0 \implies c = 1$$

$$b^2 c^2 = 1 \implies b = 1$$

$$a^2 - p^2 c^2 = 2 \implies p = \frac{1}{2}.$$

Then, by (19), (23), and (24) in the text,

$$y = x^{-3/2}[c_1 J_{1/2}(x) + c_2 J_{-1/2}(x)] = x^{-3/2}\left[c_1 \sqrt{\frac{2}{\pi x}} \sin x + c_2 \sqrt{\frac{2}{\pi x}} \cos x\right] = C_1 \frac{1}{x^2} \sin x + C_2 \frac{1}{x^2} \cos x.$$

27. (a) The recurrence relation follows from

$$-\nu J_\nu(x) + x J_{\nu-1}(x) = -\sum_{n=0}^\infty \frac{(-1)^n \nu}{n!\Gamma(1+\nu+n)} \left(\frac{x}{2}\right)^{2n+\nu} + x\sum_{n=0}^\infty \frac{(-1)^n}{n!\Gamma(\nu+n)} \left(\frac{x}{2}\right)^{2n+\nu-1}$$

$$= -\sum_{n=0}^\infty \frac{(-1)^n \nu}{n!\Gamma(1+\nu+n)} \left(\frac{x}{2}\right)^{2n+\nu} + \sum_{n=0}^\infty \frac{(-1)^n(\nu+n)}{n!\Gamma(1+\nu+n)} \cdot 2 \left(\frac{x}{2}\right) \left(\frac{x}{2}\right)^{2n+\nu-1}$$

$$= \sum_{n=0}^\infty \frac{(-1)^n(2n+\nu)}{n!\Gamma(1+\nu+n)} \left(\frac{x}{2}\right)^{2n+\nu} = x J'_\nu(x).$$

(b) The formula in part (a) is a linear first-order differential equation in $J_\nu(x)$. An integrating factor for this equation is x^ν, so

$$\frac{d}{dx}[x^\nu J_\nu(x)] = x^\nu J_{\nu-1}(x).$$

30. From (20) we obtain $J'_0(x) = -J_1(x)$, and from (21) we obtain $J'_0(x) = J_{-1}(x)$. Thus $J'_0(x) = J_{-1}(x) = -J_1(x)$.

33. Letting

$$s = \frac{2}{\alpha}\sqrt{\frac{k}{m}} e^{-\alpha t/2},$$

116

we have

$$\frac{dx}{dt} = \frac{dx}{ds}\frac{ds}{dt} = \frac{dx}{dt}\left[\frac{2}{\alpha}\sqrt{\frac{k}{m}}\left(-\frac{\alpha}{2}\right)e^{-\alpha t/2}\right] = \frac{dx}{ds}\left(-\sqrt{\frac{k}{m}}\,e^{-\alpha t/2}\right)$$

and

$$\frac{d^2x}{dt^2} = \frac{d}{dt}\left(\frac{dx}{dt}\right) = \frac{dx}{ds}\left(\frac{\alpha}{2}\sqrt{\frac{k}{m}}\,e^{-\alpha t/2}\right) + \frac{d}{dt}\left(\frac{dx}{ds}\right)\left(-\sqrt{\frac{k}{m}}\,e^{-\alpha t/2}\right)$$

$$= \frac{dx}{ds}\left(\frac{\alpha}{2}\sqrt{\frac{k}{m}}\,e^{-\alpha t/2}\right) + \frac{d^2x}{ds^2}\frac{ds}{dt}\left(-\sqrt{\frac{k}{m}}\,e^{-\alpha t/2}\right)$$

$$= \frac{dx}{ds}\left(\frac{\alpha}{2}\sqrt{\frac{k}{m}}\,e^{-\alpha t/2}\right) + \frac{d^2x}{ds^2}\left(\frac{k}{m}\,e^{-\alpha t}\right).$$

Then

$$m\frac{d^2x}{dt^2} + ke^{-\alpha t}x = ke^{-\alpha t}\frac{d^2x}{ds^2} + \frac{m\alpha}{2}\sqrt{\frac{k}{m}}\,e^{-\alpha t/2}\frac{dx}{ds} + ke^{-\alpha t}x = 0.$$

Multiplying by $2^2/\alpha^2 m$ we have

$$\frac{2^2}{\alpha^2}\frac{k}{m}e^{-\alpha t}\frac{d^2x}{ds^2} + \frac{2}{\alpha}\sqrt{\frac{k}{m}}\,e^{-\alpha t/2}\frac{dx}{ds} + \frac{2^2}{\alpha^2}\frac{k}{m}e^{-\alpha t}x = 0$$

or, since $s = (2/\alpha)\sqrt{k/m}\,e^{-\alpha t/2}$,

$$s^2\frac{d^2x}{ds^2} + s\frac{dx}{ds} + s^2x = 0.$$

36. The general solution of the differential equation is

$$y(x) = c_1 J_0(\alpha x) + c_2 Y_0(\alpha x).$$

In order to satisfy the conditions that $\lim_{x\to 0^+} y(x)$ and $\lim_{x\to 0^+} y'(x)$ are finite we are forced to define $c_2 = 0$. Thus, $y(x) = c_1 J_0(\alpha x)$. The second boundary condition, $y(2) = 0$, implies $c_1 = 0$ or $J_0(2\alpha) = 0$. In order to have a nontrivial solution we require that $J_0(2\alpha) = 0$. From Table 5.1, the first three positive zeros of J_0 are found to be

$$2\alpha_1 = 2.4048, \quad 2\alpha_2 = 5.5201, \quad 2\alpha_3 = 8.6537$$

and so $\alpha_1 = 1.2024$, $\alpha_2 = 2.7601$, $\alpha_3 = 4.3269$. The eigenfunctions corresponding to the eigenvalues $\lambda_1 = \alpha_1^2$, $\lambda_2 = \alpha_2^2$, $\lambda_3 = \alpha_3^2$ are $J_0(1.2024x)$, $J_0(2.7601x)$, and $J_0(4.3269x)$.

39. (a) We identify $m = 4$, $k = 1$, and $\alpha = 0.1$. Then

$$x(t) = c_1 J_0(10e^{-0.05t}) + c_2 Y_0(10e^{-0.05t})$$

and

$$x'(t) = -0.5c_1 J_0'(10e^{-0.05t}) - 0.5c_2 Y_0'(10e^{-0.05t}).$$

Now $x(0) = 1$ and $x'(0) = -1/2$ imply

$$c_1 J_0(10) + c_2 Y_0(10) = 1$$

$$c_1 J_0'(10) + c_2 Y_0'(10) = 1.$$

Using Cramer's rule we obtain

$$c_1 = \frac{Y_0'(10) - Y_0(10)}{J_0(10)Y_0'(10) - J_0'(10)Y_0(10)}$$

and

$$c_2 = \frac{J_0(10) - J_0'(10)}{J_0(10)Y_0'(10) - J_0'(10)Y_0(10)}.$$

117

Using $Y_0' = -Y_1$ and $J_0' = -J_1$ and Table 5.2 we find $c_1 = -4.7860$ and $c_2 = -3.1803$. Thus

$$x(t) = -4.7860J_0(10e^{-0.05t}) - 3.1803Y_0(10e^{-0.05t}).$$

(b)

42. (a) Writing the differential equation in the form $xy'' + (PL/M)y = 0$, we identify $\lambda = PL/M$. From Problem 37 the solution of this differential equation is

$$y = c_1\sqrt{x}\, J_1\left(2\sqrt{PLx/M}\right) + c_2\sqrt{x}\, Y_1\left(2\sqrt{PLx/M}\right).$$

Now $J_1(0) = 0$, so $y(0) = 0$ implies $c_2 = 0$ and

$$y = c_1\sqrt{x}\, J_1\left(2\sqrt{PLx/M}\right).$$

(b) From $y(L) = 0$ we have $y = J_1(2L\sqrt{PM}) = 0$. The first positive zero of J_1 is 3.8317 so, solving $2L\sqrt{P_1/M} = 3.8317$, we find $P_1 = 3.6705M/L^2$. Therefore,

$$y_1(x) = c_1\sqrt{x}\, J_1\left(2\sqrt{\frac{3.6705x}{L}}\right) = c_1\sqrt{x}\, J_1\left(\frac{3.8317}{\sqrt{L}}\sqrt{x}\right).$$

(c) For $c_1 = 1$ and $L = 1$ the graph of $y_1 = \sqrt{x}\, J_1(3.8317\sqrt{x})$ is shown.

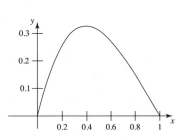

45. The recurrence relation can be written

$$P_{k+1}(x) = \frac{2k+1}{k+1}\, xP_k(x) - \frac{k}{k+1}\, P_{k-1}(x), \qquad k = 2,\, 3,\, 4,\, \dots\, .$$

$k = 1$: $\quad P_2(x) = \dfrac{3}{2}x^2 - \dfrac{1}{2}$

$k = 2$: $\quad P_3(x) = \dfrac{5}{3}x\left(\dfrac{3}{2}x^2 - \dfrac{1}{2}\right) - \dfrac{2}{3}x = \dfrac{5}{2}x^3 - \dfrac{3}{2}x$

$k = 3$: $\quad P_4(x) = \dfrac{7}{4}x\left(\dfrac{5}{2}x^3 - \dfrac{3}{2}x\right) - \dfrac{3}{4}\left(\dfrac{3}{2}x^2 - \dfrac{1}{2}\right) = \dfrac{35}{8}x^4 - \dfrac{30}{8}x^2 + \dfrac{3}{8}$

$k = 4$: $\quad P_5(x) = \dfrac{9}{5}x\left(\dfrac{35}{8}x^4 - \dfrac{30}{8}x^2 + \dfrac{3}{8}\right) - \dfrac{4}{5}\left(\dfrac{5}{2}x^3 - \dfrac{3}{2}x\right) = \dfrac{63}{8}x^5 - \dfrac{35}{4}x^3 + \dfrac{15}{8}x$

$k = 5$: $\quad P_6(x) = \dfrac{11}{6}x\left(\dfrac{63}{8}x^5 - \dfrac{35}{4}x^3 + \dfrac{15}{8}x\right) - \dfrac{5}{6}\left(\dfrac{35}{8}x^4 - \dfrac{30}{8}x^2 + \dfrac{3}{8}\right) = \dfrac{231}{16}x^6 - \dfrac{315}{16}x^4 + \dfrac{105}{16}x^2 - \dfrac{5}{16}$

$k = 6$: $\quad P_7(x) = \dfrac{13}{7}x\left(\dfrac{231}{16}x^6 - \dfrac{315}{16}x^4 + \dfrac{105}{16}x^2 - \dfrac{5}{16}\right) - \dfrac{6}{7}\left(\dfrac{63}{8}x^5 - \dfrac{35}{4}x^3 + \dfrac{15}{8}x\right)$

$$= \dfrac{429}{16}x^7 - \dfrac{693}{16}x^5 + \dfrac{315}{16}x^3 - \dfrac{35}{16}x$$

48. Using a CAS we find

$$P_1(x) = \frac{1}{2}\frac{d}{dx}\left(x^2 - 1\right)^1 = x$$

$$P_2(x) = \frac{1}{2^2 2!}\frac{d^2}{dx^2}\left(x^2 - 1\right)^2 = \frac{1}{2}(3x^2 - 1)$$

$$P_3(x) = \frac{1}{2^3 3!}\frac{d^3}{dx^3}\left(x^2 - 1\right)^3 = \frac{1}{2}(5x^3 - 3x)$$

$$P_4(x) = \frac{1}{2^4 4!}\frac{d^4}{dx^4}\left(x^2 - 1\right)^4 = \frac{1}{8}(35x^4 - 30x^2 + 3)$$

$$P_5(x) = \frac{1}{2^5 5!}\frac{d^5}{dx^5}\left(x^2 - 1\right)^5 = \frac{1}{8}(63x^5 - 70x^3 + 15x)$$

$$P_6(x) = \frac{1}{2^6 6!}\frac{d^6}{dx^6}\left(x^2 - 1\right)^6 = \frac{1}{16}(231x^6 - 315x^4 + 105x^2 - 5)$$

$$P_7(x) = \frac{1}{2^7 7!}\frac{d^7}{dx^7}\left(x^2 - 1\right)^7 = \frac{1}{16}(429x^7 - 693x^5 + 315x^3 - 35x)$$

CHAPTER 5 REVIEW EXERCISES

3. $x = -1$ is the nearest singular point to the ordinary point $x = 0$. Theorem 5.1 guarantees the existence of two power series solutions $y = \sum_{n=1}^{\infty} c_n x^n$ of the differential equation that converge at least for $-1 < x < 1$. Since $-\frac{1}{2} \leq x \leq \frac{1}{2}$ is properly contained in $-1 < x < 1$, both power series must converge for all points contained in $-\frac{1}{2} \leq x \leq \frac{1}{2}$.

6. We have

$$f(x) = \frac{\sin x}{\cos x} = \frac{x - \dfrac{x^3}{6} + \dfrac{x^5}{120} - \cdots}{1 - \dfrac{x^2}{2} + \dfrac{x^4}{24} - \cdots} = x + \frac{x^3}{3} + \frac{2x^5}{15} + \cdots.$$

9. Substituting $y = \sum_{n=0}^{\infty} c_n x^{n+r}$ into the differential equation we obtain

$$2xy'' + y' + y = \left(2r^2 - r\right)c_0 x^{r-1} + \sum_{k=1}^{\infty}[2(k+r)(k+r-1)c_k + (k+r)c_k + c_{k-1}]x^{k+r-1} = 0$$

which implies

$$2r^2 - r = r(2r - 1) = 0$$

and

$$(k+r)(2k + 2r - 1)c_k + c_{k-1} = 0.$$

The indicial roots are $r = 0$ and $r = 1/2$. For $r = 0$ the recurrence relation is

$$c_k = -\frac{c_{k-1}}{k(2k-1)}, \quad k = 1, 2, 3, \ldots,$$

so

$$c_1 = -c_0, \quad c_2 = \frac{1}{6}c_0, \quad c_3 = -\frac{1}{90}c_0.$$

119

For $r = 1/2$ the recurrence relation is

$$c_k = -\frac{c_{k-1}}{k(2k+1)}, \quad k = 1, 2, 3, \ldots,$$

so

$$c_1 = -\frac{1}{3}c_0, \quad c_2 = \frac{1}{30}c_0, \quad c_3 = -\frac{1}{630}c_0.$$

Two linearly independent solutions are

$$y_1 = 1 - x + \frac{1}{6}x^2 - \frac{1}{90}x^3 + \cdots$$

and

$$y_2 = x^{1/2}\left(1 - \frac{1}{3}x + \frac{1}{30}x^2 - \frac{1}{630}x^3 + \cdots\right).$$

12. Substituting $y = \sum_{n=0}^{\infty} c_n x^n$ into the differential equation we obtain

$$y'' - x^2 y' + xy = 2c_2 + (6c_3 + c_0)x + \sum_{k=1}^{\infty}[(k+3)(k+2)c_{k+3} - (k-1)c_k]x^{k+1} = 0$$

which implies $c_2 = 0$, $c_3 = -c_0/6$, and

$$c_{k+3} = \frac{k-1}{(k+3)(k+2)}c_k, \quad k = 1, 2, 3, \ldots.$$

Choosing $c_0 = 1$ and $c_1 = 0$ we find

$$c_3 = -\frac{1}{6}$$

$$c_4 = c_7 = c_{10} = \cdots = 0$$

$$c_5 = c_8 = c_{11} = \cdots = 0$$

$$c_6 = -\frac{1}{90}$$

and so on. For $c_0 = 0$ and $c_1 = 1$ we obtain

$$c_3 = c_6 = c_9 = \cdots = 0$$

$$c_4 = c_7 = c_{10} = \cdots = 0$$

$$c_5 = c_8 = c_{11} = \cdots = 0$$

and so on. Thus, two solutions are

$$y_1 = 1 - \frac{1}{6}x^3 - \frac{1}{90}x^6 - \cdots \quad \text{and} \quad y_2 = x.$$

15. Substituting $y = \sum_{n=0}^{\infty} c_n x^n$ into the differential equation we have

$$y'' + xy' + 2y = \underbrace{\sum_{n=2}^{\infty} n(n-1)c_n x^{n-2}}_{k=n-2} + \underbrace{\sum_{n=1}^{\infty} nc_n x^n}_{k=n} + 2\underbrace{\sum_{n=0}^{\infty} c_n x^n}_{k=n}$$

$$= \sum_{k=0}^{\infty}(k+2)(k+1)c_{k+2}x^k + \sum_{k=1}^{\infty} kc_k x^k + 2\sum_{k=0}^{\infty} c_k x^k$$

$$= 2c_2 + 2c_0 + \sum_{k=1}^{\infty}[(k+2)(k+1)c_{k+2} + (k+2)c_k]x^k = 0.$$

Thus

$$2c_2 + 2c_0 = 0$$

$$(k+2)(k+1)c_{k+2} + (k+2)c_k = 0$$

and

$$c_2 = -c_0$$

$$c_{k+2} = -\frac{1}{k+1}\,c_k, \quad k = 1, 2, 3, \ldots.$$

Choosing $c_0 = 1$ and $c_1 = 0$ we find

$$c_2 = -1$$

$$c_3 = c_5 = c_7 = \cdots = 0$$

$$c_4 = \frac{1}{3}$$

$$c_6 = -\frac{1}{15}$$

and so on. For $c_0 = 0$ and $c_1 = 1$ we obtain

$$c_2 = c_4 = c_6 = \cdots = 0$$

$$c_3 = -\frac{1}{2}$$

$$c_5 = \frac{1}{8}$$

$$c_7 = -\frac{1}{48}$$

and so on. Thus, the general solution is

$$y = C_0\left(1 - x^2 + \frac{1}{3}x^4 - \frac{1}{15}x^6 + \cdots\right) + C_1\left(x - \frac{1}{2}x^3 + \frac{1}{8}x^5 - \frac{1}{48}x^7 + \cdots\right)$$

and

$$y' = C_0\left(-2x + \frac{4}{3}x^3 - \frac{2}{5}x^5 + \cdots\right) + C_1\left(1 - \frac{3}{2}x^2 + \frac{5}{8}x^4 - \frac{7}{48}x^6 + \cdots\right).$$

Setting $y(0) = 3$ and $y'(0) = -2$ we find $c_0 = 3$ and $c_1 = -2$. Therefore, the solution of the initial-value problem is

$$y = 3 - 2x - 3x^2 + x^3 + x^4 - \frac{1}{4}x^5 - \frac{1}{5}x^6 + \frac{1}{24}x^7 + \cdots.$$

18. While we can find two solutions of the form

$$y_1 = c_0[1 + \cdots] \quad \text{and} \quad y_2 = c_1[x + \cdots],$$

the initial conditions at $x = 1$ give solutions for c_0 and c_1 in terms of infinite series. Letting $t = x - 1$ the initial-value problem becomes

$$\frac{d^2y}{dt^2} + (t+1)\frac{dy}{dt} + y = 0, \quad y(0) = -6, \ y'(0) = 3.$$

Substituting $y = \sum_{n=0}^{\infty} c_n t^n$ into the differential equation, we have

$$\frac{d^2 y}{dt^2} + (t+1)\frac{dy}{dt} + y = \underbrace{\sum_{n=2}^{\infty} n(n-1)c_n t^{n-2}}_{k=n-2} + \underbrace{\sum_{n=1}^{\infty} nc_n t^n}_{k=n} + \underbrace{\sum_{n=1}^{\infty} nc_n t^{n-1}}_{k=n-1} + \underbrace{\sum_{n=0}^{\infty} c_n t^n}_{k=n}$$

$$= \sum_{k=0}^{\infty}(k+2)(k+1)c_{k+2}t^k + \sum_{k=1}^{\infty} kc_k t^k + \sum_{k=0}^{\infty}(k+1)c_{k+1}t^k + \sum_{k=0}^{\infty} c_k t^k$$

$$= 2c_2 + c_1 + c_0 + \sum_{k=1}^{\infty}[(k+2)(k+1)c_{k+2} + (k+1)c_{k+1} + (k+1)c_k]t^k = 0.$$

Thus
$$2c_2 + c_1 + c_0 = 0$$
$$(k+2)(k+1)c_{k+2} + (k+1)c_{k+1} + (k+1)c_k = 0$$

and
$$c_2 = -\frac{c_1 + c_0}{2}$$
$$c_{k+2} = -\frac{c_{k+1} + c_k}{k+2}, \quad k = 1, 2, 3, \ldots.$$

Choosing $c_0 = 1$ and $c_1 = 0$ we find

$$c_2 = -\frac{1}{2}, \quad c_3 = \frac{1}{6}, \quad c_4 = \frac{1}{12},$$

and so on. For $c_0 = 0$ and $c_1 = 1$ we find

$$c_2 = -\frac{1}{2}, \quad c_3 = -\frac{1}{6}, \quad c_4 = \frac{1}{6},$$

and so on. Thus, the general solution is

$$y = c_0\left[1 - \frac{1}{2}t^2 + \frac{1}{6}t^3 + \frac{1}{12}t^4 + \cdots\right] + c_1\left[t - \frac{1}{2}t^2 - \frac{1}{6}t^3 + \frac{1}{6}t^4 + \cdots\right].$$

The initial conditions then imply $c_0 = -6$ and $c_1 = 3$. Thus the solution of the initial-value problem is

$$y = -6\left[1 - \frac{1}{2}(x-1)^2 + \frac{1}{6}(x-1)^3 + \frac{1}{12}(x-1)^4 + \cdots\right]$$
$$+ 3\left[(x-1) - \frac{1}{2}(x-1)^2 - \frac{1}{6}(x-1)^3 + \frac{1}{6}(x-1)^4 + \cdots\right].$$

21. Substituting $y = \sum_{n=0}^{\infty} c_n x^n$ into the differential equation we have

$$y'' + x^2 y' + 2xy = \underbrace{\sum_{n=2}^{\infty} n(n-1)c_n x^{n-2}}_{k=n-2} + \underbrace{\sum_{n=1}^{\infty} nc_n x^{n+1}}_{k=n+1} + 2\underbrace{\sum_{n=0}^{\infty} c_n x^{n+1}}_{k=n+1}$$

$$= \sum_{k=0}^{\infty}(k+2)(k+1)c_{k+2}x^k + \sum_{k=2}^{\infty}(k-1)c_{k-1}x^k + 2\sum_{k=1}^{\infty} c_{k-1}x^k$$

$$= 2c_2 + (6c_3 + 2c_0)x + \sum_{k=2}^{\infty}[(k+2)(k+1)c_{k+2} + (k+1)c_{k-1}]x^k = 5 - 2x + 10x^3.$$

Thus, equating coefficients of like powers of x gives

$$2c_2 = 5$$
$$6c_3 + 2c_0 = -2$$
$$12c_4 + 3c_1 = 0$$
$$20c_5 + 4c_2 = 10$$
$$(k+2)(k+1)c_{k+2} + (k+1)c_{k-1} = 0, \quad k = 4,5,6,\ldots,$$

and

$$c_2 = \frac{5}{2}$$
$$c_3 = -\frac{1}{3}c_0 - \frac{1}{3}$$
$$c_4 = -\frac{1}{4}c_1$$
$$c_5 = \frac{1}{2} - \frac{1}{5}c_2 = \frac{1}{2} - \frac{1}{5}\left(\frac{5}{2}\right) = 0$$
$$c_{k+2} = -\frac{1}{k+2}c_{k-1}.$$

Using the recurrence relation, we find

$$c_6 = -\frac{1}{6}c_3 = \frac{1}{3\cdot 6}(c_0 + 1) = \frac{1}{3^2\cdot 2!}c_0 + \frac{1}{3^2\cdot 2!}$$
$$c_7 = -\frac{1}{7}c_4 = \frac{1}{4\cdot 7}c_1$$
$$c_8 = c_{11} = c_{14} = \cdots = 0$$
$$c_9 = -\frac{1}{9}c_6 = -\frac{1}{3^3\cdot 3!}c_0 - \frac{1}{3^3\cdot 3!}$$
$$c_{10} = -\frac{1}{10}c_7 = -\frac{1}{4\cdot 7\cdot 10}c_1$$
$$c_{12} = -\frac{1}{12}c_9 = \frac{1}{3^4\cdot 4!}c_0 + \frac{1}{3^4\cdot 4!}$$
$$c_{13} = -\frac{1}{13}c_0 = \frac{1}{4\cdot 7\cdot 10\cdot 13}c_1$$

and so on. Thus

$$y = c_0\left[1 - \frac{1}{3}x^3 + \frac{1}{3^2\cdot 2!}x^6 - \frac{1}{3^3\cdot 3!}x^9 + \frac{1}{3^4\cdot 4!}x^{12} - \cdots\right]$$
$$+ c_1\left[x - \frac{1}{4}x^4 + \frac{1}{4\cdot 7}x^7 - \frac{1}{4\cdot 7\cdot 10}x^{10} + \frac{1}{4\cdot 7\cdot 10\cdot 13}x^{13} - \cdots\right]$$
$$+ \left[\frac{5}{2}x^2 - \frac{1}{3}x^3 + \frac{1}{3^2\cdot 2!}x^6 - \frac{1}{3^3\cdot 3!}x^9 + \frac{1}{3^4\cdot 4!}x^{12} - \cdots\right].$$

24. (a) Using formula (5) of Section 3.2 in the text, we find that a second solution of $(1-x^2)y'' - 2xy' = 0$ is

$$y_2(x) = 1\cdot\int\frac{e^{\int 2x\,dx/(1-x^2)}}{1^2}\,dx = \int e^{-\ln(1-x^2)}\,dx$$
$$= \int\frac{dx}{1-x^2} = \frac{1}{2}\ln\left(\frac{1+x}{1-x}\right),$$

where partial fractions was used to obtain the last integral.

(b) Using formula (5) of Section 3.2 in the text, we find that a second solution of $(1 - x^2)y'' - 2xy' + 2y = 0$ is

$$y_2(x) = x \cdot \int \frac{e^{\int 2x\,dx/(1-x^2)}}{x^2}\,dx = x \int \frac{e^{-\ln(1-x^2)}}{x^2}\,dx$$

$$= x \int \frac{dx}{x^2(1-x^2)}\,dx = x\left[\frac{1}{2}\ln\left(\frac{1+x}{1-x}\right) - \frac{1}{x}\right] = \frac{x}{2}\ln\left(\frac{1+x}{1-x}\right) - 1,$$

where partial fractions was used to obtain the last integral.

(c)

$$y_2(x) = \frac{1}{2}\ln\left(\frac{1+x}{1-x}\right) \qquad y_2 = \frac{x}{2}\ln\left(\frac{1+x}{1-x}\right) - 1$$

6 Numerical Solutions of Ordinary Differential Equations

3. $h = 0.1$

x_n	y_n
0.00	0.0000
0.10	0.1005
0.20	0.2030
0.30	0.3098
0.40	0.4234
0.50	0.5470

$h = 0.05$

x_n	y_n
0.00	0.0000
0.05	0.0501
0.10	0.1004
0.15	0.1512
0.20	0.2028
0.25	0.2554
0.30	0.3095
0.35	0.3652
0.40	0.4230
0.45	0.4832
0.50	0.5465

6. $h = 0.1$

x_n	y_n
0.00	0.0000
0.10	0.0050
0.20	0.0200
0.30	0.0451
0.40	0.0805
0.50	0.1266

$h = 0.05$

x_n	y_n
0.00	0.0000
0.05	0.0013
0.10	0.0050
0.15	0.0113
0.20	0.0200
0.25	0.0313
0.30	0.0451
0.35	0.0615
0.40	0.0805
0.45	0.1022
0.50	0.1266

9. $h = 0.1$

x_n	y_n
1.00	1.0000
1.10	1.0095
1.20	1.0404
1.30	1.0967
1.40	1.1866
1.50	1.3260

$h = 0.05$

x_n	y_n
1.00	1.0000
1.05	1.0024
1.10	1.0100
1.15	1.0228
1.20	1.0414
1.25	1.0663
1.30	1.0984
1.35	1.1389
1.40	1.1895
1.45	1.2526
1.50	1.3315

12. (a)

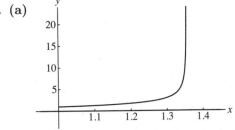

(b)

x_n	Euler	Imp. Euler
1.00	1.0000	1.0000
1.10	1.2000	1.2469
1.20	1.4938	1.6430
1.30	1.9711	2.4042
1.40	2.9060	4.5085

15. (a) Using Euler's method we obtain $y(0.1) \approx y_1 = 0.8$.

(b) Using $y'' = 5e^{-2x}$ we see that the local truncation error is

$$5e^{-2c} \frac{(0.1)^2}{2} = 0.025e^{-2c}.$$

Since e^{-2x} is a decreasing function, $e^{-2c} \leq e^0 = 1$ for $0 \leq c \leq 0.1$. Thus an upper bound for the local truncation error is $0.025(1) = 0.025$.

(c) Since $y(0.1) = 0.8234$, the actual error is $y(0.1) - y_1 = 0.0234$, which is less than 0.025.

(d) Using Euler's method with $h = 0.05$ we obtain $y(0.1) \approx y_2 = 0.8125$.

(e) The error in (d) is $0.8234 - 0.8125 = 0.0109$. With global truncation error $O(h)$, when the step size is halved we expect the error for $h = 0.05$ to be one-half the error when $h = 0.1$. Comparing 0.0109 with 0.0234 we see that this is the case.

18. (a) Using $y''' = -114e^{-3(x-1)}$ we see that the local truncation error is

$$\left| y'''(c) \frac{h^3}{6} \right| = 114e^{-3(x-1)} \frac{h^3}{6} = 19h^3 e^{-3(c-1)}.$$

(b) Since $e^{-3(x-1)}$ is a decreasing function for $1 \leq x \leq 1.5$, $e^{-3(c-1)} \leq e^{-3(1-1)} = 1$ for $1 \leq c \leq 1.5$ and

$$\left| y'''(c) \frac{h^3}{6} \right| \leq 19(0.1)^3(1) = 0.019.$$

(c) Using the improved Euler's method with $h = 0.1$ we obtain $y(1.5) \approx 2.080108$. With $h = 0.05$ we obtain $y(1.5) \approx 2.059166$.

(d) Since $y(1.5) = 2.053216$, the error for $h = 0.1$ is $E_{0.1} = 0.026892$, while the error for $h = 0.05$ is $E_{0.05} = 0.005950$. With global truncation error $O(h^2)$ we expect $E_{0.1}/E_{0.05} \approx 4$. We actually have $E_{0.1}/E_{0.05} = 4.52$.

21. Because y_{n+1}^* depends on y_n and is used to determine y_{n+1}, all of the y_n^* cannot be computed at one time independently of the corresponding y_n values. For example, the computation of y_4^* involves the value of y_3.

EXERCISES 6.2

Runge-Kutta Methods

3.

x_n	y_n
1.00	5.0000
1.10	3.9724
1.20	3.2284
1.30	2.6945
1.40	2.3163
1.50	2.0533

6.

x_n	y_n
0.00	1.0000
0.10	1.1115
0.20	1.2530
0.30	1.4397
0.40	1.6961
0.50	2.0670

9.

x_n	y_n
0.00	0.5000
0.10	0.5213
0.20	0.5358
0.30	0.5443
0.40	0.5482
0.50	0.5493

12.

x_n	y_n
0.00	0.5000
0.10	0.5250
0.20	0.5498
0.30	0.5744
0.40	0.5987
0.50	0.6225

15. (a)

x_n	$h = 0.05$	$h = 0.1$
1.00	1.0000	1.0000
1.05	1.1112	
1.10	1.2511	1.2511
1.15	1.4348	
1.20	1.6934	1.6934
1.25	1.1047	
1.30	2.9560	2.9425
1.35	7.8981	
1.40	1.0608×10^{15}	903.0282

(b)

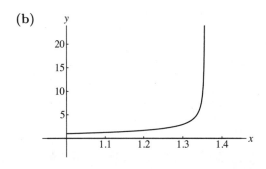

18. (a) Using $y^{(5)} = -1026e^{-3(x-1)}$ we see that the local truncation error is

$$\left| y^{(5)}(c)\, \frac{h^5}{120} \right| = 8.55 h^5 e^{-3(c-1)}.$$

(b) Since $e^{-3(x-1)}$ is a decreasing function for $1 \le x \le 1.5$, $e^{-3(c-1)} \le e^{-3(1-1)} = 1$ for $1 \le c \le 1.5$ and

$$y^{(5)}(c)\, \frac{h^5}{120} \le 8.55(0.1)^5(1) = 0.0000855.$$

(c) Using the RK4 method with $h = 0.1$ we obtain $y(1.5) \approx 2.053338827$. With $h = 0.05$ we obtain $y(1.5) \approx 2.053222989$.

21. (a) For $y' + y = 10 \sin 3x$ an integrating factor is e^x so that

$$\frac{d}{dx}[e^x y] = 10 e^x \sin 3x \implies e^x y = e^x \sin 3x - 3e^x \cos 3x + c$$

$$\implies y = \sin 3x - 3\cos 3x + ce^{-x}.$$

When $x = 0$, $y = 0$, so $0 = -3 + c$ and $c = 3$. The solution is

$$y = \sin 3x - 3\cos 3x + 3e^{-x}.$$

Using Newton's method we find that $x = 1.53235$ is the only positive root in $[0, 2]$.

(b) Using the RK4 method with $h = 0.1$ we obtain the table of values shown. These values are used to obtain an interpolating function in *Mathematica*. The graph of the interpolating function is shown. Using *Mathematica*'s root finding capability we see that the only positive root in $[0, 2]$ is $x = 1.53236$.

x_n	y_n
0.0	0.0000
0.1	0.1440
0.2	0.5448
0.3	1.1409
0.4	1.8559
0.5	2.6049
0.6	3.3019
0.7	3.8675
0.8	4.2356
0.9	4.3593
1.0	4.2147

x_n	y_n
1.0	4.2147
1.1	3.8033
1.2	3.1513
1.3	2.3076
1.4	1.3390
1.5	0.3243
1.6	−0.6530
1.7	−1.5117
1.8	−2.1809
1.9	−2.6061
2.0	−2.7539

EXERCISES 6.3

Multistep Methods

In the tables in this section "ABM" stands for Adams-Bashforth-Moulton.

3. The first predictor is $y_4^* = 0.73318477$.

x_n	y_n	
0.0	1.00000000	init. cond.
0.2	0.73280000	RK4
0.4	0.64608032	RK4
0.6	1.65851653	RK4
0.8	1.72319464	ABM

6. The first predictor for $h = 0.2$ is $y_4^* = 3.34828434$.

x_n	$h = 0.2$		$h = 0.1$	
0.0	1.00000000	init. cond.	1.00000000	init. cond.
0.1			1.21017082	RK4
0.2	1.44139950	RK4	1.44140511	RK4
0.3			1.69487942	RK4
0.4	1.97190167	RK4	1.97191536	ABM
0.5			2.27400341	ABM
0.6	2.60280694	RK4	2.60283209	ABM
0.7			2.96031780	ABM
0.8	3.34860927	ABM	3.34863769	ABM
0.9			3.77026548	ABM
1.0	4.22797875	ABM	4.22801028	ABM

EXERCISES 6.4

Higher-Order Equations and Systems

3. The substitution $y' = u$ leads to the system

$$y' = u, \qquad u' = 4u - 4y.$$

Using formula (4) in the text with x corresponding to t, y corresponding to x, and u corresponding to y, we obtain the table shown.

x_n	$h = 0.2$ y_n	$h = 0.2$ u_n	$h = 0.1$ y_n	$h = 0.1$ u_n
0.0	−2.0000	1.0000	−2.0000	1.0000
0.1			−1.8321	2.4427
0.2	−1.4928	4.4731	−1.4919	4.4753

6. Using $h = 0.1$, the RK4 method for a system, and a numerical solver, we obtain

t_n	$h = 0.2$ i_{1n}	$h = 0.2$ i_{3n}
0.0	0.0000	0.0000
0.1	2.5000	3.7500
0.2	2.8125	5.7813
0.3	2.0703	7.4023
0.4	0.6104	9.1919
0.5	−1.5619	11.4877

9.

t_n	$h = 0.2$ x_n	$h = 0.2$ y_n	$h = 0.1$ x_n	$h = 0.1$ y_n
0.0	−3.0000	5.0000	−3.0000	5.0000
0.1			−3.4790	4.6707
0.2	−3.9123	4.2857	−3.9123	4.2857

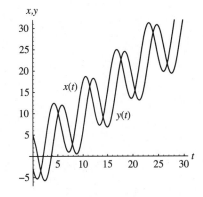

12. Solving for x' and y' we obtain the system

$$x' = \frac{1}{2}y - 3t^2 + 2t - 5$$

$$y' = -\frac{1}{2}y + 3t^2 + 2t + 5.$$

t_n	$h = 0.2$ x_n	$h = 0.2$ y_n	$h = 0.1$ x_n	$h = 0.1$ y_n
0.0	3.0000	−1.0000	3.0000	−1.0000
0.1			2.4727	−0.4527
0.2	1.9867	0.0933	1.9867	0.0933

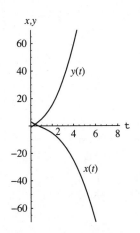

129

EXERCISES 6.5

Second-Order Boundary-Value Problems

3. We identify $P(x) = 2$, $Q(x) = 1$, $f(x) = 5x$, and $h = (1-0)/5 = 0.2$. Then the finite difference equation is

$$1.2y_{i+1} - 1.96y_i + 0.8y_{i-1} = 0.04(5x_i).$$

The solution of the corresponding linear system gives

x	0.0	0.2	0.4	0.6	0.8	1.0
y	0.0000	−0.2259	−0.3356	−0.3308	−0.2167	0.0000

6. We identify $P(x) = 5$, $Q(x) = 0$, $f(x) = 4\sqrt{x}$, and $h = (2-1)/6 = 0.1667$. Then the finite difference equation is

$$1.4167y_{i+1} - 2y_i + 0.5833y_{i-1} = 0.2778(4\sqrt{x_i}).$$

The solution of the corresponding linear system gives

x	1.0000	1.1667	1.3333	1.5000	1.6667	1.8333	2.0000
y	1.0000	−0.5918	−1.1626	−1.3070	−1.2704	−1.1541	−1.0000

9. We identify $P(x) = 1 - x$, $Q(x) = x$, $f(x) = x$, and $h = (1-0)/10 = 0.1$. Then the finite difference equation is

$$[1 + 0.05(1 - x_i)]y_{i+1} + [-2 + 0.01x_i]y_i + [1 - 0.05(1 - x_i)]y_{i-1} = 0.01x_i.$$

The solution of the corresponding linear system gives

x	0.0	0.1	0.2	0.3	0.4	0.5	0.6
y	0.0000	0.2660	0.5097	0.7357	0.9471	1.1465	1.3353

0.7	0.8	0.9	1.0
1.5149	1.6855	1.8474	2.0000

12. We identify $P(r) = 2/r$, $Q(r) = 0$, $f(r) = 0$, and $h = (4-1)/6 = 0.5$. Then the finite difference equation is

$$\left(1 + \frac{0.5}{r_i}\right)u_{i+1} - 2u_i + \left(1 - \frac{0.5}{r_i}\right)u_{i-1} = 0.$$

The solution of the corresponding linear system gives

r	1.0	1.5	2.0	2.5	3.0	3.5	4.0
u	50.0000	72.2222	83.3333	90.0000	94.4444	97.6190	100.0000

130

CHAPTER 6 REVIEW EXERCISES

3.

x_n	Euler h = 0.1	Euler h = 0.05	Imp. Euler h = 0.1	Imp. Euler h = 0.05	RK4 h = 0.1	RK4 h = 0.05
0.50	0.5000	0.5000	0.5000	0.5000	0.5000	0.5000
0.55		0.5500		0.5512		0.5512
0.60	0.6000	0.6024	0.6048	0.6049	0.6049	0.6049
0.65		0.6573		0.6609		0.6610
0.70	0.7090	0.7144	0.7191	0.7193	0.7194	0.7194
0.75		0.7739		0.7800		0.7801
0.80	0.8283	0.8356	0.8427	0.8430	0.8431	0.8431
0.85		0.8996		0.9082		0.9083
0.90	0.9559	0.9657	0.9752	0.9755	0.9757	0.9757
0.95		1.0340		1.0451		1.0452
1.00	1.0921	1.1044	1.1163	1.1168	1.1169	1.1169

6. The first predictor is $y_3^* = 1.14822731$.

x_n	y_n	
0.0	2.00000000	init. cond.
0.1	1.65620000	RK4
0.2	1.41097281	RK4
0.3	1.24645047	RK4
0.4	1.14796764	ABM

7 Vectors

EXERCISES 7.1

Vectors in 2-Space

3. (a) $\langle 12, 0 \rangle$ (b) $\langle 4, -5 \rangle$ (c) $\langle 4, 5 \rangle$ (d) $\sqrt{41}$ (e) $\sqrt{41}$

6. (a) $\langle 3, 9 \rangle$ (b) $\langle -4, -12 \rangle$ (c) $\langle 6, 18 \rangle$ (d) $4\sqrt{10}$ (e) $6\sqrt{10}$

9. (a) $\langle 4, -12 \rangle - \langle -2, 2 \rangle = \langle 6, -14 \rangle$ (b) $\langle -3, 9 \rangle - \langle -5, 5 \rangle = \langle 2, 4 \rangle$

12. (a) $\langle 8, 0 \rangle - \langle 0, -6 \rangle = \langle 8, 6 \rangle$ (b) $\langle -6, 0 \rangle - \langle 0, -15 \rangle = \langle -6, 15 \rangle$

15.

$\overrightarrow{P_1 P_2} = \langle 2, 5 \rangle$

18.

$\overrightarrow{P_1 P_2} = \langle 2, -3 \rangle$

21. $a(= -\mathbf{a})$, $b(= -\frac{1}{4}\mathbf{a})$, $c(= \frac{5}{2}\mathbf{a})$, $e(= 2\mathbf{a})$, and $f(= -\frac{1}{2}\mathbf{a})$ are parallel to $\mathbf{a}$.

24. $\langle 5, 2 \rangle$

27. $\|\mathbf{a}\| = 5$; (a) $\mathbf{u} = \frac{1}{5}\langle 0, -5 \rangle = \langle 0, -1 \rangle$; (b) $-\mathbf{u} = \langle 0, 1 \rangle$

30. $\|2\mathbf{a} - 3\mathbf{b}\| = \|\langle -5, 4 \rangle\| = \sqrt{25 + 16} = \sqrt{41}$; $\mathbf{u} = \frac{1}{\sqrt{41}}\langle -5, 4 \rangle = \langle -\frac{5}{\sqrt{41}}, \frac{4}{\sqrt{41}} \rangle$

33. $-\frac{3}{4}\mathbf{a} = \langle -3, -15/2 \rangle$

36.

$b = a + (b + c)$

39.

$b = (-c) - a$; $(b + c) + a = 0$; $a + b + c = 0$

42. From $2\mathbf{i} + 3\mathbf{j} = k_1\mathbf{b} + k_2\mathbf{c} = k_1(-2\mathbf{i} + 4\mathbf{j}) + k_2(5\mathbf{i} + 7\mathbf{j}) = (-2k_1 + 5k_2)\mathbf{i} + (4k_1 + 7k_2)\mathbf{j}$ we obtain the system of equations $-2k_1 + 5k_2 = 2$, $4k_1 + 7k_2 = 3$. Solving, we find $k_1 = \frac{1}{34}$ and $k_2 = \frac{7}{17}$.

45. (a) Since $\mathbf{F}_f = -\mathbf{F}_g$, $\|\mathbf{F}_g\| = \|\mathbf{F}_f\| = \mu\|\mathbf{F}_n\|$ and $\tan\theta = \|\mathbf{F}_g\|/\|\mathbf{F}_n\| = \mu\|\mathbf{F}_n\|/\|\mathbf{F}_n\| = \mu$.

    (b) $\theta = \tan^{-1} 0.6 \approx 31°$

48. Place one corner of the parallelogram at the origin and let two adjacent sides be $\overrightarrow{OP_1}$ and $\overrightarrow{OP_2}$. Let M be the midpoint of the diagonal connecting P_1 and P_2 and N be the midpoint of the other diagonal. Then $\overrightarrow{OM} = \frac{1}{2}(\overrightarrow{OP_1} + \overrightarrow{OP_2})$. Since $\overrightarrow{OP_1} + \overrightarrow{OP_2}$ is the main diagonal of the parallelogram and N is its midpoint, $\overrightarrow{ON} = \frac{1}{2}(\overrightarrow{OP_1} + \overrightarrow{OP_2})$. Thus, $\overrightarrow{OM} = \overrightarrow{ON}$ and the diagonals bisect each other.

EXERCISES 7.2
Vectors in 3-Space

3.

6.

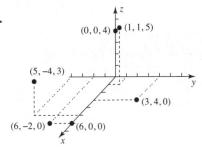

9. A line perpendicular to the xy-plane at $(2, 3, 0)$

12.

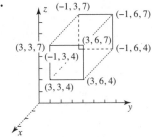

15. The union of the planes $x = 0$, $y = 0$, and $z = 0$

18. The union of the planes $x = 2$ and $z = 8$

21. $d = \sqrt{(3-6)^2 + (-1-4)^2 + (2-8)^2} = \sqrt{70}$

24. (a) 2; **(b)** $d = \sqrt{(-6)^2 + 2^2 + (-3)^2} = 7$

27. $d(P_1, P_2) = \sqrt{(4-1)^2 + (1-2)^2 + (3-3)^2} = \sqrt{10}$
$d(P_1, P_3) = \sqrt{(4-1)^2 + (6-2)^2 + (4-3)^2} = \sqrt{26}$
$d(P_2, P_3) = \sqrt{(4-4)^2 + (6-1)^2 + (4-3)^2} = \sqrt{26}$; The triangle is an isosceles triangle.

30. $d(P_1, P_2) = \sqrt{(1-2)^2 + (4-3)^2 + (4-2)^2} = \sqrt{6}$

$d(P_1, P_3) = \sqrt{(5-2)^2 + (0-3)^2 + (-4-2)^2} = 3\sqrt{6}$

$d(P_2, P_3) = \sqrt{(5-1)^2 + (0-4)^2 + (-4-4)^2} = 4\sqrt{6}$

Since $d(P_1, P_2) + d(P_1, P_3) = d(P_2, P_3)$, the points P_1, P_2, and P_3 are collinear.

33. $\left(\dfrac{1+7}{2}, \dfrac{3+(-2)}{2}, \dfrac{1/2+5/2}{2} \right) = (4, 1/2, 3/2)$

36. $(-3+(-5))/2 = x_3 = -4;$ $(4+8)/2 = y_3 = 6;$ $(1+3)/2 = z_3 = 2.$

The coordinates of P_3 are $(-4, 6, 2)$. **(a)** $\left(\dfrac{-3+(-4)}{2}, \dfrac{4+6}{2}, \dfrac{1+2}{2} \right) = (-7/2, 5, 3/2)$

(b) $\left(\dfrac{-4+(-5)}{2}, \dfrac{6+8}{2}, \dfrac{2+3}{2} \right) = (-9/2, 7, 5/2)$

39. $\overrightarrow{P_1 P_2} = \langle 2, 1, 1 \rangle$

42. $2\mathbf{a} - (\mathbf{b} - \mathbf{c}) = \langle 2, -6, 4 \rangle - \langle -3, -5, -8 \rangle = \langle 5, -1, 12 \rangle$

45. $\|\mathbf{a} + \mathbf{c}\| = \|\langle 3, 3, 11 \rangle\| = \sqrt{9 + 9 + 121} = \sqrt{139}$

48. $\|\mathbf{b}\|\mathbf{a} + \|\mathbf{a}\|\mathbf{b} = \sqrt{1+1+1}\,\langle 1, -3, 2 \rangle + \sqrt{1+9+4}\,\langle -1, 1, 1 \rangle = \langle \sqrt{3}, -3\sqrt{3}, 2\sqrt{3} \rangle + \langle -\sqrt{14}, \sqrt{14}, \sqrt{14} \rangle$

$= \langle \sqrt{3} - \sqrt{14}, -3\sqrt{3} + \sqrt{14}, 2\sqrt{3} + \sqrt{14} \rangle$

51. $\mathbf{b} = 4\mathbf{a} = 4\mathbf{i} - 4\mathbf{j} + 4\mathbf{k}$

EXERCISES 7.3

Dot Product

3. $\mathbf{a} \cdot \mathbf{b} = 2(-1) + (-3)2 + 4(5) = 12$

6. $\mathbf{a} \cdot (\mathbf{b} + \mathbf{c}) = 2(2) + (-3)8 + 4(4) = -4$

9. $\mathbf{a} \cdot \mathbf{a} = 2^2 + (-3)^2 + 4^2 = 29$

12. $(2\mathbf{a}) \cdot (\mathbf{a} - 2\mathbf{b}) = 4(4) + (-6)(-7) + 8(-6) = 10$

15. a and f, b and e, c and d

18. If $\mathbf{a}$ and $\mathbf{b}$ represent adjacent sides of the rhombus, then $\|\mathbf{a}\| = \|\mathbf{b}\|$, the diagonals of the rhombus are $\mathbf{a} + \mathbf{b}$ and $\mathbf{a} - \mathbf{b}$, and

$$(\mathbf{a} + \mathbf{b}) \cdot (\mathbf{a} - \mathbf{b}) = \mathbf{a} \cdot \mathbf{a} - \mathbf{a} \cdot \mathbf{b} + \mathbf{b} \cdot \mathbf{a} - \mathbf{b} \cdot \mathbf{b} = \mathbf{a} \cdot \mathbf{a} - \mathbf{b} \cdot \mathbf{b} = \|\mathbf{a}\|^2 - \|\mathbf{b}\|^2 = 0.$$

Thus, the diagonals are perpendicular.

21. $\mathbf{a} \cdot \mathbf{b} = 3(2) + (-1)2 = 4;$ $\|\mathbf{a}\| = \sqrt{10},$ $\|\mathbf{b}\| = 2\sqrt{2}$

$\cos\theta = \dfrac{4}{(\sqrt{10})(2\sqrt{2})} = \dfrac{1}{\sqrt{5}} \implies \theta = \cos^{-1}\dfrac{1}{\sqrt{5}} \approx 1.11 \text{ rad} \approx 63.43°$

24. $\mathbf{a} \cdot \mathbf{b} = \frac{1}{2}(2) + \frac{1}{2}(-4) + \frac{3}{2}(6) = 8;$ $\|\mathbf{a}\| = \sqrt{11}/2,$ $\|\mathbf{b}\| = 2\sqrt{14}$

$\cos\theta = \dfrac{8}{(\sqrt{11}/2)(2\sqrt{14})} = \dfrac{8}{\sqrt{154}} \implies \theta = \cos^{-1}(8/\sqrt{154}) \approx 0.87 \text{ rad} \approx 49.86°$

27. $\|\mathbf{a}\| = 2;$ $\cos\alpha = 1/2,$ $\alpha = 60°;$ $\cos\beta = 0,$ $\beta = 90°;$ $\cos\gamma = -\sqrt{3}/2,$ $\gamma = 150°$

30. If **a** and **b** are orthogonal, then $\mathbf{a} \cdot \mathbf{b} = 0$ and

$$\cos\alpha_1 \cos\alpha_2 + \cos\beta_1 \cos\beta_2 + \cos\gamma_1 \cos\gamma_2 = \frac{a_1}{\|\mathbf{a}\|}\frac{b_1}{\|\mathbf{b}\|} + \frac{a_2}{\|\mathbf{a}\|}\frac{b_2}{\|\mathbf{b}\|} + \frac{a_3}{\|\mathbf{a}\|}\frac{b_3}{\|\mathbf{b}\|}$$

$$= \frac{1}{\|\mathbf{a}\|\,\|\mathbf{b}\|}(a_1b_1 + a_2b_2 + a_3b_3) = \frac{1}{\|\mathbf{a}\|\,\|\mathbf{b}\|}(\mathbf{a}\cdot\mathbf{b}) = 0.$$

33. $\text{comp}_{\mathbf{b}}\mathbf{a} = \mathbf{a}\cdot\mathbf{b}/\|\mathbf{b}\| = \langle 1,-1,3\rangle \cdot \langle 2,6,3\rangle/7 = 5/7$

36. $\mathbf{a}+\mathbf{b} = \langle 3,5,6\rangle$; $2\mathbf{b} = \langle 4,12,6\rangle$; $\text{comp}_{2\mathbf{b}}(\mathbf{a}+\mathbf{b})\cdot 2\mathbf{b}/|2\mathbf{b}| = \langle 3,5,6\rangle \cdot \langle 4,12,6\rangle/14 = 54/7$

39. $\text{comp}_{\mathbf{b}}\mathbf{a} = \mathbf{a}\cdot\mathbf{b}/\|\mathbf{b}\| = (-5\mathbf{i}+5\mathbf{j})\cdot(-3\mathbf{i}+4\mathbf{j})/5 = 7$
$\text{proj}_{\mathbf{b}}\mathbf{a} = (\text{comp}_{\mathbf{b}}\mathbf{a})\mathbf{b}/\|\mathbf{b}\| = 7(-3\mathbf{i}+4\mathbf{j})/5 = -\frac{21}{5}\mathbf{i}+\frac{28}{5}\mathbf{j}$

42. $\text{comp}_{\mathbf{b}}\mathbf{a} = \mathbf{a}\cdot\mathbf{b}/\|\mathbf{b}\| = \langle 1,1,1\rangle \cdot \langle -2,2,-1\rangle/3 = -1/3$
$\text{proj}_{\mathbf{b}}\mathbf{a} = (\text{comp}_{\mathbf{b}}\mathbf{a})\mathbf{b}/\|\mathbf{b}\| = -\frac{1}{3}\langle -2,2,-1\rangle/3 = \langle 2/9,-2/9,1/9\rangle$

45. We identify $\|\mathbf{F}\| = 20$, $\theta = 60°$ and $\|\mathbf{d}\| = 100$. Then $W = \|\mathbf{F}\|\,\|\mathbf{d}\|\cos\theta = 20(100)(\frac{1}{2}) = 1000$ ft-lb.

48. Using $\mathbf{d} = 6\mathbf{i}+2\mathbf{j}$ and $\mathbf{F} = 3(\frac{3}{5}\mathbf{i}+\frac{4}{5}\mathbf{j})$, $W = \mathbf{F}\cdot\mathbf{d} = \langle \frac{9}{5},\frac{12}{5}\rangle \cdot \langle 6,2\rangle = \frac{78}{5}$ ft-lb.

51. $\|\mathbf{a}+\mathbf{b}\|^2 = (\mathbf{a}+\mathbf{b})\cdot(\mathbf{a}+\mathbf{b}) = \mathbf{a}\cdot\mathbf{a} + 2\mathbf{a}\cdot\mathbf{b} + \mathbf{b}\cdot\mathbf{b} = \|\mathbf{a}\|^2 + 2\mathbf{a}\cdot\mathbf{b} + \|\mathbf{b}\|^2$
$\qquad \leq \|\mathbf{a}\|^2 + 2|\mathbf{a}\cdot\mathbf{b}| + \|\mathbf{b}\|^2$ $\boxed{\text{since } x \leq |x|}$
$\qquad \leq \|\mathbf{a}\|^2 + 2\|\mathbf{a}\|\,\|\mathbf{b}\| + \|\mathbf{b}\|^2 = (\|\mathbf{a}\| + \|\mathbf{b}\|)^2$ $\boxed{\text{by Problem 50}}$
Thus, since $\|\mathbf{a}+\mathbf{b}\|$ and $\|\mathbf{a}\| + \|\mathbf{b}\|$ are positive, $\|\mathbf{a}+\mathbf{b}\| \leq \|\mathbf{a}\| + \|\mathbf{b}\|$.

EXERCISES 7.4
Cross Product

3. $\mathbf{a}\times\mathbf{b} = \begin{vmatrix} \mathbf{i} & \mathbf{j} & \mathbf{k} \\ 1 & -3 & 1 \\ 2 & 0 & 4 \end{vmatrix} = \begin{vmatrix} -3 & 1 \\ 0 & 4 \end{vmatrix}\mathbf{i} - \begin{vmatrix} 1 & 1 \\ 2 & 4 \end{vmatrix}\mathbf{j} + \begin{vmatrix} 1 & -3 \\ 2 & 0 \end{vmatrix}\mathbf{k} = \langle -12,-2,6\rangle$

6. $\mathbf{a}\times\mathbf{b} = \begin{vmatrix} \mathbf{i} & \mathbf{j} & \mathbf{k} \\ 4 & 1 & -5 \\ 2 & 3 & -1 \end{vmatrix} = \begin{vmatrix} 1 & -5 \\ 3 & -1 \end{vmatrix}\mathbf{i} - \begin{vmatrix} 4 & -5 \\ 2 & -1 \end{vmatrix}\mathbf{j} + \begin{vmatrix} 4 & 1 \\ 2 & 3 \end{vmatrix}\mathbf{k} = 14\mathbf{i} - 6\mathbf{j} + 10\mathbf{k}$

9. $\mathbf{a}\times\mathbf{b} = \begin{vmatrix} \mathbf{i} & \mathbf{j} & \mathbf{k} \\ 2 & 2 & -4 \\ -3 & -3 & 6 \end{vmatrix} = \begin{vmatrix} 2 & -4 \\ -3 & 6 \end{vmatrix}\mathbf{i} - \begin{vmatrix} 2 & -4 \\ -3 & 6 \end{vmatrix}\mathbf{j} + \begin{vmatrix} 2 & 2 \\ -3 & -3 \end{vmatrix}\mathbf{k} = \langle 0,0,0\rangle$

12. $\overrightarrow{P_1P_2} = (0,1,1)$; $\overrightarrow{P_1P_3} = (1,2,2)$; $\overrightarrow{P_1P_2}\times\overrightarrow{P_1P_3} = \begin{vmatrix} \mathbf{i} & \mathbf{j} & \mathbf{k} \\ 0 & 1 & 1 \\ 1 & 2 & 2 \end{vmatrix} = \begin{vmatrix} 1 & 1 \\ 2 & 2 \end{vmatrix}\mathbf{i} - \begin{vmatrix} 0 & 1 \\ 1 & 2 \end{vmatrix}\mathbf{j} + \begin{vmatrix} 0 & 1 \\ 1 & 2 \end{vmatrix}\mathbf{k} = \mathbf{j} - \mathbf{k}$

15. $\mathbf{a}\times\mathbf{b} = \begin{vmatrix} \mathbf{i} & \mathbf{j} & \mathbf{k} \\ 5 & -2 & 1 \\ 2 & 0 & -7 \end{vmatrix} = \begin{vmatrix} -2 & 1 \\ 0 & -7 \end{vmatrix}\mathbf{i} - \begin{vmatrix} 5 & 1 \\ 2 & -7 \end{vmatrix}\mathbf{j} + \begin{vmatrix} 5 & -2 \\ 2 & 0 \end{vmatrix}\mathbf{k} = \langle 14,37,4\rangle$

$\mathbf{a}\cdot(\mathbf{a}\times\mathbf{b}) = \langle 5,-2,-1\rangle \cdot \langle 14,37,4\rangle = 70 - 74 + 4 = 0$; $\mathbf{b}\cdot(\mathbf{a}\times\mathbf{b}) = \langle 2,0,-7\rangle \cdot \langle 14,37,4\rangle = 28 + 0 - 28 = 0$

18. (a) $\mathbf{b} \times \mathbf{c} = \begin{vmatrix} \mathbf{i} & \mathbf{j} & \mathbf{k} \\ 1 & 2 & -1 \\ -1 & 5 & 8 \end{vmatrix} = \begin{vmatrix} 2 & -1 \\ 5 & 8 \end{vmatrix} \mathbf{i} - \begin{vmatrix} 1 & -1 \\ -1 & 8 \end{vmatrix} \mathbf{j} + \begin{vmatrix} 1 & 2 \\ -1 & 5 \end{vmatrix} \mathbf{k} = 21\mathbf{i} - 7\mathbf{j} + 7\mathbf{k}$

$\mathbf{a} \times (\mathbf{b} \times \mathbf{c}) = \begin{vmatrix} \mathbf{i} & \mathbf{j} & \mathbf{k} \\ 3 & 0 & -4 \\ 21 & -7 & 7 \end{vmatrix} = \begin{vmatrix} 0 & -4 \\ -7 & 7 \end{vmatrix} \mathbf{i} - \begin{vmatrix} 3 & -4 \\ 21 & 7 \end{vmatrix} \mathbf{j} + \begin{vmatrix} 3 & 0 \\ 21 & -7 \end{vmatrix} \mathbf{k} = -28\mathbf{i} - 105\mathbf{j} - 21\mathbf{k}$

(b) $\mathbf{a} \cdot \mathbf{c} = (3\mathbf{i} - 4\mathbf{k}) \cdot (-\mathbf{i} + 5\mathbf{j} + 8\mathbf{k}) = -35;$ $(\mathbf{a} \cdot \mathbf{c})\mathbf{b} = -35(\mathbf{i} + 2\mathbf{j} - \mathbf{k}) = -35\mathbf{i} - 70\mathbf{j} + 35\mathbf{k}$

$\mathbf{a} \cdot \mathbf{b} = (3\mathbf{i} - 4\mathbf{k}) \cdot (\mathbf{i} + 2\mathbf{j} - \mathbf{k}) = 7;$ $(\mathbf{a} \cdot \mathbf{b})\mathbf{c} = 7(-\mathbf{i} + 5\mathbf{j} + 8\mathbf{k}) = -7\mathbf{i} + 35\mathbf{j} + 56\mathbf{k}$

$\mathbf{a} \times (\mathbf{b} \times \mathbf{c}) = (\mathbf{a} \cdot \mathbf{c})\mathbf{b} - (\mathbf{a} \cdot \mathbf{b})\mathbf{c} = (-35\mathbf{i} - 70\mathbf{j} + 35\mathbf{k}) - (-7\mathbf{i} + 35\mathbf{j} + 56\mathbf{k}) = -28\mathbf{i} - 105\mathbf{j} - 21\mathbf{k}$

21. $\mathbf{k} \times (2\mathbf{i} - \mathbf{j}) = \mathbf{k} \times (2\mathbf{i}) + \mathbf{k} \times (-\mathbf{j}) = 2(\mathbf{k} \times \mathbf{i}) - (\mathbf{k} \times \mathbf{j}) = 2\mathbf{j} - (-\mathbf{i}) = \mathbf{i} + 2\mathbf{j}$

24. $(2\mathbf{i} - \mathbf{j} + 5\mathbf{k}) \times \mathbf{i} = (2\mathbf{i} \times \mathbf{i}) + (-\mathbf{j} \times \mathbf{i}) + (5\mathbf{k} \times \mathbf{i}) = 2(\mathbf{i} \times \mathbf{i}) + (\mathbf{i} \times \mathbf{j}) + 5(\mathbf{k} \times \mathbf{i}) = 5\mathbf{j} + \mathbf{k}$

27. $\mathbf{k} \cdot (\mathbf{j} \times \mathbf{k}) = \mathbf{k} \cdot \mathbf{i} = 0$

30. $(\mathbf{i} \times \mathbf{j}) \cdot (3\mathbf{j} \times \mathbf{i}) = \mathbf{k} \cdot (-3\mathbf{k}) = -3(\mathbf{k} \cdot \mathbf{k}) = -3$

33. $(\mathbf{i} \times \mathbf{i}) \times \mathbf{j} = \mathbf{0} \times \mathbf{j} = \mathbf{0}$

36. $(\mathbf{i} \times \mathbf{k}) \times (\mathbf{j} \times \mathbf{i}) = (-\mathbf{j}) \times (-\mathbf{k}) = (-1)(-1)(\mathbf{j} \times \mathbf{k}) = \mathbf{j} \times \mathbf{k} = \mathbf{i}$

39. $(-\mathbf{a}) \times \mathbf{b} = -(\mathbf{a} \times \mathbf{b}) = -4\mathbf{i} + 3\mathbf{j} - 6\mathbf{k}$

42. $(\mathbf{a} \times \mathbf{b}) \cdot \mathbf{c} = 4(2) + (-3)4 + 6(-1) = -10$

45. (a) Let $A = (1, 3, 0)$, $B = (2, 0, 0)$, $C = (0, 0, 4)$, and $D = (1, -3, 4)$. Then $\overrightarrow{AB} = \mathbf{i} - 3\mathbf{j}$, $\overrightarrow{AC} = -\mathbf{i} - 3\mathbf{j} + 4\mathbf{k}$, $\overrightarrow{CD} = \mathbf{i} - 3\mathbf{j}$, and $\overrightarrow{BD} = -\mathbf{i} - 3\mathbf{j} + 4\mathbf{k}$. Since $\overrightarrow{AB} = \overrightarrow{CD}$ and $\overrightarrow{AC} = \overrightarrow{BD}$, the quadrilateral is a parallelogram.

(b) Computing

$$\overrightarrow{AB} \times \overrightarrow{AC} = \begin{vmatrix} \mathbf{i} & \mathbf{j} & \mathbf{k} \\ 1 & -3 & 0 \\ -1 & -3 & 4 \end{vmatrix} = -12\mathbf{i} - 4\mathbf{j} - 6\mathbf{k}$$

we find that the area is $\| -12\mathbf{i} - 4\mathbf{j} - 6\mathbf{k} \| = \sqrt{144 + 16 + 36} = 14$.

48. $\overrightarrow{P_1 P_2} = \mathbf{j} + 2\mathbf{k};$ $\overrightarrow{P_2 P_3} = 2\mathbf{i} + \mathbf{j} - 2\mathbf{k}$

$$\overrightarrow{P_1 P_2} \times \overrightarrow{P_2 P_3} = \begin{vmatrix} \mathbf{i} & \mathbf{j} & \mathbf{k} \\ 0 & 1 & 2 \\ 2 & 1 & -2 \end{vmatrix} = \begin{vmatrix} 1 & 2 \\ 1 & -2 \end{vmatrix} \mathbf{i} - \begin{vmatrix} 0 & 2 \\ 2 & -2 \end{vmatrix} \mathbf{j} + \begin{vmatrix} 0 & 1 \\ 2 & 1 \end{vmatrix} \mathbf{k} = -4\mathbf{i} + 4\mathbf{j} - 2\mathbf{k}$$

$A = \frac{1}{2} \| -4\mathbf{i} + 4\mathbf{j} - 2\mathbf{k} \| = 3$ sq. units

51. $\mathbf{b} \times \mathbf{c} = \begin{vmatrix} \mathbf{i} & \mathbf{j} & \mathbf{k} \\ -1 & 4 & 0 \\ 2 & 2 & 2 \end{vmatrix} = \begin{vmatrix} 4 & 0 \\ 2 & 2 \end{vmatrix} \mathbf{i} - \begin{vmatrix} -1 & 0 \\ 2 & 2 \end{vmatrix} \mathbf{j} + \begin{vmatrix} -1 & 4 \\ 2 & 2 \end{vmatrix} \mathbf{k} = 8\mathbf{i} + 2\mathbf{j} - 10\mathbf{k}$

$\mathbf{v} = |\mathbf{a} \cdot (\mathbf{b} \times \mathbf{c})| = |(\mathbf{i} + \mathbf{j}) \cdot (8\mathbf{i} + 2\mathbf{j} - 10\mathbf{k})| = |8 + 2 + 0| = 10$ cu. units

54. The four points will be coplanar if the three vectors $\overrightarrow{P_1 P_2} = \langle 3, -1, -1 \rangle$, $\overrightarrow{P_2 P_3} = \langle -3, -5, 13 \rangle$, and $\overrightarrow{P_3 P_4} = \langle -8, 7, -6 \rangle$ are coplanar.

$$\overrightarrow{P_2 P_3} \times \overrightarrow{P_3 P_4} = \begin{vmatrix} \mathbf{i} & \mathbf{j} & \mathbf{k} \\ -3 & -5 & 13 \\ -8 & 7 & -6 \end{vmatrix} = \begin{vmatrix} -5 & 13 \\ 7 & -6 \end{vmatrix} \mathbf{i} - \begin{vmatrix} -3 & 13 \\ -8 & -6 \end{vmatrix} \mathbf{j} + \begin{vmatrix} -3 & -5 \\ -8 & 7 \end{vmatrix} \mathbf{k} = \langle -61, -122, -61 \rangle$$

$\overrightarrow{P_1 P_2} \cdot (\overrightarrow{P_2 P_3} \times \overrightarrow{P_3 P_4}) = \langle 3, -1, -1 \rangle \cdot \langle -61, -122, -61 \rangle = -183 + 122 + 61 = 0$

The four points are coplanar.

57. (a) We note first that $\mathbf{a} \times \mathbf{b} = \mathbf{k}$, $\mathbf{b} \times \mathbf{c} = \frac{1}{2}(\mathbf{i} - \mathbf{k})$, $\mathbf{c} \times \mathbf{a} = \frac{1}{2}(\mathbf{j} - \mathbf{k})$, $\mathbf{a} \cdot (\mathbf{b} \times \mathbf{c}) = \frac{1}{2}$, $\mathbf{b} \cdot (\mathbf{c} \times \mathbf{a}) = \frac{1}{2}$, and $\mathbf{c} \cdot (\mathbf{a} \times \mathbf{b}) = \frac{1}{2}$. Then

$$\mathbf{A} = \frac{\frac{1}{2}(\mathbf{i} - \mathbf{k})}{\frac{1}{2}} = \mathbf{i} - \mathbf{k}, \quad \mathbf{B} = \frac{\frac{1}{2}(\mathbf{j} - \mathbf{k})}{\frac{1}{2}} = \mathbf{j} - \mathbf{k}, \quad \text{and} \quad \mathbf{C} = \frac{\mathbf{k}}{\frac{1}{2}} = 2\mathbf{k}.$$

(b) We need to compute $\mathbf{A} \cdot (\mathbf{B} \times \mathbf{C})$. Using formula (10) in the text we have

$$\mathbf{B} \times \mathbf{C} = \frac{(\mathbf{c} \times \mathbf{a}) \times (\mathbf{a} \times \mathbf{b})}{[\mathbf{b} \cdot (\mathbf{c} \times \mathbf{a})][\mathbf{c} \cdot (\mathbf{a} \times \mathbf{b})]} = \frac{[(\mathbf{c} \times \mathbf{a}) \cdot \mathbf{b}]\mathbf{a} - [(\mathbf{c} \times \mathbf{a}) \cdot \mathbf{a}]\mathbf{b}}{[\mathbf{b} \cdot (\mathbf{c} \times \mathbf{a})][\mathbf{c} \cdot (\mathbf{a} \times \mathbf{b})]}$$

$$= \frac{\mathbf{a}}{\mathbf{c} \cdot (\mathbf{a} \times \mathbf{b})} \qquad \boxed{\text{since } (\mathbf{c} \times \mathbf{a}) \cdot \mathbf{a} = 0.}$$

Then

$$\mathbf{A} \cdot (\mathbf{B} \times \mathbf{C}) = \frac{\mathbf{b} \times \mathbf{c}}{\mathbf{a} \cdot (\mathbf{b} \times \mathbf{c})} \cdot \frac{\mathbf{a}}{\mathbf{c} \cdot (\mathbf{a} \times \mathbf{b})} = \frac{1}{\mathbf{c} \cdot (\mathbf{a} \times \mathbf{b})}$$

and the volume of the unit cell of the reciprocal latrice is the reciprocal of the volume of the unit cell of the original lattice.

60. The statement is false since $\mathbf{i} \times (\mathbf{i} \times \mathbf{j}) = \mathbf{i} \times \mathbf{k} = -\mathbf{j}$ and $(\mathbf{i} \times \mathbf{i}) \times \mathbf{j} = 0 \times \mathbf{j} = 0$.

63. Since

$$\|\mathbf{a} \times \mathbf{b}\|^2 = (a_2 b_3 - a_3 b_2)^2 + (a_1 b_3 - a_3 b_1)^2 + (a_1 b_2 - a_2 b_1)^2$$

$$= a_2^2 b_3^2 - 2 a_2 b_3 a_3 b_2 + a_3^2 b_2^2 + a_1^2 b_3^2 - 2 a_1 b_3 a_3 b_1 + a_3^2 b_1^2 + a_1^2 b_2^2 - 2 a_1 b_2 a_2 b_1 + a_2^2 b_1^2$$

and

$$\|\mathbf{a}\|^2 \|\mathbf{b}\|^2 - (\mathbf{a} \cdot \mathbf{b})^2 = (a_1^2 + a_2^2 + a_3^2)(b_1^2 + b_2^2 + b_3^2) - (a_1 b_1 + a_2 b_2 + a_3 b_3)^2$$

$$= a_1^2 a_2^2 + a_1^2 b_2^2 + a_1^2 b_3^2 + a_2^2 b_1^2 + a_2^2 b_2^2 + a_2^2 b_3^2 + a_3^2 b_1^2 + a_3^2 b_2^2 + a_3^2 b_3^2$$

$$- a_1^2 b_1^2 - a_2^2 b_2^2 - a_3^2 b_3^2 - 2 a_1 b_1 a_2 b_2 - 2 a_1 b_1 a_3 b_3 - 2 a_2 b_2 a_3 b_3$$

$$= a_1^2 b_2^2 + a_1^2 b_3^2 + a_2^2 b_1^2 + a_2^2 b_3^2 + a_3^2 b_1^2 + a_3^2 b_2^2 - 2 a_1 a_2 b_1 b_2 - 2 a_1 a_3 b_1 b_3 - 2 a_2 a_3 b_2 b_3$$

we see that $\|\mathbf{a} \times \mathbf{b}\|^2 = \|\mathbf{a}\|^2 \|\mathbf{b}\|^2 - (\mathbf{a} \cdot \mathbf{b})^2$.

EXERCISES 7.5

Lines and Planes in 3-Space

The equation of a line through P_1 and P_2 in 3-space with $\mathbf{r}_1 = \overrightarrow{OP_1}$ and $\mathbf{r}_2 = \overrightarrow{OP_2}$ can be expressed as $\mathbf{r} = \mathbf{r}_1 + t(k\mathbf{a})$ or $\mathbf{r} = \mathbf{r}_2 + t(k\mathbf{a})$ where $\mathbf{a} = \mathbf{r}_2 - \mathbf{r}_1$ and k is any non-zero scalar. Thus, the form of the equation of a line is not unique. (See the alternate solution to Problem 1.)

3. $\mathbf{a} = \langle 1/2 - (-3/2), -1/2 - 5/2, 1 - (-1/2) \rangle = \langle 2, -3, 3/2 \rangle$; $\langle x, y, z \rangle = \langle 1/2, -1/2, 1 \rangle + t\langle 2, -3, 3/2 \rangle$

6. $\mathbf{a} = \langle 3 - 5/2, 2 - 1, 1 - (-2) \rangle = \langle 1/2, 1, 3 \rangle$; $\langle x, y, z \rangle = \langle 3, 2, 1 \rangle + t\langle 1/2, 1, 3 \rangle$

9. $\mathbf{a} = \langle 1 - 3, 0 - (-2), 0 - (-7) \rangle = \langle -2, 2, 7 \rangle$; $x = 1 - 2t$, $y = 2t$, $z = 7t$

12. $\mathbf{a} = \langle -3 - 4, 7 - (-8), 9 - (-1) \rangle = \langle -7, 15, 10 \rangle$; $x = -3 - 7t$, $y = 7 + 15t$, $z = 9 + 10t$

15. $a_1 = -7 - 4 = -11$, $a_2 = 2 - 2 = 0$, $a_3 = 5 - 1 = 4$; $\dfrac{x+7}{-11} = \dfrac{z-5}{4}$, $y = 2$

18. $a_1 = 5/6 - 1/3 = 1/2$; $a_2 = -1/4 - 3/8 = -5/8$; $a_3 = 1/5 - 1/10 = 1/10$

$$\frac{x - 5/6}{1/2} = \frac{y + 1/4}{-5/8} = \frac{z - 1/5}{1/10}$$

21. parametric: $x = 5t$, $y = 9t$, $z = 4t$; symmetric: $\dfrac{x}{5} = \dfrac{y}{9} = \dfrac{z}{4}$

24. A direction vector is $\langle 5, 1/3, -2\rangle$. Symmetric equations for the line are $(x-4)/5 = (y+11)/(1/3) = (z+7)/(-2)$.

27. Both lines go through the points $(0,0,0)$ and $(6,6,6)$. Since two points determine a line, the lines are the same.

30. The parametric equations for the line are $x = 1 + 2t$, $y = -2 + 3t$, $z = 4 + 2t$. In the xy-plane, $z = 4 + 2t = 0$ and $t = -2$. Then $x = 1 + 2(-2) = -3$ and $y = -2 + 3(-2) = -8$. The point is $(-3, -8, 0)$. In the xz-plane, $y = -2 + 3t = 0$ and $t = 2/3$. Then $x = 1 + 2(2/3) = 7/3$ and $z = 4 + 2(2/3) = 16/3$. The point is $(7/3, 0, 16/3)$. In the yz-plane, $x = 1 + 2t = 0$ and $t = -1/2$. Then $y = -2 + 3(-1/2) = -7/2$ and $z = 4 + 2(-1/2) = 3$. The point is $(0, -7/2, 3)$.

33. The system of equations $2 - t = 4 + s$, $3 + t = 1 + s$, $1 + t = 1 - s$, or $t + s = -2$, $t - s = -2$, $t + s = 0$ has no solution since $-2 \neq 0$. Thus, the lines do not intersect.

36. $\mathbf{a} = \langle 2, 7, -1\rangle$, $\mathbf{b} = \langle -2, 1, 4\rangle$, $\mathbf{a} \cdot \mathbf{b} = -1$, $\|\mathbf{a}\| = 3\sqrt{6}$, $\|\mathbf{b}\| = \sqrt{21}$;

$$\cos\theta = \frac{\mathbf{a} \cdot \mathbf{b}}{\|\mathbf{a}\|\,\|\mathbf{b}\|} = \frac{-1}{(3\sqrt{6})(\sqrt{21})} = -\frac{1}{9\sqrt{14}}\,; \quad \theta = \cos^{-1}\left(-\frac{1}{9\sqrt{14}}\right) \approx 91.70°$$

39. $2(x - 5) - 3(y - 1) + 4(z - 3) = 0$; $2x - 3y + 4z = 19$

42. $6x - y + 3z = 0$

45. From the points $(3, 5, 2)$ and $(2, 3, 1)$ we obtain the vector $\mathbf{u} = \mathbf{i} + 2\mathbf{j} + \mathbf{k}$. From the points $(2, 3, 1)$ and $(-1, -1, 4)$ we obtain the vector $\mathbf{v} = 3\mathbf{i} + 4\mathbf{j} - 3\mathbf{k}$. From the points $(-1, -1, 4)$ and (x, y, z) we obtain the vector $\mathbf{w} = (x + 1)\mathbf{i} + (y + 1)\mathbf{j} + (z - 4)\mathbf{k}$. Then, a normal vector is

$$\mathbf{u} \times \mathbf{v} = \begin{vmatrix} \mathbf{i} & \mathbf{j} & \mathbf{k} \\ 1 & 2 & 1 \\ 3 & 4 & -3 \end{vmatrix} = -10\mathbf{i} + 6\mathbf{j} - 2\mathbf{k}.$$

A vector equation of the plane is $-10(x + 1) + 6(y + 1) - 2(z - 4) = 0$ or $5x - 3y + z = 2$.

48. The three points are not colinear and all satisfy $x = 0$, which is the equation of the plane.

51. A normal vector to $x + y - 4z = 1$ is $\langle 1, 1, -4\rangle$. The equation of the parallel plane is $(x - 2) + (y - 3) - 4(z + 5) = 0$ or $x + y - 4z = 25$.

54. A normal vector is $\langle 0, 1, 0\rangle$. The equation of the plane is $y + 5 = 0$ or $y = -5$.

57. A direction vector for the two lines is $\langle 1, 2, 1\rangle$. Points on the lines are $(1, 1, 3)$ and $(3, 0, -2)$. Thus, another vector parallel to the plane is $\langle 1-3, 1-0, 3+2\rangle = \langle -2, 1, 5\rangle$. A normal vector to the plane is $\langle 1, 2, 1\rangle \times \langle -2, 1, 5\rangle = \langle 9, -7, 5\rangle$. Using the point $(3, 0, -2)$ in the plane, the equation of the plane is $9(x - 3) - 7(y - 0) + 5(z + 2) = 0$ or $9x - 7y + 5z = 17$.

60. A normal vector to the plane is $\langle 2 - 1, 6 - 0, -3 + 2\rangle = \langle 1, 6, -1\rangle$. The equation of the plane is $(x - 1) + 6(y - 1) - (z - 1) = 0$ or $x + 6y - z = 6$.

63. A direction vector of the line is $\langle -6, 9, 3\rangle$, and the normal vectors of the planes are **(a)** $\langle 4, 1, 2\rangle$, **(b)** $\langle 2, -3, 1\rangle$, **(c)** $\langle 10, -15, -5\rangle$, **(d)** $\langle -4, 6, 2\rangle$. Vectors **(c)** and **(d)** are multiples of the direction vector and hence the corresponding planes are perpendicular to the line.

66. Letting $y = t$ in both equations and solving $x - z = 2 - 2t$, $3x + 2z = 1 + t$, we obtain $x = 1 - \frac{3}{5}t$, $y = t$, $z = -1 + \frac{7}{5}t$ or, letting $t = 5s$, $x = 1 - 3s$, $y = 5s$, $z = -1 + 7s$.

69. Substituting the parametric equations into the equation of the plane, we obtain $2(1+2t) - 3(2-t) + 2(-3t) = -7$ or $t = -3$. Letting $t = -3$ in the equation of the line, we obtain the point of intersection $(-5, 5, 9)$.

72. Substituting the parametric equations into the equation of the plane, we obtain $4 + t - 3(2 + t) + 2(1 + 5t) = 0$ or $t = 0$. Letting $t = 0$ in the equation of the line, we obtain the point of intersection $(4, 2, 1)$.

75. In this problem the cross product of the direction vector of the line with the normal vector of the given plane will be a normal vector to the desired plane. A direction vector of the line is $\langle 3, -1, 5 \rangle$ and a normal vector to the given plane is $\langle 1, 1, 1 \rangle$. A normal vector to the desired plane is $\langle 3, -1, 5 \rangle \times \langle 1, 1, 1 \rangle = \langle -6, 2, 4 \rangle$. A point on the line, and hence in the plane, is $(4, 0, 1)$. The equation of the plane is $-6(x - 4) + 2(y - 0) + 4(z - 1) = 0$ or $3x - y - 2z = 10$.

78.

81.

EXERCISES 7.6

Vector Spaces

3. Not a vector space. Axiom (**x**) is not satisfied.

6. A vector space

9. A vector space

12. Not a subspace. Axiom (**i**) is not satisfied.

15. A subspace

18. A subspace

21. Let (x_1, y_1, z_1) and (x_2, y_2, z_2) be in S. Then

$$(x_1, y_1, z_1) + (x_2, y_2, z_2) = (at_1, bt_1, ct_1) + (at_2, bt_2, ct_2) = (a(t_1 + t_2), b(t_1 + t_2), c(t_1 + t_2))$$

is in S. Also, for (x, y, z) in S then $k(x, y, z) = (kx, ky, kz) = (a(kt), b(kt), c(kt))$ is also in S.

24. (a) The assumption $c_1 p_1 + c_2 p_2 = 0$ is equivalent to $(c_1 + c_2)x + (c_1 - c_2) = 0$. Thus $c_1 + c_2 = 0$, $c_1 - c_2 = 0$. The only solution of this system is $c_1 = 0$, $c_2 = 0$.

(b) Solving the system $c_1 + c_2 = 5$, $c_1 - c_2 = 2$ gives $c_1 = \frac{7}{2}$, $c_2 = \frac{3}{2}$. Thus $p(x) = \frac{7}{2}p_1(x) + \frac{3}{2}p_2(x)$

27. Linearly independent

30. $(x, \sin x) = \displaystyle\int_0^{2\pi} x \sin x \, dx = (-x \cos x + \sin x) \Big|_0^{2\pi} = -2\pi$

33. We need to show that $\text{Span}\{x_1, x_2, \ldots, x_n\}$ is closed under vector addition and scalar multiplication. Suppose $\mathbf{u}$ and $\mathbf{v}$ are in $\text{Span}\{x_1, x_2, \ldots, x_n\}$. Then $\mathbf{u} = a_1 x_1 + a_2 x_2 + \cdots + a_n x_n$ and $\mathbf{v} = b_1 x_1 + b_2 x_2 + \cdots + b_n x_n$, so that

$$\mathbf{u} + \mathbf{v} = (a_1 + b_1)x_1 + (a_2 + b_2)x_2 + \cdots + (a_n + b_n)x_n,$$

which is in $\text{Span}\{x_1, x_2, \ldots, x_n\}$. Also, for any real number k,

$$k\mathbf{u} = k(a_1 x_1 + a_2 x_2 + \cdots + a_n x_n) = k a_1 x_1 + k a_2 x_2 + \cdots + k a_n x_n,$$

which is in $\text{Span}\{x_1, x_2, \ldots, x_n\}$. Thus, $\text{Span}\{x_1, x_2, \ldots, x_n\}$ is a subspace of $\mathbf{V}$.

36. To show that the set of nonzero orthogonal vectors is linearly independent we set $c_1 \mathbf{v}_1 + c_2 \mathbf{v}_2 + \cdots + c_n \mathbf{v}_n = \mathbf{0}$. For $0 \leq i \leq n$,

$$(c_1 \mathbf{v}_1 + c_2 \mathbf{v}_2 + \cdots + c_i \mathbf{v}_i \cdots + c_n \mathbf{v}_n) \cdot \mathbf{v}_i = c_1 \mathbf{v}_1 \cdot \mathbf{v}_i + c_2 \mathbf{v}_2 \cdot \mathbf{v}_i + \cdots + c_i \mathbf{v}_i \cdot \mathbf{v}_i \cdots + c_n \mathbf{v}_n \cdot \mathbf{v}_i = c_i \|\mathbf{v}_i\|^2,$$

so $c_i \|\mathbf{v}_i\|^2 = 0$ because

$$(c_1 \mathbf{v}_1 + c_2 \mathbf{v}_2 + \cdots + c_i \mathbf{v}_i \cdots + c_n \mathbf{v}_n) \cdot \mathbf{v}_i = \mathbf{0} \cdot \mathbf{v}_i = 0.$$

Since $\mathbf{v}_i$ is a nonzero vector, $c_i = 0$. Thus, the assumption that $c_1 \mathbf{v}_1 + c_2 \mathbf{v}_2 + \cdots + c_n \mathbf{v}_n = \mathbf{0}$ leads to $c_1 = c_2 = \cdots = c_n = 0$, and the set is linearly independent.

EXERCISES 7.7

Gram-Schmidt Orthogonalization Process

3. Since the basis vectors in this problem are orthogonal but not orthonormal, the result of Theorem 7.5 must be slightly modified to read

$$\mathbf{u} = \frac{\mathbf{u} \cdot \mathbf{w}_1}{\|\mathbf{w}_1\|^2} \mathbf{w}_1 + \frac{\mathbf{u} \cdot \mathbf{w}_2}{\|\mathbf{w}_2\|^2} \mathbf{w}_2 + \cdots + \frac{\mathbf{u} \cdot \mathbf{w}_n}{\|\mathbf{w}_n\|^2} \mathbf{w}_n.$$

The proof is very similar to that given in the text for Theorem 7.5. Letting $\mathbf{w}_1 = \langle 1, 0, 1 \rangle$, $\mathbf{w}_2 = \langle 0, 1, 0 \rangle$, and $\mathbf{w}_3 = \langle -1, 0, 1 \rangle$ we have

$$\mathbf{w}_1 \cdot \mathbf{w}_2 = (1)(0) + (0)(1) + (1)(0) = 0$$

$$\mathbf{w}_1 \cdot \mathbf{w}_3 = (1)(-1) + (0)(0) + (1)(1) = 0$$

$$\mathbf{w}_2 \cdot \mathbf{w}_3 = (0)(-1) + (1)(0) + (0)(1) = 0$$

so the vectors are orthogonal. We also compute

$$\|\mathbf{w}_1\|^2 = 1^2 + 0^2 + 1^2 = 2$$

$$\|\mathbf{w}_2\|^2 = 0^2 + 1^2 + 0^2 = 1$$

$$\|\mathbf{w}_3\|^2 = (-1)^2 + 0^2 + 1^2 = 2$$

and, with $\mathbf{u} = \langle 10, 7, -13 \rangle$,

$$\mathbf{u} \cdot \mathbf{w}_1 = (10)(1) + (7)(0) + (-13)(1) = -3$$

$$\mathbf{u} \cdot \mathbf{w}_2 = (10)(0) + (7)(1) + (-13)(0) = 7$$

$$\mathbf{u} \cdot \mathbf{w}_3 = (10)(-1) + (7)(0) + (-13)(1) = -23.$$

140

Then, using the result given before the solution to this problem, we have

$$\mathbf{u} = -\frac{3}{2}\,\mathbf{w}_1 + 7\mathbf{w}_2 - \frac{23}{2}\,\mathbf{w}_3.$$

6. **(a)** We have $\mathbf{u}_1 = \langle -3, 4 \rangle$ and $\mathbf{u}_2 = \langle -1, 0 \rangle$. Taking $\mathbf{v}_1 = \mathbf{u}_1 = \langle -3, 4 \rangle$, and using $\mathbf{u}_2 \cdot \mathbf{v}_1 = 3$ and $\mathbf{v}_1 \cdot \mathbf{v}_1 = 25$ we obtain

$$\mathbf{v}_2 = \mathbf{u}_2 - \frac{\mathbf{u}_2 \cdot \mathbf{v}_1}{\mathbf{v}_1 \cdot \mathbf{v}_1}\,\mathbf{v}_1 = \langle -1, 0 \rangle - \frac{3}{25}\langle -3, 4 \rangle = \left\langle -\frac{16}{25}, -\frac{12}{25} \right\rangle.$$

Thus, an orthogonal basis is $\{\langle -3, 4 \rangle, \langle -\frac{16}{25}, -\frac{12}{25} \rangle\}$ and an orthonormal basis is $\{\mathbf{w}_1', \mathbf{w}_2'\}$, where

$$\mathbf{w}_1' = \frac{1}{||\langle -3, 4 \rangle||}\,\langle -3, 4 \rangle = \frac{1}{5}\,\langle -3, 4 \rangle = \left\langle -\frac{3}{5}, \frac{4}{5} \right\rangle$$

and

$$\mathbf{w}_2' = \frac{1}{||\langle -\frac{16}{25}, -\frac{12}{25} \rangle||}\,\left\langle -\frac{16}{25}, -\frac{12}{25} \right\rangle = \frac{1}{4/5}\,\left\langle -\frac{16}{25}, -\frac{12}{25} \right\rangle = \left\langle -\frac{4}{5}, -\frac{3}{5} \right\rangle.$$

(b) We have $\mathbf{u}_1 = \langle -3, 4 \rangle$ and $\mathbf{u}_2 = \langle -1, 0 \rangle$. Taking $\mathbf{v}_1 = \mathbf{u}_2 = \langle -1, 0 \rangle$, and using $\mathbf{u}_1 \cdot \mathbf{v}_1 = 3$ and $\mathbf{v}_1 \cdot \mathbf{v}_1 = 1$ we obtain

$$\mathbf{v}_2 = \mathbf{u}_1 - \frac{\mathbf{u}_1 \cdot \mathbf{v}_1}{\mathbf{v}_1 \cdot \mathbf{v}_1}\,\mathbf{v}_1 = \langle -3, 4 \rangle - \frac{3}{1}\langle -1, 0 \rangle = \langle 0, 4 \rangle.$$

Thus, an orthogonal basis is $\{\langle -1, 0 \rangle, \langle 0, 4 \rangle\}$ and an orthonormal basis is $\{\mathbf{w}_3'', \mathbf{w}_4''\}$, where

$$\mathbf{w}_3'' = \frac{1}{||\langle -1, 0 \rangle||}\,\langle -1, 0 \rangle = \frac{1}{1}\,\langle -1, 0 \rangle = \langle -1, 0 \rangle$$

and

$$\mathbf{w}_4'' = \frac{1}{||\langle 0, 4 \rangle||}\,\langle 0, 4 \rangle = \frac{1}{4}\,\langle 0, 4 \rangle = \langle 0, 1 \rangle.$$

(c)

9. We have $\mathbf{u}_1 = \langle 1, 1, 0 \rangle$, $\mathbf{u}_2 = \langle 1, 2, 2 \rangle$, and $\mathbf{u}_3 = \langle 2, 2, 1 \rangle$. Taking $\mathbf{v}_1 = \mathbf{u}_1 = \langle 1, 1, 0 \rangle$ and using $\mathbf{u}_2 \cdot \mathbf{v}_1 = 3$ and $\mathbf{v}_1 \cdot \mathbf{v}_1 = 2$ we obtain

$$\mathbf{v}_2 = \mathbf{u}_2 - \frac{\mathbf{u}_2 \cdot \mathbf{v}_1}{\mathbf{v}_1 \cdot \mathbf{v}_1}\,\mathbf{v}_1 = \langle 1, 2, 2 \rangle - \frac{3}{2}\langle 1, 1, 0 \rangle = \left\langle -\frac{1}{2}, \frac{1}{2}, 2 \right\rangle.$$

Next, using $\mathbf{u}_3 \cdot \mathbf{v}_1 = 4$, $\mathbf{u}_3 \cdot \mathbf{v}_2 = 2$, and $\mathbf{v}_2 \cdot \mathbf{v}_2 = \frac{9}{2}$, we obtain

$$\mathbf{v}_3 = \mathbf{u}_3 - \frac{\mathbf{u}_3 \cdot \mathbf{v}_1}{\mathbf{v}_1 \cdot \mathbf{v}_1}\,\mathbf{v}_1 - \frac{\mathbf{u}_3 \cdot \mathbf{v}_2}{\mathbf{v}_2 \cdot \mathbf{v}_2}\,\mathbf{v}_2 = \langle 2, 2, 1 \rangle - \frac{4}{2}\,\langle 1, 1, 0 \rangle - \frac{2}{9/2}\,\left\langle -\frac{1}{2}, \frac{1}{2}, 2 \right\rangle = \left\langle \frac{2}{9}, -\frac{2}{9}, \frac{1}{9} \right\rangle.$$

Thus, an orthogonal basis is

$$B' = \left\{ \langle 1, 1, 0 \rangle, \left\langle -\frac{1}{2}, \frac{1}{2}, 2 \right\rangle, \left\langle \frac{2}{9}, -\frac{2}{9}, \frac{1}{9} \right\rangle \right\},$$

7.7 Gram-Schmidt Orthogonalization Process

and an orthonormal basis is

$$B'' = \left\{ \left\langle \frac{1}{\sqrt{2}}, \frac{1}{\sqrt{2}}, 0 \right\rangle, \left\langle -\frac{1}{3\sqrt{2}}, \frac{1}{3\sqrt{2}}, \frac{4}{3\sqrt{2}} \right\rangle, \left\langle \frac{2}{3}, -\frac{2}{3}, \frac{1}{3} \right\rangle \right\}.$$

12. We have $\mathbf{u}_1 = \langle 1,1,1 \rangle, \mathbf{u}_2 = \langle 9,-1,1 \rangle$, and $\mathbf{u}_3 = \langle -1,4,-2 \rangle$. Taking $\mathbf{v}_1 = \mathbf{u}_1 = \langle 1,1,1 \rangle$ and using $\mathbf{u}_2 \cdot \mathbf{v}_1 = 9$ and $\mathbf{v}_1 \cdot \mathbf{v}_1 = 3$ we obtain

$$\mathbf{v}_2 = \mathbf{u}_2 - \frac{\mathbf{u}_2 \cdot \mathbf{v}_1}{\mathbf{v}_1 \cdot \mathbf{v}_1} \mathbf{v}_1 = \langle 9,-1,1 \rangle - \frac{9}{3} \langle 1,1,1 \rangle = \langle 6,-4,-2 \rangle.$$

Next, using $\mathbf{u}_3 \cdot \mathbf{v}_1 = 1, \mathbf{u}_3 \cdot \mathbf{v}_2 = -18$, and $\mathbf{v}_2 \cdot \mathbf{v}_2 = 56$, we obtain

$$\mathbf{v}_3 = \mathbf{u}_3 - \frac{\mathbf{u}_3 \cdot \mathbf{v}_1}{\mathbf{v}_1 \cdot \mathbf{v}_1} \mathbf{v}_1 - \frac{\mathbf{u}_3 \cdot \mathbf{v}_2}{\mathbf{v}_2 \cdot \mathbf{v}_2} \mathbf{v}_2 = \langle -1,4,-2 \rangle - \frac{1}{3} \langle 1,1,1 \rangle - \frac{-18}{56} \langle 6,-4,-2 \rangle = \left\langle \frac{25}{42}, \frac{50}{21}, -\frac{125}{42} \right\rangle.$$

Thus, an orthogonal basis is

$$B' = \left\{ \langle 1,1,1 \rangle, \langle 6,-4,-2 \rangle, \left\langle \frac{25}{42}, \frac{50}{21}, -\frac{125}{42} \right\rangle \right\},$$

and an orthonormal basis is

$$B'' = \left\{ \left\langle \frac{1}{\sqrt{3}}, \frac{1}{\sqrt{3}}, \frac{1}{\sqrt{3}} \right\rangle, \left\langle \frac{3}{\sqrt{14}}, -\frac{2}{\sqrt{14}}, -\frac{1}{\sqrt{14}} \right\rangle, \left\langle \frac{1}{\sqrt{42}}, \frac{4}{\sqrt{42}}, -\frac{5}{\sqrt{42}} \right\rangle \right\}.$$

15. We have $\mathbf{u}_1 = \langle 1,-1,1,-1 \rangle$, and $\mathbf{u}_2 = \langle 1,3,0,1 \rangle$. Taking $\mathbf{v}_1 = \mathbf{u}_1 = \langle 1,-1,1,-1 \rangle$ and using $\mathbf{u}_2 \cdot \mathbf{v}_1 = -3$ and $\mathbf{v}_1 \cdot \mathbf{v}_1 = 4$ we obtain

$$\mathbf{v}_2 = \mathbf{u}_2 - \frac{\mathbf{u}_2 \cdot \mathbf{v}_1}{\mathbf{v}_1 \cdot \mathbf{v}_1} \mathbf{v}_1 = \langle 1,3,0,1 \rangle - \frac{-3}{4} \langle 1,-1,1,-1 \rangle = \left\langle \frac{7}{4}, \frac{9}{4}, \frac{3}{4}, \frac{1}{4} \right\rangle.$$

Thus, an orthogonal basis is $B' = \left\{ \langle 1,-1,1,-1 \rangle, \left\langle \frac{7}{4}, \frac{9}{4}, \frac{3}{4}, \frac{1}{4} \right\rangle \right\}$, and an orthonormal basis is

$$B'' = \left\{ \left\langle \frac{1}{2}, -\frac{1}{2}, \frac{1}{2}, -\frac{1}{2} \right\rangle, \left\langle \frac{7}{2\sqrt{35}}, \frac{9}{2\sqrt{35}}, \frac{3}{2\sqrt{35}}, \frac{1}{2\sqrt{35}} \right\rangle \right\}.$$

18. We have $u_1 = x^2 - x, u_2 = x^2 + 1$, and $u_3 = 1 - x^2$. Taking $v_1 = u_1 = x^2 - x$ and using

$$(u_2, v_1) = \int_{-1}^{1} (x^2+1)(x^2-x)dx = \frac{16}{15} \quad \text{and} \quad (v_1, v_1) = \int_{-1}^{1} (x^2-x)(x^2-x)dx = \frac{16}{15}$$

we obtain

$$v_2 = u_2 - \frac{(u_2, v_1)}{(v_1, v_1)} v_1 = x^2 + 1 - \frac{16/15}{16/15}(x^2-x) = x+1.$$

Next, using

$$(u_3, v_1) = \int_{-1}^{1} (1-x^2)(x^2-x)dx = \frac{4}{15}, \quad (u_3, v_2) = \int_{-1}^{1} (1-x^2)(x+1)dx = \frac{4}{3},$$

and

$$(v_2, v_2) = \int_{-1}^{1} (x+1)(x+1)dx = \frac{8}{3},$$

we obtain

$$v_3 = u_3 - \frac{(u_3, v_1)}{(v_1, v_1)} v_1 - \frac{(u_3, v_2)}{(v_2, v_2)} v_2 = 1 - x^2 - \frac{4/15}{16/15}(x^2-x) - \frac{4/3}{8/3}(x+1) = -\frac{5}{4}x^3 - \frac{1}{4}x + \frac{1}{2}.$$

Thus, an orthogonal basis is $B' = \left\{ x^2 - x, x+1, -\frac{5}{4}x^3 - \frac{1}{4}x + \frac{1}{2} \right\}$.

21. Using $w_1 = 1/\sqrt{2}$, $w_2 = 3x/\sqrt{6}$, and $w_3 = (15/2\sqrt{10})(x^2 - 1/3)$, and computing

$$(p, w_1) = \int_{-1}^{1} (9x^2 - 6x + 5)\frac{1}{\sqrt{2}}\, dx = 8\sqrt{2},$$

$$(p, w_2) = \int_{-1}^{1} (9x^2 - 6x + 5)\frac{3}{\sqrt{6}}\, x\, dx = -2\sqrt{6}$$

$$(p, w_3) = \int_{-1}^{1} (9x^2 - 6x + 5)\left[\frac{15}{2\sqrt{10}}\left(x^2 - \frac{1}{3}\right)\right]\, dx = \frac{12}{\sqrt{10}},$$

we find from Theorem 7.5

$$p(x) = 9x^2 - 6x + 5 = (p, w_1)w_1 + (p, w_2)w_2 + (p, w_3)w_3 = 8\sqrt{2}\, w_1 - 2\sqrt{6}\, w_2 + \frac{12}{\sqrt{10}}\, w_3.$$

CHAPTER 7 REVIEW EXERCISES

3. False; since a normal to the plane is $\langle 2, 3, -4 \rangle$ which is not a multiple of the direction vector $\langle 5, -2, 1 \rangle$ of the line.

6. True

9. True

12. orthogonal

15. $\sqrt{(-12)^2 + 4^2 + 6^2} = 14$

18. The coordinates of $(1, -2, -10)$ satisfy the given equation.

21. $x_2 - 2 = 3$, $x_2 = 5$; $y_2 - 1 = 5$, $y_2 = 6$; $z_2 - 7 = -4$, $z_2 = 3$; $P_2 = (5, 6, 3)$

24. $2\mathbf{b} = \langle -2, 4, 2 \rangle$; $4\mathbf{c} = \langle 0, -8, 8 \rangle$; $\mathbf{a} \cdot (2\mathbf{b} + 4\mathbf{c}) = \langle 3, 1, 0 \rangle \cdot \langle -2, -4, 10 \rangle = -10$

27. $A = \frac{1}{2}|5\mathbf{i} - 4\mathbf{j} - 7\mathbf{k}| = \frac{3\sqrt{10}}{2}$

30. parallel: $-2c = 5$, $c = -5/2$; orthogonal: $1(-2) + 3(-6) + c(5) = 0$, $c = 4$

33. $\text{comp}_\mathbf{b}\mathbf{a} = \mathbf{a} \cdot \mathbf{b}/\|\mathbf{b}\| = \langle 1, 2, -2 \rangle \cdot \langle 4, 3, 0 \rangle / 5 = 2$

36. $\mathbf{a} - \mathbf{b} = \langle 1, 2, -2 \rangle - \langle 4, 3, 0 \rangle = \langle -3, -1, -2 \rangle$
   $\text{comp}_\mathbf{b}(\mathbf{a} - \mathbf{b}) = (\mathbf{a} - \mathbf{b}) \cdot \mathbf{b}/\sqrt{16 + 9} = \frac{1}{5}(\mathbf{a} \cdot \mathbf{b} - \mathbf{b} \cdot \mathbf{b}) = \frac{1}{5}[(4 + 6 + 0) - (16 + 9)] = -3$
   $\text{proj}_\mathbf{b}(\mathbf{a} - \mathbf{b}) = [\text{comp}_\mathbf{b}(\mathbf{a} - \mathbf{b})](\mathbf{b}/\|\mathbf{b}\|) = -3\langle \frac{4}{5}, \frac{3}{5}, 0 \rangle = \langle -\frac{12}{5}, -\frac{9}{5}, 0 \rangle$

39. A direction vector of the given line is $\langle 4, -2, 6 \rangle$. A parallel line containing $(7, 3, -5)$ is $(x - 7)/4 = (y - 3)/(-2) = (z + 5)/6$.

42. Vectors in the plane are $\langle 2, 3, 1 \rangle$ and $\langle 1, 0, 2 \rangle$. A normal vector is $\langle 2, 3, 1 \rangle \times \langle 1, 0, 2 \rangle = \langle 6, -3, -3 \rangle = 3\langle 2, -1, -1 \rangle$. An equation of the plane is $2x - y - z = 0$

45. $\mathbf{F} = 10\frac{\mathbf{a}}{\|\mathbf{a}\|} = \frac{10}{\sqrt{2}}(\mathbf{i} + \mathbf{j}) = 5\sqrt{2}\,\mathbf{i} + 5\sqrt{2}\,\mathbf{j}$; $\mathbf{d} = \langle 7, 4, 0 \rangle - \langle 4, 1, 0 \rangle = 3\mathbf{i} + 3\mathbf{j}$

   $W = \mathbf{F} \cdot \mathbf{d} = 15\sqrt{2} + 15\sqrt{2} = 30\sqrt{2}$ N-m

48. Let $\|\mathbf{F}_1\| = F_1$ and $\|\mathbf{F}_2\| = F_2$. Then $\mathbf{F}_1 = F_1[(\cos 45°)\mathbf{i} + (\sin 45°)\mathbf{j}]$ and $\mathbf{F}_2 = F_2[(\cos 120°)\mathbf{i} + (\sin 120°)\mathbf{j}]$, or $\mathbf{F}_1 = F_1(\frac{1}{\sqrt{2}}\mathbf{i} + \frac{1}{\sqrt{2}}\mathbf{j})$ and $\mathbf{F}_2 = F_2(-\frac{1}{2}\mathbf{i} + \frac{\sqrt{3}}{2}\mathbf{j})$. Since $\mathbf{w} + \mathbf{F}_1 + \mathbf{F}_2 = \mathbf{0}$,

$$F_1\left(\frac{1}{\sqrt{2}}\mathbf{i} + \frac{1}{\sqrt{2}}\mathbf{j}\right) + F_2\left(-\frac{1}{2}\mathbf{i} + \frac{\sqrt{3}}{2}\mathbf{j}\right) = 50\mathbf{j}, \qquad \left(\frac{1}{\sqrt{2}}F_1 - \frac{1}{2}F_2\right)\mathbf{i} + \left(\frac{1}{\sqrt{2}}F_1 + \frac{\sqrt{3}}{2}F_2\right)\mathbf{j} = 50\mathbf{j}$$

and

$$\frac{1}{\sqrt{2}}F_1 - \frac{1}{2}F_2 = 0, \qquad \frac{1}{\sqrt{2}}F_1 + \frac{\sqrt{3}}{2}F_2 = 50.$$

Solving, we obtain $F_1 = 25(\sqrt{6} - \sqrt{2}) \approx 25.9$ lb and $F_2 = 50(\sqrt{3} - 1) \approx 36.6$ lb.

51. Let p_1 and p_2 be in P_n such that $\dfrac{d^2p_1}{dx^2} = 0$ and $\dfrac{d^2p_2}{dx^2} = 0$. Since

$$0 = \frac{d^2p_1}{dx^2} + \frac{d^2p_2}{dx^2} = \frac{d^2}{dx^2}(p_1 + p_2) \quad \text{and} \quad 0 = k\frac{d^2p_1}{dx^2} = \frac{d^2}{dx^2}(kp_1)$$

we conclude that the set of polynomials with the given property is a subspace of P_n. A basis for the subspace is $1, x$.

8 Matrices

3. 3×3

6. 8×1

9. Not equal

12. Solving $x^2 = 9$, $y = 4x$ we obtain $x = 3$, $y = 12$ and $x = -3$, $t = -12$.

15. (a) $\mathbf{A} + \mathbf{B} = \begin{pmatrix} 4-2 & 5+6 \\ -6+8 & 9-10 \end{pmatrix} = \begin{pmatrix} 2 & 11 \\ 2 & -1 \end{pmatrix}$

(b) $\mathbf{B} - \mathbf{A} = \begin{pmatrix} -2-4 & 6-5 \\ 8+6 & -10-9 \end{pmatrix} = \begin{pmatrix} -6 & 1 \\ 14 & -19 \end{pmatrix}$

(c) $2\mathbf{A} + 3\mathbf{B} = \begin{pmatrix} 8 & 10 \\ -12 & 18 \end{pmatrix} + \begin{pmatrix} -6 & 18 \\ 24 & -30 \end{pmatrix} = \begin{pmatrix} 2 & 28 \\ 12 & -12 \end{pmatrix}$

18. (a) $\mathbf{AB} = \begin{pmatrix} -4+4 & 6-12 & -3+8 \\ -20+10 & 30-30 & -15+20 \\ -32+12 & 48-36 & -24+24 \end{pmatrix} = \begin{pmatrix} 0 & -6 & 5 \\ -10 & 0 & 5 \\ -20 & 12 & 0 \end{pmatrix}$

(b) $\mathbf{BA} = \begin{pmatrix} -4+30-24 & -16+60-36 \\ 1-15+16 & 4-30+24 \end{pmatrix} = \begin{pmatrix} 2 & 8 \\ 2 & -2 \end{pmatrix}$

21. (a) $\mathbf{A}^T\mathbf{A} = \begin{pmatrix} 4 & 8 & -10 \end{pmatrix} \begin{pmatrix} 4 \\ 8 \\ -10 \end{pmatrix} = (180)$

(b) $\mathbf{B}^T\mathbf{B} = \begin{pmatrix} 2 \\ 4 \\ 5 \end{pmatrix} \begin{pmatrix} 2 & 4 & 5 \end{pmatrix} = \begin{pmatrix} 4 & 8 & 10 \\ 8 & 16 & 20 \\ 10 & 20 & 25 \end{pmatrix}$

(c) $\mathbf{A} + \mathbf{B}^T = \begin{pmatrix} 4 \\ 8 \\ -10 \end{pmatrix} + \begin{pmatrix} 2 \\ 4 \\ 5 \end{pmatrix} = \begin{pmatrix} 6 \\ 12 \\ -5 \end{pmatrix}$

24. (a) $\mathbf{A}^T + \mathbf{B} = \begin{pmatrix} 5 & -4 \\ 9 & 6 \end{pmatrix} + \begin{pmatrix} -3 & 11 \\ -7 & 2 \end{pmatrix} = \begin{pmatrix} 2 & 7 \\ 2 & 8 \end{pmatrix}$

(b) $2\mathbf{A} + \mathbf{B}^T = \begin{pmatrix} 10 & 18 \\ -8 & 12 \end{pmatrix} + \begin{pmatrix} -3 & -7 \\ 11 & 2 \end{pmatrix} = \begin{pmatrix} 7 & 11 \\ 3 & 14 \end{pmatrix}$

145

27. $\begin{pmatrix} -19 \\ 18 \end{pmatrix} - \begin{pmatrix} 19 \\ 20 \end{pmatrix} = \begin{pmatrix} -38 \\ -2 \end{pmatrix}$

30. 3×2

33. $(\mathbf{AB})^T = \begin{pmatrix} 16 & 40 \\ -8 & -20 \end{pmatrix}^T = \begin{pmatrix} 16 & -8 \\ 40 & -20 \end{pmatrix};$ $\quad \mathbf{B}^T\mathbf{A}^T = \begin{pmatrix} 4 & 2 \\ 10 & 5 \end{pmatrix}\begin{pmatrix} 2 & -3 \\ 4 & 2 \end{pmatrix} = \begin{pmatrix} 16 & -8 \\ 40 & -20 \end{pmatrix}$

36. Using Problem 33 we have $(\mathbf{AA}^T)^T = (\mathbf{A}^T)^T\mathbf{A}^T = \mathbf{AA}^T$, so that $\mathbf{AA}^T$ is symmetric.

39. Since $(\mathbf{A}+\mathbf{B})^2 = (\mathbf{A}+\mathbf{B})(\mathbf{A}+\mathbf{B}) \neq \mathbf{A}^2+\mathbf{AB}+\mathbf{BA}+\mathbf{B}^2$, and $\mathbf{AB} \neq \mathbf{BA}$ in general, $(\mathbf{A}+\mathbf{B})^2 \neq \mathbf{A}^2+2\mathbf{AB}+\mathbf{B}^2$.

42. $\begin{pmatrix} 2 & 6 & 1 \\ 1 & 2 & -1 \\ 5 & 7 & -4 \end{pmatrix}\begin{pmatrix} x_1 \\ x_2 \\ x_3 \end{pmatrix} = \begin{pmatrix} 7 \\ -1 \\ 9 \end{pmatrix}$

45. (a) $M_Y\begin{pmatrix} x \\ y \\ z \end{pmatrix} = \begin{pmatrix} \cos\gamma & \sin\gamma & 0 \\ -\sin\gamma & \cos\gamma & 0 \\ 0 & 0 & 1 \end{pmatrix}\begin{pmatrix} x \\ y \\ z \end{pmatrix} = \begin{pmatrix} x\cos\gamma + y\sin\gamma \\ -x\sin\gamma + y\cos\gamma \\ z \end{pmatrix} = \begin{pmatrix} x_Y \\ y_Y \\ z_Y \end{pmatrix}$

(b) $M_R = \begin{pmatrix} \cos\beta & 0 & -\sin\beta \\ 0 & 1 & 0 \\ \sin\beta & 0 & \cos\beta \end{pmatrix};$ $\quad M_P\begin{pmatrix} 1 & 0 & 0 \\ 0 & \cos\alpha & \sin\alpha \\ 0 & -\sin\alpha & \cos\alpha \end{pmatrix}$

(c) $M_P\begin{pmatrix} 1 \\ 1 \\ 1 \end{pmatrix} = \begin{pmatrix} 1 & 0 & 0 \\ 0 & \cos 30° & \sin 30° \\ 0 & -\sin 30° & \cos 30° \end{pmatrix}\begin{pmatrix} 1 \\ 1 \\ 1 \end{pmatrix} = \begin{pmatrix} 1 & 0 & 0 \\ 0 & \frac{\sqrt{3}}{2} & \frac{1}{2} \\ 0 & -\frac{1}{2} & \frac{\sqrt{3}}{2} \end{pmatrix}\begin{pmatrix} 1 \\ 1 \\ 1 \end{pmatrix} = \begin{pmatrix} 1 \\ \frac{1}{2}(\sqrt{3}+1) \\ \frac{1}{2}(\sqrt{3}-1) \end{pmatrix}$

$M_RM_P\begin{pmatrix} 1 \\ 1 \\ 1 \end{pmatrix} = \begin{pmatrix} \cos 45° & 0 & -\sin 45° \\ 0 & 1 & 0 \\ \sin 45° & 0 & \cos 45° \end{pmatrix}\begin{pmatrix} 1 \\ \frac{1}{2}(\sqrt{3}+1) \\ \frac{1}{2}(\sqrt{3}-1) \end{pmatrix} = \begin{pmatrix} \frac{\sqrt{2}}{2} & 0 & -\frac{\sqrt{2}}{2} \\ 0 & 1 & 0 \\ \frac{\sqrt{2}}{2} & 0 & \frac{\sqrt{2}}{2} \end{pmatrix}\begin{pmatrix} 1 \\ \frac{1}{2}(\sqrt{3}+1) \\ \frac{1}{2}(\sqrt{3}-1) \end{pmatrix}$

$= \begin{pmatrix} \frac{1}{4}(3\sqrt{2}-\sqrt{6}) \\ \frac{1}{2}(\sqrt{3}+1) \\ \frac{1}{4}(\sqrt{2}+\sqrt{6}) \end{pmatrix}$

$M_YM_RM_P\begin{pmatrix} 1 \\ 1 \\ 1 \end{pmatrix} = \begin{pmatrix} \cos 60° & \sin 60° & 0 \\ -\sin 60° & \cos 60° & 0 \\ 0 & 0 & 1 \end{pmatrix}\begin{pmatrix} \frac{1}{4}(3\sqrt{2}-\sqrt{6}) \\ \frac{1}{2}(\sqrt{3}+1) \\ \frac{1}{4}(\sqrt{2}+\sqrt{6}) \end{pmatrix}$

$= \begin{pmatrix} \frac{1}{2} & \frac{\sqrt{3}}{2} & 0 \\ -\frac{\sqrt{3}}{2} & \frac{1}{2} & 0 \\ 0 & 0 & 1 \end{pmatrix}\begin{pmatrix} \frac{1}{4}(3\sqrt{2}-\sqrt{6}) \\ \frac{1}{2}(\sqrt{3}+1) \\ \frac{1}{4}(\sqrt{2}+\sqrt{6}) \end{pmatrix} = \begin{pmatrix} \frac{1}{8}(3\sqrt{2}-\sqrt{6}+6+2\sqrt{3}) \\ \frac{1}{8}(-3\sqrt{6}+3\sqrt{2}+2\sqrt{3}+2) \\ \frac{1}{4}(\sqrt{2}+\sqrt{6}) \end{pmatrix}$

EXERCISES 8.2

Systems of Linear Algebraic Equations

3. $\begin{pmatrix} 9 & 3 & | & -5 \\ 2 & -1 & | & -1 \end{pmatrix} \xrightarrow{\frac{1}{9}R_1} \begin{pmatrix} 1 & \frac{1}{3} & | & -\frac{5}{9} \\ 2 & 1 & | & -1 \end{pmatrix} \xrightarrow{-2R_1+R_2} \begin{pmatrix} 1 & \frac{1}{3} & | & -\frac{5}{9} \\ 0 & \frac{1}{3} & | & \frac{1}{9} \end{pmatrix} \xrightarrow{3R_2} \begin{pmatrix} 1 & \frac{1}{3} & | & -\frac{5}{9} \\ 0 & 1 & | & \frac{1}{3} \end{pmatrix}$

$\xrightarrow{-\frac{1}{3}R_2+R_1} \begin{pmatrix} 1 & 0 & | & -\frac{2}{3} \\ 0 & 1 & | & \frac{1}{3} \end{pmatrix}$

The solution is $x_1 = -\frac{2}{3}$, $x_2 = \frac{1}{3}$.

6. $\begin{pmatrix} 1 & 2 & -1 & | & 0 \\ 2 & 1 & 2 & | & 9 \\ 1 & -1 & 1 & | & 3 \end{pmatrix} \xrightarrow[-R_1+R_3]{-2R_1+R_2} \begin{pmatrix} 1 & 2 & -1 & | & 0 \\ 0 & -3 & 4 & | & 9 \\ 0 & -3 & 2 & | & 3 \end{pmatrix} \xrightarrow{-\frac{1}{3}R_2} \begin{pmatrix} 1 & 2 & -1 & | & 0 \\ 0 & 1 & -\frac{4}{3} & | & -3 \\ 0 & -3 & 2 & | & 3 \end{pmatrix}$

$\xrightarrow[3R_2+R_3]{-2R_2+R_1} \begin{pmatrix} 1 & 0 & \frac{5}{3} & | & 6 \\ 0 & 1 & -\frac{4}{3} & | & -3 \\ 0 & 0 & -2 & | & -6 \end{pmatrix} \xrightarrow{-\frac{1}{2}R_3} \begin{pmatrix} 1 & 0 & \frac{5}{3} & | & 6 \\ 0 & 1 & -\frac{4}{3} & | & -3 \\ 0 & 0 & 1 & | & 3 \end{pmatrix} \xrightarrow[\frac{4}{3}R_3+R_2]{-\frac{5}{3}R_3+R_1} \begin{pmatrix} 1 & 0 & 0 & | & 1 \\ 0 & 1 & 0 & | & 1 \\ 0 & 0 & 1 & | & 3 \end{pmatrix}$

The solution is $x_1 = 1$, $x_2 = 1$, $x_3 = 3$.

9. $\begin{pmatrix} 1 & -1 & -1 & | & 8 \\ 1 & -1 & 1 & | & 3 \\ -1 & 1 & 1 & | & 4 \end{pmatrix} \xrightarrow[\text{operations}]{\text{row}} \begin{pmatrix} 1 & -1 & -1 & | & 8 \\ 0 & 0 & 2 & | & -5 \\ 0 & 0 & 0 & | & 12 \end{pmatrix}$

Since the bottom row implies $0 = 12$, the system is inconsistent.

12. $\begin{pmatrix} 1 & -1 & -2 & | & 0 \\ 2 & 4 & 5 & | & 0 \\ 6 & 0 & -3 & | & 0 \end{pmatrix} \xrightarrow[\text{operations}]{\text{row}} \begin{pmatrix} 1 & -1 & -2 & | & 0 \\ 0 & 1 & \frac{3}{2} & | & 0 \\ 0 & 0 & 0 & | & 0 \end{pmatrix}$

The solution is $x_1 = \frac{1}{2}t$, $x_2 = -\frac{3}{2}t$, $x_3 = t$.

15. $\begin{pmatrix} 1 & 1 & 1 & | & 3 \\ 1 & -1 & -1 & | & -1 \\ 3 & 1 & 1 & | & 5 \end{pmatrix} \xrightarrow[\text{operations}]{\text{row}} \begin{pmatrix} 1 & 1 & 1 & | & 3 \\ 0 & 1 & 1 & | & 2 \\ 0 & 0 & 0 & | & 0 \end{pmatrix}$

If $x_3 = t$ the solution is $x_1 = 1$, $x_2 = 2 - t$, $x_3 = t$.

18. $\begin{pmatrix} 2 & 1 & 1 & 0 & | & 3 \\ 3 & 1 & 1 & 1 & | & 4 \\ 1 & 2 & 2 & 3 & | & 3 \\ 4 & 5 & -2 & 1 & | & 16 \end{pmatrix} \xrightarrow[\text{operations}]{\text{row}} \begin{pmatrix} 1 & \frac{1}{2} & \frac{1}{2} & 0 & | & \frac{3}{2} \\ 0 & 1 & 1 & -2 & | & 1 \\ 0 & 0 & 1 & -1 & | & -1 \\ 0 & 0 & 0 & 1 & | & 0 \end{pmatrix}$

The solution is $x_1 = 1$, $x_2 = 2$, $x_3 = -1$, $x_4 = 0$.

21. $\begin{pmatrix} 1 & 1 & 1 & | & 4.280 \\ 0.2 & -0.1 & -0.5 & | & -1.978 \\ 4.1 & 0.3 & 0.12 & | & 1.686 \end{pmatrix} \xrightarrow[\text{operations}]{\text{row}} \begin{pmatrix} 1 & 1 & 1 & | & 4.28 \\ 0 & 1 & 2.333 & | & 9.447 \\ 0 & 0 & 1 & | & 4.1 \end{pmatrix}$

The solution is $x_1 = 0.3$, $x_2 = -0.12$, $x_3 = 4.1$.

147

24. From $x_1KClO_3 \rightarrow x_2KCl + x_3O_2$ we obtain the system $x_1 = x_2$, $x_1 = x_2$, $3x_1 = 2x_3$. Letting $x_3 = t$ we see that a solution of the system is $x_1 = x_2 = \frac{2}{3}t$, $x_3 = t$. Taking $t = 3$ we obtain the balanced equation

$$2KClO_3 \rightarrow 2KCl + 3O_2.$$

27. From $x_1Cu + x_2HNO_3 \rightarrow x_3Cu(NO_3)_2 + x_4H_2O + x_5NO$ we obtain the system

$$x_1 = 3, \qquad x_2 = 2x_4, \qquad x_2 = 2x_3 + x_5, \qquad 3x_2 = 6x_3 + x_4 + x_5.$$

Letting $x_4 = t$ we see that $x_2 = 2t$ and

$$
\begin{aligned}
2t &= 2x_3 + x_5 \\
6t &= 6x_3 + t + x_5
\end{aligned}
\qquad \text{or} \qquad
\begin{aligned}
2x_3 + x_5 &= 2t \\
6x_3 + x_5 &= 5t.
\end{aligned}
$$

Then $x_3 = \frac{3}{4}t$ and $x_5 = \frac{1}{2}t$. Finally, $x_1 = x_3 = \frac{3}{4}t$. Taking $t = 4$ we obtain the balanced equation

$$3Cu + 8HNO_3 \rightarrow 3Cu(NO_3)_2 + 4H_2O + 2NO.$$

30. The system of equations is

$$
\begin{aligned}
i_1 - i_2 - i_3 &= 0 \\
52 - i_1 - 5i_2 &= 0 \\
-10i_3 + 5i_2 &= 0
\end{aligned}
\qquad \text{or} \qquad
\begin{aligned}
i_1 - i_2 - i_3 &= 0 \\
i_1 + 5i_2 &= 52 \\
5i_2 - 10i_3 &= 0
\end{aligned}
$$

Gaussian elimination gives

$$
\begin{pmatrix}
1 & -1 & -1 & | & 0 \\
1 & 5 & 0 & | & 52 \\
0 & 5 & -10 & | & 0
\end{pmatrix}
\xrightarrow[\text{operations}]{\text{row}}
\begin{pmatrix}
1 & -1 & -1 & | & 0 \\
0 & 1 & 1/6 & | & 26/3 \\
0 & 0 & 1 & | & 4
\end{pmatrix}.
$$

The solution is $i_1 = 12$, $i_2 = 8$, $i_3 = 4$.

33. Add c times row 2 to row 3 in $\mathbf{I}_3$.

36. $\mathbf{EA} = \begin{pmatrix} a_{11} & a_{12} & a_{13} \\ a_{21} & a_{22} & a_{23} \\ ca_{31} & ca_{32} & ca_{33} \end{pmatrix}$

39. The system is equivalent to

$$
\begin{pmatrix} 1 & 0 \\ \frac{1}{2} & 1 \end{pmatrix}
\begin{pmatrix} 2 & -2 \\ 0 & 3 \end{pmatrix}
\mathbf{X} = \begin{pmatrix} 2 \\ 6 \end{pmatrix}.
$$

Letting

$$
\mathbf{Y} = \begin{pmatrix} y_1 \\ y_2 \end{pmatrix} = \begin{pmatrix} 2 & -2 \\ 0 & 3 \end{pmatrix} \mathbf{X}
$$

we have

$$
\begin{pmatrix} 1 & 0 \\ \frac{1}{2} & 1 \end{pmatrix}
\begin{pmatrix} y_1 \\ y_2 \end{pmatrix} = \begin{pmatrix} 2 \\ 6 \end{pmatrix}.
$$

This implies $y_1 = 2$ and $\frac{1}{2}y_1 + y_2 = 1 + y_2 = 6$ or $y_2 = 5$. Then

$$
\begin{pmatrix} 2 & -2 \\ 0 & 3 \end{pmatrix}
\begin{pmatrix} x_1 \\ x_2 \end{pmatrix} = \begin{pmatrix} 2 \\ 5 \end{pmatrix},
$$

which implies $3x_2 = 5$ or $x_2 = \frac{5}{3}$ and $2x_1 - 2x_2 = 2x_1 - \frac{10}{3} = 2$ or $x_1 = \frac{8}{3}$. The solution is $\mathbf{X} = \left(\frac{8}{3}, \frac{5}{3}\right)$.

42. The system is equivalent to

$$\begin{pmatrix} 1 & 0 & 0 \\ 3 & 1 & 0 \\ 1 & 1 & 1 \end{pmatrix} \begin{pmatrix} 1 & 1 & 1 \\ 0 & -2 & -1 \\ 0 & 0 & 1 \end{pmatrix} \mathbf{X} = \begin{pmatrix} 0 \\ 1 \\ 4 \end{pmatrix}.$$

Letting

$$\mathbf{Y} = \begin{pmatrix} y_1 \\ y_2 \\ y_3 \end{pmatrix} = \begin{pmatrix} 1 & 1 & 1 \\ 0 & -2 & -1 \\ 0 & 0 & 1 \end{pmatrix} \mathbf{X}$$

we have

$$\begin{pmatrix} 1 & 0 & 0 \\ 3 & 1 & 0 \\ 1 & 1 & 1 \end{pmatrix} \begin{pmatrix} y_1 \\ y_2 \\ y_3 \end{pmatrix} = \begin{pmatrix} 0 \\ 1 \\ 4 \end{pmatrix}.$$

This implies $y_1 = 0$, $3y_1 + y_2 = y_2 = 1$, and $y_1 + y_2 + y_3 = 0 + 1 + y_3 = 4$ or $y_3 = 3$. Then

$$\begin{pmatrix} 1 & 1 & 1 \\ 0 & -2 & -1 \\ 0 & 0 & 1 \end{pmatrix} \begin{pmatrix} x_1 \\ x_2 \\ x_3 \end{pmatrix} = \begin{pmatrix} 0 \\ 1 \\ 3 \end{pmatrix},$$

which implies $x_3 = 3$, $-2x_2 - x_3 = -2x_2 - 3 = 1$ or $x_2 = -2$, and $x_1 + x_2 + x_3 = x_1 - 2 + 3 = 0$ or $x_1 = -1$. The solution is $\mathbf{X} = (-1, -2, 3)$.

45. Using the **Solve** function in *Mathematica* we find $x_1 = -3.76993$, $x_2 = -1.09071$, $x_3 = -4.50461$, $x_4 = -3.12221$

EXERCISES 8.3

Rank of a Matrix

3. $\begin{pmatrix} 2 & 1 & 3 \\ 6 & 3 & 9 \\ -1 & -\frac{1}{2} & -\frac{3}{2} \end{pmatrix} \xrightarrow[\text{operations}]{\text{row}} \begin{pmatrix} 1 & \frac{1}{2} & \frac{3}{2} \\ 0 & 0 & 0 \\ 0 & 0 & 0 \end{pmatrix};$ The rank is 1.

6. $\begin{pmatrix} 3 & -1 & 2 & 0 \\ 6 & 2 & 4 & 5 \end{pmatrix} \xrightarrow[\text{operations}]{\text{row}} \begin{pmatrix} 1 & -\frac{1}{3} & \frac{2}{3} & 0 \\ 0 & 1 & 0 & \frac{5}{4} \end{pmatrix};$ The rank is 2.

9. $\begin{pmatrix} 0 & 2 & 4 & 2 & 2 \\ 4 & 1 & 0 & 5 & 1 \\ 2 & 1 & \frac{2}{3} & 3 & \frac{1}{3} \\ 6 & 6 & 6 & 12 & 0 \end{pmatrix} \xrightarrow[\text{operations}]{\text{row}} \begin{pmatrix} 1 & \frac{1}{2} & \frac{1}{3} & \frac{3}{2} & \frac{1}{6} \\ 0 & 1 & \frac{4}{3} & 1 & -\frac{1}{3} \\ 0 & 0 & 1 & 0 & 2 \\ 0 & 0 & 0 & 0 & 0 \end{pmatrix};$ The rank is 3.

12. $\begin{pmatrix} 2 & 6 & 3 \\ 1 & -1 & 4 \\ 3 & 2 & 1 \\ 2 & 5 & 4 \end{pmatrix} \xrightarrow[\text{operations}]{\text{row}} \begin{pmatrix} 1 & -1 & 4 \\ 0 & 1 & -\frac{5}{8} \\ 0 & 0 & 1 \\ 0 & 0 & 0 \end{pmatrix}$

Since the rank of the matrix is 3 and there are 4 vectors, the vectors are linearly dependent.

15. Since the number of unknowns is $n = 8$ and the rank of the coefficient matrix is $r = 3$, the solution of the system has $n - r = 5$ parameters.

18. Since the rank of $\mathbf{A}$ is $r = 3$ and the number of equations is $n = 6$, the solution of the system has $n - r = 3$ parameters. Thus, the solution of the system is not unique.

EXERCISES 8.4
Determinants

3. $C_{13} = (-1)^{1+3} \begin{vmatrix} 1 & -1 \\ -2 & 3 \end{vmatrix} = 1$

6. $M_{41} = \begin{vmatrix} 2 & 4 & 0 \\ 2 & -2 & 3 \\ 1 & 0 & -1 \end{vmatrix} = 24$

9. -7

12. $-1/2$

15. $\begin{vmatrix} 0 & 2 & 0 \\ 3 & 0 & 1 \\ 0 & 5 & 8 \end{vmatrix} = -3 \begin{vmatrix} 2 & 0 \\ 5 & 8 \end{vmatrix} = -48$

18. $\begin{vmatrix} 1 & -1 & -1 \\ 2 & 2 & -2 \\ 1 & 1 & 9 \end{vmatrix} = \begin{vmatrix} 2 & -2 \\ 1 & 9 \end{vmatrix} - 2 \begin{vmatrix} -1 & -1 \\ 1 & 9 \end{vmatrix} + \begin{vmatrix} -1 & -1 \\ 2 & -2 \end{vmatrix} = 20 - 2(-8) + 4 = 40$

21. $\begin{vmatrix} -2 & -1 & 4 \\ -3 & 6 & 1 \\ -3 & 4 & 8 \end{vmatrix} = -2 \begin{vmatrix} 6 & 1 \\ 4 & 8 \end{vmatrix} + 3 \begin{vmatrix} -1 & 4 \\ 4 & 8 \end{vmatrix} - 3 \begin{vmatrix} -1 & 4 \\ 6 & 1 \end{vmatrix} = -2(44) + 3(-24) - 3(-25) = -85$

24. $\begin{vmatrix} 1 & 1 & 1 \\ x & y & z \\ 2+x & 3+y & 4+z \end{vmatrix} = \begin{vmatrix} y & z \\ 3+y & 4+z \end{vmatrix} - \begin{vmatrix} x & z \\ 2+x & 4+z \end{vmatrix} + \begin{vmatrix} x & y \\ 2+x & 3+y \end{vmatrix}$

$$= (4y + yz - 3z - yz) - (4x + xz - 2z - xz) + (3x + xy - 2y - xy) = -x + 2y - z$$

27. Expanding along the first column in the original matrix and each succeeding minor, we obtain $3(1)(2)(4)(2) = 48$.

30. Solving $-\lambda^3 + 3\lambda^2 - 2\lambda = -\lambda(\lambda - 2)(\lambda - 1) = 0$ we obtain $\lambda = 0, 1$, and 2.

EXERCISES 8.5

Properties of Determinants

3. Theorem 8.14

6. Theorem 8.11 (twice)

9. Theorem 8.8

12. $\det \mathbf{B} = 2(3)(5) = 30$

15. $\det \mathbf{A} = 6(\frac{2}{3})(-4)(-5) = 80$

18. $\det \mathbf{D} = 4(7)(-2) = -56$

21. $\det \mathbf{AB} = \begin{vmatrix} 0 & -2 & 2 \\ 10 & 7 & 23 \\ 8 & 4 & 16 \end{vmatrix} = -80 = 20(-4) = \det \mathbf{A} \det \mathbf{B}$

24. Using Theorems 8.14 and 8.9,

$$\det \mathbf{A} = \begin{vmatrix} 1 & 1 & 1 \\ x & y & z \\ x+y+z & x+y+z & x+y+z \end{vmatrix} = (x+y+z) \begin{vmatrix} 1 & 1 & 1 \\ x & y & z \\ 1 & 1 & 1 \end{vmatrix} = 0.$$

27. $\begin{vmatrix} -1 & 2 & 3 \\ 4 & -5 & -2 \\ 9 & -9 & 6 \end{vmatrix} = \begin{vmatrix} -1 & 2 & 3 \\ 0 & 3 & 10 \\ 0 & 9 & 33 \end{vmatrix} = \begin{vmatrix} -1 & 2 & 3 \\ 0 & 3 & 10 \\ 0 & 0 & 3 \end{vmatrix} = -1(3)(3) = -9$

30. $\begin{vmatrix} 0 & 1 & 4 & 5 \\ 2 & 5 & 0 & 1 \\ 1 & 2 & 2 & 0 \\ 3 & 1 & 3 & 2 \end{vmatrix} = -\begin{vmatrix} 1 & 2 & 2 & 0 \\ 2 & 5 & 0 & 1 \\ 0 & 1 & 4 & 5 \\ 3 & 1 & 3 & 2 \end{vmatrix} = -\begin{vmatrix} 1 & 2 & 2 & 0 \\ 0 & 1 & -4 & 1 \\ 0 & 1 & 4 & 5 \\ 0 & -5 & -3 & 2 \end{vmatrix} = -\begin{vmatrix} 1 & 2 & 2 & 0 \\ 0 & 1 & -4 & 1 \\ 0 & 0 & 8 & 4 \\ 0 & 0 & -23 & 7 \end{vmatrix} = -\begin{vmatrix} 1 & 2 & 2 & 0 \\ 0 & 1 & -4 & 1 \\ 0 & 0 & 8 & 4 \\ 0 & 0 & -23 & \frac{37}{2} \end{vmatrix}$

$$= -(1)(1)(8)(\frac{37}{2}) = -148$$

33. We first use the second row to reduce the third row. Then we use the first row to reduce the second row.

$$\begin{vmatrix} 1 & 1 & 1 \\ a & b & c \\ 0 & b^2 - ab & c^2 - ac \end{vmatrix} = \begin{vmatrix} 1 & 1 & 1 \\ 0 & b-a & c-a \\ 0 & b(b-a) & c(c-a) \end{vmatrix} = (b-a)(c-a) \begin{vmatrix} 1 & 1 & 1 \\ 0 & 1 & 1 \\ 0 & b & c \end{vmatrix}.$$

Expanding along the first row gives $(b-a)(c-a)(c-b)$.

36. Since $C_{11} + -7$, $C_{12} = -8$, and $C_{13} = -10$ we have $a_{21}C_{11} + a_{22}C_{12} + a_{23}C_{13} = -2(-7) + 3(-8) - 1(-10) = 0$. Since $C_{12} = -8$, $C_{22} = -19$, and $C_{32} = -7$ we have $a_{13}C_{12} + a_{23}C_{22} + a_{33}C_{32} = 5(-8) - 1(-19) - 3(-7) = 0$.

39. Factoring -1 out of each row we see that $\det(-\mathbf{A}) = (-1)^5 \det \mathbf{A} = -\det \mathbf{A}$. Then $-\det \mathbf{A} = \det(-\mathbf{A}) = \det \mathbf{A}^T = \det \mathbf{A}$ and $\det \mathbf{A} = 0$.

EXERCISES 8.6

Inverse of a Matrix

3. $\det \mathbf{A} = 9.$ $\mathbf{A}$ is nonsingular. $\mathbf{A}^{-1} = \dfrac{1}{9}\begin{pmatrix} 1 & 1 \\ -4 & 5 \end{pmatrix} = \begin{pmatrix} \frac{1}{9} & \frac{1}{9} \\ -\frac{4}{9} & \frac{5}{9} \end{pmatrix}$

6. $\det \mathbf{A} = -3\pi^2.$ $\mathbf{A}$ is nonsingular. $\mathbf{A}^{-1} = -\dfrac{1}{3\pi^2}\begin{pmatrix} \pi & \pi \\ \pi & -2\pi \end{pmatrix} = \begin{pmatrix} -\frac{1}{3\pi} & -\frac{1}{3\pi} \\ -\frac{1}{3\pi} & \frac{2}{3\pi} \end{pmatrix}$

9. $\det \mathbf{A} = -30.$ $\mathbf{A}$ is nonsingular. $\mathbf{A}^{-1} = -\dfrac{1}{30}\begin{pmatrix} -14 & 13 & 16 \\ -2 & 4 & -2 \\ -4 & -7 & -4 \end{pmatrix} = \begin{pmatrix} \frac{7}{15} & -\frac{13}{30} & -\frac{8}{15} \\ \frac{1}{15} & -\frac{2}{15} & \frac{1}{15} \\ \frac{2}{15} & \frac{7}{30} & \frac{2}{15} \end{pmatrix}$

12. $\det \mathbf{A} = 16.$ $\mathbf{A}$ is nonsingular. $\mathbf{A}^{-1} = \dfrac{1}{16}\begin{pmatrix} 0 & 0 & 2 \\ 8 & 0 & 0 \\ 0 & 16 & 0 \end{pmatrix} = \begin{pmatrix} 0 & 0 & \frac{1}{8} \\ \frac{1}{2} & 0 & 0 \\ 0 & 1 & 0 \end{pmatrix}$

15. $\begin{pmatrix} 6 & -2 & | & 1 & 0 \\ 0 & 4 & | & 0 & 1 \end{pmatrix} \xrightarrow[\frac{1}{4}R_2]{\frac{1}{6}R_1} \begin{pmatrix} 1 & -\frac{1}{3} & | & \frac{1}{6} & 0 \\ 0 & 1 & | & 0 & \frac{1}{4} \end{pmatrix} \xrightarrow{\frac{1}{3}R_2 + R_1} \begin{pmatrix} 1 & 0 & | & \frac{1}{6} & \frac{1}{12} \\ 0 & 1 & | & 0 & \frac{1}{4} \end{pmatrix};$ $\mathbf{A}^{-1} = \begin{pmatrix} \frac{1}{6} & \frac{1}{12} \\ 0 & \frac{1}{4} \end{pmatrix}$

18. $\begin{pmatrix} 2 & -3 & | & 1 & 0 \\ -2 & 4 & | & 0 & 1 \end{pmatrix} \xrightarrow{\frac{1}{2}R_1} \begin{pmatrix} 1 & -\frac{3}{2} & | & \frac{1}{2} & 0 \\ -2 & 4 & | & 0 & 1 \end{pmatrix} \xrightarrow{2R_1 + R_2} \begin{pmatrix} 1 & -\frac{3}{2} & | & \frac{1}{2} & 0 \\ 0 & 1 & | & 1 & 1 \end{pmatrix} \xrightarrow{\frac{3}{2}R_2 + R_1} \begin{pmatrix} 1 & 0 & | & 2 & \frac{3}{2} \\ 0 & 1 & | & 1 & 1 \end{pmatrix};$

$\mathbf{A}^{-1} = \begin{pmatrix} 2 & \frac{3}{2} \\ 1 & 1 \end{pmatrix}$

21. $\begin{pmatrix} 4 & 2 & 3 & | & 1 & 0 & 0 \\ 2 & 1 & 0 & | & 0 & 1 & 0 \\ -1 & -2 & 0 & | & 0 & 0 & 1 \end{pmatrix} \xrightarrow{R_{13}} \begin{pmatrix} -1 & -2 & 0 & | & 0 & 0 & 1 \\ 2 & 1 & 0 & | & 0 & 1 & 0 \\ 4 & 2 & 3 & | & 1 & 0 & 0 \end{pmatrix} \xrightarrow[\text{operations}]{\text{row}} \begin{pmatrix} 1 & 0 & 0 & | & 0 & \frac{2}{3} & \frac{1}{3} \\ 0 & 1 & 0 & | & 0 & -\frac{1}{3} & -\frac{2}{3} \\ 0 & 0 & 1 & | & \frac{1}{3} & -\frac{2}{3} & 0 \end{pmatrix};$

$\mathbf{A}^{-1} = \begin{pmatrix} 0 & \frac{2}{3} & \frac{1}{3} \\ 0 & -\frac{1}{3} & -\frac{2}{3} \\ \frac{1}{3} & -\frac{2}{3} & 0 \end{pmatrix}$

24. $\begin{pmatrix} 1 & 2 & 3 & | & 1 & 0 & 0 \\ 0 & 1 & 4 & | & 0 & 1 & 0 \\ 0 & 0 & 8 & | & 0 & 0 & 1 \end{pmatrix} \xrightarrow[\text{operations}]{\text{row}} \begin{pmatrix} 1 & 0 & 0 & | & 1 & -2 & \frac{5}{8} \\ 0 & 1 & 0 & | & 0 & 1 & -\frac{1}{2} \\ 0 & 0 & 1 & | & 0 & 0 & \frac{1}{8} \end{pmatrix};$ $\mathbf{A}^{-1} = \begin{pmatrix} 1 & -2 & \frac{5}{8} \\ 0 & 1 & -\frac{1}{2} \\ 0 & 0 & \frac{1}{8} \end{pmatrix}$

27. $(\mathbf{AB})^{-1} = \mathbf{B}^{-1}\mathbf{A}^{-1} = \begin{pmatrix} -\frac{1}{3} & \frac{1}{3} \\ -1 & \frac{10}{3} \end{pmatrix}$

30. $\mathbf{A}^T = \begin{pmatrix} 1 & 2 \\ 4 & 10 \end{pmatrix};$ $(\mathbf{A}^T)^{-1} = \begin{pmatrix} 5 & -1 \\ -2 & \frac{1}{2} \end{pmatrix};$ $\mathbf{A}^{-1} = \begin{pmatrix} 5 & -2 \\ -1 & \frac{1}{2} \end{pmatrix};$ $(\mathbf{A}^{-1})^T = \begin{pmatrix} 5 & -1 \\ -2 & \frac{1}{2} \end{pmatrix}$

33. (a) $\mathbf{A}^T = \begin{pmatrix} \sin\theta & -\cos\theta \\ \cos\theta & \sin\theta \end{pmatrix} = \mathbf{A}^{-1}$
   (b) $\mathbf{A}^T = \begin{pmatrix} \frac{1}{\sqrt{3}} & \frac{1}{\sqrt{3}} & \frac{1}{\sqrt{3}} \\ 0 & \frac{1}{\sqrt{2}} & -\frac{1}{\sqrt{2}} \\ -\frac{2}{\sqrt{6}} & \frac{1}{\sqrt{6}} & \frac{1}{\sqrt{6}} \end{pmatrix} = \mathbf{A}^{-1}$

36. Suppose **A** is singular. Then $\det \mathbf{A} = 0$, $\det \mathbf{AB} = \det \mathbf{A} \cdot \det \mathbf{B} = 0$, and **AB** is singular.

39. If **A** is nonsingular, then $\mathbf{A}^{-1}$ exists, and $\mathbf{AB} = \mathbf{0}$ implies $\mathbf{A}^{-1}\mathbf{AB} = \mathbf{A}^{-1}\mathbf{0}$, so $\mathbf{B} = \mathbf{0}$.

42. **A** is nonsingular if $a_{11}a_{22}a_{33} = 0$ or a_{11}, a_{22}, and a_{33} are all nonzero.

$$\mathbf{A}^{-1} = \begin{pmatrix} 1/a_{11} & 0 & 0 \\ 0 & 1/a_{22} & 0 \\ 0 & 0 & 1/a_{33} \end{pmatrix}$$

For any diagonal matrix, the inverse matrix is obtaining by taking the reciprocals of the diagonal entries and leaving all other entries 0.

45. $\mathbf{A}^{-1} = \begin{pmatrix} \frac{1}{16} & \frac{3}{8} \\ -\frac{1}{8} & \frac{1}{4} \end{pmatrix}$; $\mathbf{A}^{-1}\begin{pmatrix} 6 \\ 1 \end{pmatrix} = \begin{pmatrix} \frac{3}{4} \\ -\frac{1}{2} \end{pmatrix}$; $x_1 = \frac{3}{4}$, $x_2 = -\frac{1}{2}$

48. $\mathbf{A}^{-1} = \begin{pmatrix} \frac{5}{12} & -\frac{1}{12} & \frac{1}{4} \\ -\frac{2}{3} & \frac{1}{3} & 0 \\ -\frac{1}{12} & \frac{5}{12} & -\frac{1}{4} \end{pmatrix}$; $\mathbf{A}^{-1}\begin{pmatrix} 1 \\ 2 \\ -3 \end{pmatrix} = \begin{pmatrix} -\frac{1}{2} \\ 0 \\ \frac{3}{2} \end{pmatrix}$; $x_1 = -\frac{1}{2}$, $x_2 = 0$, $x_3 = \frac{3}{2}$

51. $\begin{pmatrix} 7 & -2 \\ 3 & 2 \end{pmatrix}\begin{pmatrix} x_1 \\ x_2 \end{pmatrix} = \begin{pmatrix} b_1 \\ b_2 \end{pmatrix}$; $\mathbf{A}^{-1} = \begin{pmatrix} \frac{1}{10} & \frac{1}{10} \\ -\frac{3}{20} & \frac{7}{20} \end{pmatrix}$; $\mathbf{X} = \mathbf{A}^{-1}\begin{pmatrix} 5 \\ 4 \end{pmatrix} = \begin{pmatrix} \frac{9}{10} \\ \frac{13}{20} \end{pmatrix}$; $\mathbf{X} = \mathbf{A}^{-1}\begin{pmatrix} 10 \\ 50 \end{pmatrix} = \begin{pmatrix} 6 \\ 16 \end{pmatrix}$;

$\mathbf{X} = \mathbf{A}^{-1}\begin{pmatrix} 0 \\ -20 \end{pmatrix} = \begin{pmatrix} -2 \\ -7 \end{pmatrix}$

54. $\det \mathbf{A} = 0$, so the system has a nontrivial solution.

57. (a) $\begin{pmatrix} 1 & 1 & 1 \\ -R_1 & R_2 & 0 \\ 0 & -R_2 & R_3 \end{pmatrix}\begin{pmatrix} i_1 \\ i_2 \\ i_3 \end{pmatrix} = \begin{pmatrix} 0 \\ E_2 - E_1 \\ E_3 - E_2 \end{pmatrix}$

(b) $\det \mathbf{A} = R_1R_2 + R_1R_3 + R_2R_3 > 0$, so **A** is nonsingular.

(c) $\mathbf{A}^{-1} = \frac{1}{R_1R_2 + R_1R_3 + R_2R_3}\begin{pmatrix} R_2R_3 & -R_2 - R_3 & -R_2 \\ R_1R_3 & R_3 & -R_1 \\ R_1R_2 & R_2 & R_1 + R_2 \end{pmatrix}$;

$\mathbf{A}^{-1}\begin{pmatrix} 0 \\ E_2 - E_1 \\ E_3 - E_2 \end{pmatrix} = \frac{1}{R_1R_2 + R_1R_3 + R_2R_3}\begin{pmatrix} R_2E_1 - R_2E_3 + R_3E_1 - R_3E_2 \\ R_1E_2 - R_1E_3 - R_3E_1 + R_3E_2 \\ -R_1E_2 + R_1E_3 - R_2E_1 + R_2E_3 \end{pmatrix}$

153

EXERCISES 8.7

Cramer's Rule

3. $\det \mathbf{A} = 0.3$, $\det \mathbf{A}_1 = 0.03$, $\det \mathbf{A}_2 = -0.09$; $x_1 = \frac{0.03}{0.3} = 0.1$, $x_2 = \frac{-0.09}{0.3} = -0.3$

6. $\det \mathbf{A} = -70$, $\det \mathbf{A}_1 = -14$, $\det \mathbf{A}_2 = 35$; $r = \frac{-14}{-70} = \frac{1}{5}$, $s = \frac{35}{-70} = -\frac{1}{2}$

9. $\det \mathbf{A} = -12$, $\det \mathbf{A}_1 = -48$, $\det \mathbf{A}_2 = -18$, $\det \mathbf{A}_3 = -12$; $u = \frac{48}{12} = 4$, $v = \frac{18}{12} = \frac{3}{2}$, $w = 1$

12. (a) $\det \mathbf{A} = \epsilon - 1$, $\det \mathbf{A}_1 = \epsilon - 2$, $\det \mathbf{A}_2 = 1$; $x_1 = \dfrac{\epsilon - 2}{\epsilon - 1} = \dfrac{\epsilon - 1 - 1}{\epsilon - 1} = 1 - \dfrac{1}{\epsilon - 1}$, $x_2 = \dfrac{1}{\epsilon - 1}$

(b) When $\epsilon = 1.01$, $x_1 = -99$ and $x_2 = 100$. When $\epsilon = 0.99$, $x_1 = 101$ and $x_2 = -100$.

15. The system is

$$i_1 + i_2 - i_3 = 0$$

$$r_1 i_1 - r_2 i_2 = E_1 - E_2$$

$$r_2 i_2 + R i_3 = E_2$$

$\det \mathbf{A} = -r_1 R - r_2 R - r_1 r_2$, $\det A_3 = -r_1 E_2$, $-r_2 E_1$; $i_3 = \dfrac{r_1 E_2 + r_2 E_1}{r_1 R + r_2 R + r_1 r_2}$

EXERCISES 8.8

The Eigenvalue Problem

3. $\mathbf{K}_3$ since $\begin{pmatrix} 6 & 3 \\ 2 & 1 \end{pmatrix} \begin{pmatrix} -5 \\ 10 \end{pmatrix} = \begin{pmatrix} 0 \\ 0 \end{pmatrix} = 0 \begin{pmatrix} -5 \\ 10 \end{pmatrix}$; $\quad \lambda = 0$

6. $\mathbf{K}_2$ since $\begin{pmatrix} -1 & 1 & 0 \\ 1 & 2 & 1 \\ 0 & 3 & -1 \end{pmatrix} \begin{pmatrix} 1 \\ 4 \\ 3 \end{pmatrix} = \begin{pmatrix} 3 \\ 12 \\ 9 \end{pmatrix} = 3 \begin{pmatrix} 1 \\ 4 \\ 3 \end{pmatrix}$; $\quad \lambda = 3$

9. We solve $\det(\mathbf{A} - \lambda \mathbf{I}) = \begin{vmatrix} -8 - \lambda & -1 \\ 16 & -\lambda \end{vmatrix} = (\lambda + 4)^2 = 0$.

For $\lambda_1 = \lambda_2 = -4$ we have $\begin{pmatrix} -4 & -1 & | & 0 \\ 16 & 4 & | & 0 \end{pmatrix} \Longrightarrow \begin{pmatrix} 1 & 1/4 & | & 0 \\ 0 & 0 & | & 0 \end{pmatrix}$

so that $k_1 = -\frac{1}{4} k_2$. If $k_2 = 4$ then $\mathbf{K}_1 = \begin{pmatrix} -1 \\ 4 \end{pmatrix}$.

12. We solve $\det(\mathbf{A} - \lambda \mathbf{I}) = \begin{vmatrix} 1 - \lambda & -1 \\ 1 & 1 - \lambda \end{vmatrix} = \lambda^2 - 2\lambda + 2 = 0$.

For $\lambda_1 = 1 - i$ we have $\begin{pmatrix} i & -1 & | & 0 \\ 1 & i & | & 0 \end{pmatrix} \Longrightarrow \begin{pmatrix} i & -1 & | & 0 \\ 0 & 0 & | & 0 \end{pmatrix}$

154

so that $k_1 = -ik_2$. If $k_2 = 1$ then $\mathbf{K}_1 = \begin{pmatrix} -i \\ 1 \end{pmatrix}$ and $\mathbf{K}_2 = \overline{\mathbf{K}}_1 = \begin{pmatrix} i \\ 1 \end{pmatrix}$.

15. We solve $\quad \det(\mathbf{A} - \lambda\mathbf{I}) = \begin{vmatrix} 5-\lambda & -1 & 0 \\ 0 & -5-\lambda & 9 \\ 5 & -1 & -\lambda \end{vmatrix} = \begin{vmatrix} 4-\lambda & -1 & 0 \\ 4-\lambda & -5-\lambda & 9 \\ 4-\lambda & -1 & -\lambda \end{vmatrix} = \lambda(4-\lambda)(\lambda+4) = 0.$

For $\lambda_1 = 0$ we have
$$\begin{pmatrix} 5 & -1 & 0 & | & 0 \\ 0 & -5 & 9 & | & 0 \\ 5 & -1 & 0 & | & 0 \end{pmatrix} \Longrightarrow \begin{pmatrix} 1 & 0 & -9/25 & | & 0 \\ 0 & 1 & -9/5 & | & 0 \\ 0 & 0 & 0 & | & 0 \end{pmatrix}$$

so that $k_1 = \frac{9}{25}k_3$ and $k_2 = \frac{9}{5}k_3$. If $k_3 = 25$ then $\mathbf{K}_1 = \begin{pmatrix} 9 \\ 45 \\ 25 \end{pmatrix}$. If $\lambda_2 = 4$ then

$$\begin{pmatrix} 1 & -1 & 0 & | & 0 \\ 0 & -9 & 9 & | & 0 \\ 5 & -1 & -4 & | & 0 \end{pmatrix} \Longrightarrow \begin{pmatrix} 1 & 0 & -1 & | & 0 \\ 0 & 1 & -1 & | & 0 \\ 0 & 0 & 0 & | & 0 \end{pmatrix}$$

so that $k_1 = k_3$ and $k_2 = k_3$. If $k_3 = 1$ then $\mathbf{K}_2 = \begin{pmatrix} 1 \\ 1 \\ 1 \end{pmatrix}$. If $\lambda_3 = -4$ then

$$\begin{pmatrix} 9 & -1 & 0 & | & 0 \\ 0 & -1 & 9 & | & 0 \\ 5 & -1 & 4 & | & 0 \end{pmatrix} \Longrightarrow \begin{pmatrix} 1 & 0 & -1 & | & 0 \\ 0 & 1 & -9 & | & 0 \\ 0 & 0 & 0 & | & 0 \end{pmatrix}$$

so that $k_1 = k_3$ and $k_2 = 9k_3$. If $k_3 = 1$ then $\mathbf{K}_3 = \begin{pmatrix} 1 \\ 9 \\ 1 \end{pmatrix}$.

18. We solve $\quad \det(\mathbf{A} - \lambda\mathbf{I}) = \begin{vmatrix} 1-\lambda & 6 & 0 \\ 0 & 2-\lambda & 1 \\ 0 & 1 & 2-\lambda \end{vmatrix} = \begin{vmatrix} 1-\lambda & 6 & 0 \\ 0 & 3-\lambda & 3-\lambda \\ 0 & 1 & 2-\lambda \end{vmatrix} = (3-\lambda)(1-\lambda)^2 = 0.$

For $\lambda_1 = 3$ we have
$$\begin{pmatrix} -2 & 6 & 0 & | & 0 \\ 0 & 0 & 0 & | & 0 \\ 0 & 1 & -1 & | & 0 \end{pmatrix} \Longrightarrow \begin{pmatrix} 1 & 0 & -3 & | & 0 \\ 0 & 1 & -1 & | & 0 \\ 0 & 0 & 0 & | & 0 \end{pmatrix}$$

so that $k_1 = 3k_3$ and $k_2 = k_3$. If $k_3 = 1$ then $\mathbf{K}_1 = \begin{pmatrix} 3 \\ 1 \\ 1 \end{pmatrix}$. For $\lambda_2 = \lambda_3 = 1$ we have

$$\begin{pmatrix} 0 & 6 & 0 & | & 0 \\ 0 & 1 & 1 & | & 0 \\ 0 & 1 & 1 & | & 0 \end{pmatrix} \Longrightarrow \begin{pmatrix} 0 & 1 & 0 & | & 0 \\ 0 & 0 & 1 & | & 0 \\ 0 & 0 & 0 & | & 0 \end{pmatrix}$$

so that $k_2 = 0$ and $k_3 = 0$. If $k_1 = 1$ then $\mathbf{K}_2 = \begin{pmatrix} 1 \\ 0 \\ 0 \end{pmatrix}$.

21. We solve
$$\det(\mathbf{A} - \lambda\mathbf{I}) = \begin{vmatrix} 1-\lambda & 2 & 3 \\ 0 & 5-\lambda & 6 \\ 0 & 0 & -7-\lambda \end{vmatrix} = -(\lambda-1)(\lambda-5)(\lambda+7) = 0.$$

For $\lambda_1 = 1$ we have
$$\begin{pmatrix} 0 & 2 & 3 & | & 0 \\ 0 & 4 & 6 & | & 0 \\ 0 & 0 & -6 & | & 0 \end{pmatrix} \implies \begin{pmatrix} 0 & 1 & 0 & | & 0 \\ 0 & 0 & 1 & | & 0 \\ 0 & 0 & 0 & | & 0 \end{pmatrix}$$

so that $k_2 = k_3 = 0$. If $k_1 = 1$ then $\mathbf{K}_1 = \begin{pmatrix} 1 \\ 0 \\ 0 \end{pmatrix}$. For $\lambda_2 = 5$ we have

$$\begin{pmatrix} -4 & 2 & 3 & | & 0 \\ 0 & 0 & 6 & | & 0 \\ 0 & 0 & -12 & | & 0 \end{pmatrix} \implies \begin{pmatrix} 1 & -\frac{1}{2} & 0 & | & 0 \\ 0 & 0 & 1 & | & 0 \\ 0 & 0 & 0 & | & 0 \end{pmatrix}$$

so that $k_3 = 0$ and $k_2 = 2k_1$. If $k_1 = 1$ then $\mathbf{K}_2 = \begin{pmatrix} 1 \\ 2 \\ 0 \end{pmatrix}$. For $\lambda_3 = -7$ we have

$$\begin{pmatrix} 8 & 2 & 3 & | & 0 \\ 0 & 12 & 6 & | & 0 \\ 0 & 0 & 0 & | & 0 \end{pmatrix} \implies \begin{pmatrix} 1 & 0 & \frac{1}{4} & | & 0 \\ 0 & 1 & \frac{1}{2} & | & 0 \\ 0 & 0 & 0 & | & 0 \end{pmatrix}$$

so that $k_1 = -\frac{1}{4}k_3$ and $k_2 = -\frac{1}{2}k_3$. If $k_3 = 4$ then $\mathbf{K}_3 = \begin{pmatrix} -1 \\ -2 \\ 4 \end{pmatrix}$.

24. The eigenvalues and eigenvectors of $\mathbf{A} = \begin{pmatrix} 1 & 2 & -1 \\ 1 & 0 & 1 \\ 4 & -4 & 5 \end{pmatrix}$ are

$$\lambda_1 = 1, \quad \lambda_2 = 2, \quad \lambda_3 = 3, \quad \mathbf{K}_1 = \begin{pmatrix} -1 \\ 1 \\ 2 \end{pmatrix}, \quad \mathbf{K}_2 = \begin{pmatrix} -2 \\ 1 \\ 4 \end{pmatrix}, \quad \mathbf{K}_3 = \begin{pmatrix} -1 \\ 1 \\ 4 \end{pmatrix}.$$

and the eigenvalues and eigenvectors of $\mathbf{A}^{-1} = \frac{1}{6}\begin{pmatrix} 4 & -6 & 2 \\ -1 & 9 & -2 \\ -4 & 12 & -2 \end{pmatrix}$ are

$$\lambda_1 = 1, \quad \lambda_2 = \frac{1}{2}, \quad \lambda_3 = \frac{1}{3}, \quad \mathbf{K}_1 = \begin{pmatrix} -1 \\ 1 \\ 2 \end{pmatrix}, \quad \mathbf{K}_2 = \begin{pmatrix} -2 \\ 1 \\ 4 \end{pmatrix}, \quad \mathbf{K}_3 = \begin{pmatrix} -1 \\ 1 \\ 4 \end{pmatrix}.$$

27. (a) Since $p + 1 - p = 1$ and $q + 1 - q = 1$, the first matrix $\mathbf{A}$ is stochastic. Since $\frac{1}{2} + \frac{1}{4} + \frac{1}{4} = 1$, $\frac{1}{3} + \frac{1}{3} + \frac{1}{3} = 1$, and $\frac{1}{6} + \frac{1}{3} + \frac{1}{2} = 1$, the second matrix $\mathbf{A}$ is stochastic.

(b) The matrix from part (a) is shown with its eigenvalues and corresponding eigenvectors.

$$\begin{pmatrix} \frac{1}{2} & \frac{1}{4} & \frac{1}{4} \\ \frac{1}{3} & \frac{1}{3} & \frac{1}{3} \\ \frac{1}{6} & \frac{1}{3} & \frac{1}{2} \end{pmatrix}; \quad \text{eigenvalues: } 1, \frac{1}{6} - \frac{1}{12}\sqrt{2}, \frac{1}{6} + \frac{1}{12}\sqrt{2};$$

eigenvectors: $(1, 1, 1)$, $\left(-\frac{3(-1+\sqrt{2})}{-6+\sqrt{2}}, \frac{2(2+\sqrt{2})}{-6+\sqrt{2}}, 1\right)$, $\left(-\frac{3(1+\sqrt{2})}{6+\sqrt{2}}, \frac{2(-2+\sqrt{2})}{6+\sqrt{2}}, 1\right)$

Further examples indicate that 1 is always an eigenvalue with corresponding eigenvector $(1, 1, 1)$. To prove this, let $\mathbf{A}$ be a stochastic matrix and $\mathbf{K} = (1, 1, 1)$. Then

$$\mathbf{AK} = \begin{pmatrix} a_{11} & \cdots & a_{1n} \\ \vdots & & \vdots \\ a_{n1} & \cdots & a_{nn} \end{pmatrix} \begin{pmatrix} 1 \\ \vdots \\ 1 \end{pmatrix} = \begin{pmatrix} a_{11} + \cdots + a_{1n} \\ \vdots \\ a_{n1} + \cdots + a_{nn} \end{pmatrix} = \begin{pmatrix} 1 \\ \vdots \\ 1 \end{pmatrix} = 1\mathbf{K},$$

and 1 is an eigenvalue of $\mathbf{A}$ with corresponding eigenvector $(1, 1, 1)$.

(c) For the 3×3 matrix in part (a) we have

$$\mathbf{A}^2 = \begin{pmatrix} \frac{3}{8} & \frac{7}{24} & \frac{1}{3} \\ \frac{1}{3} & \frac{11}{36} & \frac{13}{36} \\ \frac{5}{18} & \frac{23}{72} & \frac{29}{72} \end{pmatrix}, \quad \mathbf{A}^3 = \begin{pmatrix} \frac{49}{144} & \frac{29}{96} & \frac{103}{288} \\ \frac{71}{216} & \frac{11}{36} & \frac{79}{216} \\ \frac{5}{16} & \frac{67}{216} & \frac{163}{432} \end{pmatrix}.$$

These powers of $\mathbf{A}$ are also stochastic matrices. To prove that this is true in general for 2×2 matrices, we prove the more general theorem that any product of 2×2 stochastic matrices is stochastic. Let

$$\mathbf{A} = \begin{pmatrix} a_{11} & a_{12} \\ a_{21} & a_{22} \end{pmatrix} \quad \text{and} \quad \mathbf{B} = \begin{pmatrix} b_{11} & b_{12} \\ b_{21} & b_{22} \end{pmatrix}$$

be stochastic matrices. Then

$$\mathbf{AB} = \begin{pmatrix} a_{11}b_{11} + a_{12}b_{21} & a_{11}b_{12} + a_{12}b_{22} \\ a_{21}b_{11} + a_{22}b_{21} & a_{21}b_{12} + a_{22}b_{22} \end{pmatrix}.$$

The sums of the rows are

$$a_{11}b_{11} + a_{12}b_{21} + a_{11}b_{12} + a_{12}b_{22} = a_{11}(b_{11} + b_{12}) + a_{12}(b_{21} + b_{22})$$
$$= a_{11}(1) + a_{12}(1) = a_{11} + a_{12} = 1$$
$$a_{21}b_{11} + a_{22}b_{21} + a_{21}b_{12} + a_{22}b_{22} = a_{21}(b_{11} + b_{12}) + a_{22}(b_{21} + b_{22})$$
$$= a_{21}(1) + a_{22}(1) = a_{21} + a_{22} = 1.$$

Thus, the product matrix $\mathbf{AB}$ is stochastic. It follows that any power of a 2×2 matrix is stochastic. The proof in the case of an $n \times n$ matrix is very similar.

EXERCISES 8.9
Powers of Matrices

3. The characteristic equation is $\lambda^2 - 3\lambda - 10 = 0$, with eigenvalues -2 and 5. Substituting the eigenvalues into $\lambda^m = c_0 + c_1\lambda$ generates

$$(-2)^m = c_0 - 2c_1$$
$$5^m = c_0 + 5c_1.$$

Solving the system gives

$$c_0 = \frac{1}{7}[5(-2)^m + 2(5)^m], \qquad c_1 = \frac{1}{7}[-(-2)^m + 5^m].$$

Thus

$$\mathbf{A}^m = c_0\mathbf{I} + c_1\mathbf{A} = \begin{pmatrix} \frac{1}{7}[3(-1)^m2^{m+1}+5^m] & \frac{3}{7}[-(-2)^m+5^m] \\ \frac{2}{7}[-(-2)^m+5^m] & \frac{1}{7}[(-2)^m+6(5)^m] \end{pmatrix}$$

and

$$\mathbf{A}^3 = \begin{pmatrix} 11 & 57 \\ 38 & 106 \end{pmatrix}.$$

6. The characteristic equation is $\lambda^2 + 4\lambda + 3 = 0$, with eigenvalues -3 and -1. Substituting the eigenvalues into $\lambda^m = c_0 + c_1\lambda$ generates

$$(-3)^m = c_0 - 3c_1$$
$$(-1)^m = c_0 - c_1.$$

Solving the system gives

$$c_0 = \frac{1}{2}[-(-3)^m + 3(-1)^m], \qquad c_1 = \frac{1}{2}[-(-3)^m + (-1)^m].$$

Thus

$$\mathbf{A}^m = c_0\mathbf{I} + c_1\mathbf{A} = \begin{pmatrix} (-1)^m & -(-3)^m + (-1)^m \\ 0 & (-3)^m \end{pmatrix}$$

and

$$\mathbf{A}^6 = \begin{pmatrix} 1 & -728 \\ 0 & 729 \end{pmatrix}.$$

9. The characteristic equation is $-\lambda^3 + 3\lambda^2 + 6\lambda - 8 = 0$, with eigenvalues -2, 1, and 4. Substituting the eigenvalues into $\lambda^m = c_0 + c_1\lambda + c_2\lambda^2$ generates

$$(-2)^m = c_0 - 2c_1 + 4c_2$$
$$1 = c_0 + c_1 + c_2$$
$$4^m = c_0 + 4c_1 + 16c_2.$$

Solving the system gives

$$c_0 = \frac{1}{9}[8 + (-1)^m2^{m+1} - 4^m],$$

$$c_1 = \frac{1}{18}[4 - 5(-2)^m + 4^m],$$

$$c_2 = \frac{1}{18}[-2 + (-2)^m + 4^m].$$

Thus

$$\mathbf{A}^m = c_0\mathbf{I} + c_1\mathbf{A} + c_2\mathbf{A}^2 = \begin{pmatrix} \frac{1}{9}[(-2)^m + (-1)^m2^{m+1} + 3\cdot2^{2m+1}] & \frac{1}{3}[-(-2)^m + 4^m] & 0 \\ -\frac{2}{3}[(-2)^m - 4^m] & \frac{1}{3}[(-1)^m2^{m+1} + 4^m] & 0 \\ \frac{1}{3}[-3 + (-2)^m + 2^{2m+1}] & \frac{1}{3}[-(-2)^m + 4^m] & 1 \end{pmatrix}$$

and

$$\mathbf{A}^{10} = \begin{pmatrix} 699392 & 349184 & 0 \\ 698368 & 350208 & 0 \\ 699391 & 349184 & 1 \end{pmatrix}.$$

12. The characteristic equation is $-\lambda^3 - \lambda^2 + 21\lambda + 45 = 0$, with eigenvalues -3, -3, and 5. Substituting the eigenvalues into $\lambda^m = c_0 + c_1\lambda + c_2\lambda^2$ generates

$$(-3)^m = c_0 - 3c_1 + 9c_2$$
$$(-3)^{m-1}m = c_1 - 6c_2$$
$$5^m = c_0 + 5c_1 + 25c_2.$$

Solving the system gives

$$c_0 = \frac{1}{64}[73(-3)^m - 2(-1)^m 3^{m+2} + 9 \cdot 5^m - 40(-3)^m m],$$

$$c_1 = \frac{1}{96}[-(-1)^m 3^{m+2} + 9 \cdot 5^m - 8(-3)^m m],$$

$$c_2 = \frac{1}{64}[-(-3)^m + 5^m - 8(-3)^{m-1}m].$$

Thus

$$\mathbf{A}^m = c_0\mathbf{I} + c_1\mathbf{A} + c_2\mathbf{A}^2$$

$$= \begin{pmatrix} \frac{1}{32}[31(-3)^m - (-1)^m 3^{m+1} + 4 \cdot 5^m] & \frac{1}{16}[-(-3)^m - (-1)^m 3^{m+1} + 4 \cdot 5^m] & \frac{1}{32}[(-3)^m + (-1)^m 3^{m+1} - 4 \cdot 5^m] \\ \frac{1}{16}[-(-3)^m - (-1)^m 3^{m+1} + 4 \cdot 5^m] & \frac{1}{8}[7(-3)^m - (-1)^m 3^{m+1} + 4 \cdot 5^m] & \frac{1}{16}[(-3)^m + (-1)^m 3^{m+1} - 4 \cdot 5^m] \\ \frac{3}{32}[(-3)^m + (-1)^m 3^{m+1} - 4 \cdot 5^m] & \frac{3}{16}[(-3)^m + (-1)^m 3^{m+1} - 4 \cdot 5^m] & \frac{1}{32}[29(-3)^m - (-1)^m 3^{m+2} + 12 \cdot 5^m] \end{pmatrix}$$

and

$$\mathbf{A}^5 = \begin{pmatrix} 178 & 842 & -421 \\ 842 & 1441 & -842 \\ -1263 & -2526 & 1020 \end{pmatrix}.$$

15. The characteristic equation of $\mathbf{A}$ is $\lambda^2 - 5\lambda + 10 = 0$, so $\mathbf{A}^2 - 5\mathbf{A} + 10\mathbf{I} = \mathbf{0}$ and $\mathbf{I} = -\frac{1}{10}\mathbf{A}^2 + \frac{1}{2}\mathbf{A}$. Multiplying by $\mathbf{A}^{-1}$ we find

$$\mathbf{A}^{-1} = -\frac{1}{10}\mathbf{A} + \frac{1}{2}\mathbf{I} = -\frac{1}{10}\begin{pmatrix} 2 & -4 \\ 1 & 3 \end{pmatrix} + \frac{1}{2}\begin{pmatrix} 1 & 0 \\ 0 & 1 \end{pmatrix} = \begin{pmatrix} \frac{3}{10} & \frac{2}{5} \\ -\frac{1}{10} & \frac{1}{5} \end{pmatrix}.$$

18. (a) If $\mathbf{A}^m = \mathbf{0}$ for some m, then $(\det \mathbf{A})^m = \det \mathbf{A}^m = \det \mathbf{0} = 0$, and $\mathbf{A}$ is a singular matrix.

(b) By (1) of Section 8.8 we have $\mathbf{A}\mathbf{K} = \lambda\mathbf{K}$, $\mathbf{A}^2\mathbf{K} = \lambda\mathbf{A}\mathbf{K} = \lambda^2\mathbf{K}$, $\mathbf{A}^3\mathbf{K} = \lambda^2\mathbf{A}\mathbf{K} = \lambda^3\mathbf{K}$, and, in general, $\mathbf{A}^m\mathbf{K} = \lambda^m\mathbf{K}$. If $\mathbf{A}$ is nilpotent with index m, then $\mathbf{A}^m = \mathbf{0}$ and $\lambda^m = 0$.

EXERCISES 8.10

Orthogonal Matrices

3. (a)–(b) $\begin{pmatrix} 5 & 13 & 0 \\ 13 & 5 & 0 \\ 0 & 0 & -8 \end{pmatrix} \begin{pmatrix} \frac{\sqrt{2}}{2} \\ \frac{\sqrt{2}}{2} \\ 0 \end{pmatrix} = \begin{pmatrix} 9\sqrt{2} \\ 9\sqrt{2} \\ 0 \end{pmatrix} = 18 \begin{pmatrix} \frac{\sqrt{2}}{2} \\ \frac{\sqrt{2}}{2} \\ 0 \end{pmatrix}; \quad \lambda_1 = 18$

$\begin{pmatrix} 5 & 13 & 0 \\ 13 & 5 & 0 \\ 0 & 0 & -8 \end{pmatrix} \begin{pmatrix} \frac{\sqrt{3}}{3} \\ -\frac{\sqrt{3}}{3} \\ \frac{\sqrt{3}}{3} \end{pmatrix} = \begin{pmatrix} -\frac{8\sqrt{2}}{3} \\ \frac{8\sqrt{3}}{3} \\ -\frac{8\sqrt{3}}{3} \end{pmatrix} = (-8) \begin{pmatrix} \frac{\sqrt{3}}{3} \\ -\frac{\sqrt{3}}{3} \\ \frac{\sqrt{3}}{3} \end{pmatrix}; \quad \lambda_2 = -8$

$\begin{pmatrix} 5 & 13 & 0 \\ 13 & 5 & 0 \\ 0 & 0 & -8 \end{pmatrix} \begin{pmatrix} \frac{\sqrt{6}}{6} \\ -\frac{\sqrt{6}}{6} \\ -\frac{\sqrt{6}}{3} \end{pmatrix} = \begin{pmatrix} -\frac{8\sqrt{6}}{6} \\ \frac{8\sqrt{6}}{6} \\ \frac{8\sqrt{6}}{3} \end{pmatrix} = (-8) \begin{pmatrix} \frac{\sqrt{6}}{6} \\ -\frac{\sqrt{6}}{6} \\ -\frac{\sqrt{6}}{3} \end{pmatrix}; \quad \lambda_3 = -8$

(c) $\mathbf{K}_1^T \mathbf{K}_2 = \begin{pmatrix} \frac{\sqrt{2}}{2} & \frac{\sqrt{2}}{2} & 0 \end{pmatrix} \begin{pmatrix} \frac{\sqrt{3}}{3} \\ -\frac{\sqrt{3}}{3} \\ \frac{\sqrt{3}}{3} \end{pmatrix} = \frac{\sqrt{6}}{6} - \frac{\sqrt{6}}{6} = 0;$

$\mathbf{K}_1^T \mathbf{K}_3 = \begin{pmatrix} \frac{\sqrt{2}}{2} & \frac{\sqrt{2}}{2} & 0 \end{pmatrix} \begin{pmatrix} \frac{\sqrt{6}}{6} \\ -\frac{\sqrt{6}}{6} \\ -\frac{\sqrt{6}}{3} \end{pmatrix} = \frac{\sqrt{12}}{12} - \frac{\sqrt{12}}{12} = 0$

$\mathbf{K}_2^T \mathbf{K}_3 = \begin{pmatrix} \frac{\sqrt{3}}{3} & -\frac{\sqrt{3}}{3} & \frac{\sqrt{3}}{3} \end{pmatrix} \begin{pmatrix} \frac{\sqrt{6}}{6} \\ -\frac{\sqrt{6}}{6} \\ -\frac{\sqrt{6}}{3} \end{pmatrix} = \frac{\sqrt{18}}{18} + \frac{\sqrt{18}}{18} - \frac{\sqrt{18}}{9} = 0$

6. Not orthogonal. Columns one and three are not unit vectors.

9. Not orthogonal. Columns are not unit vectors.

12. $\lambda_1 = 7, \; \lambda_2 = 4, \; \mathbf{K}_1 = \begin{pmatrix} 1 \\ 0 \end{pmatrix}, \; \mathbf{K}_2 = \begin{pmatrix} 0 \\ 1 \end{pmatrix}, \; \mathbf{P} = \begin{pmatrix} 1 & 0 \\ 0 & 1 \end{pmatrix}$

15. $\lambda_1 = 0, \; \lambda_2 = 2, \; \lambda_3 = 1, \; \mathbf{K}_1 = \begin{pmatrix} -1 \\ 0 \\ 1 \end{pmatrix}, \; \mathbf{K}_2 = \begin{pmatrix} 1 \\ 0 \\ 1 \end{pmatrix}, \; \mathbf{K}_3 = \begin{pmatrix} 0 \\ 1 \\ 0 \end{pmatrix}, \; \mathbf{P} = \begin{pmatrix} -\frac{1}{\sqrt{2}} & \frac{1}{\sqrt{2}} & 0 \\ 0 & 0 & 1 \\ \frac{1}{\sqrt{2}} & \frac{1}{\sqrt{2}} & 0 \end{pmatrix}$

18. $\lambda_1 = -18, \; \lambda_2 = 0, \; \lambda_3 = 9, \; \mathbf{K}_1 = \begin{pmatrix} 1 \\ -2 \\ 2 \end{pmatrix}, \; \mathbf{K}_2 = \begin{pmatrix} -2 \\ 1 \\ 2 \end{pmatrix}, \; \mathbf{K}_3 = \begin{pmatrix} 2 \\ 2 \\ 1 \end{pmatrix}, \; \mathbf{P} = \begin{pmatrix} \frac{1}{3} & -\frac{2}{3} & \frac{2}{3} \\ -\frac{2}{3} & \frac{1}{3} & \frac{2}{3} \\ \frac{2}{3} & \frac{2}{3} & \frac{1}{3} \end{pmatrix}$

160

21. (a)–(b) We compute

$$\mathbf{AK}_1 = \begin{pmatrix} 0 & 2 & 2 \\ 2 & 0 & 2 \\ 2 & 2 & 0 \end{pmatrix} \begin{pmatrix} 1 \\ -1 \\ 0 \end{pmatrix} = \begin{pmatrix} -2 \\ 2 \\ 0 \end{pmatrix} = -2 \begin{pmatrix} 1 \\ -1 \\ 0 \end{pmatrix} = -2\mathbf{K}_1$$

$$\mathbf{AK}_2 = \begin{pmatrix} 0 & 2 & 2 \\ 2 & 0 & 2 \\ 2 & 2 & 0 \end{pmatrix} \begin{pmatrix} 1 \\ 0 \\ -1 \end{pmatrix} = \begin{pmatrix} -2 \\ 0 \\ 2 \end{pmatrix} = -2 \begin{pmatrix} 1 \\ 0 \\ -1 \end{pmatrix} = -2\mathbf{K}_2$$

$$\mathbf{AK}_3 = \begin{pmatrix} 0 & 2 & 2 \\ 2 & 0 & 2 \\ 2 & 2 & 0 \end{pmatrix} \begin{pmatrix} 1 \\ 1 \\ 1 \end{pmatrix} = \begin{pmatrix} 4 \\ 4 \\ 4 \end{pmatrix} = 4 \begin{pmatrix} 1 \\ 1 \\ 1 \end{pmatrix} = 4\mathbf{K}_3,$$

and observe that $\mathbf{K}_1$ is an eigenvector with corresponding eigenvalue -2, $\mathbf{K}_2$ is an eigenvector with corresponding eigenvalue -2, and $\mathbf{K}_3$ is an eigenvector with corresponding eigenvalue 4.

(c) Since $\mathbf{K}_1 \cdot \mathbf{K}_2 = 1 \neq 0$, $\mathbf{K}_1$ and $\mathbf{K}_2$ are not orthogonal, while $\mathbf{K}_1 \cdot \mathbf{K}_3 = 0$ and $\mathbf{K}_2 \cdot \mathbf{K}_3 = 0$ so $\mathbf{K}_3$ is orthogonal to both $\mathbf{K}_1$ and $\mathbf{K}_2$, To transform $\{\mathbf{K}_1, \mathbf{K}_2\}$ into an orthogonal set we let $\mathbf{V}_1 = \mathbf{K}_1$ and compute $\mathbf{K}_2 \cdot \mathbf{V}_1 = 1$ and $\mathbf{V}_1 \cdot \mathbf{V}_1 = 2$. Then

$$\mathbf{V}_2 = \mathbf{K}_2 - \frac{\mathbf{K}_2 \cdot \mathbf{V}_1}{\mathbf{V}_1 \cdot \mathbf{V}_1} \mathbf{V}_1 = \begin{pmatrix} 1 \\ 0 \\ -1 \end{pmatrix} - \frac{1}{2} \begin{pmatrix} 1 \\ -1 \\ 0 \end{pmatrix} = \begin{pmatrix} \frac{1}{2} \\ \frac{1}{2} \\ -1 \end{pmatrix}.$$

Now, $\{\mathbf{V}_1, \mathbf{V}_2, \mathbf{K}_3\}$ is an orthogonal set of eigenvectors with

$$\|\mathbf{V}_1\| = \sqrt{2}, \quad \|\mathbf{V}_2\| = \frac{3}{\sqrt{6}}, \quad \text{and} \quad \|\mathbf{K}_3\| = \sqrt{3}.$$

An orthonormal set of vectors is

$$\begin{pmatrix} \frac{1}{\sqrt{2}} \\ -\frac{1}{\sqrt{2}} \\ 0 \end{pmatrix}, \quad \begin{pmatrix} \frac{1}{\sqrt{6}} \\ \frac{1}{\sqrt{6}} \\ -\frac{2}{\sqrt{6}} \end{pmatrix}, \quad \text{and} \quad \begin{pmatrix} \frac{1}{\sqrt{3}} \\ \frac{1}{\sqrt{3}} \\ \frac{1}{\sqrt{3}} \end{pmatrix},$$

and so the matrix

$$\mathbf{P} = \begin{pmatrix} \frac{1}{\sqrt{2}} & \frac{1}{\sqrt{6}} & \frac{1}{\sqrt{3}} \\ -\frac{1}{\sqrt{2}} & \frac{1}{\sqrt{6}} & \frac{1}{\sqrt{3}} \\ 0 & -\frac{2}{\sqrt{6}} & \frac{1}{\sqrt{3}} \end{pmatrix}$$

is orthogonal.

24. The eigenvalues and corresponding eigenvectors of $\mathbf{A}$ are

$$\lambda_1 = \lambda_2 = -1, \quad \lambda_3 = \lambda_4 = 3, \quad \text{and} \quad \mathbf{K}_1 = \begin{pmatrix} -1 \\ 1 \\ 0 \\ 0 \end{pmatrix}, \quad \mathbf{K}_2 = \begin{pmatrix} 0 \\ 0 \\ -1 \\ 1 \end{pmatrix}, \quad \mathbf{K}_3 = \begin{pmatrix} 0 \\ 0 \\ 1 \\ 1 \end{pmatrix}, \quad \mathbf{K}_4 = \begin{pmatrix} 1 \\ 1 \\ 0 \\ 0 \end{pmatrix}.$$

Since $\mathbf{K}_1 \cdot \mathbf{K}_2 = \mathbf{K}_1 \cdot \mathbf{K}_3 = \mathbf{K}_1 \cdot \mathbf{K}_4 = \mathbf{K}_2 \cdot \mathbf{K}_3 = \mathbf{K}_2 \cdot \mathbf{K}_4 = \mathbf{K}_3 \cdot \mathbf{K}_4 = 0$, the vectors are orthogonal. Using $\|\mathbf{K}_1\| = \|\mathbf{K}_2\| = \|\mathbf{K}_3\| = \|\mathbf{K}_4\| = \sqrt{2}$, we construct the orthogonal matrix

$$\mathbf{P} = \begin{pmatrix} -\frac{1}{\sqrt{2}} & 0 & 0 & \frac{1}{\sqrt{2}} \\ \frac{1}{\sqrt{2}} & 0 & 0 & \frac{1}{\sqrt{2}} \\ 0 & -\frac{1}{\sqrt{2}} & \frac{1}{\sqrt{2}} & 0 \\ 0 & \frac{1}{\sqrt{2}} & \frac{1}{\sqrt{2}} & 0 \end{pmatrix}.$$

161

EXERCISES 8.11

Approximation of Eigenvalues

3. Taking $\mathbf{X}_0 = \begin{pmatrix} 1 \\ 1 \end{pmatrix}$ and computing $\mathbf{AX}_0 = \begin{pmatrix} 6 \\ 16 \end{pmatrix}$, we define $\mathbf{X}_1 = \dfrac{1}{16} \begin{pmatrix} 6 \\ 16 \end{pmatrix} = \begin{pmatrix} 0.375 \\ 1 \end{pmatrix}$. Continuing in this manner we obtain

$$\mathbf{X}_2 = \begin{pmatrix} 0.3363 \\ 1 \end{pmatrix}, \quad \mathbf{X}_3 = \begin{pmatrix} 0.3335 \\ 1 \end{pmatrix}, \quad \mathbf{X}_4 = \begin{pmatrix} 0.3333 \\ 1 \end{pmatrix}.$$

We conclude that a dominant eigenvector is $\mathbf{K} = \begin{pmatrix} 0.3333 \\ 1 \end{pmatrix}$ with corresponding eigenvalue $\lambda = 14$.

6. Taking $\mathbf{X}_0 = \begin{pmatrix} 1 \\ 1 \\ 1 \end{pmatrix}$ and computing $\mathbf{AX}_0 = \begin{pmatrix} 5 \\ 2 \\ 2 \end{pmatrix}$, we define $\mathbf{X}_1 = \dfrac{1}{5} \begin{pmatrix} 5 \\ 2 \\ 2 \end{pmatrix} = \begin{pmatrix} 1 \\ 0.4 \\ 0.4 \end{pmatrix}$. Continuing in this manner we obtain

$$\mathbf{X}_2 = \begin{pmatrix} 1 \\ 0.2105 \\ 0.2105 \end{pmatrix}, \quad \mathbf{X}_3 = \begin{pmatrix} 1 \\ 0.1231 \\ 0.1231 \end{pmatrix}, \quad \mathbf{X}_4 = \begin{pmatrix} 1 \\ 0.0758 \\ 0.0758 \end{pmatrix}, \quad \mathbf{X}_5 = \begin{pmatrix} 1 \\ 0.0481 \\ 0.0481 \end{pmatrix}.$$

At this point if we restart with $\mathbf{X}_0 = \begin{pmatrix} 1 \\ 0 \\ 0 \end{pmatrix}$ we see that $\mathbf{K} = \begin{pmatrix} 1 \\ 0 \\ 0 \end{pmatrix}$ is a dominant eigenvector with corresponding eigenvalue $\lambda = 3$.

9. Taking $\mathbf{X}_0 = \begin{pmatrix} 1 \\ 1 \\ 1 \end{pmatrix}$ and using scaling we obtain

$$\mathbf{X}_1 = \begin{pmatrix} 1 \\ 0 \\ 1 \end{pmatrix}, \quad \mathbf{X}_2 = \begin{pmatrix} 1 \\ -0.6667 \\ 1 \end{pmatrix}, \quad \mathbf{X}_3 = \begin{pmatrix} 1 \\ -0.9091 \\ 1 \end{pmatrix}, \quad \mathbf{X}_4 = \begin{pmatrix} 1 \\ -0.9767 \\ 1 \end{pmatrix}, \quad \mathbf{X}_5 = \begin{pmatrix} 1 \\ -0.9942 \\ 1 \end{pmatrix}.$$

Taking $\mathbf{K} = \begin{pmatrix} 1 \\ -1 \\ 1 \end{pmatrix}$ as the dominant eigenvector we find $\lambda_1 = 4$. Now the normalized eigenvector is

$$\mathbf{K}_1 = \begin{pmatrix} 0.5774 \\ -0.5774 \\ 0.5774 \end{pmatrix} \text{ and } \mathbf{B} = \begin{pmatrix} 1.6667 & 0.3333 & -1.3333 \\ 0.3333 & 0.6667 & 0.3333 \\ -1.3333 & 0.3333 & 1.6667 \end{pmatrix}. \text{ If } \mathbf{X}_0 = \begin{pmatrix} 1 \\ 1 \\ 1 \end{pmatrix} \text{ is now chosen only one more}$$

eigenvalue is found. Thus, try $\mathbf{X}_0 = \begin{pmatrix} 1 \\ 1 \\ 0 \end{pmatrix}$. Using scaling we obtain

$$\mathbf{X}_1 = \begin{pmatrix} 1 \\ 0.5 \\ -0.5 \end{pmatrix}, \quad \mathbf{X}_2 = \begin{pmatrix} 1 \\ 0.2 \\ -0.8 \end{pmatrix}, \quad \mathbf{X}_3 = \begin{pmatrix} 1 \\ 0.0714 \\ -0.9286 \end{pmatrix}, \quad \mathbf{X}_4 = \begin{pmatrix} 1 \\ 0.0244 \\ -0.9756 \end{pmatrix}, \quad \mathbf{X}_5 = \begin{pmatrix} 1 \\ 0.0082 \\ -0.9918 \end{pmatrix}.$$

Taking $\mathbf{K} = \begin{pmatrix} 1 \\ 0 \\ -1 \end{pmatrix}$ as the eigenvector we find $\lambda_2 = 3$. The normalized eigenvector in this case is

$$\mathbf{K}_2 = \begin{pmatrix} 0.7071 \\ 0 \\ -0.7071 \end{pmatrix} \quad \text{and} \quad \mathbf{C} = \begin{pmatrix} 0.1667 & 0.3333 & 0.1667 \\ 0.3333 & 0.6667 & 0.3333 \\ 0.1667 & 0.3333 & 0.1667 \end{pmatrix}. \text{ If } \mathbf{X}_0 = \begin{pmatrix} 1 \\ 1 \\ 1 \end{pmatrix} \text{ is chosen, and scaling is used we}$$

obtain $\mathbf{X}_1 = \begin{pmatrix} 0.5 \\ 1 \\ 0.5 \end{pmatrix}, \mathbf{X}_2 = \begin{pmatrix} 0.5 \\ 1 \\ 0.5 \end{pmatrix}$. Taking $\mathbf{K} = \begin{pmatrix} 0.5 \\ 1 \\ 0.5 \end{pmatrix}$ we find $\lambda_3 = 1$. The eigenvalues are 4, 3, and 1.

The difficulty in choosing $\mathbf{X}_0 = \begin{pmatrix} 1 \\ 1 \\ 1 \end{pmatrix}$ to find the second eigenvector results from the fact that this vector is a

linear combination of the eigenvectors corresponding to the other two eigenvalues, with 0 contribution from the second eigenvector. When this occurs the development of the power method, shown in the text, breaks down.

12. The inverse matrix is $\begin{pmatrix} 1 & 3 \\ 4 & 2 \end{pmatrix}$. Taking $\mathbf{X}_0 = \begin{pmatrix} 1 \\ 1 \end{pmatrix}$ and using scaling we obtain

$$\mathbf{X}_1 = \begin{pmatrix} 0.6667 \\ 1 \end{pmatrix}, \quad \mathbf{X}_2 = \begin{pmatrix} 0.7857 \\ 1 \end{pmatrix}, \dots, \mathbf{X}_{10} = \begin{pmatrix} 0.75 \\ 1 \end{pmatrix}.$$

Using $\mathbf{K} = \begin{pmatrix} 0.75 \\ 1 \end{pmatrix}$ we find $\lambda = 5$. The minimum eigenvalue of $\begin{pmatrix} -0.2 & 0.3 \\ 0.4 & -0.1 \end{pmatrix}$ is $1/5 = 0.2$

15. (a) $\mathbf{A}^{10} = \begin{pmatrix} 67{,}745{,}349 & -43{,}691{,}832 & 8{,}258{,}598 \\ -43{,}691{,}832 & 28{,}182{,}816 & -5{,}328{,}720 \\ 8{,}258{,}598 & -5{,}328{,}720 & 1{,}008{,}180 \end{pmatrix}$

(b) $\mathbf{X}_{10} = \mathbf{A}^{10} \begin{pmatrix} 1 \\ 0 \\ 0 \end{pmatrix} = \begin{pmatrix} 67{,}745{,}349 \\ -43{,}691{,}832 \\ 8{,}258{,}598 \end{pmatrix} \approx 67{,}745{,}349 \begin{pmatrix} 1 \\ -0.644942 \\ 0.121906 \end{pmatrix}$

$$\mathbf{X}_{12} = \mathbf{A}^{12} \begin{pmatrix} 1 \\ 0 \\ 0 \end{pmatrix} = \begin{pmatrix} 2{,}680{,}201{,}629 \\ -1{,}728{,}645{,}624 \\ 326{,}775{,}222 \end{pmatrix} \approx 2{,}680{,}201{,}629 \begin{pmatrix} 1 \\ -0.644968 \\ 0.121922 \end{pmatrix}.$$

The vectors appear to be approaching scalar multiples of $\mathbf{K} = (1, -0.644968, 0.121922)$, which approximates the dominant eigenvector.

(c) The dominant eigenvalue is $\lambda_1 = (\mathbf{AK} \cdot \mathbf{K})/(\mathbf{K} \cdot \mathbf{K}) = 6.28995$.

EXERCISES 8.12

Diagonalization

3. For $\lambda_1 = \lambda_2 = 1$ we obtain the single eigenvector $\mathbf{K}_1 = \begin{pmatrix} 1 \\ 1 \end{pmatrix}$. Hence $\mathbf{A}$ is not diagonalizable.

6. Distinct eigenvalues $\lambda_1 = -4$, $\lambda_2 = 10$ imply $\mathbf{A}$ is diagonalizable.

$$\mathbf{P} = \begin{pmatrix} -3 & 1 \\ 1 & -5 \end{pmatrix}, \quad \mathbf{D} = \begin{pmatrix} -4 & 0 \\ 0 & 10 \end{pmatrix}$$

9. Distinct eigenvalues $\lambda_1 = -i$, $\lambda_2 = i$ imply $\mathbf{A}$ is diagonalizable.

$$\mathbf{P} = \begin{pmatrix} 1 & 1 \\ -i & i \end{pmatrix}, \quad \mathbf{D} = \begin{pmatrix} -i & 0 \\ 0 & i \end{pmatrix}$$

12. Distinct eigenvalues $\lambda_1 = 3$, $\lambda_2 = 4$, $\lambda_3 = 5$ imply $\mathbf{A}$ is diagonalizable.

$$\mathbf{P} = \begin{pmatrix} 1 & 2 & 0 \\ 0 & 2 & 1 \\ 1 & 1 & -1 \end{pmatrix}, \quad \mathbf{D} = \begin{pmatrix} 3 & 0 & 0 \\ 0 & 4 & 0 \\ 0 & 0 & 5 \end{pmatrix}$$

15. The eigenvalues are $\lambda_1 = \lambda_2 = 1$, $\lambda_3 = 2$. For $\lambda_1 = \lambda_2 = 1$ we obtain the single eigenvector $\mathbf{K}_1 = \begin{pmatrix} 1 \\ 0 \\ 0 \end{pmatrix}$.

Hence $\mathbf{A}$ is not diagonalizable.

18. For $\lambda_1 = \lambda_2 = \lambda_3 = 1$ we obtain the single eigenvector $\mathbf{K}_1 = \begin{pmatrix} 1 \\ -2 \\ 1 \end{pmatrix}$. Hence $\mathbf{A}$ is not diagonalizable.

21. $\lambda_1 = 0$, $\lambda_2 = 2$, $\mathbf{K}_1 = \begin{pmatrix} 1 \\ -1 \end{pmatrix}$, $\mathbf{K}_2 = \begin{pmatrix} 1 \\ 1 \end{pmatrix}$, $\mathbf{P} = \begin{pmatrix} \frac{1}{\sqrt{2}} & \frac{1}{\sqrt{2}} \\ -\frac{1}{\sqrt{2}} & \frac{1}{\sqrt{2}} \end{pmatrix}$, $\mathbf{D} = \begin{pmatrix} 0 & 0 \\ 0 & 2 \end{pmatrix}$

24. $\lambda_1 = -1$, $\lambda_2 = 3$, $\mathbf{K}_1 = \begin{pmatrix} 1 \\ 1 \end{pmatrix}$, $\mathbf{K}_2 = \begin{pmatrix} 1 \\ -1 \end{pmatrix}$, $\mathbf{P} = \begin{pmatrix} \frac{1}{\sqrt{2}} & \frac{1}{\sqrt{2}} \\ \frac{1}{\sqrt{2}} & -\frac{1}{\sqrt{2}} \end{pmatrix}$, $\mathbf{D} = \begin{pmatrix} -1 & 0 \\ 0 & 3 \end{pmatrix}$

27. $\lambda_1 = 3$, $\lambda_2 = 6$, $\lambda_3 = 9$, $\mathbf{K}_1 = \begin{pmatrix} 2 \\ 2 \\ 1 \end{pmatrix}$, $\mathbf{K}_2 = \begin{pmatrix} 2 \\ -1 \\ -2 \end{pmatrix}$, $\mathbf{K}_3 = \begin{pmatrix} 1 \\ -2 \\ 2 \end{pmatrix}$, $\mathbf{P} = \begin{pmatrix} \frac{2}{3} & \frac{2}{3} & \frac{1}{3} \\ \frac{2}{3} & -\frac{1}{3} & -\frac{2}{3} \\ \frac{1}{3} & -\frac{2}{3} & \frac{2}{3} \end{pmatrix}$, $\mathbf{D} = \begin{pmatrix} 3 & 0 & 0 \\ 0 & 6 & 0 \\ 0 & 0 & 9 \end{pmatrix}$

30. $\lambda_1 = \lambda_2 = 0$, $\lambda_3 = -2$, $\lambda_4 = 2$, $\mathbf{K}_1 = \begin{pmatrix} -1 \\ 0 \\ 1 \\ 0 \end{pmatrix}$, $\mathbf{K}_2 = \begin{pmatrix} 0 \\ -1 \\ 0 \\ 1 \end{pmatrix}$, $\mathbf{K}_3 = \begin{pmatrix} 1 \\ -1 \\ 1 \\ -1 \end{pmatrix}$, $\mathbf{K}_4 = \begin{pmatrix} 1 \\ 1 \\ 1 \\ 1 \end{pmatrix}$

$$\mathbf{P} = \begin{pmatrix} -\frac{1}{\sqrt{2}} & 0 & \frac{1}{2} & \frac{1}{2} \\ 0 & -\frac{1}{\sqrt{2}} & -\frac{1}{2} & \frac{1}{2} \\ \frac{1}{\sqrt{2}} & 0 & \frac{1}{2} & \frac{1}{2} \\ 0 & \frac{1}{\sqrt{2}} & -\frac{1}{2} & \frac{1}{2} \end{pmatrix}, \quad \mathbf{D} = \begin{pmatrix} 0 & 0 & 0 & 0 \\ 0 & 0 & 0 & 0 \\ 0 & 0 & -2 & 0 \\ 0 & 0 & 0 & 2 \end{pmatrix}$$

33. The given equation can be written as $\mathbf{X}^T\mathbf{A}\mathbf{X} = 20$: $(x \quad y)\begin{pmatrix} -3 & 4 \\ 4 & 3 \end{pmatrix}\begin{pmatrix} x \\ y \end{pmatrix} = 20$. Using

$\lambda_1 = 5, \ \lambda_2 = -5, \ \mathbf{K}_1 = \begin{pmatrix} 1 \\ 2 \end{pmatrix}, \ \mathbf{K}_2 = \begin{pmatrix} -2 \\ 1 \end{pmatrix}, \ \mathbf{P} = \begin{pmatrix} \frac{1}{\sqrt{5}} & -\frac{2}{\sqrt{5}} \\ \frac{2}{\sqrt{5}} & \frac{1}{\sqrt{5}} \end{pmatrix}$ and $\mathbf{X} = \mathbf{P}\mathbf{X}'$ we find

$$(X \quad Y)\begin{pmatrix} 5 & 0 \\ 0 & -5 \end{pmatrix}\begin{pmatrix} X \\ Y \end{pmatrix} = 20 \quad \text{or} \quad 5X^2 - 5Y^2 = 20.$$

The conic section is a hyperbola. Now from $\mathbf{X}' = \mathbf{P}^T\mathbf{X}$ we see that the XY-coordinates of $(1,2)$ and $(-2,1)$ are $(\sqrt{5},0)$ and $(0,\sqrt{5})$, respectively. From this we conclude that the X-axis and Y-axis are as shown in the accompanying figure.

36. Since eigenvectors are mutually orthogonal we use an orthogonal matrix $\mathbf{P}$ and $\mathbf{A} = \mathbf{P}\mathbf{D}\mathbf{P}^T$.

$$\mathbf{A} = \begin{pmatrix} \frac{1}{\sqrt{3}} & \frac{1}{\sqrt{2}} & \frac{1}{\sqrt{6}} \\ -\frac{1}{\sqrt{3}} & 0 & \frac{2}{\sqrt{6}} \\ \frac{1}{\sqrt{3}} & -\frac{1}{\sqrt{2}} & \frac{1}{\sqrt{6}} \end{pmatrix}\begin{pmatrix} 1 & 0 & 0 \\ 0 & 3 & 0 \\ 0 & 0 & 5 \end{pmatrix}\begin{pmatrix} \frac{1}{\sqrt{3}} & -\frac{1}{\sqrt{3}} & \frac{1}{\sqrt{3}} \\ \frac{1}{\sqrt{2}} & 0 & -\frac{1}{\sqrt{2}} \\ \frac{1}{\sqrt{6}} & \frac{2}{\sqrt{6}} & \frac{1}{\sqrt{6}} \end{pmatrix} = \begin{pmatrix} \frac{8}{3} & \frac{4}{3} & -\frac{1}{3} \\ \frac{4}{3} & \frac{11}{3} & \frac{4}{3} \\ -\frac{1}{3} & \frac{4}{3} & \frac{8}{3} \end{pmatrix}$$

39. $\lambda_1 = 2, \ \lambda_2 = -1, \ \mathbf{K}_1 = \begin{pmatrix} 1 \\ 1 \end{pmatrix}, \ \mathbf{K}_2 = \begin{pmatrix} -1 \\ 2 \end{pmatrix}, \ \mathbf{P} = \begin{pmatrix} 1 & -1 \\ 1 & 2 \end{pmatrix}, \ \mathbf{P}^{-1} = \begin{pmatrix} \frac{2}{3} & \frac{1}{3} \\ -\frac{1}{3} & \frac{1}{3} \end{pmatrix}$

$$\mathbf{A}^5 = \begin{pmatrix} 1 & -1 \\ 1 & 2 \end{pmatrix}\begin{pmatrix} 32 & 0 \\ 0 & -1 \end{pmatrix}\begin{pmatrix} \frac{2}{3} & \frac{1}{3} \\ -\frac{1}{3} & \frac{1}{3} \end{pmatrix} = \begin{pmatrix} 21 & 11 \\ 22 & 10 \end{pmatrix}$$

EXERCISES 8.13

Cryptography

3. (a) The message is $\mathbf{M} = \begin{pmatrix} 16 & 8 & 15 & 14 & 5 \\ 0 & 8 & 15 & 13 & 5 \end{pmatrix}$. The encoded message is

$$\mathbf{B} = \mathbf{A}\mathbf{M} = \begin{pmatrix} 3 & 5 \\ 2 & 3 \end{pmatrix}\begin{pmatrix} 16 & 8 & 15 & 14 & 5 \\ 0 & 8 & 15 & 13 & 5 \end{pmatrix} = \begin{pmatrix} 48 & 64 & 120 & 107 & 40 \\ 32 & 40 & 75 & 67 & 25 \end{pmatrix}.$$

(b) The decoded message is

$$\mathbf{M} = \mathbf{A}^{-1}\mathbf{B} = \begin{pmatrix} -3 & 5 \\ 2 & -3 \end{pmatrix}\begin{pmatrix} 48 & 64 & 120 & 107 & 40 \\ 32 & 40 & 75 & 67 & 25 \end{pmatrix} = \begin{pmatrix} 16 & 8 & 15 & 14 & 5 \\ 0 & 8 & 15 & 13 & 5 \end{pmatrix}.$$

6. (a) The message is $\mathbf{M} = \begin{pmatrix} 4 & 18 & 0 & 10 & 15 & 8 \\ 14 & 0 & 9 & 19 & 0 & 20 \\ 8 & 5 & 0 & 19 & 16 & 25 \end{pmatrix}$. The encoded message is

$$\mathbf{B} = \mathbf{A}\mathbf{M} = \begin{pmatrix} 5 & 3 & 0 \\ 4 & 3 & -1 \\ 5 & 2 & 2 \end{pmatrix}\begin{pmatrix} 4 & 18 & 0 & 10 & 15 & 8 \\ 14 & 0 & 9 & 19 & 0 & 20 \\ 8 & 5 & 0 & 19 & 16 & 25 \end{pmatrix} = \begin{pmatrix} 62 & 90 & 27 & 107 & 75 & 100 \\ 50 & 67 & 27 & 78 & 44 & 67 \\ 64 & 100 & 18 & 126 & 107 & 130 \end{pmatrix}.$$

(b) The decoded message is

$$\mathbf{M} = \mathbf{A}^{-1}\mathbf{B} = \begin{pmatrix} 8 & -6 & -3 \\ -13 & 10 & 5 \\ -7 & 5 & 3 \end{pmatrix} \begin{pmatrix} 62 & 90 & 27 & 107 & 75 & 100 \\ 50 & 67 & 27 & 78 & 44 & 67 \\ 64 & 100 & 18 & 126 & 107 & 130 \end{pmatrix} = \begin{pmatrix} 4 & 18 & 0 & 10 & 15 & 8 \\ 14 & 0 & 9 & 19 & 0 & 20 \\ 8 & 5 & 0 & 19 & 16 & 25 \end{pmatrix}.$$

9. The decoded message is

$$\mathbf{M} = \mathbf{A}^{-1}\mathbf{B} = \begin{pmatrix} 0 & 0 & 1 \\ 0 & 1 & 0 \\ 1 & 0 & -1 \end{pmatrix} \begin{pmatrix} 31 & 21 & 21 & 22 & 20 & 9 \\ 19 & 0 & 9 & 13 & 16 & 15 \\ 13 & 1 & 20 & 8 & 0 & 9 \end{pmatrix} = \begin{pmatrix} 13 & 1 & 20 & 8 & 0 & 9 \\ 19 & 0 & 9 & 13 & 16 & 15 \\ 18 & 20 & 1 & 14 & 20 & 0 \end{pmatrix}.$$

From correspondence (1) we obtain: MATH_IS_IMPORTANT.

12. (a) $\mathbf{M}^T = \begin{pmatrix} 22 & 8 & 19 & 27 & 21 & 3 & 3 & 27 & 21 & 18 & 21 \\ 13 & 3 & 21 & 22 & 3 & 25 & 27 & 6 & 7 & 14 & 23 \\ 2 & 27 & 21 & 7 & 27 & 5 & 21 & 17 & 2 & 25 & 7 \end{pmatrix}$

(b) $\mathbf{B}^T = \mathbf{M} = \begin{pmatrix} 1 & 1 & 0 \\ 1 & 0 & 1 \\ 1 & 1 & -1 \end{pmatrix} = \begin{pmatrix} 37 & 38 & 61 & 56 & 51 & 33 & 51 & 50 & 30 & 57 & 51 \\ 24 & 35 & 40 & 34 & 48 & 8 & 24 & 44 & 23 & 43 & 28 \\ 11 & -24 & 0 & 15 & -24 & 20 & 6 & -11 & 5 & -11 & 16 \end{pmatrix}$

(c) $\mathbf{B}\mathbf{A}^{-1} = \mathbf{B}\begin{pmatrix} -1 & 1 & 1 \\ 2 & -1 & -1 \\ 1 & 0 & -1 \end{pmatrix} = \mathbf{M}$

EXERCISES 8.14
An Error-Correcting Code

3. $(0 \ \ 0 \ \ 0 \ \ 1 \ \ 1)$

6. $(0 \ \ 1 \ \ 1 \ \ 0 \ \ 1 \ \ 0 \ \ 1 \ \ 0)$

9. Parity error

12. Parity error

15. In this problem $\mathbf{D} = (\, c_1 \ \ c_2 \ \ c_3 \,)$ and $\mathbf{P} = \begin{pmatrix} 1 & 1 & 0 & 1 \\ 1 & 0 & 1 & 1 \\ 0 & 1 & 1 & 1 \end{pmatrix}$.

$\mathbf{D}^T = \mathbf{P}(0 \ \ 1 \ \ 0 \ \ 1)^T = (0 \ \ 1 \ \ 0)^T$; $\quad \mathbf{C} = (0 \ \ 1 \ \ 0 \ \ 0 \ \ 1 \ \ 0 \ \ 1)$

18. In this problem $\mathbf{D} = (\, c_1 \ \ c_2 \ \ c_3 \,)$ and $\mathbf{P} = \begin{pmatrix} 1 & 1 & 0 & 1 \\ 1 & 0 & 1 & 1 \\ 0 & 1 & 1 & 1 \end{pmatrix}$.

$\mathbf{D}^T = \mathbf{P}(1 \ \ 1 \ \ 0 \ \ 0)^T = (0 \ \ 1 \ \ 1)^T$; $\quad \mathbf{C} = (0 \ \ 1 \ \ 1 \ \ 1 \ \ 1 \ \ 0 \ \ 0)$

21. In this problem $\mathbf{W}$ represents the correctly decoded message.

$\mathbf{S} = \mathbf{HR}^T = \mathbf{H}(1 \quad 1 \quad 0 \quad 1 \quad 1 \quad 0 \quad 1) = (1 \quad 0 \quad 1)^T$; not a code word. The error is in the fifth bit.
$\mathbf{W} = (0 \quad 0 \quad 0 \quad 1)$

24. In this problem $\mathbf{W}$ represents the correctly decoded message.

$\mathbf{S} = \mathbf{HR}^T = \mathbf{H}(1 \quad 1 \quad 0 \quad 0 \quad 1 \quad 1 \quad 0) = (0 \quad 0 \quad 0)^T$; a code word. $\mathbf{W} = (0 \quad 1 \quad 1 \quad 0)$

27. In this problem $\mathbf{W}$ represents the correctly decoded message.

$\mathbf{S} = \mathbf{HR}^T = \mathbf{H}(1 \quad 0 \quad 1 \quad 1 \quad 0 \quad 1 \quad 1) = (1 \quad 1 \quad 1)^T$; not a code word. The error is in the seventh bit.
$\mathbf{W} = (1 \quad 0 \quad 1 \quad 0)$

30. (a) $c_4 = 0$, $c_3 = 1$, $c_2 = 1$, $c_1 = 0$; $(0 \quad 1 \quad 1 \quad 0 \quad 0 \quad 1 \quad 1 \quad 0)$

(b) $\mathbf{H} = \begin{pmatrix} 0 & 0 & 0 & 0 & 1 & 1 & 1 & 1 \\ 0 & 0 & 1 & 1 & 0 & 0 & 1 & 1 \\ 0 & 1 & 0 & 1 & 0 & 1 & 0 & 1 \\ 1 & 1 & 1 & 1 & 1 & 1 & 1 & 1 \end{pmatrix}$ (c) $\mathbf{S} = \mathbf{HR}^T = \mathbf{H}(0 \quad 0 \quad 1 \quad 1 \quad 1 \quad 1 \quad 0 \quad 0)^T = (0 \quad 0 \quad 0 \quad 0)^T$

EXERCISES 8.15

Method of Least Squares

3. We have $\mathbf{Y}^T = (1 \quad 1.5 \quad 3 \quad 4.5 \quad 5)$ and $\mathbf{A}^T = \begin{pmatrix} 1 & 2 & 3 & 4 & 5 \\ 1 & 1 & 1 & 1 & 1 \end{pmatrix}$.

Now $\mathbf{A}^T\mathbf{A} = \begin{pmatrix} 55 & 15 \\ 15 & 5 \end{pmatrix}$ and $(\mathbf{A}^T\mathbf{A})^{-1} = \frac{1}{50}\begin{pmatrix} 5 & -15 \\ -15 & 55 \end{pmatrix}$

so $\mathbf{X} = (\mathbf{A}^T\mathbf{A})^{-1}\mathbf{A}^T\mathbf{Y} = \begin{pmatrix} 1.1 \\ -0.3 \end{pmatrix}$ and the least squares line is $y = 1.1x - 0.3$.

6. We have $\mathbf{Y}^T = (2 \quad 2.5 \quad 1 \quad 1.5 \quad 2 \quad 3.2 \quad 5)$ and $\mathbf{A}^T = \begin{pmatrix} 1 & 2 & 3 & 4 & 5 & 6 & 7 \\ 1 & 1 & 1 & 1 & 1 & 1 & 1 \end{pmatrix}$.

Now $\mathbf{A}^T\mathbf{A} = \begin{pmatrix} 140 & 28 \\ 28 & 7 \end{pmatrix}$ and $(\mathbf{A}^T\mathbf{A})^{-1} = \frac{1}{196}\begin{pmatrix} 7 & -28 \\ -28 & 140 \end{pmatrix}$

so $\mathbf{X} = (\mathbf{A}^T\mathbf{A})^{-1}\mathbf{A}^T\mathbf{Y} = \begin{pmatrix} 0.407143 \\ 0.828571 \end{pmatrix}$ and the least squares line is $y = 0.407143x + 0.828571$.

EXERCISES 8.16

3. (a) In this problem we use the fact that the element τ_{ij} in the transfer matrix $\mathbf{T}$ is the rate of transfer from compartment j to compartment i, and the fact that the sum of each column in $\mathbf{T}$ is 1.

The initial state and the transfer matrix are

$$\mathbf{X}_0 = \begin{pmatrix} 100 \\ 0 \\ 0 \end{pmatrix} \quad \text{and} \quad \mathbf{T} = \begin{pmatrix} 0.2 & 0.5 & 0 \\ 0.3 & 0.1 & 0 \\ 0.5 & 0.4 & 1 \end{pmatrix}.$$

(b) We have

$$\mathbf{X}_1 = \mathbf{T}\mathbf{X}_0 = \begin{pmatrix} 20 \\ 30 \\ 50 \end{pmatrix} \quad \text{and} \quad \mathbf{X}_2 = \mathbf{T}\mathbf{X}_1 = \begin{pmatrix} 19 \\ 9 \\ 72 \end{pmatrix}.$$

(c) From $\mathbf{T}\hat{\mathbf{X}} - \hat{\mathbf{X}} = (\mathbf{T} - \mathbf{I})\hat{\mathbf{X}} = \mathbf{0}$ and the fact that the system is closed we obtain

$$\begin{aligned} -0.8x_1 + 0.5x_2 &= 0 \\ 0.3x_1 - 0.9x_2 &= 0 \\ x_1 + x_2 + x_3 &= 100. \end{aligned}$$

The solution is $x_1 = x_2 = 0$, $x_3 = 100$, so the equilibrium state is $\hat{\mathbf{X}} = \begin{pmatrix} 0 \\ 0 \\ 100 \end{pmatrix}$.

6. (a) The initial state and the transfer matrix are

$$\mathbf{X}_0 = \begin{pmatrix} 0 \\ 100 \\ 0 \end{pmatrix} \quad \text{and} \quad \mathbf{T} = \begin{pmatrix} 0.88 & 0.02 & 0 \\ 0.06 & 0.97 & 0.05 \\ 0.06 & 0.01 & 0.95 \end{pmatrix}.$$

(b)

Year	Phytoplankton	Water	Zooplankton
0	0.00	100.00	0.00
1	2.00	97.00	1.00
2	3.70	94.26	2.04
3	5.14	91.76	3.10
4	6.36	89.47	4.17
5	7.39	87.37	5.24
6	8.25	85.46	6.30
7	8.97	83.70	7.33
8	9.56	82.10	8.34
9	10.06	80.62	9.32
10	10.46	79.28	10.26
11	10.79	78.04	11.17
12	11.06	76.90	12.04

CHAPTER 8 REVIEW EXERCISES

3. $\mathbf{AB} = \begin{pmatrix} 3 & 4 \\ 6 & 8 \end{pmatrix}$; $\quad \mathbf{BA} = (11)$

6. True

9. 0

12. True

15. False; if the characteristic equation of an $n \times n$ matrix has repeated roots, there may not be n linearly independent eigenvectors.

18. True

21. $\mathbf{A} = \frac{1}{2}(\mathbf{A} + \mathbf{A}^T) + \frac{1}{2}(\mathbf{A} - \mathbf{A}^T)$ where $\frac{1}{2}(\mathbf{A} + \mathbf{A}^T)$ is symmetric and $\frac{1}{2}(\mathbf{A} - \mathbf{A}^T)$ is skew-symmetric.

24. (a) $\sigma_x \sigma_y = \begin{pmatrix} i & 0 \\ 0 & -i \end{pmatrix} = -\sigma_y \sigma_x$; $\quad \sigma_x \sigma_z = \begin{pmatrix} 0 & -1 \\ 1 & 0 \end{pmatrix} = -\sigma_z \sigma_x$; $\quad \sigma_y \sigma_z = \begin{pmatrix} 0 & i \\ i & 0 \end{pmatrix} = -\sigma_z \sigma_y$

(b) We first note that for anticommuting matrices $\mathbf{AB} = -\mathbf{BA}$, so $\mathbf{C} = 2\mathbf{AB}$. Then $\mathbf{C}_{xy} = \begin{pmatrix} 2i & 0 \\ 0 & -2i \end{pmatrix}$,

$\mathbf{C}_{yz} = \begin{pmatrix} 0 & 2i \\ 2i & 0 \end{pmatrix}$, and $\mathbf{C}_{zx} = \begin{pmatrix} 0 & 2 \\ -2 & 0 \end{pmatrix}$.

27. Multiplying the second row by abc we obtain the third row. Thus the determinant is 0.

30. $(-3)(6)(9)(1) = -162$

33. From $x_1 I_2 + x_2 HNO_3 \rightarrow x_3 HIO_3 + x_4 NO_2 + x_5 H_2O$ we obtain the system $2x_1 = x_3$, $x_2 = x_3 + 2x_5$, $x_2 = x_4$, $3x_2 = 3x_3 + 2x_4 + x_5$. Letting $x_4 = x_2$ in the fourth equation we obtain $x_2 = 3x_3 + x_5$. Taking $x_1 = t$ we see that

$x_3 = 2t$, $x_2 = 2t + 2x_5$, and $x_2 = 6t + x_5$. From the latter two equations we get $x_5 = 4t$. Taking $t = 1$ we have $x_1 = 1$, $x_2 = 10$, $x_3 = 2$, $x_4 = 10$, and $x_5 = 4$. The balanced equation is $I_2 + 10HNO_3 \rightarrow 2HIO_3 + 10NO_2 + 4H_2O$.

36. det $= 4$, det $\mathbf{A}_1 = 16$, det $\mathbf{A}_2 = -4$, det $\mathbf{A}_3 = 0$; $x_1 = \dfrac{16}{4} = 4$, $x_2 = \dfrac{-4}{4} = -1$, $x_3 = \dfrac{0}{4} = 0$

39. $\mathbf{AX} = \mathbf{B}$ is $\begin{pmatrix} 2 & 3 & -1 \\ 1 & -2 & 0 \\ -2 & 0 & 1 \end{pmatrix} \begin{pmatrix} x_1 \\ x_2 \\ x_3 \end{pmatrix} = \begin{pmatrix} 6 \\ -3 \\ 9 \end{pmatrix}$. Since $\mathbf{A}^{-1} = -\dfrac{1}{3} \begin{pmatrix} -2 & -3 & -2 \\ -1 & 0 & -1 \\ -4 & -6 & -7 \end{pmatrix}$, we have

$\mathbf{X} = \mathbf{A}^{-1}\mathbf{B} = \begin{pmatrix} 7 \\ 5 \\ 23 \end{pmatrix}$.

42. From the characteristic equation $\lambda^2 = 0$ we see that the eigenvalues are $\lambda_1 = \lambda_2 = 0$. For $\lambda_1 = \lambda_2 = 0$ we have $4k_1 = 0$ and $\mathbf{K}_1 = \begin{pmatrix} 0 \\ 1 \end{pmatrix}$ is a single eigenvector.

45. From the characteristic equation $-\lambda^3 - \lambda^2 + 21\lambda + 45 = -(\lambda + 3)^2(\lambda - 5) = 0$ we see that the eigenvalues are $\lambda_1 = \lambda_2 = -3$ and $\lambda_3 = 5$. For $\lambda_1 = \lambda_2 = -3$ we have

$$\begin{pmatrix} 1 & 2 & -3 & | & 0 \\ 2 & 4 & -6 & | & 0 \\ -1 & -2 & 3 & | & 0 \end{pmatrix} \xrightarrow[\text{operations}]{\text{row}} \begin{pmatrix} 1 & 2 & -3 & | & 0 \\ 0 & 0 & 0 & | & 0 \\ 0 & 0 & 0 & | & 0 \end{pmatrix}.$$

Thus $\mathbf{K}_1 = (-2 \ \ 1 \ \ 0)^T$ and $\mathbf{K}_2 = (3 \ \ 0 \ \ 1)^T$. For $\lambda_3 = 5$ we have

$$\begin{pmatrix} -7 & 2 & -3 & | & 0 \\ 2 & -4 & -6 & | & 0 \\ -1 & -2 & -5 & | & 0 \end{pmatrix} \xrightarrow[\text{operations}]{\text{row}} \begin{pmatrix} 1 & -\frac{2}{7} & \frac{3}{7} & | & 0 \\ 0 & 1 & 2 & | & 0 \\ 0 & 0 & 0 & | & 0 \end{pmatrix}.$$

Thus $\mathbf{K}_3 = (-1 \ \ -2 \ \ 1)^T$.

48. (a) Eigenvalues are $\lambda_1 = \lambda_2 = 0$ and $\lambda_3 = 5$ with corresponding eigenvectors $\mathbf{K}_1 = (0 \ \ 1 \ \ 0)^T$, $\mathbf{K}_2 = (2 \ \ 0 \ \ 1)^T$, and $\mathbf{K}_3 = (-1 \ \ 0 \ \ 2)^T$. Since $\|\mathbf{K}_1\| = 1$, $\|\mathbf{K}_2\| = \sqrt{5}$, and $\|\mathbf{K}_3\| = \sqrt{5}$, we have

$$\mathbf{P} = \begin{pmatrix} 0 & \frac{2}{\sqrt{5}} & -\frac{1}{\sqrt{5}} \\ 1 & 0 & 0 \\ 0 & \frac{1}{\sqrt{5}} & \frac{2}{\sqrt{5}} \end{pmatrix} \quad \text{and} \quad \mathbf{P}^{-1} = \mathbf{P}^T = \begin{pmatrix} 0 & 1 & 0 \\ \frac{2}{\sqrt{5}} & 0 & \frac{1}{\sqrt{5}} \\ -\frac{1}{\sqrt{5}} & 0 & \frac{2}{\sqrt{5}} \end{pmatrix}.$$

(b) $\mathbf{P}^{-1}\mathbf{AP} = \begin{pmatrix} 0 & 0 & 0 \\ 0 & 0 & 0 \\ 0 & 0 & 5 \end{pmatrix}$

51. The encoded message is

$$\mathbf{B} = \mathbf{AM} = \begin{pmatrix} 10 & 1 \\ 9 & 1 \end{pmatrix} \begin{pmatrix} 19 & 1 & 20 & 5 & 12 & 12 & 9 & 20 & 5 & 0 & 12 & 1 & 21 \\ 14 & 3 & 8 & 5 & 4 & 0 & 15 & 14 & 0 & 6 & 18 & 9 & 0 \end{pmatrix}$$

$$= \begin{pmatrix} 204 & 13 & 208 & 55 & 124 & 120 & 105 & 214 & 50 & 6 & 138 & 19 & 210 \\ 185 & 12 & 188 & 50 & 112 & 108 & 96 & 194 & 45 & 6 & 126 & 18 & 189 \end{pmatrix}.$$

54. The decoded message is

$$\mathbf{M} = \mathbf{A}^{-1}\mathbf{B} = \begin{pmatrix} -3 & 2 & -1 \\ 1 & 0 & 0 \\ 2 & -1 & 1 \end{pmatrix} \begin{pmatrix} 5 & 2 & 21 \\ 27 & 17 & 40 \\ 21 & 13 & -2 \end{pmatrix} = \begin{pmatrix} 18 & 15 & 19 \\ 5 & 2 & 21 \\ 4 & 0 & 0 \end{pmatrix}.$$

From correspondence (1) we obtain: ROSEBUD_ _.

9 Vector Calculus

3.

6.

9.

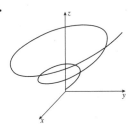

Note: the scale is distorted in this graph. For $t = 0$, the graph starts at $(1, 0, 1)$. The upper loop shown intersects the xz-plane at about $(286751, 0, 286751)$.

12. $x = t$, $y = 2t$, $z = \pm\sqrt{t^2 + 4t^2 + 1} = \pm\sqrt{5t^2 - 1}$; $\mathbf{r}(t) = t\mathbf{i} + 2t\mathbf{j} \pm \sqrt{5t^2 - 1}\,\mathbf{k}$

15. $\mathbf{r}(t) = \dfrac{\sin 2t}{t}\mathbf{i} + (t-2)^5\mathbf{j} + \dfrac{\ln t}{1/t}\mathbf{k}$. Using L'Hôpital's Rule,

$$\lim_{t \to 0^+} \mathbf{r}(t) = \left[\frac{2\cos 2t}{1}\mathbf{i} + (t-2)^5\mathbf{j} + \frac{1/t}{-1/t^2}\mathbf{k}\right] = 2\mathbf{i} - 32\mathbf{j}.$$

18. $\mathbf{r}'(t) = \langle -t\sin t, 1 - \sin t\rangle$; $\mathbf{r}''(t) = \langle -t\cos t - \sin t, -\cos t\rangle$

171

21. $\mathbf{r}'(t) = -2\sin t\,\mathbf{i} + 6\cos t\,\mathbf{j}$

$\mathbf{r}'(\pi/6) = -\mathbf{i} + 3\sqrt{3}\,\mathbf{j}$

24. $\mathbf{r}'(t) = -3\sin t\,\mathbf{i} + 3\cos t\,\mathbf{j} + 2\mathbf{k}$

$\mathbf{r}'(\pi/4) = \dfrac{-3\sqrt{2}}{2}\,\mathbf{i} + \dfrac{3\sqrt{2}}{2}\,\mathbf{j} + 2\mathbf{k}$

27. $\dfrac{d}{dt}[\mathbf{r}(t) \times \mathbf{r}'(t)] = \mathbf{r}(t) \times \mathbf{r}''(t) + \mathbf{r}'(t) \times \mathbf{r}'(t) = \mathbf{r}(t) \times \mathbf{r}''(t)$

30. $\dfrac{d}{dt}[\mathbf{r}_1(t) \times (\mathbf{r}_2(t) \times \mathbf{r}_3(t))] = \mathbf{r}_1(t) \times \dfrac{d}{dt}(\mathbf{r}_2(t) \times \mathbf{r}_3(t)) + \mathbf{r}_1'(t) \times (\mathbf{r}_2(t) \times \mathbf{r}_3(t))$

$\qquad = \mathbf{r}_1(t) \times (\mathbf{r}_2(t) \times \mathbf{r}_3'(t) + \mathbf{r}_2'(t) \times \mathbf{r}_3(t)) + \mathbf{r}_1'(t) \times (\mathbf{r}_2(t) \times \mathbf{r}_3(t))$

$\qquad = \mathbf{r}_1(t) \times (\mathbf{r}_2(t) \times \mathbf{r}_3'(t)) + \mathbf{r}_1(t) \times (\mathbf{r}_2'(t) \times \mathbf{r}_3(t)) + \mathbf{r}_1(t) \times (\mathbf{r}_2(t) \times \mathbf{r}_3(t))$

33. $\displaystyle\int_{-1}^{2} \mathbf{r}(t)\,dt = \left[\int_{-1}^{2} t\,dt\right]\mathbf{i} + \left[\int_{-1}^{2} 3t^2\,dt\right]\mathbf{j} + \left[\int_{-1}^{2} 4t^3\,dt\right]\mathbf{k} = \dfrac{1}{2}t^2 \Big|_{-1}^{2}\mathbf{i} + t^3 \Big|_{-1}^{2}\mathbf{j} + t^4 \Big|_{-1}^{2}\mathbf{k} = \dfrac{3}{2}\mathbf{i} + 9\mathbf{j} + 15\mathbf{k}$

36. $\displaystyle\int \mathbf{r}(t)\,dt = \left[\int \dfrac{1}{1+t^2}\,dt\right]\mathbf{i} + \left[\int \dfrac{t}{1+t^2}\,dt\right]\mathbf{j} + \left[\int \dfrac{t^2}{1+t^2}\,dt\right]\mathbf{k}$

$\qquad = [\tan^{-1} t + c_1]\mathbf{i} + \left[\dfrac{1}{2}\ln(1+t^2) + c_2\right]\mathbf{j} + \left[\int \left(1 - \dfrac{1}{1+t^2}\right)dt\right]\mathbf{k}$

$\qquad = [\tan^{-1} t + c_1]\mathbf{i} + \left[\dfrac{1}{2}\ln(1+t^2) + c_2\right]\mathbf{j} + [t - \tan^{-1} t + c_3]\mathbf{k}$

$\qquad = \tan^{-1} t\,\mathbf{i} + \dfrac{1}{2}\ln(1+t^2)\mathbf{j} + (t - \tan^{-1} t)\mathbf{k} + \mathbf{c},$

where $\mathbf{c} = c_1\mathbf{i} + c_2\mathbf{j} + c_3\mathbf{k}$.

39. $\mathbf{r}'(t) = \displaystyle\int \mathbf{r}''(t)\,dt = \left[\int 12t\,dt\right]\mathbf{i} + \left[\int -3t^{-1/2}\,dt\right]\mathbf{j} + \left[\int 2\,dt\right]\mathbf{k} = [6t^2 + c_1]\mathbf{i} + [-6t^{1/2} + c_2]\mathbf{j} + [2t + c_3]\mathbf{k}$

Since $\mathbf{r}'(1) = \mathbf{j} = (6 + c_1)\mathbf{i} + (-6 + c_2)\mathbf{j} + (2 + c_3)\mathbf{k}$, $c_1 = -6$, $c_2 = 7$, and $c_3 = -2$. Thus,

$$\mathbf{r}'(t) = (6t^2 - 6)\mathbf{i} + (-6t^{1/2} + 7)\mathbf{j} + (2t - 2)\mathbf{k}.$$

$\mathbf{r}(t) = \displaystyle\int \mathbf{r}'(t)\,dt = \left[\int (6t^2 - 6)\,dt\right]\mathbf{i} + \left[\int (-6t^{1/2} + 7)\,dt\right]\mathbf{j} + \left[\int (2t - 2)\,dt\right]\mathbf{k}$

$\qquad = [2t^3 - 6t + c_4]\mathbf{i} + [-4t^{3/2} + 7t + c_5]\mathbf{j} + [t^2 - 2t + c_6]\mathbf{k}.$

Since

$$\mathbf{r}(1) = 2\mathbf{i} - \mathbf{k} = (-4 + c_4)\mathbf{i} + (3 + c_5)\mathbf{j} + (-1 + c_6)\mathbf{k},$$

$c_4 = 6$, $c_5 = -3$, and $c_6 = 0$. Thus,

$$\mathbf{r}(t) = (2t^3 - 6t + 6)\mathbf{i} + (-4t^{3/2} + 7t - 3)\mathbf{j} + (t^2 - 2t)\mathbf{k}.$$

42. $\mathbf{r}'(t) = \mathbf{i} + (\cos t - t\sin t)\mathbf{j} + (\sin t + t\cos t)\mathbf{k}$

$\|\mathbf{r}'(t)\| = \sqrt{1^2 + (\cos t - t\sin t)^2 + (\sin t + t\cos t)^2} = \sqrt{2 + t^2}$

$s = \displaystyle\int_0^{\pi} \sqrt{2 + t^2}\,dt = \left(\dfrac{t}{2}\sqrt{2 + t^2} + \ln\left|t + \sqrt{2 + t^2}\right|\right)\Big|_0^{\pi} = \dfrac{\pi}{2}\sqrt{2 + \pi^2} + \ln(\pi + \sqrt{2 + \pi^2}) - \ln\sqrt{2}$

45. $\mathbf{r}'(t) = -a\sin t\mathbf{i} + a\cos t\mathbf{j}$; $\|\mathbf{r}'(t)\| = \sqrt{a^2\sin^2 t + a^2\cos^2 t} = a$, $a > 0$; $s = \int_0^t a\,du = at$

$\mathbf{r}(s) = a\cos(s/a)\mathbf{i} + a\sin(s/a)\mathbf{j}$; $\mathbf{r}'(s) = -\sin(s/a)\mathbf{i} + \cos(s/a)\mathbf{j}$

$\|\mathbf{r}'(s)\| = \sqrt{\sin^2(s/a) + \cos^2(s/a)} = 1$

48. Since $\|\mathbf{r}(t)\|$ is the length of $\mathbf{r}(t)$, $\|\mathbf{r}(t)\| = c$ represents a curve lying on a sphere of radius c centered at the origin.

51. $\dfrac{d}{dt}[\mathbf{r}_1(t) \times \mathbf{r}_2(t)] = \lim\limits_{h\to 0} \dfrac{\mathbf{r}_1(t+h) \times \mathbf{r}_2(t+h) - \mathbf{r}_1(t) \times \mathbf{r}_2(t)}{h}$

$= \lim\limits_{h\to 0} \dfrac{\mathbf{r}_1(t+h) \times \mathbf{r}_2(t+h) - \mathbf{r}_1(t+h) \times \mathbf{r}_2(t) + \mathbf{r}_1(t+h) \times \mathbf{r}_2(t) - \mathbf{r}_1(t) \times \mathbf{r}_2(t)}{h}$

$= \lim\limits_{h\to 0} \dfrac{\mathbf{r}_1(t+h) \times [\mathbf{r}_2(t+h) - \mathbf{r}_2(t)]}{h} + \lim\limits_{h\to 0} \dfrac{[\mathbf{r}_1(t+h) - \mathbf{r}_1(t)] \times \mathbf{r}_2(t)}{h}$

$= \mathbf{r}_1(t) \times \left(\lim\limits_{h\to 0} \dfrac{\mathbf{r}_2(t+h) - \mathbf{r}_2(t)}{h}\right) + \left(\lim\limits_{h\to 0} \dfrac{\mathbf{r}_1(t+h) - \mathbf{r}_1(t)}{h}\right) \times \mathbf{r}_2(t)$

$= \mathbf{r}_1(t) \times \mathbf{r}_2'(t) + \mathbf{r}_1'(t) \times \mathbf{r}_2(t)$

EXERCISES 9.2

Motion on a Curve

3. $\mathbf{v}(t) = -2\sinh 2t\mathbf{i} + 2\cosh 2t\mathbf{j}$; $\mathbf{v}(0) = 2\mathbf{j}$; $\|\mathbf{v}(0)\| = 2$;
$\mathbf{a}(t) = -4\cosh 2t\mathbf{i} + 4\sinh 2t\mathbf{j}$; $\mathbf{a}(0) = -4\mathbf{i}$

6. $\mathbf{v}(t) = \mathbf{i}+\mathbf{j}+3t^2\mathbf{k}$; $\mathbf{v}(2) = \mathbf{i}+\mathbf{j}+12\mathbf{k}$; $\|\mathbf{v}(2)\| = \sqrt{1+1+144} = \sqrt{146}$; $\mathbf{a}(t) = 6t\mathbf{k}$;
$\mathbf{a}(2) = 12\mathbf{k}$

9. The particle passes through the xy-plane when $z(t) = t^2 - 5t = 0$ or $t = 0, 5$ which gives us the points $(0,0,0)$ and $(25, 115, 0)$. $\mathbf{v}(t) = 2t\mathbf{i}+(3t^2-2)\mathbf{j}+(2t-5)\mathbf{k}$; $\mathbf{v}(0) = -2\mathbf{j}-5\mathbf{k}$; $\mathbf{v}(5) = 10\mathbf{i}+73\mathbf{j}+5\mathbf{k}$; $\mathbf{a}(t) = 2\mathbf{i}+6t\mathbf{j}+2\mathbf{k}$;
$\mathbf{a}(0) = 2\mathbf{i} + 2\mathbf{k}$, $\mathbf{a}(5) = 2\mathbf{i} + 30\mathbf{j} + 2\mathbf{k}$

12. Initially we are given $\mathbf{s}_0 = 1600\mathbf{j}$ and $\mathbf{v}_0 = (480\cos 30°)\mathbf{i} + (480\sin 30°)\mathbf{j} = 240\sqrt{3}\,\mathbf{i} + 240\mathbf{j}$. Using $\mathbf{a}(t) = -32\mathbf{j}$ we find

$$\mathbf{v}(t) = \int \mathbf{a}(t)\,dt = -32t\mathbf{j} + \mathbf{c}$$

$$240\sqrt{3}\,\mathbf{i} + 240\mathbf{j} = \mathbf{v}(0) = \mathbf{c}$$

$$\mathbf{v}(t) = -32t\mathbf{j} + 240\sqrt{3}\,\mathbf{i} + 240\mathbf{j} = 240\sqrt{3}\,\mathbf{i} + (240 - 32t)\mathbf{j}$$

$$\mathbf{r}(t) = \int \mathbf{v}(t)\, dt = 240\sqrt{3}\, t\mathbf{i} + (240t - 16t^2)\mathbf{j} + \mathbf{b}$$

$$1600\mathbf{j} = \mathbf{r}(0) = \mathbf{b}.$$

(a) The shell's trajectory is given by $\mathbf{r}(t) = 240\sqrt{3}\, t\mathbf{i} + (240t - 16t^2 + 1600)\mathbf{j}$ or $x = 240\sqrt{3}\, t$, $y = 240t - 16t^2 + 1600$.

(b) Solving $dy/dt = 240 - 32t = 0$, we see that y is maximum when $t = 15/2$. The maximum altitude is $y(15/2) = 2500$ ft.

(c) Solving $y(t) = -16t^2 + 240t + 1600 = -16(t-20)(t+5) = 0$, we see that the shell hits the ground when $t = 20$. The range of the shell is $x(20) = 4800\sqrt{3} \approx 8314$ ft.

(d) From (c), impact is when $t = 20$. The speed at impact is

$$\|\mathbf{v}(20)\| = |240\sqrt{3}\,\mathbf{i} + (240 - 32\cdot 20)\mathbf{j}| = \sqrt{240^2\cdot 3 + (-400)^2} = 160\sqrt{13} \approx 577 \text{ ft/s}.$$

15. Let s be the initial speed. Then $\mathbf{v}(0) = s\cos 45°\mathbf{i} + s\sin 45°\mathbf{j} = \dfrac{s\sqrt{2}}{2}\mathbf{i} + \dfrac{s\sqrt{2}}{2}\mathbf{j}$. Using $\mathbf{a}(t) = -32\mathbf{j}$, we have

$$\mathbf{v}(t) = \int \mathbf{a}(t)\, dt = -32t\mathbf{j} + \mathbf{c}$$

$$\frac{s\sqrt{2}}{2}\mathbf{i} + \frac{s\sqrt{2}}{2}\mathbf{j} = \mathbf{v}(0) = \mathbf{c}$$

$$\mathbf{v}(t) = \frac{s\sqrt{2}}{2}\mathbf{i} + \left(\frac{s\sqrt{2}}{2} - 32t\right)\mathbf{j}$$

$$\mathbf{r}(t) = \frac{s\sqrt{2}}{2}\, t\mathbf{i} + \left(\frac{s\sqrt{2}}{2}\, t - 16t^2\right)\mathbf{j} + \mathbf{b}.$$

Since $\mathbf{r}(0) = \mathbf{0}$, $\mathbf{b} = \mathbf{0}$ and

$$\mathbf{r}(t) = \frac{s\sqrt{2}}{2}\, t\mathbf{i} + \left(\frac{s\sqrt{2}}{2}\, t - 16t^2\right)\mathbf{j}.$$

Setting $y(t) = s\sqrt{2}\, t/2 - 16t^2 = t(s\sqrt{2}/2 - 16t) = 0$ we see that the ball hits the ground when $t = \sqrt{2}\, s/32$. Thus, using $x(t) = s\sqrt{2}\, t/2$ and the fact that 100 yd $= 300$ ft, $300 = x(t) = \dfrac{s\sqrt{2}}{2}(\sqrt{2}\, s/32) = \dfrac{s^2}{32}$ and $s = \sqrt{9600} \approx 97.98$ ft/s.

18. The initial angle is $\theta = 0$, the initial height is 1024 ft, and the initial speed is $s = 180(5280)/3600 = 264$ ft/s. Then $x(t) = 264t$ and $y(t) = -16t^2 + 1024$. Solving $y(t) = 0$ we see that the pack hits the ground at $t = 8$ seconds The horizontal distance travelled is $x(8) = 2112$ feet. From the figure in the text, $\tan\alpha = 1024/2112 = 16/33$ and $\alpha \approx 0.45$ radian or $25.87°$.

21. By Problem 19, $a = v^2/r_0 = 1530^2/(4000\cdot 5280) \approx 0.1108$. We are given $mg = 192$, so $m = 192/32$ and $w_e = 192 - (192/32)(0.1108) \approx 191.33$ lb.

24. Since the projectile is launched from ground level, $s_0 = 0$. To find the maximum height we maximize $y(t) = -\frac{1}{2}gt^2 + (v_0\sin\theta)t$. Solving $y'(t) = -gt + v_0\sin\theta = 0$, we see that $t = (v_0/g)\sin\theta$ is a critical point. Since $y''(t) = -g < 0$,

$$H = y\left(\frac{v_0\sin\theta}{g}\right) = -\frac{1}{2}g\frac{v_0^2\sin^2\theta}{g^2} + v_0\sin\theta\frac{v_0\sin\theta}{g} = \frac{v_0^2\sin^2\theta}{2g}$$

is the maximum height. To find the range we solve $y(t) = -\frac{1}{2}gt^2 + (v_0\sin\theta)t = t(v_0\sin\theta - \frac{1}{2}gt) = 0$. The positive solution of this equation is $t = (2v_0\sin\theta)/g$. The range is thus

$$x(t) = (v_0\cos\theta)\frac{2v_0\sin\theta}{g} = \frac{v_0^2\sin 2\theta}{g}.$$

174

27. (a) Since $\mathbf{F}$ is directed along $\mathbf{r}$ we have $\mathbf{F} = c\mathbf{r}$ for some constant c. Then

$$\tau = \mathbf{r} \times \mathbf{F} = \mathbf{r} \times (c\mathbf{r}) = c(\mathbf{r} \times \mathbf{r}) = 0.$$

(b) If $\tau = 0$ then $d\mathbf{L}/dt = 0$ and $\mathbf{L}$ is constant.

EXERCISES 9.3

Curvature and Components of Acceleration

3. We assume $a > 0$. $\mathbf{r}'(t) = -a\sin t\mathbf{i} + a\cos t\mathbf{j} + c\mathbf{k}$; $|\mathbf{r}'(t)| = \sqrt{a^2\sin^2 t + a^2\cos^2 t + c^2} = \sqrt{a^2 + c^2}$;

$$\mathbf{T}(t) = -\frac{a\sin t}{\sqrt{a^2 + c^2}}\mathbf{i} + \frac{a\cos t}{\sqrt{a^2 + c^2}}\mathbf{j} + \frac{c}{\sqrt{a^2 + c^2}}\mathbf{k}; \quad \frac{d\mathbf{T}}{dt} = -\frac{a\cos t}{\sqrt{a^2 + c^2}}\mathbf{i} - \frac{a\sin t}{\sqrt{a^2 + c^2}}\mathbf{j},$$

$$\left|\frac{d\mathbf{T}}{dt}\right| = \sqrt{\frac{a^2\cos^2 t}{a^2 + c^2} + \frac{a^2\sin^2 t}{a^2 + c^2}} = \frac{a}{\sqrt{a^2 + c^2}}; \quad \mathbf{N} = -\cos t\mathbf{i} - \sin t\mathbf{j};$$

$$\mathbf{B} = \mathbf{T} \times \mathbf{N} = \begin{vmatrix} \mathbf{i} & \mathbf{j} & \mathbf{k} \\ -\dfrac{a\sin t}{\sqrt{a^2 + c^2}} & \dfrac{a\cos t}{\sqrt{a^2 + c^2}} & \dfrac{c}{\sqrt{a^2 + c^2}} \\ -\cos t & -\sin t & 0 \end{vmatrix} = \frac{c\sin t}{\sqrt{a^2 + c^2}}\mathbf{i} - \frac{c\cos t}{\sqrt{a^2 + c^2}}\mathbf{j} + \frac{a}{\sqrt{a^2 + c^2}}\mathbf{k};$$

$$\kappa = \frac{|d\mathbf{T}/dt|}{|\mathbf{r}'(t)|} = \frac{a/\sqrt{a^2 + c^2}}{\sqrt{a^2 + c^2}} = \frac{a}{a^2 + c^2}$$

6. From Problem 4, a normal to the osculating plane is $\mathbf{B}(1) = \frac{1}{\sqrt{6}}(\mathbf{i} - 2\mathbf{j} + \mathbf{k})$. The point on the curve when $t = 1$ is $(1, 1/2, 1/3)$. An equation of the plane is $(x - 1) - 2(y - 1/2) + (z - 1/3) = 0$ or $x - 2y + z = 1/3$.

9. $\mathbf{v}(t) = 2t\mathbf{i} + 2t\mathbf{j} + 4t\mathbf{k}$, $|\mathbf{v}(t)| = 2\sqrt{6}\,t$, $t > 0$; $\mathbf{a}(t) = 2\mathbf{i} + 2\mathbf{j} + 4\mathbf{k}$; $\mathbf{v} \cdot \mathbf{a} = 24t$, $\mathbf{v} \times \mathbf{a} = 0$;

$$a_T = \frac{24t}{2\sqrt{6}\,t} = 2\sqrt{6}, \quad a_N = 0, \ t > 0$$

12. $\mathbf{v}(t) = \dfrac{1}{1 + t^2}\mathbf{i} + \dfrac{t}{1 + t^2}\mathbf{j}$, $|\mathbf{v}(t)| = \dfrac{\sqrt{1 + t^2}}{1 + t^2}$; $\mathbf{a}(t) = -\dfrac{2t}{(1 + t^2)^2}\mathbf{i} + \dfrac{1 - t^2}{(1 + t^2)^2}\mathbf{j}$;

$$\mathbf{v} \cdot \mathbf{a} = -\frac{2t}{(1 + t^2)^3} + \frac{t - t^3}{(1 + t^2)^3} = -\frac{t}{(1 + t^2)^2}; \quad \mathbf{v} \times \mathbf{a} = \frac{1}{(1 + t^2)^2}\mathbf{k}, \ |\mathbf{v} \times \mathbf{a}| = \frac{1}{(1 + t^2)^2};$$

$$a_T = -\frac{t/(1 + t^2)^2}{\sqrt{1 + t^2}/(1 + t^2)} = -\frac{t}{(1 + t^2)^{3/2}}, \quad a_N = \frac{1/(1 + t^2)^2}{\sqrt{1 + t^2}/(1 + t^2)} = \frac{1}{(1 + t^2)^{3/2}}$$

15. $\mathbf{v}(t) = -e^{-t}(\mathbf{i} + \mathbf{j} + \mathbf{k})$, $|\mathbf{v}(t)| = \sqrt{3}\,e^{-t}$; $\mathbf{a}(t) = e^{-t}(\mathbf{i} + \mathbf{j} + \mathbf{k})$; $\mathbf{v} \cdot \mathbf{a} = -3e^{-2t}$; $\mathbf{v} \times \mathbf{a} = 0$, $|\mathbf{v} \times \mathbf{a}| = 0$; $a_T = -\sqrt{3}\,e^{-t}$, $a_N = 0$

18. (a) $\mathbf{v}(t) = -a\sin t\mathbf{i} + b\cos t\mathbf{j}$, $|\mathbf{v}(t)| = \sqrt{a^2\sin^2 t + b^2\cos^2 t}$; $\mathbf{a}(t) = -a\cos t\mathbf{i} - b\sin t\mathbf{j}$;

$$\mathbf{v} \times \mathbf{a} = ab\mathbf{k}; \quad |\mathbf{v} \times \mathbf{a}| = ab; \quad \kappa = \frac{ab}{(a^2\sin^2 t + b^2\cos^2 t)^{3/2}}$$

(b) When $a = b$, $|\mathbf{v}(t)| = a$, $|\mathbf{v} \times \mathbf{a}| = a^2$, and $\kappa = a^2/a^3 = 1/a$.

21. $\mathbf{v}(t) = f'(t)\mathbf{i} + g'(t)\mathbf{j}, \quad |\mathbf{v}(t)| = \sqrt{[f'(t)]^2 + [g'(t)]^2}; \quad \mathbf{a}(t) = f''(t)\mathbf{i} + g''(t)\mathbf{j};$

$\mathbf{v} \times \mathbf{a} = [f'(t)g''(t) - g'(t)f''(t)]\mathbf{k}, \quad |\mathbf{v} \times \mathbf{a}| = |f'(t)g''(t) - g'(t)f''(t)|;$

$\kappa = \dfrac{|\mathbf{v} \times \mathbf{a}|}{|\mathbf{v}|^3} = \dfrac{|f'(t)g''(t) - g'(t)f''(t)|}{([f'(t)]^2 + [g'(t)]^2)^{3/2}}$

24. $F(x) = x^3, \quad F(-1) = -1, \quad F(1/2) = 1/8; \quad F'(x) = 3x^2, \quad F'(-1) = 3, \quad F'(1/2) = 3/4; \quad F''(x) = 6x,$

$F''(-1) = -6, \quad F''(1/2) = 3; \quad \kappa(-1) = \dfrac{|-6|}{(1+3^2)^{3/2}} = \dfrac{6}{10\sqrt{10}} = \dfrac{3}{5\sqrt{10}} \approx 0.19;$

$\rho(-1) = \dfrac{5\sqrt{10}}{3} \approx 5.27; \quad \kappa\left(\dfrac{1}{2}\right) = \dfrac{3}{[1+(3/4)^2]^{3/2}} = \dfrac{3}{125/64} = \dfrac{192}{125} \approx 1.54; \quad \rho\left(\dfrac{1}{2}\right) = \dfrac{125}{192} \approx 0.65$

Since $1.54 > 0.19$, the curve is "sharper" at $(1/2, 1/8)$.

EXERCISES 9.4

Partial Derivatives

3. $x^2 - y^2 = 1 + c^2$

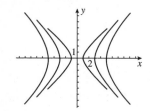

6. $y = x + \tan c, \quad -\pi/x < c < \pi/2$

9. $x^2 + 3y^2 + 6z^2 = c$; ellipsoid

12. Setting $x = -4$, $y = 2$, and $z = -3$ in $x^2/16 + y^2/4 + z^2/9 = c$ we obtain $c = 3$. The equation of the surface is $x^2/16 + y^2/4 + z^2/9 = 3$. Setting $y = z = 0$ we find the x-intercepts are $\pm 4\sqrt{3}$. Similarly, the y-intercepts are $\pm 2\sqrt{3}$ and the z-intercepts are $\pm 3\sqrt{3}$.

15. $z_x = 20x^3y^3 - 2xy^6 + 30x^4; \quad z_y = 15x^4y^2 - 6x^2y^5 - 4$

18. $z_x = 12x^2 - 10x + 8; \quad z_y = 0$

21. $z_x = 2(\cos 5x)(-\sin 5x)(5) = -10\sin 5x \cos 5x; \quad z_y = 2(\sin 5y)(\cos 5y)(5) = 10\sin 5y \cos 5y$

24. $f_\theta = \phi^2\left(\cos\dfrac{\theta}{\phi}\right)\left(\dfrac{1}{\phi}\right) = \phi\cos\dfrac{\theta}{\phi}; \quad f_\phi = \phi^2\left(\cos\dfrac{\theta}{\phi}\right)\left(-\dfrac{\theta}{\phi^2}\right) + 2\phi\sin\dfrac{\theta}{\phi} = -\theta\cos\dfrac{\theta}{\phi} + 2\phi\sin\dfrac{\theta}{\phi}$

27. $g_u = \dfrac{8u}{4u^2 + 5v^3}; \quad g_v = \dfrac{15v^2}{4u^2 + 5v^3}$

30. $w_x = xy\left(\dfrac{1}{x}\right) + (\ln xz)y = y + y\ln xz; \quad w_y = x\ln xz; \quad w_z = \dfrac{xy}{z}$

33. $\dfrac{\partial z}{\partial x} = \dfrac{2x}{x^2 + y^2}, \quad \dfrac{\partial^2 z}{\partial x^2} = \dfrac{(x^2 + y^2)2 - 2x(2x)}{(x^2 + y^2)^2} = \dfrac{2y^2 - 2x^2}{(x^2 + y^2)^2}; \quad \dfrac{\partial z}{\partial y} = \dfrac{2y}{x^2 + y^2},$

$\dfrac{\partial^2 z}{\partial y^2} = \dfrac{(x^2 + y^2)2 - 2y(2y)}{(x^2 + y^2)^2} = \dfrac{2x^2 - 2y^2}{(x^2 + y^2)^2}; \quad \dfrac{\partial^2 z}{\partial x^2} + \dfrac{\partial^2 z}{\partial y^2} = \dfrac{2y^2 - 2x^2 + 2x^2 - 2y^2}{(x^2 + y^2)^2} = 0$

36. $\dfrac{\partial u}{\partial x} = -\sin(x + at) + \cos(x - at),\quad \dfrac{\partial^2 u}{\partial x^2} = -\cos(x + at) - \sin(x - at);$

$\dfrac{\partial u}{\partial t} = -a\sin(x + at) - a\cos(x - at),\quad \dfrac{\partial^2 u}{\partial t^2} = -a^2\cos(x + at) - a^2\sin(x - at);$

$a^2\,\dfrac{\partial^2 u}{\partial x^2} = -a^2\cos(x + at) - a^2\sin(x - at) = \dfrac{\partial^2 u}{\partial t^2}$

39. $z_x = v^2 e^{uv^2}(3x^2) + 2uve^{uv^2}(1) = 3x^2 v^2 e^{uv^2} + 2uve^{uv^2};\quad z_y = v^2 e^{uv^2}(0) + 2uve^{uv^2}(-2y) = -4yuve^{uv^2}$

42. $z_u = \dfrac{2y}{(x + y)^2}\left(\dfrac{1}{v}\right) + \dfrac{-2x}{(x + y)^2}\left(-\dfrac{v^2}{u^2}\right) = \dfrac{2y}{v(x + y)^2} + \dfrac{2xv^2}{u^2(x + y)^2}$

$z_v = \dfrac{2y}{(x + y)^2}\left(-\dfrac{u}{v^2}\right) + \dfrac{-2x}{(x + y)^2}\left(\dfrac{2v}{u}\right) = -\dfrac{2yu}{v^2(x + y)^2} - \dfrac{4xv}{u(x + y)^2}$

45. $R_u = s^2 t^4(e^{v^2}) + 2rst^4(-2uve^{-u^2}) + 4rs^2 t^3(2uv^2 e^{u^2 v^2}) = s^2 t^4 e^{v^2} - 4uvrst^4 e^{-u^2} + 8uv^2 rs^2 t^3 e^{u^2 v^2}$

$R_v = s^2 t^4(2uve^{v^2}) + 2rst^4(e^{-u^2}) + 4rs^2 t^3(2u^2 ve^{u^2 v^2}) = 2s^2 t^4 uve^{v^2} + 2rst^4 e^{-u^2} + 8rs^2 t^3 u^2 ve^{u^2 v^2}$

48. $s_\phi = 2pe^{3\theta} + 2q[-\sin(\phi + \theta)] - 2r\theta^2 + 4(2) = 2pe^{3\theta} - 2q\sin(\phi + \theta) - 2r\theta^2 + 8$

$s_\theta = 2p(3\phi e^{3\theta}) + 2q[-\sin(\phi + \theta)] - 2r(2\phi\theta) + 4(8) = 6p\phi e^{3\theta} - 2q\sin(\phi + \theta) - 4r\phi\theta + 32$

51. $\dfrac{dw}{dt} = -3\sin(3u + 4v)(2) - 4\sin(3u + 4v)(-1);\quad u(\pi) = 5\pi/2,\ v(\pi) = -5\pi/4$

$\dfrac{dw}{dt}\bigg|_\pi = -6\sin\left(\dfrac{15\pi}{2} - 5\pi\right) + 4\sin\left(\dfrac{15\pi}{2} - 5\pi\right) = -2\sin\dfrac{5\pi}{2} = -2$

54. $\dfrac{dP}{dt} = \dfrac{(V - 0.0427)(0.08)dT/dt}{(V - 0.0427)^2} - \dfrac{0.08T(dV/dt)}{(V - 0.0427)^2} + \dfrac{3.6}{V^3}\dfrac{dV}{dt}$

$\qquad = \dfrac{0.08}{V - 0.0427}\dfrac{dT}{dt} + \left(\dfrac{3.6}{V^3} - \dfrac{0.08T}{(V - 0.0427)^2}\right)\dfrac{dV}{dt}$

57. Since the height of the triangle is $x\sin\theta$, the area is given by $A = \frac{1}{2}xy\sin\theta$. Then

$$\dfrac{dA}{dt}\dfrac{\partial A}{\partial x}\dfrac{dx}{dt} + \dfrac{\partial A}{\partial y}\dfrac{dy}{dt} + \dfrac{\partial A}{\partial \theta}\dfrac{d\theta}{dt} = \dfrac{1}{2}y\sin\theta\dfrac{dx}{dt} + \dfrac{1}{2}x\sin\theta\dfrac{dy}{dt} + \dfrac{1}{2}xy\cos\theta\dfrac{d\theta}{dt}.$$

When $x = 10$, $y = 8$, $\theta = \pi/6$, $dx/dt = 0.3$, $dy/dt = 0.5$, and $d\theta/dt = 0.1$,

$$\dfrac{dA}{dt} = \dfrac{1}{2}(8)\left(\dfrac{1}{2}\right)(0.3) + \dfrac{1}{2}(10)\left(\dfrac{1}{2}\right)(0.5) + \dfrac{1}{2}(10)(8)\left(\dfrac{\sqrt{3}}{2}\right)(0.1)$$

$$= 0.6 + 1.25 + 2\sqrt{3} = 1.85 + 2\sqrt{3} \approx 5.31 \text{ cm}^2/\text{s}.$$

EXERCISES 9.5

Directional Derivative

3. $\nabla F = \dfrac{y^2}{z^3}\mathbf{i} + \dfrac{2xy}{z^3}\mathbf{j} - \dfrac{3xy^2}{z^4}\mathbf{k}$

6. $\nabla f = \dfrac{3x^2}{2\sqrt{x^3y - y^4}}\mathbf{i} + \dfrac{x^3 - 4y^3}{2\sqrt{x^3y - y^4}}\mathbf{j}; \quad \nabla f(3,2) = \dfrac{27}{\sqrt{38}}\mathbf{i} - \dfrac{5}{2\sqrt{38}}\mathbf{j}$

9. $D_{\mathbf{u}}f(x,y) = \lim\limits_{h \to 0} \dfrac{f(x + h\sqrt{3}/2, y + h/2) - f(x,y)}{h} = \lim\limits_{h \to 0} \dfrac{(x + h\sqrt{3}/2)^2 + (y + h/2)^2 - x^2 - y^2}{h}$

$\qquad = \lim\limits_{h \to 0} \dfrac{h\sqrt{3}\,x + 3h^2/4 + hy + h^2/4}{h} = \lim\limits_{h \to 0}(\sqrt{3}\,x + 3h/4 + y + h/4) = \sqrt{3}\,x + y$

12. $\mathbf{u} = \dfrac{\sqrt{2}}{2}\mathbf{i} + \dfrac{\sqrt{2}}{2}\mathbf{j}; \quad \nabla f = (4 + y^2)\mathbf{i} + (2xy - 5)\mathbf{j}; \quad \nabla f(3,-1) = 5\mathbf{i} - 11\mathbf{j};$

$\qquad D_{\mathbf{u}}f(3,-1) = \dfrac{5\sqrt{2}}{2} - \dfrac{11\sqrt{2}}{2} = -3\sqrt{2}$

15. $\mathbf{u} = (2\mathbf{i} + \mathbf{j})/\sqrt{5}; \quad \nabla f = 2y(xy + 1)\mathbf{i} + 2x(xy + 1)\mathbf{j}; \quad \nabla f(3,2) = 28\mathbf{i} + 42\mathbf{j}$

$\qquad D_{\mathbf{u}}f(3,2) = \dfrac{2(28)}{\sqrt{5}} + \dfrac{42}{\sqrt{5}} = \dfrac{98}{\sqrt{5}}$

18. $\mathbf{u} = \dfrac{1}{\sqrt{6}}\mathbf{i} - \dfrac{2}{\sqrt{6}}\mathbf{j} + \dfrac{1}{\sqrt{6}}\mathbf{k}; \quad \nabla F = \dfrac{2x}{z^2}\mathbf{i} - \dfrac{2y}{z^2}\mathbf{j} + \dfrac{2y^2 - 2x^2}{z^3}\mathbf{k}; \quad \nabla F(2,4,-1) = 4\mathbf{i} - 8\mathbf{j} - 24\mathbf{k}$

$\qquad D_{\mathbf{u}}F(2,4,-1) = \dfrac{4}{\sqrt{6}} - \dfrac{16}{\sqrt{6}} - \dfrac{24}{\sqrt{6}} = -6\sqrt{6}$

21. $\mathbf{u} = (-4\mathbf{i} - \mathbf{j})/\sqrt{17}; \quad \nabla f = 2(x - y)\mathbf{i} - 2(x - y)\mathbf{j}; \quad \nabla f(4,2) = 4\mathbf{i} - 4\mathbf{j}; \quad D_{\mathbf{u}}F(4,2) = -\dfrac{16}{\sqrt{17}}\dfrac{4}{\sqrt{17}} = -\dfrac{12}{\sqrt{17}}$

24. $\nabla f = (xye^{x-y} + ye^{x-y})\mathbf{i} + (-xye^{x-y} + xe^{x-y})\mathbf{j}; \quad \nabla f(5,5) = 30\mathbf{i} - 20\mathbf{j}$

The maximum $D_{\mathbf{u}}$ is $[30^2 + (-20)^2]^{1/2} = 10\sqrt{13}$ in the direction $30\mathbf{i} - 20\mathbf{j}$.

27. $\nabla f = 2x\sec^2(x^2 + y^2)\mathbf{i} + 2y\sec^2(x^2 + y^2)\mathbf{j};$

$\qquad \nabla f(\sqrt{\pi/6}, \sqrt{\pi/6}) = 2\sqrt{\pi/6}\sec^2(\pi/3)(\mathbf{i} + \mathbf{j}) = 8\sqrt{\pi/6}\,(\mathbf{i} + \mathbf{j})$

The minimum $D_{\mathbf{u}}$ is $-8\sqrt{\pi/6}\,(1^2 + 1^2)^{1/2} = -8\sqrt{\pi/3}$ in the direction $-(\mathbf{i} + \mathbf{j})$.

30. $\nabla F = \dfrac{1}{x}\mathbf{i} + \dfrac{1}{y}\mathbf{j} - \dfrac{1}{z}\mathbf{k}; \quad \nabla F(1/2, 1/6, 1/3) = 2\mathbf{i} + 6\mathbf{j} - 3\mathbf{k}$

The minimum $D_{\mathbf{u}}$ is $-[2^2 + 6^2 + (-3)^2]^{1/2} = -7$ in the direction $-2\mathbf{i} - 6\mathbf{j} + 3\mathbf{k}$.

33. (a) Vectors perpendicular to $4\mathbf{i} + 3\mathbf{j}$ are $\pm(3\mathbf{i} - 4\mathbf{j})$. Take $\mathbf{u} = \pm\left(\dfrac{3}{5}\mathbf{i} - \dfrac{4}{5}\mathbf{j}\right)$.

(b) $\mathbf{u} = (4\mathbf{i} + 3\mathbf{j})/\sqrt{16 + 9} = \dfrac{4}{5}\mathbf{i} + \dfrac{3}{5}\mathbf{j}$

(c) $\mathbf{u} = -\dfrac{4}{5}\mathbf{i} - \dfrac{3}{5}\mathbf{j}$

36. $\nabla U = \dfrac{Gmx}{(x^2 + y^2)^{3/2}}\mathbf{i} + \dfrac{Gmy}{(x^2 + y^2)^{3/2}}\mathbf{j} = \dfrac{Gm}{(x^2 + y^2)^{3/2}}(x\mathbf{i} + y\mathbf{j})$

178

The maximum and minimum values of $D_{\mathbf{u}}U(x,y)$ are obtained when $\mathbf{u}$ is in the directions ∇U and $-\nabla U$, respectively. Thus, at a point (x,y), not $(0,0)$, the directions of maximum and minimum increase in U are $x\mathbf{i}+y\mathbf{j}$ and $-x\mathbf{i}-y\mathbf{j}$, respectively. A vector at (x,y) in the direction $\pm(x\mathbf{i}+y\mathbf{j})$ lies on a line through the origin.

39. $\nabla T = 4x\mathbf{i} + 2y\mathbf{j}$; $\nabla T(4,2) = 16\mathbf{i} + 4\mathbf{j}$. The minimum change in temperature (that is, the maximum decrease in temperature) is in the direction $-\nabla T(4,3) = -16\mathbf{i} - 4\mathbf{j}$.

42. Substituting $x = 0$, $y = 0$, $z = 1$, and $T = 500$ into $T = k/(x^2 + y^2 + z^2)$ we see that $k = 500$ and $T(x,y,z) = 500/(x^2 + y^2 + z^2)$. To find the rate of change of T at $\langle 2,3,3 \rangle$ in the direction of $\langle 3,1,1 \rangle$ we first compute $\langle 3,1,1 \rangle - \langle 2,3,3 \rangle = \langle 1,-2,-2 \rangle$. Then $\mathbf{u} = \frac{1}{3}\langle 1,-2,-2 \rangle = \frac{1}{3}\mathbf{i} - \frac{2}{3}\mathbf{j} - \frac{2}{3}\mathbf{k}$. Now

$$\nabla T = -\frac{1000x}{(x^2+y^2+z^2)^2}\mathbf{i} - \frac{1000y}{(x^2+y^2+z^2)^2}\mathbf{j} - \frac{1000z}{(x^2+y^2+z^2)^2}\mathbf{k} \quad\text{and}\quad \nabla T(2,3,3) = -\frac{500}{121}\mathbf{i} - \frac{750}{121}\mathbf{j} - \frac{750}{121}\mathbf{k},$$

so

$$D_{\mathbf{u}}T(2,3,3) = \frac{1}{3}\left(-\frac{500}{121}\right) - \frac{2}{3}\left(-\frac{750}{121}\right) - \frac{2}{3}\left(-\frac{750}{121}\right) = \frac{2500}{363}.$$

The direction of maximum increase is $\nabla T(2,3,3) = -\frac{500}{121}\mathbf{i} - \frac{750}{121}\mathbf{j} - \frac{750}{121}\mathbf{k} = \frac{250}{121}(-2\mathbf{i}-3\mathbf{j}-3\mathbf{k})$, and the maximum rate of change of T is $|\nabla T(2,3,3)| = \frac{250}{121}\sqrt{4+9+9} = \frac{250}{121}\sqrt{22}$.

45. $\nabla(cf) = \dfrac{\partial}{\partial x}(cf)\mathbf{i} + \dfrac{\partial}{\partial y}(cf)\mathbf{j} = cf_x\mathbf{i} + cf_y\mathbf{j} = c(f_x\mathbf{i} + f_y\mathbf{j}) = c\nabla f$

48. $\nabla(f/g) = [(gf_x - fg_x)/g^2]\mathbf{i} + [(gf_y - fg_y)/g^2]\mathbf{j} = g(f_x\mathbf{i} + f_y\mathbf{j})/g^2 - f(g_x\mathbf{i} + g_y\mathbf{j})/g^2$
$= g\nabla f/g^2 - f\nabla g/g^2 = (g\nabla f - f\nabla g)/g^2$

EXERCISES 9.6

Tangent Planes and Normal Lines

3. Since $f(2,5) = 1$, the level curve is $y = x^2 + 1$.
$\nabla f = -2x\mathbf{i} + \mathbf{j}$; $\nabla f(2,5) = -10\mathbf{i} + \mathbf{j}$

6. Since $f(2,2) = 2$, the level curve is $y^2 = 2x$, $x \neq 0$. $\nabla f = -\dfrac{y^2}{x^2}\mathbf{i} + \dfrac{2y}{x}\mathbf{j}$;
$\nabla f(2,2) = -\mathbf{i} + 2\mathbf{j}$

9. Since $F(3,1,1) = 2$, the level surface is $y + z = 2$. $\nabla F = \mathbf{j} + \mathbf{k}$;
$\nabla F(3,1,1) = \mathbf{j} + \mathbf{k}$

12. Since $F(0,-1,1) = 0$, the level surface is $x^2 - y^2 + z = 0$ or $z = y^2 - x^2$.
$\nabla F = 2x\mathbf{i} - 2y\mathbf{j} + \mathbf{k}$; $\nabla F(0,-1,1) = 2\mathbf{j} + \mathbf{k}$

15. $F(x, y, z) = x^2 + y^2 + z^2$; $\nabla F = 2x\mathbf{i} + 2y\mathbf{j} + 2z\mathbf{k}$. $\nabla F(-2, 2, 1) = -4\mathbf{i} + 4\mathbf{j} + 2\mathbf{k}$. The equation of the tangent plane is $-4(x+2) + 4(y-2) + 2(z-1) = 0$ or $-2x + 2y + z = 9$.

18. $F(x, y, z) = xy + yz + zx$; $\nabla F = (y+z)\mathbf{i} + (x+z)\mathbf{j} + (y+x)\mathbf{k}$; $\nabla F(1, -3, -5) = -8\mathbf{i} - 4\mathbf{j} - 2\mathbf{k}$. The equation of the tangent plane is $-8(x-1) - 4(y+3) - 2(z+5) = 0$ or $4x + 2y + z = -7$.

21. $F(x, y, z) = \cos(2x+y) - z$; $\nabla F = -2\sin(2x+y)\mathbf{i} - \sin(2x+y)\mathbf{j} - \mathbf{k}$; $\nabla F(\pi/2, \pi/4, -1/\sqrt{2}) = \sqrt{2}\mathbf{i} + \dfrac{\sqrt{2}}{2}\mathbf{j} - \mathbf{k}$.

The equation of the tangent plane is $\sqrt{2}\left(x - \dfrac{\pi}{2}\right) + \dfrac{\sqrt{2}}{2}\left(y - \dfrac{\pi}{4}\right) - \left(z + \dfrac{1}{\sqrt{2}}\right) = 0$,

$2\left(x - \dfrac{\pi}{2}\right) + \left(y - \dfrac{\pi}{4}\right) - \sqrt{2}\left(z + \dfrac{1}{\sqrt{2}}\right) = 0$, or $2x + y - \sqrt{2}z = \dfrac{5\pi}{4} + 1$.

24. $F(x, y, z) = 8e^{-2y}\sin 4x - z$; $\nabla F = 32e^{-2y}\cos 4x\mathbf{i} - 16e^{-2y}\sin 4x\mathbf{j} - \mathbf{k}$; $\nabla F(\pi/24, 0, 4) = 16\sqrt{3}\mathbf{i} - 8\mathbf{j} - \mathbf{k}$. The equation of the tangent plane is

$$16\sqrt{3}(x - \pi/24) - 8(y - 0) - (z - 4) = 0 \quad \text{or} \quad 16\sqrt{3}x - 8y - z = \dfrac{2\sqrt{3}\pi}{3} - 4.$$

27. The gradient of $F(x, y, z) = x^2 + 4x + y^2 + z^2 - 2z$ is $\nabla F = (2x + 4)\mathbf{i} + 2y\mathbf{j} + (2z - 2)\mathbf{k}$, so a normal to the surface at (x_0, y_0, z_0) is $(2x_0 + 4)\mathbf{i} + 2y_0\mathbf{j} + (2z_0 - 2)\mathbf{k}$. A horizontal plane has normal $c\mathbf{k}$ for $c \neq 0$. Thus, we want $2x_0 + 4 = 0$, $2y_0 = 0$, $2z_0 - 2 = c$ or $x_0 = -2$, $y_0 = 0$, $z_0 = c + 1$. since (x_0, y_0, z_0) is on the surface, $(-2)^2 + 4(-2) + (c+1)^2 - 2(c+1) = c^2 - 5 = 11$ and $c = \pm 4$. The points on the surface are $(-2, 0, 5)$ and $(-2, 0, -3)$.

30. If (x_0, y_0, z_0) is on $x^2/a^2 - y^2/b^2 + z^2/c^2 = 1$, then $x_0^2/a^2 - y_0^2/b^2 + z_0^2/c^2 = 1$ and (x_0, y_0, z_0) is on the plane $xx_0/a^2 - yy_0/b^2 + zz_0/c^2 = 1$. A normal to the surface at (x_0, y_0, z_0) is

$$\nabla F(x_0, y_0, z_0) = (2x_0/a^2)\mathbf{i} - (2y_0/b^2)\mathbf{j} + (2z_0/c^2)\mathbf{k}.$$

A normal to the plane is $(x_0/a^2)\mathbf{i} - (y_0/b^2)\mathbf{j} + (z_0/c^2)\mathbf{k}$. Since the normal to the surface is a multiple of the normal to the plane, the normal vectors are parallel, and the plane is tangent to the surface.

33. $F(x, y, z) = x^2 + 2y^2 + z^2$; $\nabla F = 2x\mathbf{i} + 4y\mathbf{j} + 2z\mathbf{k}$; $\nabla F(1, -1, 1) = 2\mathbf{i} - 4\mathbf{j} + 2\mathbf{k}$. Parametric equations of the line are $x = 1 + 2t$, $y = -1 - 4t$, $z = 1 + 2t$.

36. $F(x, y, z) = x^2 + y^2 - z^2$; $\nabla F = 2x\mathbf{i} + 2y\mathbf{j} - 2z\mathbf{k}$; $\nabla F(3, 4, 5) = 6\mathbf{i} + 8\mathbf{j} - 10\mathbf{k}$. Symmetric equations of the line are $\dfrac{x-3}{6} = \dfrac{y-4}{8} = \dfrac{z-5}{-10}$.

39. Let $F(x, y, z) = x^2 + y^2 + z^2 - 25$ and $G(x, y, z) = -x^2 + y^2 + z^2$. Then

$$F_x G_x + F_y G_y + F_z G_z = (2x)(-2x) + (2y)(2y) + (2z)(2z) = 4(-x^2 + y^2 + z^2).$$

For (x, y, z) on both surfaces, $F(x, y, z) = G(x, y, z) = 0$. Thus, $F_x G_x + F_y G_y + F_z G_z = 4(0) = 0$ and the surfaces are orthogonal at points of intersection.

EXERCISES 9.7

Divergence and Curl

3.

6.

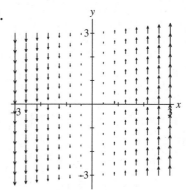

9. curl $\mathbf{F} = \mathbf{0}$; div $\mathbf{F} = 4y + 8z$

12. curl $\mathbf{F} = -x^3 z\mathbf{i} + (3x^2 yz - z)\mathbf{j} + \left(\frac{3}{2}x^2 y^2 - y - 15y^2\right)\mathbf{k}$; div $\mathbf{F} = (x^3 y - x) - (x^3 y - x) = 0$

15. curl $\mathbf{F} = (xy^2 e^y + 2xy e^y + x^3 y e^z + x^3 yz e^z)\mathbf{i} - y^2 e^y \mathbf{j} + (-3x^2 yz e^z - xe^x)\mathbf{k}$; div $\mathbf{F} = xy e^x + ye^x - x^3 z e^z$

18. curl $\mathbf{r} = \begin{vmatrix} \mathbf{i} & \mathbf{j} & \mathbf{k} \\ \partial/\partial x & \partial/\partial y & \partial/\partial z \\ x & y & z \end{vmatrix} = 0\mathbf{i} - 0\mathbf{j} + 0\mathbf{k} = \mathbf{0}$

21. $\nabla \cdot (\mathbf{a} \times \mathbf{r}) = \begin{vmatrix} \partial/\partial x & \partial/\partial y & \partial/\partial z \\ a_1 & a_2 & a_3 \\ x & y & z \end{vmatrix} = \dfrac{\partial}{\partial x}(a_2 z - a_3 y) - \dfrac{\partial}{\partial y}(a_1 z - a_3 x) + \dfrac{\partial}{\partial z}(a_1 y - a_2 x) = 0$

24. $\mathbf{r} \cdot \mathbf{a} = a_1 x + a_2 y + a_3 z$; $\mathbf{r} \cdot \mathbf{r} = x^2 + y^2 + z^2$; $\nabla \cdot [(\mathbf{r} \cdot \mathbf{r})\mathbf{a}] = 2x a_1 + 2y a_2 + 2z a_3 = 2(\mathbf{r} \cdot \mathbf{a})$

27. $\nabla \cdot (f\mathbf{F}) = \nabla \cdot (fP\mathbf{i} + fQ\mathbf{j} + fR\mathbf{k}) = fP_x + Pf_x + fQ_y + Qf_y + fR_z + Rf_z$

$= f(P_x + Q_y + R_z) + (Pf_x + Qf_y + Rf_z) = f(\nabla \cdot \mathbf{F}) + \mathbf{F} \cdot (\nabla f)$

30. Assuming continuous second partial derivatives,

$$\text{div (curl } \mathbf{F}) = \nabla \cdot [(R_y - Q_z)\mathbf{i} - (R_x - P_z)\mathbf{j} + (Q_x - P_y)\mathbf{k}]$$
$$= (R_{yx} - Q_{zx} - (R_{xy} - P_{zy}) + (Q_{xz} - P_{yz}) = 0.$$

33. $\nabla \cdot \nabla f = \nabla \cdot (f_x \mathbf{i} + f_y \mathbf{j} + f_z \mathbf{k}) = f_{xx} + f_{yy} + f_{zz}$

36. (a) For $\mathbf{F} = P\mathbf{i} + Q\mathbf{j} + R\mathbf{k}$,

$$\text{curl (curl } \mathbf{F}) = (Q_{xy} - P_{yy} - P_{zz} + R_{xz})\mathbf{i} + (R_{yz} - Q_{zz} - Q_{xx} + P_{yx})\mathbf{j}$$
$$+ (P_{zx} - R_{xx} - R_{yy} + Q_{zy})\mathbf{k}$$

and

181

$$-\nabla^2 \mathbf{F} + \text{grad } (\text{div } \mathbf{F}) = -(P_{xx} + P_{yy} + P_{zz})\mathbf{i} - (Q_{xx} + Q_{yy} + Q_{zz})\mathbf{j} - (R_{xx} + R_{yy} + R_{zz})\mathbf{k}$$
$$+ \text{grad } (P_x + Q_y + R_z)$$
$$= -P_{xx}\mathbf{i} - Q_{yy}\mathbf{j} - R_{zz}\mathbf{k} + (-P_{yy} - P_{zz})\mathbf{i} + (-Q_{xx} - Q_{zz})\mathbf{j}$$
$$+ (-R_{xx} - R_{yy})\mathbf{k} + (P_{xx} + Q_{yx} + R_{zx})\mathbf{i} + (P_{xy} + Q_{yy} + R_{zy})\mathbf{j}$$
$$+ (P_{xz} + Q_{yz} + R_{zz})\mathbf{k}$$
$$= (-P_{yy} - P_{zz} + Q_{yx} + R_{zx})\mathbf{i} + (-Q_{xx} - Q_{zz} + P_{xy} + R_{zy})\mathbf{j}$$
$$+ (-R_{xx} - R_{yy} + P_{xz} + Q_{yz})\mathbf{k}.$$

Thus, curl (curl $\mathbf{F}$) $= -\nabla^2 \mathbf{F} + \text{grad } (\text{div } \mathbf{F})$.

(b) For $\mathbf{F} = xy\mathbf{i} + 4yz^2\mathbf{j} + 2xz\mathbf{k}$, $\nabla^2\mathbf{F} = 0\mathbf{i} + 8y\mathbf{j} + 0\mathbf{k}$, div $\mathbf{F} = y + 4z^2 + 2x$, and grad (div $\mathbf{F}$) $= 2\mathbf{i} + \mathbf{j} + 8z\mathbf{k}$. Then curl (curl $\mathbf{F}$) $= -8y\mathbf{j} + 2\mathbf{i} + \mathbf{j} + 8z\mathbf{k} = 2\mathbf{i} + (1 - 8y)\mathbf{j} + 8z\mathbf{k}$.

39. curl $\mathbf{F} = -Gm_1m_2 \begin{vmatrix} \mathbf{i} & \mathbf{j} & \mathbf{k} \\ \partial/\partial x & \partial/\partial y & \partial/\partial z \\ x/|\mathbf{r}|^3 & y/|\mathbf{r}|^3 & z/|\mathbf{r}|^3 \end{vmatrix}$

$$= -Gm_1m_2[(-3yz/|\mathbf{r}|^5 + 3yz/|\mathbf{r}|^5)\mathbf{i} - (-3xz/|\mathbf{r}|^5 + 3xz/|\mathbf{r}|^5)\mathbf{j} + (-3xy/|\mathbf{r}|^5 + 3xy/|\mathbf{r}|^5)\mathbf{k}]$$

$$= \mathbf{0}$$

$$\text{div } \mathbf{F} = -Gm_1m_2 \left[\frac{-2x^2 + y^2 + z^2}{|\mathbf{r}|^{5/2}} + \frac{x^2 - 2y^2 + z^2}{|\mathbf{r}|^{5/2}} + \frac{x^2 + y^2 - 2z^2}{|\mathbf{r}|^{5/2}} \right] = 0$$

42. Recall that $\mathbf{a} \cdot (\mathbf{a} \times \mathbf{b}) = 0$. Then, using Problems 31, 29, and 28,

$$\nabla \cdot \mathbf{F} = \text{div } (\nabla f \times f\nabla g) = f\nabla g \cdot (\text{curl } \nabla f) - \nabla f \cdot (\text{curl } f\nabla g) = f\nabla g \cdot \mathbf{0} - \nabla f \cdot (\nabla \times f\nabla g)$$
$$= -\nabla f \cdot [f(\nabla \times \nabla g) + (\nabla f \times \nabla g)] = -\nabla f \cdot [f\text{curl } \nabla g + (\nabla f \times \nabla g]$$
$$= -\nabla f \cdot [f\mathbf{0} + (\nabla f \times \nabla g)] = -\nabla f \cdot (\nabla f \times \nabla g) = 0.$$

45. We note that div $\mathbf{F} = 2xyz - 2xyz + 1 = 1 \neq 0$. If $\mathbf{F} = \text{curl } \mathbf{G}$, then div(curl $\mathbf{G}$) = div $\mathbf{F} = 1$. But, by Problem 30, for any vector field $\mathbf{G}$, div(curl $\mathbf{G}$) = 0. Thus, $\mathbf{F}$ cannot be the curl of $\mathbf{G}$.

EXERCISES 9.8
Line Integrals

3. $\displaystyle\int_C (3x^2 + 6y^2)\, dx = \int_{-1}^{0} [3x^2 + 6(2x + 1)^2]\, dx = \int_{-1}^{0} (27x^2 + 24x + 6)\, dx = (9x^3 + 12x^2 + 6x) \Big|_{-1}^{0}$

$$= -(-9 + 12 - 6) = 3$$

$\displaystyle\int_C (3x^2 + 6y^2)\, dy = \int_{-1}^{0} [3x^2 + 6(2x + 1)^2]2\, dx = 6$

$\displaystyle\int_C (3x^2 + 6y^2)\, ds = \int_{-1}^{0} [3x^2 + 6(2x + 1)^2]\sqrt{1 + 4}\, dx = 3\sqrt{5}$

6. $\displaystyle\int_C 4xyz\, dx = \int_0^1 4\left(\frac{1}{3}t^3\right)(t^2)(2t)t^2\, dt = \frac{8}{3}\int_0^1 t^8\, dt = \frac{8}{27}t^9 \Big|_0^1 = \frac{8}{27}$

$$\int_C 4xyz\, dy = \int_0^1 4\left(\frac{1}{3}t^3\right)(t^2)(2t)2t\, dt = \frac{16}{3}\int_0^1 t^7\, dt = \frac{2}{3}t^8\Big|_0^1 = \frac{2}{3}$$

$$\int_C 4xyz\, dz = \int_0^1 4\left(\frac{1}{3}t^3\right)(t^2)(2t)2\, dt = \frac{16}{3}\int_0^1 t^6\, dt = \frac{16}{21}t^7\Big|_0^1 = \frac{16}{21}$$

$$\int_C 4xyz\, ds = \int_0^1 4\left(\frac{1}{3}t^3\right)(t^2)(2t)\sqrt{t^4+4t^2+4}\, dt = \frac{8}{3}\int_0^1 t^6(t^2+2)\, dt = \frac{8}{3}\left(\frac{1}{9}t^9 + \frac{2}{7}t^7\right)\Big|_0^1 = \frac{200}{189}$$

9. From $(-1,2)$ to $(2,2)$ we use x as a parameter with $y=2$ and $dy=0$. From $(2,2)$ to $(2,5)$ we use y as a parameter with $x=2$ and $dx=0$.

$$\int_C (2x+y)\, dx + xy\, dy = \int_{-1}^2 (2x+2)\, dx + \int_2^5 2y\, dy = (x^2+2x)\Big|_{-1}^2 + y^2\Big|_2^5 = 9 + 21 = 30$$

12. Using x as the parameter, $dy = dx$.

$$\int_C y\, dx + x\, dy = \int_0^1 x\, dx + \int_0^1 x\, dx = \int_0^1 2x\, dx = x^2\Big|_0^1 = 1$$

15. $$\int_C (6x^2 + 2y^2)\, dx + 4xy\, dy = \int_4^9 (6t + 2t^2)\frac{1}{2}t^{-1/2}\, dt + \int_4^9 4\sqrt{t}\, t\, dt = \int_4^9 (3t^{1/2} + 5t^{3/2})\, dt$$

$$= (2t^{3/2} + 2t^{5/2})\Big|_4^9 = 460$$

18. $$\int_C 4x\, dx + 2y\, dy = \int_{-1}^2 4(y^3+1)3y^2\, dy + \int_{-1}^2 2y\, dy = \int_{-1}^2 (12y^5 + 12y^2 + 2y)\, dy$$

$$= (2y^6 + 4y^3 + y^2)\Big|_{-1}^2 = 165$$

21. From $(1,1)$ to $(-1,1)$ and $(-1,-1)$ to $(1,-1)$ we use x as a parameter with $y=1$ and $y=-1$, respectively, and $dy=0$. From $(-1,1)$ to $(-1,-1)$ and $(1,-1)$ to $(1,1)$ we use y as a parameter with $x=-1$ and $x=1$, respectively, and $dx=0$.

$$\oint_C x^2 y^3\, dx - xy^2\, dy = \int_1^{-1} x^2(1)\, dx + \int_1^{-1} -(-1)y^2\, dy + \int_{-1}^1 x^2(-1)^3\, dx + \int_{-1}^1 -(1)y^2\, dy$$

$$= \frac{1}{3}x^3\Big|_1^{-1} + \frac{1}{3}y^3\Big|_1^{-1} - \frac{1}{3}x^3\Big|_{-1}^1 - \frac{1}{3}y^3\Big|_{-1}^1 = -\frac{8}{3}$$

24. $$\oint_C y\, dx - x\, dy = \int_0^\pi 3\sin t(-2\sin t)\, dt - \int_0^\pi 2\cos t(3\cos t)\, dt = -6\int_0^\pi (\sin^2 t + \cos^2 t)\, dt = -6\int_0^\pi dt = -6\pi$$

Thus, $\int_{-C} y\, dx - x\, dy = 6\pi$.

27. From $(0,0,0)$ to $(6,0,0)$ we use x as a parameter with $y=dy=0$ and $z=dz=0$. From $(6,0,0)$ to $(6,0,5)$ we use z as a parameter with $x=6$ and $dx=0$ and $y=dy=0$. From $(6,0,5)$ to $(6,8,5)$ we use y as a parameter with $x=6$ and $dx=0$ and $z=5$ and $dz=0$.

$$\int_C y\, dx + z\, dy + x\, dz = \int_0^6 0 + \int_0^5 6\, dz + \int_0^8 5\, dy = 70$$

30. $\mathbf{F} = e^t\mathbf{i} + te^{t^3}\mathbf{j} + t^3 e^{t^6}\mathbf{k}$; $d\mathbf{r} = (\mathbf{i} + 2t\mathbf{j} + 3t^2\mathbf{k})\, dt$;

$$\int_C \mathbf{F} \cdot d\mathbf{r} = \int_0^1 (e^t + 2t^2 e^{t^3} + 3t^5 e^{t^6})\, dt = \left(e^t + \frac{2}{3}e^{t^3} + \frac{1}{2}e^{t^6}\right)\Big|_0^1 = \frac{13}{6}(e-1)$$

33. Let $\mathbf{r}_1 = (1+2t)\mathbf{i} + \mathbf{j}$, $\mathbf{r}_2 = 3\mathbf{i} + (1+t)\mathbf{j}$, and $\mathbf{r}_3 = (3-2t)\mathbf{i} + (2-t)\mathbf{j}$ for $0 \le t \le 1$. Then

$$d\mathbf{r}_1 = 2\mathbf{i}, \qquad d\mathbf{r}_2 = \mathbf{j}, \qquad d\mathbf{r}_3 = -2\mathbf{i} - \mathbf{j},$$

$$\mathbf{F}_1 = (1+2t+2)\mathbf{i} + (6-2-4t)\mathbf{j} = (3+2t)\mathbf{i} + (4-4t)\mathbf{j},$$

$$\mathbf{F}_2 = (3+2+2t)\mathbf{i} + (6+6t-6)\mathbf{j} = (5+2t)\mathbf{i} + 6t\mathbf{j},$$

$$\mathbf{F}_3 = (3-2t+4-2t)\mathbf{i} + (12-6t-6+4t)\mathbf{j} = (7-4t)\mathbf{i} + (6-2t)\mathbf{j},$$

and

$$W = \int_{C_1} \mathbf{F}_1 \cdot d\mathbf{r}_1 + \int_{C_2} \mathbf{F}_2 \cdot d\mathbf{r}_2 + \int_{C_3} \mathbf{F}_3 \cdot d\mathbf{r}_3$$

$$= \int_0^1 (6+4t)\,dt + \int_0^1 6t\,dt + \int_0^1 (-14+8t-6+2t)\,dt$$

$$= \int_0^1 (-14+20t)\,dt = (-14t+10t^2)\,\Big|_0^1 = -4.$$

36. Let $\mathbf{r} = t\mathbf{i} + t\mathbf{j} + t\mathbf{k}$ for $1 \le t \le 3$. Then $d\mathbf{r} = \mathbf{i} + \mathbf{j} + \mathbf{k}$, and

$$\mathbf{F} = \frac{c}{|\mathbf{r}|^3}(t\mathbf{i} + t\mathbf{j} + t\mathbf{k}) = \frac{ct}{(\sqrt{3t^2})^3}(\mathbf{i}+\mathbf{j}+\mathbf{k}) = \frac{c}{3\sqrt{3}\,t^2}(\mathbf{i}+\mathbf{j}+\mathbf{k}),$$

$$W = \int_C \mathbf{F} \cdot d\mathbf{r} = \int_1^3 \frac{c}{3\sqrt{3}\,t^2}(1+1+1)\,dt = \frac{c}{\sqrt{3}}\int_1^3 \frac{1}{t^2}\,dt = \frac{c}{\sqrt{3}}\left(-\frac{1}{t}\right)\Big|_1^3 = \frac{c}{\sqrt{3}}\left(-\frac{1}{3}+1\right) = \frac{2c}{3\sqrt{3}}.$$

39. Since $\mathbf{v} \cdot \mathbf{v} = v^2$, $\dfrac{d}{dt}v^2 = \dfrac{d}{dt}(\mathbf{v}\cdot\mathbf{v}) = \mathbf{v}\cdot\dfrac{d\mathbf{v}}{dt} + \dfrac{d\mathbf{v}}{dt}\cdot\mathbf{v} = 2\dfrac{d\mathbf{v}}{dt}\cdot\mathbf{v}$. Then

$$W = \int_C \mathbf{F} \cdot d\mathbf{r} = \int_a^b m\mathbf{a} \cdot \left(\frac{d\mathbf{r}}{dt}\,dt\right) = m\int_a^b \frac{d\mathbf{v}}{dt}\cdot\mathbf{v}\,dt = m\int_a^b \frac{1}{2}\left(\frac{d}{dt}v^2\right)dt$$

$$= \frac{1}{2}m(v^2)\,\Big|_a^b = \frac{1}{2}m[v(b)]^2 - \frac{1}{2}m[v(a)]^2.$$

42. On C_1, $\mathbf{T} = \mathbf{i}$ and $\mathbf{F}\cdot\mathbf{T} = \text{comp}_\mathbf{T}\mathbf{F} \approx 1$. On C_2, $\mathbf{T} = -\mathbf{j}$ and $\mathbf{F}\cdot\mathbf{T} = \text{comp}_\mathbf{T}\mathbf{F} \approx 2$. On C_3, $\mathbf{T} = -\mathbf{i}$ and $\mathbf{F}\cdot\mathbf{T} = \text{comp}_\mathbf{T}\mathbf{F} \approx 1.5$. Using the fact that the lengths of C_1, C_2, and C_3 are 4, 5, and 5, respectively, we have

$$W = \int_C \mathbf{F}\cdot\mathbf{T}\,ds = \int_{C_1} \mathbf{F}\cdot\mathbf{T}\,ds + \int_{C_2} \mathbf{F}\cdot\mathbf{T}\,ds + \int_{C_3} \mathbf{F}\cdot\mathbf{T}\,ds \approx 1(4) + 2(5) + 1.5(5) = 21.5 \text{ ft-lb.}$$

EXERCISES 9.9

Independence of Path

3. (a) $P_y = 2 = Q_x$ and the integral is independent of path. $\phi_x = x + 2y$, $\phi = \dfrac{1}{2}x^2 + 2xy + g(y)$,

$$\phi_y = 2x + g'(y) = 2x - y, \quad g(y) = -\frac{1}{2}y^2, \quad \phi = \frac{1}{2}x^2 + 2xy - \frac{1}{2}y^2,$$

$$\int_{(1,0)}^{(3,2)} (x+2y)\,dx + (2x-y)\,dy = \left(\frac{1}{2}x^2 + 2xy - \frac{1}{2}y^2\right)\Bigg|_{(1,0)}^{(3,2)} = 14$$

(b) Use $y = x - 1$ for $1 \leq x \leq 3$.

$$\int_{(1,0)}^{(3,2)} (x + 2y)\,dx + (2x - y)\,dy = \int_1^3 [x + 2(x - 1) + 2x - (x - 1)]\,dx$$

$$= \int_1^3 (4x - 1)\,dx = (2x^2 - x)\Big|_1^3 = 14$$

6. (a) $P_y = -xy(x^2 + y^2)^{-3/2} = Q_x$ and the integral is independent of path. $\phi_x = \dfrac{x}{\sqrt{x^2 + y^2}}$,

$\phi = \sqrt{x^2 + y^2} + g(y)$, $\phi_y = \dfrac{y}{\sqrt{x^2 + y^2}} + g'(y) = \dfrac{y}{\sqrt{x^2 + y^2}}$, $g(y) = 0$, $\phi = \sqrt{x^2 + y^2}$,

$$\int_{(1,0)}^{(3,4)} \frac{x\,dx + y\,dy}{\sqrt{x^2 + y^2}} = \sqrt{x^2 + y^2}\;\Big|_{(1,0)}^{(3,4)} = 4$$

(b) Use $y = 2x - 2$ for $1 \leq x \leq 3$.

$$\int_{(1,0)}^{(3,4)} \frac{x\,dx + y\,dy}{\sqrt{x^2 + y^2}} = \int_1^3 \frac{x + (2x - 2)2}{\sqrt{x^2 + (2x - 2)^2}}\,dx = \int_1^3 \frac{5x - 4}{\sqrt{5x^2 - 8x + 4}} = \sqrt{5x^2 - 8x + 4}\;\Big|_1^3 = 4$$

9. (a) $P_y = 3y^2 + 3x^2 = Q_x$ and the integral is independent of path. $\phi_x = y^3 + 3x^2 y$,

$\phi = xy^3 + x^3 y + g(y)$, $\phi_y = 3xy^2 + x^3 + g'(y) = x^3 + 3y^2 x + 1$, $g(y) = y$, $\phi = xy^3 + x^3 y + y$,

$$\int_{(0,0)}^{(2,8)} (y^3 + 3x^2 y)\,dx + (x^3 + 3y^2 x + 1)\,dy = (xy^3 + x^3 y + y)\;\Big|_{(0,0)}^{(2,8)} = 1096$$

(b) Use $y = 4x$ for $0 \leq x \leq 2$.

$$\int_{(0,0)}^{(2,8)} (y^3 + 3x^2 y)\,dx + (x^3 + 3y^2 x + 1)\,dy = \int_0^2 [(64x^3 + 12x^3) + (x^3 + 48x^3 + 1)(4)]\,dx$$

$$= \int_0^2 (272x^3 + 4)\,dx = (68x^4 + 4x)\;\Big|_0^2 = 1096$$

12. $P_y = 6xy^2 = Q_x$ and the vector field is a gradient field. $\phi_x = 2xy^3$, $\phi = x^2 y^3 + g(y)$,

$\phi_y = 3x^2 y^2 + g'(y) = 3x^2 y^2 + 3y^2$, $g(y) = y^3$, $\phi = x^2 y^3 + y^3$

15. $P_y = 1 = Q_x$ and the vector field is a gradient field. $\phi_x = x^3 + y$, $\phi = \dfrac{1}{4}x^4 + xy + g(y)$, $\phi_y = x + g'(y) = x + y^3$,

$g(y) - \dfrac{1}{4}y^4$, $\phi = \dfrac{1}{4}x^4 + xy + \dfrac{1}{4}y^4$

18. Since $P_y = -e^{-y} = Q_x$, $\mathbf{F}$ is conservative and $\displaystyle\int_C \mathbf{F} \cdot d\mathbf{r}$ is independent of the path. Thus, instead of the given curve we may use the simpler curve C_1: $y = 0$, $-2 \leq -x \leq 2$. Then $dy = 0$ and

$$W = \int_{C_1} (2x + e^{-y})\,dx + (4y - xe^{-y})\,dy = \int_2^{-2} (2x + 1)\,dx = (x^2 + x)\;\Big|_2^{-2} = (4 - 2) - (4 + 2) = -4.$$

21. $P_y = 2x \cos y = Q_x$, $Q_z = 0 = R_y$, $R_x = 3e^{3z} = P_z$, and the integral is independent of path. Integrating $\phi_x = 2x \sin y + e^{3z}$ we find $\phi = x^2 \sin y + xe^{3z} + g(y, z)$. Then $\phi_y = x^2 \cos y + g_y = Q = x^2 \cos y$, so $g_y = 0$, $g(y, z) = h(z)$, and $\phi = x^2 \sin y + xe^{3z} + h(z)$. Now $\phi_z = 3xe^{3z} + h'(z) = R = 3xe^{3z} + 5$, so $h'(z) = 5$ and $h(z) = 5z$. Thus $\phi = x^2 \sin y + xe^{3z} + 5z$ and

$$\int_{(1,0,0)}^{(2,\pi/2,1)} (2x \sin y + e^{3z})\,dx + x^2 \cos y\,dy + (3xe^{3z} + 5)\,dz$$

$$= (x^2 \sin y + xe^{3z} + 5z)\;\Big|_{(1,0,0)}^{(2,\pi/2,1)} = [4(1) + 2e^3 + 5] - [0 + 1 + 0] = 8 + 2e^3.$$

24. $P_y = 0 = Q_x$, $Q_z = 2y = R_y$, $R_x = 2x = P_z$ and the integral is independent of path. Parameterize the line segment between the points by $x = -2(1-t)$, $y = 3(1-t)$, $z = 1-t$, for $0 \le t \le 1$. Then $dx = 2\,dt$, $dy = -3\,dt$, $dz = -dt$, and

$$\int_{(-2,3,1)}^{(0,0,0)} 2xz\,dx + 2yz\,dy + (x^2 + y^2)\,dz$$

$$= \int_0^1 [-4(1-t)^2(2) + 6(1-t)^2(-3) + 4(1-t)^2(-1) + 9(1-t)^2(-1)]\,dt$$

$$= \int_0^1 -39(1-t)^2\,dt = 13(1-t)^3 \Big|_0^1 = -13.$$

27. Since $P_y = Gm_1m_2(2xy/|\mathbf{r}|^5) = Q_x$, $Q_z = Gm_1m_2(2yz/|\mathbf{r}|^5) = R_y$, and $R_x = Gm_1m_2(2xz/|\mathbf{r}|^5) = P_z$, the force field is conservative.

$$\phi_x = -Gm_1m_2\frac{x}{(x^2+y^2+z^2)^{3/2}}, \quad \phi = Gm_1m_2(x^2+y^2+z^2)^{-1/2} + g(y,z),$$

$$\phi_y = -Gm_1m_2\frac{y}{(x^2+y^2+z^2)^{3/2}} + g_y(y,z) = -Gm_1m_2\frac{y}{(x^2+y^2+z^2)^{3/2}}, \quad g(y,z) = h(z),$$

$$\phi = Gm_1m_2(x^2+y^2+z^2)^{-1/2} + h(z),$$

$$\phi_z = -Gm_1m_2\frac{z}{(x^2+y^2+z^2)^{3/2}} + h'(z) = -Gm_1m_2\frac{z}{(x^2+y^2+z^2)^{3/2}},$$

$$h(z) = 0, \quad \phi = \frac{Gm_1m_2}{\sqrt{x^2+y^2+z^2}} = \frac{Gm_1m_2}{|\mathbf{r}|}$$

30. From $\mathbf{F} = (x^2 + y^2)^{n/2}(x\mathbf{i} + y\mathbf{j})$ we obtain $P_y = nxy(x^2 + y^2)^{n/2-1} = Q_x$, so that $\mathbf{F}$ is conservative. From $\phi_x = x(x^2 + y^2)^{n/2}$ we obtain the potential function $\phi = (x^2 + y^2)^{(n+2)/2}/(n+2)$. Then

$$W = \int_{(x_1,y_1)}^{(x_2,y_2)} \mathbf{F} \cdot d\mathbf{r} = \left(\frac{(x^2 + y^2)^{(n+2)/2}}{n+2} \right) \Bigg|_{(x_1,y_1)}^{(x_2,y_2)} = \frac{1}{n+2} \left[(x_2^2 + y_2^2)^{(n+2)/2} - (x_1^2 + y_1^2)^{(n+2)/2} \right].$$

EXERCISES 9.10
Double Integrals

3. $\displaystyle\int_1^{3x} x^3 e^{xy}\,dy = x^2 e^{xy} \Big|_1^{3x} = x^2(e^{3x^2} - e^x)$

6. $\displaystyle\int_{x^3}^x e^{2y/x}\,dy = \frac{x}{2} e^{2y/x} \Big|_{x^3}^x = \frac{x}{2}(e^{2x/x} - e^{2x^3/x}) = \frac{x}{2}(e^2 - e^{2x^2})$

9.

12.

15. $\displaystyle\iint_R (2x+4y+1)\,dA = \int_0^1 \int_{x^3}^{x^2} (2x+4y+1)\,dy\,dx = \int_0^1 (2xy+2y^2+y)\,\Big|_{x^3}^{x^2}\,dx$

$$= \int_0^1 [(2x^3+2x^4+x^2)-(2x^4+2x^6+x^3)]\,dx$$

$$= \int_0^1 (x^3+x^2-2x^6)\,dx = \left(\frac{1}{4}x^4+\frac{1}{3}x^3-\frac{2}{7}x^7\right)\Big|_0^1$$

$$= \frac{1}{4}+\frac{1}{3}-\frac{2}{7} = \frac{25}{84}$$

18. $\displaystyle\iint_R \frac{x}{\sqrt{y}}\,dA = \int_{-1}^1 \int_{x^2+1}^{3-x^2} xy^{-1/2}\,dy\,dx = \int_{-1}^1 2x\sqrt{y}\,\Big|_{x^2+1}^{3-x^2}\,dx$

$$= 2\int_{-1}^1 (x\sqrt{3-x^2}-x\sqrt{x^2+1})\,dx$$

$$= 2\left[-\frac{1}{3}(3-x^2)^{3/2}-\frac{1}{3}(x^2+1)^{3/2}\right]\Big|_{-1}^1$$

$$= -\frac{2}{3}[(2^{3/2}+2^{3/2})-(2^{3/2}+2^{3/2})] = 0$$

21. $\displaystyle\iint_R \sqrt{x^2+1}\,dA = \int_0^{\sqrt{3}} \int_{-x}^{x} \sqrt{x^2+1}\,dy\,dx = \int_0^{\sqrt{3}} y\sqrt{x^2+1}\,\Big|_{-x}^{x}\,dx$

$$= \int_0^{\sqrt{3}} (x\sqrt{x^2+1}+x\sqrt{x^2+1})\,dx = \int_0^{\sqrt{3}} 2x\sqrt{x^2+1}\,dx$$

$$= \frac{2}{3}(x^2+1)^{3/2}\,\Big|_0^{\sqrt{3}} = \frac{2}{3}(4^{3/2}-1^{3/2}) = \frac{14}{3}$$

24. The correct integral is **(b)**.

$$V = 8\int_0^2 \int_0^{\sqrt{4-y^2}} (4-y^2)^{1/2}dx\,dy = 8\int_0^2 (4-y^2)^{1/2}x\,\Big|_0^{\sqrt{4-y^2}}\,dy = 8\int_0^2 (4-y^2)\,dy$$

$$= 8\left(4y-\frac{1}{3}y^3\right)\Big|_0^2 = \frac{128}{3}$$

27. Solving for z, we have $x = 2-\frac{1}{2}x+\frac{1}{2}y$. Setting $z=0$, we see that this surface (plane) intersects the xy-plane in the line $y = x-4$. since $z(0,0) = 2 > 0$, the surface lies above the xy-plane over the quarter-circular region.

$$V = \int_0^2 \int_0^{\sqrt{4-x^2}} \left(2-\frac{1}{2}x+\frac{1}{2}y\right)\,dy\,dx = \int_0^2 \left(2y-\frac{1}{2}xy+\frac{1}{4}y^2\right)\Big|_0^{\sqrt{4-x^2}}\,dx$$

$$= \int_0^2 \left(2\sqrt{4-x^2}-\frac{1}{2}x\sqrt{4-x^2}+1-\frac{1}{4}x^2\right)dx = \left[x\sqrt{4-x^2}+4\sin^{-1}\frac{x}{2}+\frac{1}{6}(4-x^2)^{3/2}+x-\frac{1}{12}x^3\right]\Big|_0^2$$

$$= \left(2\pi+2-\frac{2}{3}\right)-\frac{4}{3} = 2\pi$$

30. In the first octant, $z = x+y$ is nonnegative. Then

$$V = \int_0^3 \int_0^{\sqrt{9-x^2}} (x+y)\,dy\,dx = \int_0^3 \left(xy+\frac{1}{2}y^2\right)\Big|_0^{\sqrt{9-x^2}}\,dx$$

$$= \int_0^3 \left(x\sqrt{9-x^2}+\frac{9}{2}-\frac{1}{2}x^2\right)dx = \left[-\frac{1}{3}(9-x^2)^{3/2}+\frac{9}{2}x-\frac{1}{6}x^3\right]\Big|_0^3 = \left(\frac{27}{2}-\frac{9}{2}\right)-(-9) = 18.$$

33. Note that $z = 4 - y^2$ is positive for $|y| \leq 1$. Using symmetry,

$$V = 2 \int_0^2 \int_0^{\sqrt{2x-x^2}} (4 - y^2)\, dy\, dx = 2 \int_0^2 \left(4y - \frac{1}{3}y^3\right)\Big|_0^{\sqrt{2x-x^2}}\, dx$$

$$= 2 \int_0^2 \left[4\sqrt{2x - x^2} - \frac{1}{3}(2x - x^2)\sqrt{2x - x^2}\right] dx$$

$$= 2 \int_0^2 \left(4\sqrt{1 - (x-1)^2} - \frac{1}{3}[1 - (x-1)^2]\sqrt{1 - (x-1)^2}\right) dx \qquad \boxed{u = x - 1,\ du = dx}$$

$$= 2 \int_{-1}^1 \left[4\sqrt{1 - u^2} - \frac{1}{3}(1 - u^2)\sqrt{1 - u^2}\right] du = 2 \int_{-1}^1 \left(\frac{11}{3}\sqrt{1 - u^2} + \frac{1}{3}u^2\sqrt{1 - u^2}\right) du$$

$$\boxed{\text{Trig substitution}}$$

$$= 2 \left[\frac{11}{6}u\sqrt{1 - u^2} + \frac{11}{6}\sin u + \frac{1}{24}x(2x^2 - 1)\sqrt{1 - u^2} + \frac{1}{24}\sin^{-1} u\right]\Big|_{-1}^1$$

$$= 2\left[\left(\frac{11}{6}\frac{\pi}{2} + \frac{1}{24}\frac{\pi}{2}\right) - \left(-\frac{11}{6}\frac{\pi}{2} - \frac{1}{24}\frac{\pi}{2}\right)\right] = \frac{15\pi}{4}.$$

36.

$$\int_0^1 \int_{2y}^2 e^{-y/x}\, dx\, dy = \int_0^2 \int_0^{x/2} e^{-y/x}\, dy\, dx = \int_0^2 -xe^{-y/x}\Big|_0^{x/2} = \int_0^2 (-xe^{-1/2} + x)\, dx$$

$$= \int_0^2 (1 - e^{-1/2})x\, dx = \frac{1}{2}(1 - e^{-1/2})x^2\Big|_0^2 = 2(1 - e^{-1/2})$$

39. $$\int_0^1 \int_x^1 \frac{1}{1 + y^4}\, dy\, dx = \int_0^1 \int_0^y \frac{1}{1 + y^4}\, dx\, dy = \int_0^1 \frac{x}{1 + y^4}\Big|_0^y\, dy = \int_0^1 \frac{y}{1 + y^4}\, dy$$

$$= \frac{1}{2}\tan^{-1} y^2\Big|_0^1 = \frac{\pi}{8}$$

42.

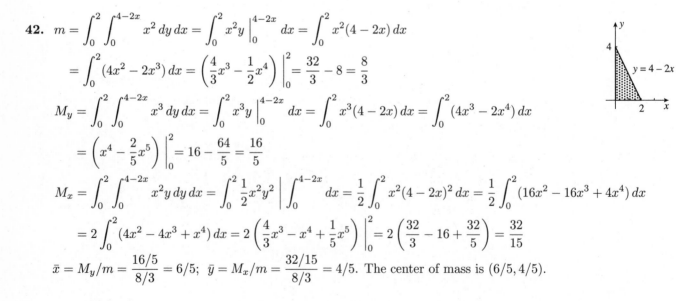

$$m = \int_0^2 \int_0^{4-2x} x^2\, dy\, dx = \int_0^2 x^2 y\Big|_0^{4-2x}\, dx = \int_0^2 x^2(4 - 2x)\, dx$$

$$= \int_0^2 (4x^2 - 2x^3)\, dx = \left(\frac{4}{3}x^3 - \frac{1}{2}x^4\right)\Big|_0^2 = \frac{32}{3} - 8 = \frac{8}{3}$$

$$M_y = \int_0^2 \int_0^{4-2x} x^3\, dy\, dx = \int_0^2 x^3 y\Big|_0^{4-2x}\, dx = \int_0^2 x^3(4 - 2x)\, dx = \int_0^2 (4x^3 - 2x^4)\, dx$$

$$= \left(x^4 - \frac{2}{5}x^5\right)\Big|_0^2 = 16 - \frac{64}{5} = \frac{16}{5}$$

$$M_x = \int_0^2 \int_0^{4-2x} x^2 y\, dy\, dx = \int_0^2 \frac{1}{2}x^2 y^2 \left|\int_0^{4-2x}\right. dx = \frac{1}{2}\int_0^2 x^2(4 - 2x)^2\, dx = \frac{1}{2}\int_0^2 (16x^2 - 16x^3 + 4x^4)\, dx$$

$$= 2\int_0^2 (4x^2 - 4x^3 + x^4)\, dx = 2\left(\frac{4}{3}x^3 - x^4 + \frac{1}{5}x^5\right)\Big|_0^2 = 2\left(\frac{32}{3} - 16 + \frac{32}{5}\right) = \frac{32}{15}$$

$\bar{x} = M_y/m = \dfrac{16/5}{8/3} = 6/5$; $\quad \bar{y} = M_x/m = \dfrac{32/15}{8/3} = 4/5$. The center of mass is $(6/5, 4/5)$.

45. $m = \displaystyle\int_0^1 \int_0^{x^2} (x+y)\,dy\,dx = \int_0^1 \left(xy + \frac{1}{2}y^2\right)\Big|_0^{x^2}\,dx = \int_0^1 \left(x^3 + \frac{1}{2}x^4\right)dx$

$\qquad = \left(\dfrac{1}{4}x^4 + \dfrac{1}{10}x^5\right)\Big|_0^1 = \dfrac{7}{20}$

$M_y = \displaystyle\int_0^1 \int_0^{x^2} (x^2+xy)\,dy\,dx = \int_0^1 \left(x^2 y + \frac{1}{2}xy^2\right)\Big|_0^{x^2}\,dx = \int_0^1 \left(x^4 + \frac{1}{2}x^5\right)dx$

$\qquad = \left(\dfrac{1}{5}x^5 + \dfrac{1}{12}x^6\right)\Big|_0^1 = \dfrac{17}{60}$

$M_x = \displaystyle\int_0^1 \int_0^{x^2} (xy+y^2)\,dy\,dx = \int_0^1 \left(\frac{1}{2}xy^2 + \frac{1}{3}y^3\right)\Big|_0^{x^2} = \int_0^1 \left(\frac{1}{2}x^5 + \frac{1}{3}x^6\right)dx = \left(\dfrac{1}{12}x^6 + \dfrac{1}{21}x^7\right)\Big|_0^1 = \dfrac{11}{84}$

$\bar{x} = M_y/m = \dfrac{17/60}{7/20} = 17/21; \quad \bar{y} = M_x/m = \dfrac{11/84}{7/20} = 55/147.$ The center of mass is $(17/21, 55/147)$.

48. The density is $\rho = kx$.

$m = \displaystyle\int_0^{\pi} \int_0^{\sin x} kx\,dy\,dx = \int_0^{\pi} kxy\Big|_0^{\sin x}\,dx = \int_0^{\pi} kx \sin x\,dx$

$\boxed{\text{Integration by parts}}$

$\qquad = k(\sin x - x\cos x)\Big|_0^{\pi} = k\pi$

$M_y = \displaystyle\int_0^{\pi} \int_0^{\sin x} kx^2\,dy\,dx = \int_0^{\pi} kx^2 y\Big|_0^{\sin x}\,dx = \int_0^{\pi} kx^2 \sin x\,dx \qquad \boxed{\text{Integration by parts}}$

$\qquad = k(-x^2 \cos x + 2\cos x + 2x\sin x)\Big|_0^{\pi} = k[(\pi^2 - 2) - 2] = k(\pi^2 - 4)$

$M_x = \displaystyle\int_0^{\pi} \int_0^{\sin x} kxy\,dy\,dx = \int_0^{\pi} \frac{1}{2}kxy^2\Big|_0^{\sin x}\,dx = \int_0^{\pi} \frac{1}{2}kx\sin^2 x\,dx = \int_0^{\pi} \frac{1}{4}kx(1 - \cos 2x)\,dx$

$\qquad = \dfrac{1}{4}k\left[\displaystyle\int_0^{\pi} x\,dx - \int_0^{\pi} x\cos 2x\,dx\right] \qquad \boxed{\text{Integration by parts}}$

$\qquad = \dfrac{1}{4}k\left[\dfrac{1}{2}x^2\Big|_0^{\pi} - \dfrac{1}{4}(\cos 2x + 2x\sin 2x)\Big|_0^{\pi}\right] = \dfrac{1}{4}k\left(\dfrac{1}{2}\pi^2\right) = \dfrac{1}{8}k\pi^2$

$\bar{x} = M_y/m = \dfrac{k(\pi^2 - 4)}{k\pi} = \pi - 4/\pi; \quad \bar{y} = M_x/m = \dfrac{k\pi^2/8}{k\pi} = \pi/8.$ The center of mass is $(\pi - 4/\pi, \pi/8)$.

51. $I_x = \displaystyle\int_0^1 \int_0^{y-y^2} 2xy^2\,dx\,dy = \int_0^1 x^2 y^2\Big|_0^{y-y^2}\,dy = \int_0^1 (y-y^2)^2\,y^2\,dy$

$\qquad = \displaystyle\int_0^1 (y^4 - 2y^5 + y^6)\,dy = \left(\dfrac{1}{5}y^5 - \dfrac{1}{3}y^6 + \dfrac{1}{7}y^7\right)\Big|_0^1 = \dfrac{1}{105}$

54. $I_x = \displaystyle\int_0^2 \int_0^{\sqrt{4-x^2}} y^3\,dy\,dx = \int_0^2 \frac{1}{4}y^4\Big|_0^{\sqrt{4-x^2}}\,dx = \frac{1}{4}\int_0^2 (4-x^2)^2\,dx$

$\qquad = \dfrac{1}{4}\displaystyle\int_0^2 (16 - 8x^2 + x^4)\,dx = \dfrac{1}{4}\left(16x - \dfrac{8}{3}x^3 + \dfrac{1}{5}x^5\right)\Big|_0^2 = \dfrac{1}{4}\left(32 - \dfrac{64}{3} + \dfrac{32}{5}\right)$

$\qquad = 8\left(1 - \dfrac{2}{3} + \dfrac{1}{5}\right) = \dfrac{64}{15}$

57. $I_y = \int_0^1 \int_y^3 (4x^3 + 3x^2y)\, dx\, dy = \int_0^1 (x^4 + x^3y)\Big|_y^3 \, dy = \int_0^1 (81 + 27y - 2y^4)\, dy$

$= \left(81y + \dfrac{27}{2}y^2 - \dfrac{2}{5}y^5\right)\Big|_0^1 = \dfrac{941}{10}$

60. $m = \int_0^a \int_0^{a-x} k\, dy\, dx = \int_0^a ky\Big|_0^{a-x}\, dx = k\int_0^a (a-x)\, dx = k\left(ax - \dfrac{1}{2}x^2\right)\Big|_0^a = \dfrac{1}{2}ka^2$

$I_x = \int_0^a \int_0^{a-x} ky^2\, dy\, dx = \int_0^a \dfrac{1}{3}ky^3\Big|_0^{a-x}\, dx = \dfrac{1}{3}k\int_0^a (a-x)^3\, dx$

$= \dfrac{1}{3}k\int_0^a (a^3 - 3a^2x + 3ax^2 - x^3)\, dx = \dfrac{1}{3}k\left(a^3x - \dfrac{3}{2}a^2x^2 + ax^3 - \dfrac{1}{4}x^4\right)\Big|_0^a = \dfrac{1}{12}ka^4$

$R_g = \sqrt{\dfrac{I_x}{m}} = \sqrt{\dfrac{ka^4/12}{ka^2/2}} = \sqrt{\dfrac{1}{6}}\, a$

63. From the solution to Problem 60, $m = \dfrac{1}{2}ka^2$ and $I_x = \dfrac{1}{12}ka^4$.

$I_y = \int_0^a \int_0^{a-x} kx^2\, dy\, dx = \int_0^a kx^2y\Big|_0^{a-x}\, dx = k\int_0^a x^2(a-x)\, dx$

$= k\left(\dfrac{1}{3}ax^3 - \dfrac{1}{4}x^4\right)\Big|_0^a = \dfrac{1}{12}ka^4$

$I_0 = I_x + I_y = \dfrac{1}{12}ka^4 + \dfrac{1}{12}ka^4 = \dfrac{1}{6}ka^4$

66. $I_0 = \int_0^3 \int_y^4 k(x^2 + y^2)\, dx\, dy = k\int_0^3 \left(\dfrac{1}{3}x^3 + xy^2\right)\Big|_y^4\, dy$

$= k\int_0^3 \left(\dfrac{64}{3} + 4y^2 - \dfrac{1}{3}y^3 - y^3\right)\, dy = k\left(\dfrac{64}{3}y + \dfrac{4}{3}y^3 - \dfrac{1}{3}y^4\right)\Big|_0^3 = 73k$

EXERCISES 9.11

Double Integrals in Polar Coordinates

3. Solving $r = 2\sin\theta$ and $r = 1$, we obtain $\sin\theta = 1/2$ or $\theta = \pi/6$. Using symmetry,

$A = 2\int_0^{\pi/6} \int_0^{2\sin\theta} r\, dr\, d\theta + 2\int_{\pi/6}^{\pi/2} \int_0^1 r\, dr\, d\theta$

$= 2\int_0^{\pi/6} \dfrac{1}{2}r^2 \Big|_0^{2\sin\theta}\, d\theta + 2\int_{\pi/6}^{\pi/2} \dfrac{1}{2}r^2 \Big|_0^1\, d\theta = \int_0^{\pi/6} 4\sin^2\theta\, d\theta + \int_{\pi/6}^{\pi/2} d\theta$

$= (2\theta - \sin 2\theta)\Big|_0^{\pi/6} + \left(\dfrac{\pi}{2} - \dfrac{\pi}{6}\right) = \dfrac{\pi}{3} - \dfrac{\sqrt{3}}{2} + \dfrac{\pi}{3} = \dfrac{4\pi - 3\sqrt{3}}{6}$

6. $V = \int_0^{2\pi} \int_0^2 \sqrt{9 - r^2}\, r\, dr\, d\theta = \int_0^{2\pi} -\dfrac{1}{3}(9 - r^2)^{3/2}\Big|_0^2\, d\theta$

$= -\dfrac{1}{3}\int_0^{2\pi} (5^{3/2} - 27)\, d\theta = \dfrac{1}{3}(27 - 5^{3/2})2\pi = \dfrac{2\pi(27 - 5\sqrt{5})}{3}$

9. $V = \int_0^{\pi/2} \int_0^{1+\cos\theta} (r\sin\theta) r \, dr \, d\theta = \int_0^{\pi/2} \frac{1}{3} r^3 \sin\theta \Big|_0^{1+\cos\theta} d\theta$

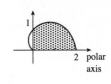

$= \frac{1}{3} \int_0^{\pi/2} (1+\cos\theta)^3 \sin\theta \, d\theta = \frac{1}{3}\left[-\frac{1}{4}(1+\cos\theta)^4 \right] \Big|_0^{\pi/2} = -\frac{1}{12}(1-2^4) = \frac{5}{4}$

12. The interior of the upper-half circle is traced from $\theta = 0$ to $\pi/2$. The density is kr.
Since both the region and the density are symmetric about the polar axis, $\bar{y} = 0$.

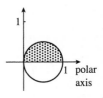

$m = \int_0^{\pi/2} \int_0^{\cos\theta} kr^2 \, dr \, d\theta = k \int_0^{\pi/2} \frac{1}{3} r^3 \Big|_0^{\cos\theta} d\theta = \frac{k}{3} \int_0^{\pi/2} \cos^3\theta \, d\theta$

$= \frac{k}{3} \left(\frac{2}{3} + \frac{1}{3}\cos^2\theta \right) \sin\theta \Big|_0^{\pi/2} = \frac{2k}{9}$

$M_y = k \int_0^{\pi/2} \int_0^{\cos\theta} (r\cos\theta)(r)(r\,dr\,d\theta) = k \int_0^{\pi/2} \int_0^{\cos\theta} r^3 \cos\theta \, dr \, d\theta = k \int_0^{\pi/2} \frac{1}{4} r^4 \cos\theta \Big|_0^{\cos\theta} d\theta$

$= \frac{k}{4} \int_0^{\pi/2} \cos^5\theta \, d\theta = \frac{k}{4}\left(\sin\theta - \frac{2}{3}\sin^3\theta + \frac{1}{5}\sin^5\theta \right) \Big|_0^{\pi/2} = \frac{2k}{15}$

Thus, $\bar{x} = \dfrac{2k/15}{2k/9} = 3/5$ and the center of mass is $(3/5, 0)$.

15. The density is $\rho = k/r$.

$m = \int_0^{\pi/2} \int_2^{2+2\cos\theta} \frac{k}{r} r \, dr \, d\theta = k \int_0^{\pi/2} \int_2^{2+2\cos\theta} dr \, d\theta$

$= k \int_0^{\pi/2} 2\cos\theta \, d\theta = 2k(\sin\theta) \Big|_0^{\pi/2} = 2k$

$M_y = \int_0^{\pi/2} \int_2^{2+2\cos\theta} x \frac{k}{r} r \, dr \, d\theta = k \int_0^{\pi/2} \int_2^{2+2\cos\theta} r \cos\theta \, dr \, d\theta = k \int_0^{\pi/2} \frac{1}{2} r^2 \Big|_2^{2+2\cos\theta} \cos\theta \, d\theta$

$= \frac{1}{2} k \int_0^{\pi/2} (8\cos\theta + 4\cos^2\theta) \cos\theta \, d\theta = 2k \int_0^{\pi/2} (2\cos^2\theta + \cos\theta - \sin^2\theta \cos\theta) \, d\theta$

$= 2k \left(\theta + \frac{1}{2}\sin 2\theta + \sin\theta - \frac{1}{3}\sin^3\theta \right) \Big|_0^{\pi/2} = 2k\left(\frac{\pi}{2} + \frac{2}{3} \right) = \frac{3\pi+4}{3} k$

$M_x = \int_0^{\pi/2} \int_2^{2+2\cos\theta} y \frac{k}{r} r \, dr \, d\theta = k \int_0^{\pi/2} \int_2^{2+2\cos\theta} r \sin\theta \, dr \, d\theta = k \int_0^{\pi/2} \frac{1}{2} r^2 \Big|_2^{2+2\cos\theta} \sin\theta \, d\theta$

$= \frac{1}{2} k \int_0^{\pi/2} (8\cos\theta + 4\cos^2\theta) \sin\theta \, d\theta = \frac{1}{2} k \left(-4\cos^2\theta - \frac{4}{3}\cos^3\theta \right) \Big|_0^{\pi/2} = \frac{1}{2} k \left[-\left(-4 - \frac{4}{3} \right) \right] = \frac{8}{3} k$

$\bar{x} = M_y/m = \dfrac{(3\pi+4)k/3}{2k} = \dfrac{3\pi+4}{6}; \quad \bar{y} = M_x/m = \dfrac{8k/3}{2k} = \dfrac{4}{3}.$ The center of mass is $((3\pi+4)/6, 4/3)$.

18. $I_x = \int_0^{2\pi} \int_0^a y^2 \frac{1}{1+r^4} r \, dr \, d\theta = \int_0^{2\pi} \int_0^a \frac{r^3}{1+r^4} \sin^2\theta \, dr \, d\theta$

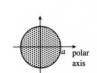

$= \int_0^{2\pi} \frac{1}{4} \ln(1+r^4) \Big|_0^a \sin^2\theta \, d\theta = \frac{1}{4} \ln(1+a^4) \left(\frac{1}{2}\theta - \frac{1}{4}\sin 2\theta \right) \Big|_0^{2\pi} = \frac{\pi}{4} \ln(1+a^4)$

21. From the solution to Problem 17, $I_x = k\pi a^4/4$. By symmetry, $I_y = I_x$. Thus $I_0 = k\pi a^4/2$.

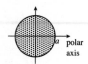

24. $I_0 = \displaystyle\int_0^\pi \int_0^{2a\cos\theta} r^2 kr\, dr\, d\theta = k\int_0^\pi \frac{1}{4}r^4 \Big|_0^{2a\cos\theta}\, d\theta = 4ka^4 \int_0^\pi \cos^4\theta\, d\theta$

$$= 4ka^4 \left(\frac{3}{8}\theta + \frac{1}{4}\sin 2\theta + \frac{1}{32}\sin 4\theta\right)\Big|_0^\pi = 4ka^4\left(\frac{3\pi}{8}\right) = \frac{3k\pi a^4}{2}$$

27. $\displaystyle\int_0^1 \int_0^{\sqrt{1-y^2}} e^{x^2+y^2}\, dx\, dy = \int_0^{\pi/2}\int_0^1 e^{r^2} r\, dr\, d\theta = \int_0^{\pi/2} \frac{1}{2}e^{r^2}\Big|_0^1 d\theta$

$$= \frac{1}{2}\int_0^{\pi/2}(e-1)\, d\theta = \frac{\pi(e-1)}{4}$$

30. $\displaystyle\int_0^1 \int_0^{\sqrt{2y-y^2}}(1-x^2-y^2)\, dx\, dy$

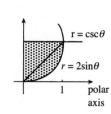

$$= \int_0^{\pi/4}\int_0^{2\sin\theta}(1-r^2)r\, dr\, d\theta + \int_{\pi/4}^{\pi/2}\int_0^{\csc\theta}(1-r^2)r\, dr\, d\theta$$

$$= \int_0^{\pi/4}\left(\frac{1}{2}r^2 - \frac{1}{4}r^4\right)\Big|_0^{2\sin\theta} d\theta + \int_{\pi/4}^{\pi/2}\left(\frac{1}{2}r^2 - \frac{1}{4}r^4\right)\Big|_0^{\csc\theta} d\theta$$

$$= \int_0^{\pi/4}(2\sin^2\theta - 4\sin^4\theta)\, d\theta + \int_{\pi/4}^{\pi/2}\left(\frac{1}{2}\csc^2\theta - \frac{1}{4}\csc^4\theta\right) d\theta$$

$$= \left[\theta - \frac{1}{2}\sin 2\theta - \left(\frac{3}{2}\theta - \sin 2\theta + \frac{1}{8}\sin 4\theta\right)\right] + \left[-\frac{1}{2}\cot\theta - \frac{1}{4}\left(-\cot\theta - \frac{1}{3}\cot^3\theta\right)\right]\Big|_{\pi/4}^{\pi/2}$$

$$= \left(-\frac{\pi}{8} + \frac{1}{2}\right) + \left[0 - \left(-\frac{1}{4} + \frac{1}{12}\right)\right] = \frac{16-3\pi}{24}$$

33. The volume of the cylindrical portion of the tank is $V_c = \pi(4.2)^2 19.3 \approx 1069.56$ m^3. We take the equation of the ellipsoid to be

$$\frac{x^2}{(4.2)^2} + \frac{z^2}{(5.15)^2} = 1 \quad\text{or}\quad z = \pm\frac{5.15}{4.2}\sqrt{(4.2)^2 - x^2 - y^2}\,.$$

The volume of the ellipsoid is

$$V_e = 2\left(\frac{5.15}{4.2}\right)\iint_R \sqrt{(4.2)^2 - x^2 - y^2}\, dx\, dy = \frac{10.3}{4.2}\int_0^{2\pi}\int_0^{4.2}[(4.2)^2 - r^2]^{1/2} r\, dr\, d\theta$$

$$= \frac{10.3}{4.2}\int_0^{2\pi}\left[\left(-\frac{1}{2}\right)\frac{2}{3}[(4.2)^2 - r^2]^{3/2}\Big|_0^{4.2}\right] d\theta = \frac{10.3}{4.2}\frac{1}{3}\int_0^{2\pi}(4.2)^3\, d\theta$$

$$= \frac{2\pi}{3}\frac{10.3}{4.2}(4.2)^3 \approx 380.53.$$

The volume of the tank is approximately $1069.56 + 380.53 = 1450.09$ m^3.

EXERCISES 9.12

Green's Theorem

3. $\oint_C - y^2\, dx + x^2\, dy = \int_0^{2\pi} (-9\sin^2 t)(-3\sin t)\, dt + \int_0^{2\pi} 9\cos^2 t(3\cos t)\, dt$

$$= 27\int_0^{2\pi} [(1 - \cos^2 t)\sin t + (1 - \sin^2 t)\cos t]\, dt$$

$$= 27\left(-\cos t + \frac{1}{3}\cos^3 t + \sin t - \frac{1}{3}\sin^3 t\right)\bigg|_0^{2\pi} = 27(0) = 0$$

$\iint_R (2x + 2y)\, dA = 2\int_0^{2\pi} \int_0^3 (r\cos\theta + r\sin\theta)r\, dr\, d\theta = 2\int_0^{2\pi} \int_0^3 r^2(\cos\theta + \sin\theta)\, dr\, d\theta$

$$= 2\int_0^{2\pi} \left[\frac{1}{3}r^3(\cos\theta + \sin\theta)\right]\bigg|_0^3 d\theta = 18\int_0^{2\pi} (\cos\theta + \sin\theta)\, d\theta$$

$$= 18(\sin\theta - \cos\theta)\bigg|_0^{2\pi} = 18(0) = 0$$

6. $P = x + y^2,\ P_y = 2y,\ Q = 2x^2 - y,\ Q_x = 4x$

$\oint_C (x + y^2)\, dx + (2x^2 - y)\, dy = \iint_R (4x - 2y)\, dA = \int_{-2}^2 \int_{x^2}^4 (4x - 2y)\, dy\, dx$

$$= \int_{-2}^2 (4xy - y^2)\bigg|_{x^2}^4 dx = \int_{-2}^2 (16x - 16 - 4x^3 + x^4)\, dx$$

$$= \left(8x^2 - 16x - x^4 + \frac{1}{5}x^5\right)\bigg|_{-2}^2 = -\frac{96}{5}$$

9. $P = 2xy,\ P_y = 2x,\ Q = 3xy^2,\ Q_x = 3y^2$

$\oint_C 2xy\, dx + 3xy^2\, dy = \iint_R (3y^2 - 2x)\, dA = \int_1^2 \int_2^{2x} (3y^2 - 2x)\, dy\, dx$

$$= \int_1^2 (y^3 - 2xy)\bigg|_2^{2x} dx = \int_1^2 (8x^3 - 4x^2 - 8 + 4x)\, dx$$

$$= \left(2x^4 - \frac{4}{3}x^3 - 8x + 2x^2\right)\bigg|_1^2 = \frac{40}{3} - \left(-\frac{16}{3}\right) = \frac{56}{3}$$

12. $P = e^{x^2},\ P_y = 0,\ Q = 2\tan^{-1} x,\ Q_x = \dfrac{2}{1 + x^2}$

$\oint_C e^{x^2}\, dx + 2\tan^{-1} x\, dy = \iint_R \dfrac{2}{1 + x^2}\, dA = \int_{-1}^0 \int_{-x}^1 \dfrac{2}{1 + x^2}\, dy\, dx$

$$= \int_{-1}^0 \left(\frac{2y}{1 + x^2}\right)\bigg|_{-x}^1 dx = \int_{-1}^0 \left(\frac{2}{1 + x^2} + \frac{2x}{1 + x^2}\right)\, dx$$

$$= [2\tan^{-1} x + \ln(1 + x^2)]\bigg|_{-1}^0 = 0 - \left(-\frac{\pi}{2} + \ln 2\right) = \frac{\pi}{2} - \ln 2$$

15. $P = ay,\ P_y = a,\ Q = bx,\ Q_x = b.\ \oint_C ay\, dx + bx\, dy = \iint_R (b - a)\, dA = (b - a) \times (\text{area bounded by } C)$

18. $P = -y,\ P_y = -1,\ Q = x,\ Q_x = 1.\ \dfrac{1}{2}\oint_C - y\, dx + x\, dy = \dfrac{1}{2}\iint_R 2\, dA = \iint_R dA = \text{area of } R$

21. (a) Parameterize C by $x = x_1 + (x_2 - x_1)t$ and $y = y_1 + (y_2 - y_1)t$ for $0 \le t \le 1$. Then

$$\int_C -y\,dx + x\,dy = \int_0^1 -[y_1 + (y_2 - y_1)t](x_2 - x_1)\,dt + \int_0^1 [x_1 + (x_2 - x_1)t](y_2 - y_1)\,dt$$

$$= -(x_2 - x_1)\left[y_1 t + \frac{1}{2}(y_2 - y_1)t^2\right]\Big|_0^1 + (y_2 - y_1)\left[x_1 t + \frac{1}{2}(x_2 - x_1)t^2\right]\Big|_0^1$$

$$= -(x_2 - x_1)\left[y_1 + \frac{1}{2}(y_2 - y_1)\right] + (y_2 - y_1)\left[x_1 + \frac{1}{2}(x_2 - x_1)\right] = x_1 y_2 - x_2 y_1.$$

(b) Let C_i be the line segment from (x_i, y_i) to (x_{i+1}, y_{i+1}) for $i = 1, 2, \ldots, n-1$, and C_2 the line segment from (x_n, y_n) to (x_1, y_1). Then

$$A = \frac{1}{2}\oint_C -y\,dx + x\,dy \qquad \boxed{\text{Using Problem 18}}$$

$$= \frac{1}{2}\left[\int_{C_1} -y\,dx + x\,dy + \int_{C_2} -y\,dx + x\,dy + \cdots + \int_{C_{n-1}} -y\,dx + x\,dy + \int_{C_n} -y\,dx + x\,dy\right]$$

$$= \frac{1}{2}(x_1 y_2 - x_2 y_1) + \frac{1}{2}(x_2 y_3 - x_3 y_2) + \frac{1}{2}(x_{n-1}y_n - x_n y_{n-1}) + \frac{1}{2}(x_n y_1 - x_1 y_n).$$

24. $P = \cos x^2 - y$, $P_y = -1$; $Q = \sqrt{y^3 + 1}$, $Q_x = 0$

$$\oint_C (\cos x^2 - y)\,dx + \sqrt{y^3 + 1}\,dy = \iint_R (0 + 1)\,dA \iint_R dA = (6\sqrt{2})^2 - \pi(2)(4) = 72 - 8\pi$$

27. Writing $\iint_R x^2\,dA = \iint_R (Q_x - P_y)\,dA$ we identify $Q = 0$ and $P = -x^2 y$. Then, with C: $x = 3\cos t$, $y = 2\sin t$, $0 \le t \le 2\pi$, we have

$$\iint_R x^2\,dA = \oint_C P\,dx + Q\,dy = \oint_C -x^2 y\,dx = -\int_0^{2\pi} 9\cos^2 t (2\sin t)(-3\sin t)\,dt$$

$$= \frac{54}{4}\int_0^{2\pi} 4\sin^2 t \cos^2 t\,dt = \frac{27}{2}\int_0^{2\pi} \sin^2 2t\,dt = \frac{27}{4}\int_0^{2\pi}(1 - \cos 4t)\,dt$$

$$= \frac{27}{4}\left(t - \frac{1}{4}\sin 4t\right)\Big|_0^{2\pi} = \frac{27\pi}{2}.$$

30. $P = -xy^2$, $P_y = -2xy$, $Q = x^2 y$, $Q_x = 2xy$. Using polar coordinates,

$$W = \oint_C \mathbf{F}\cdot d\mathbf{r} = \iint_R 4xy\,dA = \int_0^{\pi/2}\int_1^2 4(r\cos\theta)(r\sin\theta)r\,dr\,d\theta = \int_0^{\pi/2}(r^4\cos\theta\sin\theta)\Big|_1^2\,d\theta$$

$$= 15\int_0^{\pi/2}\sin\theta\cos\theta\,d\theta = \frac{15}{2}\sin^2\theta\Big|_0^{\pi/2} = \frac{15}{2}.$$

33. Using Green's Theorem,

$$W = \oint_C \mathbf{F}\cdot d\mathbf{r} = \oint_C -y\,dx + x\,dy = \iint_R 2\,dA = 2\int_0^{2\pi}\int_0^{1+\cos\theta} r\,dr\,d\theta$$

$$= 2\int_0^{2\pi}\left(\frac{1}{2}r^2\right)\Big|_0^{1+\cos\theta}\,d\theta = \int_0^{2\pi}(1 + 2\cos\theta + \cos^2\theta)\,d\theta$$

$$= \left(\theta + 2\sin\theta + \frac{1}{2}\theta + \frac{1}{4}\sin 2\theta\right)\Big|_0^{2\pi} = 3\pi.$$

EXERCISES 9.13

Surface Integrals

3. Using $f(x, y) = z = \sqrt{16 - x^2}$ we see that for $0 \le x \le 2$ and $0 \le y \le 5$, $z > 0$. Thus, the surface is entirely above the region. Now $f_x = -\dfrac{x}{\sqrt{16 - x^2}}$, $f_y = 0$,

$$1 + f_x^2 + f_y^2 = 1 + \frac{x^2}{16 - x^2} = \frac{16}{16 - x^2} \text{ and}$$

$$A = \int_0^5 \int_0^2 \frac{4}{\sqrt{16 - x^2}} \, dx \, dy = 4 \int_0^5 \sin^{-1} \frac{x}{4} \Big|_0^2 \, dy = 4 \int_0^5 \frac{\pi}{6} \, dy = \frac{10\pi}{3} \, .$$

6. The surfaces $x^2 + y^2 + z^2 = 2$ and $z^2 = x^2 + y^2$ intersect on the cylinder $2x^2 + 2y^2 = 2$ or $x^2 + y^2 = 1$. There are portions of the sphere within the cone both above and below the xy-plane. Using $f(x, y) = \sqrt{2 - x^2 - y^2}$ we have $f_x = -\dfrac{x}{\sqrt{2 - x^2 - y^2}}$,

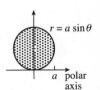

$$f_y = -\frac{y}{\sqrt{2 - x^2 - y^2}}, \quad 1 + f_x^2 + f_y^2 = \frac{2}{2 - x^2 - y^2} \, . \text{ Then}$$

$$A = 2 \left[\int_0^{2\pi} \int_0^1 \frac{\sqrt{2}}{\sqrt{2 - r^2}} \, r \, dr \, d\theta \right] = 2\sqrt{2} \int_0^{2\pi} -\sqrt{2 - r^2} \, \Big|_0^1 \, d\theta$$

$$= 2\sqrt{2} \int_0^{2\pi} (\sqrt{2} - 1) d\theta = 4\pi\sqrt{2}(\sqrt{2} - 1).$$

9. There are portions of the sphere within the cylinder both above and below the xy-plane.

Using $f(x, y) = z = \sqrt{a^2 - x^2 - y^2}$ we have $f_x = -\dfrac{x}{\sqrt{1^2 - x^2 - y^2}}$, $f_y = -\dfrac{y}{\sqrt{a^2 - x^2 - y^2}}$,

$1 + f_x^2 + f_y^2 = \dfrac{a^2}{a^2 - x^2 - y^2}$. Then, using symmetry,

$$A = 2 \left[2 \int_0^{\pi/2} \int_0^{a \sin \theta} \frac{a}{\sqrt{a^2 - r^2}} \, r \, dr \, d\theta \right] = 4a \int_0^{\pi/2} -\sqrt{a^2 - r^2} \, \Big|_0^{a \sin \theta} \, d\theta$$

$$= 4a \int_0^{\pi/2} (a - a\sqrt{1 - \sin^2 \theta}) d\theta = 4a^2 \int_0^{\pi/2} (1 - \cos \theta) d\theta$$

$$= 4a^2 (\theta - \sin \theta) \Big|_0^{\pi/2} = 4a^2 \left(\frac{\pi}{2} - 1 \right) = 2a^2 (\pi - 2).$$

12. From Example 1, the area of the portion of the hemisphere within $x^2 + y^2 = b^2$ is $2\pi a(a - \sqrt{a^2 - b^2})$. Thus, the area of the sphere is $A = 2 \lim_{b \to a} 2\pi a(a - \sqrt{a^2 - b^2}) = 2(2\pi a^2) = 4\pi a^2$.

15. $z_x = -2x$, $z_y = 0$; $dS = \sqrt{1 + 4x^2} \, dA$

$$\iint_S x \, dS = \int_0^4 \int_0^{\sqrt{2}} x\sqrt{1 + 4x^2} \, dx \, dy = \int_0^4 \frac{1}{12} (1 + 4x^2)^{3/2} \Big|_0^{\sqrt{2}} \, dy$$

$$= \int_0^4 \frac{13}{6} \, dy = \frac{26}{3}$$

18. $z_x = \dfrac{x}{\sqrt{x^2 + y^2}}$, $z_y = \dfrac{y}{\sqrt{x^2 + y^2}}$; $dS = \sqrt{2}\, dA$.

Using polar coordinates,

$$\iint_S (x + y + z)\, dS = \iint_R (x + y + \sqrt{x^2 + y^2}\,)\sqrt{2}\, dA$$

$$= \sqrt{2} \int_0^{2\pi} \int_1^4 (r\cos\theta + r\sin\theta + r)r\, dr\, d\theta$$

$$= \sqrt{2} \int_0^{2\pi} \int_1^4 r^2(1 + \cos\theta + \sin\theta)\, dr\, d\theta = \sqrt{2} \int_0^{2\pi} \left.\frac{1}{3}r^3(1 + \cos\theta + \sin\theta)\right|_1^4 d\theta$$

$$= \frac{63\sqrt{2}}{3} \int_0^{2\pi} (1 + \cos\theta + \sin\theta)\, d\theta = 21\sqrt{2}(\theta + \sin\theta - \cos\theta)\Big|_0^{2\pi} = 42\sqrt{2}\,\pi.$$

21. $z_x = -x,\ z_y = -y;\ dS = \sqrt{1 + x^2 + y^2}\, dA$

$$\iint_S xy\, dS = \int_0^1 \int_0^1 xy\sqrt{1 + x^2 + y^2}\, dx\, dy = \int_0^1 \left.\frac{1}{3}y(1 + x^2 + y^2)^{3/2}\right|_0^1 dy$$

$$= \int_0^1 \left[\frac{1}{3}y(2 + y^2)^{3/2} - \frac{1}{3}y(1 + y^2)^{3/2}\right] dy$$

$$= \left[\frac{1}{15}(2 + y^2)^{5/2} - \frac{1}{15}(1 + y^2)^{5/2}\right]\Big|_0^1 = \frac{1}{15}(3^{5/2} - 2^{7/2} + 1)$$

24. $x_y = -2y,\ x_z = -2z;\ dS = \sqrt{1 + 4y^2 + 4z^2}\, dA$

Using polar coordinates,

$$\iint_S (1 + 4y^2 + 4z^2)^{1/2}\, dS = \int_0^{\pi/2} \int_1^2 (1 + 4r^2)r\, dr\, d\theta$$

$$= \int_0^{\pi/2} \left.\frac{1}{16}(1 + 4r^2)^2\right|_1^2 d\theta = \frac{1}{16} \int_0^{\pi/2} 12\, d\theta = \frac{3\pi}{8}\,.$$

27. The density is $\rho = kx^2$. The surface is $z = 1 - x - y$. Then $z_x = -1,\ z_y = -1;\ dS = \sqrt{3}\, dA$.

$$m = \iint_S kx^2\, dS = k \int_0^1 \int_0^{1-x} x^2\sqrt{3}\, dy\, dx = \sqrt{3}\,k \int_0^1 \left.\frac{1}{3}x^3\right|_0^{1-x} dx$$

$$= \frac{\sqrt{3}}{3}k \int_0^1 (1 - x)^3\, dx = \frac{\sqrt{3}}{3}k\left[-\frac{1}{4}(1 - x)^4\right]\Big|_0^1 = \frac{\sqrt{3}}{12}k$$

30. The surface is $g(x, y, z) = x^2 + y^2 + z - 5 = 0$. $\nabla g = 2x\mathbf{i} + 2y\mathbf{j} + \mathbf{k}$,

$$|\nabla g| = \sqrt{1 + 4x^2 + 4y^2}\,;\quad \mathbf{n} = \frac{2x\mathbf{i} + 2y\mathbf{j} + \mathbf{k}}{\sqrt{1 + 4x^2 + 4y^2}}\,;\quad \mathbf{F} \cdot \mathbf{n} = \frac{z}{\sqrt{1 + 4x^2 + 4y^2}}\,;$$

$z_x = -2x,\ z_y = -2y,\ dS = \sqrt{1 + 4x^2 + 4y^2}\, dA$. Using polar coordinates,

$$\text{Flux} = \iint_S \mathbf{F} \cdot \mathbf{n}\, dS = \iint_R \frac{z}{\sqrt{1 + 4x^2 + 4y^2}} \sqrt{1 + 4x^2 + 4y^2}\, dA = \iint_R (5 - x^2 - y^2)\, dA$$

$$= \int_0^{2\pi} \int_0^2 (5 - r^2)r\, dr\, d\theta = \int_0^{2\pi} \left(\frac{5}{2}r^2 - \frac{1}{4}r^4\right)\Big|_0^2 d\theta = \int_0^{2\pi} 6\, d\theta = 12\pi.$$

33. The surface is $g(x, y, z) = x^2 + y^2 + z - 4$. $\nabla g = 2x\mathbf{i} + 2y\mathbf{j} + \mathbf{k}$,

$$|\nabla g| = \sqrt{4x^2 + 4y^2 + 1}\,; \quad \mathbf{n} = \frac{2x\mathbf{i} + 2y\mathbf{j} + \mathbf{k}}{\sqrt{4x^2 + 4y^2 + 1}}\,; \quad \mathbf{F} \cdot \mathbf{n} = \frac{x^3 + y^3 + z}{\sqrt{4x^2 + 4y^2 + 1}}\,;$$

$z_x = -2x$, $z_y = -2y$, $dS = \sqrt{1 + 4x^2 + 4y^2}\,dA$. Using polar coordinates,

$$\text{Flux} = \iint_S \mathbf{F} \cdot \mathbf{n}\, dS = \iint_R (x^3 + y^3 + z)\, dA = \iint_R (4 - x^2 - y^2 + x^3 + y^3)\, dA$$

$$= \int_0^{2\pi} \int_0^2 (4 - r^2 + r^3 \cos^3 \theta + r^3 \sin^3 \theta)\, r\, dr\, d\theta$$

$$= \int_0^{2\pi} \left(2r^2 - \frac{1}{4}r^4 + \frac{1}{5}r^5 \cos^3 \theta + \frac{1}{5}r^5 \sin^3 \theta \right) \Big|_0^2 \, d\theta$$

$$= \int_0^{2\pi} \left(4 + \frac{32}{5} \cos^3 \theta + \frac{32}{5} \sin^3 \theta \right) d\theta = 4\theta \Big|_0^{2\pi} + 0 + 0 = 8\pi.$$

36. For S_1: $g(x, y, z) = x^2 + y^2 + z - 4$, $\nabla g = 2x\mathbf{i} + 2y\mathbf{j} + \mathbf{k}$, $|\nabla g| = \sqrt{4x^2 + 4y^2 + 1}$;

$$\mathbf{n}_1 = \frac{2x\mathbf{i} + 2y\mathbf{j} + \mathbf{k}}{\sqrt{4x^2 + 4y^2 + 1}}\,; \quad \mathbf{F} \cdot \mathbf{n}_1 = 6z^2/\sqrt{4x^2 + 4y^2 + 1}\,; \quad z_x = -2x,\ z_y = -2y,$$

$dS_1 = \sqrt{1 + 4x^2 + 4y^2}\,dA$. For S_2: $g(x, y, z) = x^2 + y^2 - z$, $\nabla g = 2x\mathbf{i} + 2y\mathbf{j} - \mathbf{k}$,

$$|\nabla g| = \sqrt{4x^2 + 4y^2 + 1}\,; \quad \mathbf{n}_2 = \frac{2x\mathbf{i} + 2y\mathbf{j} - \mathbf{k}}{\sqrt{4x^2 + y^2 + 1}}\,; \quad \mathbf{F} \cdot \mathbf{n}_2 = -6z^2/\sqrt{4x^2 + 4y^2 + 1}\,; \quad z_x = 2x,\ z_y = 2y,$$

$dS_2 = \sqrt{1 + 4x^2 + 4y^2}\,dA$. Using polar coordinates and R: $x^2 + y^2 \leq 2$ we have

$$\text{Flux} = \iint_{S_1} \mathbf{F} \cdot \mathbf{n}_1\, dS_1 + \iint_{S_1} \mathbf{F} \cdot \mathbf{n}_2\, dS_2 = \iint_R 6z^2\, dA + \iint -6z^2\, dA$$

$$= \iint_R [6(4 - x^2 - y^2)^2 - 6(x^2 + y^2)^2]\, dA = 6 \int_0^{2\pi} \int_0^{\sqrt{2}} [(4 - r^2)^2 - r^4]\, r\, dr\, d\theta$$

$$= 6 \int_0^{2\pi} \left[-\frac{1}{6}(4 - r^2)^3 - \frac{1}{6}r^6 \right] \Big|_0^{\sqrt{2}} \, d\theta = -\int_0^{2\pi} [(2^3 - 4^3) + (\sqrt{2})^6]\, d\theta = \int_0^{2\pi} 48\, d\theta = 96\pi.$$

39. Refering to the solution to Problem 37, we find $\mathbf{n} = \dfrac{x\mathbf{i} + y\mathbf{j} + z\mathbf{k}}{\sqrt{x^2 + y^2 + z^2}}$ and $dS = \dfrac{a}{\sqrt{a^2 - x^2 - y^2}}\, dA$.

Now

$$\mathbf{F} \cdot \mathbf{n} = kq\, \frac{\mathbf{r}}{|\mathbf{r}|^3} \cdot \frac{\mathbf{r}}{|\mathbf{r}|} = \frac{kq}{|\mathbf{r}|^4} |\mathbf{r}|^2 = \frac{kq}{|\mathbf{r}|^2} = \frac{kq}{x^2 + y^2 + z^2} = \frac{kq}{a^2}$$

and

$$\text{Flux} = \iint_S \mathbf{F} \cdot \mathbf{n}\, dS = \iint_S \frac{kq}{a^2}\, dS = \frac{kq}{a^2} \times \text{area} = \frac{kq}{a^2}(4\pi a^2) = 4\pi kq.$$

42. The area of the hemisphere is $A(s) = 2\pi a^2$. By symmetry, $\bar{x} = \bar{y} = 0$.

$$z_x = -\frac{x}{\sqrt{a^2 - x^2 - y^2}}, \ z_y = -\frac{y}{\sqrt{a^2 - x^2 - y^2}};$$

$$dS = \sqrt{1 + \frac{x^2}{a^2 - x^2 - y^2} + \frac{y^2}{a^2 - x^2 - y^2}} \, dA = \frac{a}{\sqrt{a^2 - x^2 - y^2}} \, dA$$

Using polar coordinates,

$$z = \iint_S \frac{z \, dS}{2\pi a^2} = \frac{1}{2\pi a^2} \iint_R \sqrt{a^2 - x^2 - y^2} \ \frac{a}{\sqrt{a^2 - x^2 - y^2}} \, dA = \frac{1}{2\pi a} \int_0^{2\pi} \int_0^a r \, dr \, d\theta$$

$$= \frac{1}{2\pi a} \int_0^{2\pi} \frac{1}{2} r^2 \bigg|_0^a \, d\theta = \frac{1}{2\pi a} \int_0^{2\pi} \frac{1}{2} s^2 \, d\theta = \frac{a}{2}.$$

The centroid is $(0, 0, a/2)$.

EXERCISES 9.14

Stokes' Theorem

3. Surface Integral: $\text{curl } \mathbf{F} = \mathbf{i} + \mathbf{j} + \mathbf{k}$. Letting $g(x, y, z) = 2x + y + 2z - 6$, we have $\nabla g = 2\mathbf{i} + \mathbf{j} + 2\mathbf{k}$ and $\mathbf{n} = (2\mathbf{i} + \mathbf{j} + 2\mathbf{k})/3$. Then $\iint_S (\text{curl } \mathbf{F}) \cdot \mathbf{n} \, dS = \iint_S \frac{5}{3} \, dS$. Letting the surface be $z = 3 - \frac{1}{2}y - x$ we have $z_x = -1$, $z_y = -\frac{1}{2}$, and $dS = \sqrt{1 + (-1)^2 + (-\frac{1}{2})^2} \, dA = \frac{3}{2} \, dA$. Then

$$\iint_S (\text{curl } \mathbf{F}) \cdot \mathbf{n} \, dS = \iint_R \frac{5}{3}\left(\frac{3}{2}\right) dA = \frac{5}{2} \times (\text{area of } R) = \frac{5}{2}(9) = \frac{45}{2}.$$

Line Integral: C_1: $z = 3 - x$, $0 \leq x \leq 3$, $y = 0$; C_2: $y = 6 - 2x$, $3 \geq x \geq 0$, $z = 0$; C_3: $z = 3 - y/2$, $6 \geq y \geq 0$, $x = 0$.

$$\oint_C z \, dx + x \, dy + y \, dz = \iint_{C_1} z \, dx + \int_{C_2} x \, dy + \int_{C_3} y \, dz$$

$$= \int_0^3 (3 - x) \, dx + \int_3^0 x(-2 \, dx) + \int_6^0 y(-dy/2)$$

$$= \left(3x - \frac{1}{2}x^2\right)\bigg|_0^3 - x^2 \bigg|_3^0 - \frac{1}{4}y^2 \bigg|_6^0 = \frac{9}{2} - (0 - 9) - \frac{1}{4}(0 - 36) = \frac{45}{2}$$

6. $\text{curl } \mathbf{F} = -2xz\mathbf{i} + z^2\mathbf{k}$. A unit vector normal to the plane is $\mathbf{n} = (\mathbf{j} + \mathbf{k})/\sqrt{2}$. From $z = 1 - y$, we have $z_x = 0$ and $z_y = -1$. Thus, $dS = \sqrt{1 + 1} \, dA = \sqrt{2} \, dA$ and

$$\oint_C \mathbf{F} \cdot d\mathbf{r} = \iint_S (\text{curl } \mathbf{F}) \cdot \mathbf{n} \, dS = \iint_R \frac{1}{\sqrt{2}} z^2\sqrt{2} \, dA = \iint_R (1 - y)^2 \, dA$$

$$= \int_0^2 \int_0^1 (1 - y)^2 \, dy \, dx = \int_0^2 -\frac{1}{3}(1 - y)^3 \bigg|_0^1 \, dx = \int_0^2 \frac{1}{3} \, dx = \frac{2}{3}.$$

9. curl $\mathbf{F} = (-3x^2 - 3y^2)\mathbf{k}$. A unit vector normal to the plane is $\mathbf{n} = (\mathbf{i}+\mathbf{j}+\mathbf{k})/\sqrt{3}$. From $z = 1 - x - y$, we have $z_x = z_y = -1$ and $dS = \sqrt{3}\,dA$. Then, using polar coordinates,

$$\oint_C \mathbf{F} \cdot d\mathbf{r} = \iint_S (\text{curl } \mathbf{F}) \cdot \mathbf{n}\,dS = \iint_R (-\sqrt{3}\,x^2 - \sqrt{3}\,y^2)\sqrt{3}\,dA$$

$$= 3 \iint_R (-x^2 - y^2)\,dA = 3 \int_0^{2\pi} \int_0^1 (-r^2)r\,dr\,d\theta$$

$$= 3 \int_0^{2\pi} -\frac{1}{4}r^4 \Big|_0^1 d\theta = 3 \int_0^{2\pi} -\frac{1}{4}\,d\theta = -\frac{3\pi}{2}.$$

12. curl $\mathbf{F} = \mathbf{i} + \mathbf{j} + \mathbf{k}$. Taking the surface S bounded by C to be the portion of the plane $x + y + z = 0$ inside C, we have $\mathbf{n} = (\mathbf{i}+\mathbf{j}+\mathbf{k})/\sqrt{3}$ and $dS = \sqrt{3}\,dA$.

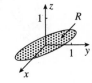

$$\oint_C \mathbf{F} \cdot d\mathbf{r} = \iint_S (\text{curl } \mathbf{F}) \cdot \mathbf{n}\,dS = \iint_S \sqrt{3}\,dS = \sqrt{3} \iint_R \sqrt{3}\,dA = 3 \times (\text{area of } R)$$

The region R is obtained by eliminating z from the equations of the plane and the sphere. This gives $x^2 + xy + y^2 = \frac{1}{2}$. Rotating axes, we see that R is enclosed by the ellipse $X^2/(1/3) + Y^2/1 = 1$ in a rotated coordinate system. Thus,

$$\oint_C \mathbf{F} \cdot d\mathbf{r} = 3 \times (\text{area of } R) = 3\left(\pi \frac{1}{\sqrt{3}} 1\right) = \sqrt{3}\,\pi.$$

15. Parameterize C by C_1: $x = 0$, $z = 0$, $2 \geq y \geq 0$; C_2: $z = x$, $y = 0$, $0 \leq x \leq 2$; C_3: $x = 2$, $z = 2$, $0 \leq y \leq 2$; C_4: $z = x$, $y = 2$, $2 \geq x \geq 0$. Then

$$\iint_S (\text{curl } \mathbf{F}) \cdot \mathbf{n}\,dS = \oint_C \mathbf{F} \cdot \mathbf{r} = \oint_C 3x^2\,dx + 8x^3y\,dy + 3x^2y\,dz$$

$$= \int_{C_1} 0\,dx + 0\,dy + 0\,dz + \int_{C_2} 3x^2\,dx + \int_{C_3} 64\,dy + \int_{C_4} 3x^2\,dx + 6x^2\,dx$$

$$= \int_0^2 3x^2\,dx + \int_0^2 64\,dy + \int_2^0 9x^2\,dx = x^3 \Big|_0^2 + 64y \Big|_0^2 + 3x^3 \Big|_2^0 = 112.$$

18. (a) curl $\mathbf{F} = xz\mathbf{i} - yz\mathbf{j}$. A unit vector normal to the surface is $\mathbf{n} = \dfrac{2x\mathbf{i} + 2y\mathbf{j} + \mathbf{k}}{\sqrt{4x^2 + 4y^2 + 1}}$ and $dS = \sqrt{1 + 4x^2 + 4y^2}\,dA$. Then, using $x = \cos t$, $y = \sin t$, $0 \leq t \leq 2\pi$, we have

$$\iint_S (\text{curl } \mathbf{F}) \cdot \mathbf{n}\,dS = \iint_R (2x^2 z - 2y^2 z)\,dA = \iint_R (2x^2 - 2y^2)(1 - x^2 - y^2)\,dA$$

$$= \iint_R (2x^2 - 2y^2 - 2x^4 + 2y^4)\,dA$$

$$= \int_0^{2\pi} \int_0^1 (2r^2 \cos^2\theta - 2r^2 \sin^2\theta - 2r^4 \cos^4\theta + 2r^4 \cos^4\theta)\,r\,dr\,d\theta$$

$$= 2 \int_0^{2\pi} \int_0^1 [r^3 \cos 2\theta - r^5(\cos^2\theta - \sin^2\theta)(\cos^2\theta + \sin^2\theta)]\,dr\,d\theta$$

$$= 2 \int_0^{2\pi} \int_0^1 (r^3 \cos 2\theta - r^5 \cos 2\theta)\,dr\,d\theta = 2 \int_0^{2\pi} \cos 2\theta \left(\frac{1}{4}r^4 - \frac{1}{6}r^6\right) \Big|_0^1 d\theta$$

$$= \frac{1}{6} \int_0^{2\pi} \cos 2\theta\,d\theta = 0.$$

(b) We take the surface to be $z = 0$. Then $\mathbf{n} = \mathbf{k}$, curl $\mathbf{F} \cdot \mathbf{n} = $ curl $\mathbf{F} \cdot \mathbf{k} = 0$ and $\iint_S (\text{curl } \mathbf{F}) \cdot \mathbf{n} \, dS = 0$.

(c) By Stokes' Theorem, using $z = 0$, we have

$$\iint_S (\text{curl } \mathbf{F}) \cdot \mathbf{n} \, dS = \oint_C \mathbf{F} \cdot d\mathbf{r} = \oint_C xyz \, dz = \oint_C xy(0) \, dz = 0.$$

EXERCISES 9.15

Triple Integrals

3. $\displaystyle\int_0^6 \int_0^{6-x} \int_0^{6-x-z} dy \, dz \, dx = \int_0^6 \int_0^{6-x} (6 - x - z) dz \, dx = \int_0^6 \left(6z - xz - \frac{1}{2}z^2 \right) \Big|_0^{6-x} dx$

$\displaystyle = \int_0^6 \left[6(6-x) - x(6-x) - \frac{1}{2}(6-x)^2 \right] dx = \int_0^6 \left(18 - 6x + \frac{1}{2}x^2 \right) dx$

$\displaystyle = \left(18x - 3x^2 + \frac{1}{6}x^3 \right) \Big|_0^6 = 36$

6. $\displaystyle\int_0^{\sqrt{2}} \int_{\sqrt{y}}^2 \int_0^{e^{x^2}} x \, dz \, dx \, dy = \int_0^{\sqrt{2}} \int_{\sqrt{y}}^2 xe^{x^2} dx \, dy = \int_0^{\sqrt{2}} \frac{1}{2}e^{x^2} \Big|_{\sqrt{y}}^2 dy = \frac{1}{2}\int_0^{\sqrt{2}} (e^4 - e^y) dy$

$\displaystyle = \frac{1}{2}(ye^4 - e^y) \Big|_0^{\sqrt{2}} = \frac{1}{2}[(e^4\sqrt{2} - e^{\sqrt{2}}) - (-1)] = \frac{1}{2}(1 + e^4\sqrt{2} - e^{\sqrt{2}})$

9. $\displaystyle\iiint_D z \, dV = \int_0^5 \int_1^3 \int_y^{y+2} z \, dx \, dy \, dz = \int_0^5 \int_1^3 xz \Big|_y^{y+2} dy \, dz = \int_0^5 \int_1^3 2z \, dy \, dz$

$\displaystyle = \int_0^5 2yz \Big|_1^3 dz = \int_0^5 4z \, dz = 2z^2 \Big|_0^5 = 50$

12. The other five integrals are $\displaystyle\int_0^3 \int_0^{\sqrt{36-4y^2}/3} \int_1^3 F(x, y, z) \, dz \, dx \, dy,$

$\displaystyle\int_1^3 \int_0^2 \int_0^{\sqrt{36-9x^2}/2} F(x, y, z) \, dy \, dx \, dz, \quad \int_1^3 \int_0^3 \int_0^{\sqrt{36-4y^2}/3} F(x, y, z) \, dx \, dy \, dz,$

$\displaystyle\int_0^3 \int_1^3 \int_0^{\sqrt{36-4y^2}/3} F(x, y, z) \, dx \, dz \, dy, \quad \int_0^2 \int_1^3 \int_0^{\sqrt{36-9x^2}/2} F(x, y, z) \, dy \, dz \, dx.$

15.

18.

21. Solving $x = y^2$ and $4 - x = y^2$, we obtain $x = 2$, $y = \pm\sqrt{2}$. Using symmetry,

$$V = 2 \int_0^3 \int_0^{\sqrt{2}} \int_{y^2}^{4-y^2} dx\, dy\, dz = 2 \int_0^3 \int_0^{\sqrt{2}} (4 - 2y^2)\, dy\, dz$$

$$= 2 \int_0^3 \left(4y - \frac{2}{3}y^3 \right) \Big|_0^{\sqrt{2}} dz = 2 \int_0^3 \frac{8\sqrt{2}}{3}\, dz = 16\sqrt{2}.$$

24. Solving $x = 2$, $y = x$, and $z = x^2 + y^2$, we obtain the point $(2, 2, 8)$.

$$V = \int_0^2 \int_0^x \int_0^{x^2+y^2} dz\, dy\, dx = \int_0^2 \int_0^x (x^2 + y^2)\, dy\, dx = \int_0^2 \left(x^2 y + \frac{1}{3}y^3 \right) \Big|_0^x dx$$

$$= \int_0^2 \frac{4}{3}x^3\, dx = \frac{1}{3}x^4 \Big|_0^2 = \frac{16}{3}.$$

27. The density is $\rho(x, y, z) = ky$. Since both the region and the density function are symmetric with respect to the xy- and yz-planes, $\bar{x} = \bar{z} = 0$. Using symmetry,

$$m = 4 \int_0^3 \int_0^2 \int_0^{\sqrt{4-x^2}} ky\, dz\, dx\, dy = 4k \int_0^3 \int_0^2 yz \Big|_0^{\sqrt{4-x^2}} dx\, dy = 4k \int_0^3 \int_0^2 y\sqrt{4 - x^2}\, dx\, dy$$

$$= 4k \int_0^3 y \left(\frac{x}{2}\sqrt{4 - x^2} + 2\sin^{-1}\frac{x}{2} \right) \Big|_0^2 dy = 4k \int_0^3 \pi y\, dy = 4\pi k \left(\frac{1}{2}y^2 \right) \Big|_0^3 = 18\pi k$$

$$M_{xz} = 4 \int_0^3 \int_0^2 \int_0^{\sqrt{4-x^2}} ky^2\, dz\, dx\, dy = 4k \int_0^3 \int_0^2 y^2 z \Big|_0^{\sqrt{4-x^2}} dx\, dy = 4k \int_0^3 \int_0^2 y^2\sqrt{4 - x^2}\, dx\, dy$$

$$= 4k \int_0^3 y^2 \left(\frac{x}{2}\sqrt{4 - x^2} + 2\sin^{-1}\frac{x}{2} \right) \Big|_0^2 dy = 4k \int_0^3 \pi y^2\, dy = 4\pi k \left(\frac{1}{3}y^3 \right) \Big|_0^3 = 36\pi k.$$

$\bar{y} = M_{xz}/m = \dfrac{36\pi k}{18\pi k} = 2.$ The center of mass is $(0, 2, 0)$.

30. Both the region and the density function are symmetric with respect to the xz- and yz-planes. Thus, $m = 4 \displaystyle\int_{-1}^2 \int_0^{\sqrt{1+z^2}} \int_0^{\sqrt{1+z^2-y^2}} z^2\, dx\, dy\, dz.$

33. $I_z = k \displaystyle\int_0^1 \int_0^{1-x} \int_0^{1-x-y} (x^2 + y^2)\, dz\, dy\, dx = k \int_0^1 \int_0^{1-x} (x^2 + y^2)(1 - x - y)\, dy\, dx$

$$= k \int_0^1 \int_0^{1-x} (x^2 - x^3 - x^2 y + y^2 - xy^2 - y^3)\, dy\, dx$$

$$= k \int_0^1 \left[(x^2 - x^3)y - \frac{1}{2}x^2 y^2 + \frac{1}{3}(1 - x)y^3 - \frac{1}{4}y^4 \right] \Big|_0^{1-x} dx$$

$$= k \int_0^1 \left[\frac{1}{2}x^2 - x^3 + \frac{1}{2}x^4 + \frac{1}{12}(1 - x)^4 \right] dx = k \left[\frac{1}{6}x^6 - \frac{1}{4}x^4 + \frac{1}{10}x^5 - \frac{1}{60}(1 - x)^5 \right] \Big|_0^1 = \frac{k}{30}$$

36. $x = 2\cos 5\pi/6 = -\sqrt{3};\ \ y = 2\sin 5\pi/6 = 1;\ \ (-\sqrt{3}, 1, -3)$

39. With $x = 1$ and $y = -1$ we have $r^2 = 2$ and $\tan\theta = -1$. The point is $(\sqrt{2}, -\pi/4, -9)$.

42. With $x = 1$ and $y = 2$ we have $r^2 = 5$ and $\tan\theta = 2$. The point is $(\sqrt{5}, \tan^{-1} 2, 7)$.

45. $r^2 - z^2 = 1$ **48.** $z = 2y$

51. The equations are $r^2 = 4$, $r^2 + z^2 = 16$, and $z = 0$.

$$V = \int_0^{2\pi} \int_0^2 \int_0^{\sqrt{16-r^2}} r\,dz\,dr\,d\theta = \int_0^{2\pi} \int_0^2 r\sqrt{16-r^2}\,dr\,d\theta$$

$$= \int_0^{2\pi} -\frac{1}{3}(16-r^2)^{3/2}\Big|_0^2\,d\theta = \int_0^{2\pi} \frac{1}{3}(64-24\sqrt{3})\,d\theta = \frac{2\pi}{3}(64-24\sqrt{3})$$

54. Substituting the first equation into the second, we see that the surfaces intersect in the plane $y = 4$. Using polar coordinates in the xz-plane, the equations of the surfaces become $y = r^2$ and $y = \frac{1}{2}r^2 + 2$.

$$V = \int_0^{2\pi} \int_0^2 \int_{r^2}^{r^2/2+2} r\,dy\,dr\,d\theta = \int_0^{2\pi} \int_0^2 r\left(\frac{r^2}{2} + 2 - r^2\right)dr\,d\theta$$

$$= \int_0^{2\pi} \int_0^2 \left(2r - \frac{1}{2}r^3\right)dr\,d\theta = \int_0^{2\pi} \left(r^2 - \frac{1}{8}r^4\right)\Big|_0^2\,d\theta = \int_0^{2\pi} 2\,d\theta = 4\pi$$

57. The equation is $z = \sqrt{9 - r^2}$ and the density is $\rho = k/r^2$. When $z = 2$, $r = \sqrt{5}$.

$$I_z = \int_0^{2\pi} \int_0^{\sqrt{5}} \int_2^{\sqrt{9-r^2}} r^2(k/r^2)r\,dz\,dr\,d\theta = k\int_0^{2\pi} \int_0^{\sqrt{5}} rz\Big|_2^{\sqrt{9-r^2}}\,dr\,d\theta$$

$$= k\int_0^{2\pi} \int_0^{\sqrt{5}} (r\sqrt{9-r^2} - 2r)\,dr\,d\theta = k\int_0^{2\pi} \left[-\frac{1}{3}(9-r^2)^{3/2} - r^2\right]\Big|_0^{\sqrt{5}}\,d\theta$$

$$= k\int_0^{2\pi} \frac{4}{3}\,d\theta = \frac{8}{3}\pi k$$

60. (a) $x = 5\sin(5\pi/4)\cos(2\pi/3) = 5\sqrt{2}/4$; $y = 5\sin(5\pi/4)\sin(2\pi/3) = -5\sqrt{6}/4$;
$z = 5\cos(5\pi/4) = -5\sqrt{2}/2$; $(5\sqrt{2}/4, -5\sqrt{6}/4, -5\sqrt{2}/2)$

(b) With $x = 5\sqrt{2}/4$ and $y = -5\sqrt{6}/4$ we have $r^2 = 25/2$ and $\tan\theta = -\sqrt{3}$.
The point is $(5/\sqrt{2}, 2\pi/3, -5\sqrt{2}/2)$.

63. With $x = -5$, $y = -5$, and $z = 0$, we have $\rho^2 = 50$, $\tan\theta = 1$, and $\cos\phi = 0$. The point is $(5\sqrt{2}, \pi/2, 5\pi/4)$.

66. With $x = -\sqrt{3}/2$, $y = 0$, and $z = -1/2$, we have $\rho^2 = 1$, $\tan\theta = 0$, and $\cos\phi = -1/2$. The point is $(1, 2\pi/3, 0)$.

69. $4z^2 = 3x^2 + 3y^2 + 3z^2$; $4\rho^2\cos^2\phi = 3\rho^2$; $\cos\phi = \pm\sqrt{3}/2$; $\phi = \pi/6$ or equivalently, $\phi = 5\pi/6$

72. $\cos\phi = 1/2$; $\rho^2\cos^2\phi = \rho^2/4$; $4z^2 = x^2 + y^2 + z^2$; $x^2 + y^2 = 3z^2$

75. The equations are $\phi = \pi/4$ and $\rho = 3$.

$$V = \int_0^{2\pi} \int_0^{\pi/4} \int_0^3 \rho^2\sin\phi\,d\rho\,d\phi\,d\theta = \int_0^{2\pi} \int_0^{\pi/4} \frac{1}{3}\rho^3\sin\phi\Big|_0^3\,d\phi\,d\theta = \int_0^{2\pi} \int_0^{\pi/4} 9\sin\phi\,d\phi\,d\theta$$

$$= \int_0^{2\pi} -9\cos\phi\Big|_0^{\pi/4}\,d\theta = -9\int_0^{2\pi}\left(\frac{\sqrt{2}}{2} - 1\right)d\theta = 9\pi(2 - \sqrt{2})$$

78. The equations are $\rho = 1$ and $\phi = \pi/4$. We find the volume above the xy-plane and double.

$$V = 2\int_0^{2\pi} \int_{\pi/4}^{\pi/2} \int_0^1 \rho^2\sin\phi\,d\rho\,d\phi\,d\theta = 2\int_0^{2\pi} \int_{\pi/4}^{\pi/2} \frac{1}{3}\rho^3\sin\phi\Big|_0^1\,d\phi\,d\theta$$

$$= \frac{2}{3}\int_0^{2\pi} \int_{\pi/4}^{\pi/2} \sin\phi\,d\phi\,d\theta = \frac{2}{3}\int_0^{2\pi} -\cos\phi\Big|_{\pi/4}^{\pi/2}\,d\theta = \frac{2}{3}\int_0^{2\pi} \frac{\sqrt{2}}{2}\,d\theta = \frac{2\pi\sqrt{2}}{3}$$

81. We are given density $= k/\rho$.

$$m = \int_0^{2\pi} \int_0^{\cos^{-1} 4/5} \int_{4\sec\phi}^5 \frac{k}{\rho} \rho^2 \sin\phi \, d\rho \, d\phi \, d\theta = k \int_0^{2\pi} \int_0^{\cos^{-1} 4/5} \frac{1}{2} \rho^2 \sin\phi \Big|_{4\sec\phi}^5 d\phi \, d\theta$$

$$= \frac{1}{2} k \int_0^{2\pi} \int_0^{\cos^{-1} 4/5} (25\sin\phi - 16\tan\phi\sec\phi) \, d\phi \, d\theta$$

$$= \frac{1}{2} k \int_0^{2\pi} (-25\cos\phi - 16\sec\phi) \Big|_0^{\cos^{-1} 4/5} d\theta = \frac{1}{2} k \int_0^{2\pi} [-25(4/5) - 16(5/4) - (-25 - 16)] \, d\theta$$

$$= \frac{1}{2} k \int_0^{2\pi} d\theta = k\pi$$

EXERCISES 9.16

Divergence Theorem

3. div $\mathbf{F} = 3x^2 + 3y^2 + 3z^2$. Using spherical coordinates,

$$\iint_S \mathbf{F} \cdot \mathbf{n} \, dS = \iiint_D 3(x^2 + y^2 + z^2) \, dV = \int_0^{2\pi} \int_0^{\pi} \int_0^a 3\rho^2 \rho^2 \sin\phi \, d\rho \, d\phi \, d\theta$$

$$= \int_0^{2\pi} \int_0^{\pi} \frac{3}{5} \rho^5 \sin\phi \Big|_0^a d\phi \, d\theta = \frac{3a^5}{5} \int_0^{2\pi} \int_0^{\pi} \sin\phi \, d\phi \, d\theta$$

$$= \frac{3a^5}{5} \int_0^{2\pi} -\cos\phi \Big|_0^{\pi} d\theta = \frac{6a^5}{5} \int_0^{2\pi} d\theta = \frac{12\pi a^5}{5}.$$

6. div $\mathbf{F} = 2x + 2z + 12z^2$.

$$\iint_S \mathbf{F} \cdot \mathbf{n} \, dS = \iiint_D \text{div } \mathbf{F} \, dV = \int_0^3 \int_0^2 \int_0^1 (2x + 2z + 12z^2) \, dx \, dy \, dz$$

$$= \int_0^3 \int_0^2 (x^2 + 2xz + 12xz^2) \Big|_0^1 dy \, dz = \int_0^3 \int_0^2 (1 + 2z + 12z^2) \, dy \, dz$$

$$= \int_0^3 2(1 + 2z + 12z^2) \, dz = (2z + 2z^2 + 8z^3) \Big|_0^3 = 240$$

9. div $\mathbf{F} = \dfrac{1}{x^2 + y^2 + z^2}$. Using spherical coordinates,

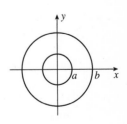

$$\iint_S \mathbf{F} \cdot \mathbf{n} \, dS = \iiint_D \text{div } \mathbf{F} \, dV = \int_0^{2\pi} \int_0^{\pi} \int_a^b \frac{1}{\rho^2} \rho^2 \sin\phi \, d\rho \, d\phi \, d\theta$$

$$= \int_0^{2\pi} \int_0^{\pi} (b - a)\sin\phi \, d\phi \, d\theta = (b - a) \int_0^{2\pi} -\cos\phi \Big|_0^{\pi} d\theta$$

$$= (b - a) \int_0^{2\pi} 2 \, d\theta = 4\pi(b - a).$$

12. div $\mathbf{F} = 30xy$.

$$\iint_S \mathbf{F} \cdot \mathbf{n}\, dS = \iiint_D 30xy\, dV = \int_0^2 \int_0^{2-x} \int_{x+y}^3 30xy\, dz\, dy\, dx$$

$$= \int_0^2 \int_0^{2-x} 30xyz\, \Big|_{x+y}^3 \, dy\, dx$$

$$= \int_0^2 \int_0^{2-x} (90xy - 30x^2y - 30xy^2)\, dy\, dx$$

$$= \int_0^2 (45xy^2 - 15x^2y^2 - 10xy^3)\, \Big|_0^{2-x}\, dx$$

$$= \int_0^2 (-5x^4 + 45x^3 - 120x^2 + 100x)\, dx = \left(-x^5 + \frac{45}{4}x^4 - 40x^3 + 50x^2\right)\Big|_0^2 = 28$$

15. (a) div $\mathbf{E} = q\left[\dfrac{-2x^2 + y^2 + z^2}{(x^2 + y^2 + z^2)^{5/2}} + \dfrac{x^2 - 2y^2 + z^2}{(x^2 + y^2 + z^2)^{5/2}} + \dfrac{x^2 + y^2 - 2z^2}{(x^2 + y^2 + z^2)^{5/2}}\right] = 0$

$$\iint_{S \cup S_a} (\mathbf{E} \cdot \mathbf{n})\, dS = \iiint_D \text{div } \mathbf{E}\, dV = \iiint_D 0\, dV = 0$$

(b) From (a), $\displaystyle\iint_S (\mathbf{E} \cdot \mathbf{n})\, dS + \iint_{S_a} (\mathbf{E} \cdot \mathbf{n})\, dS = 0$ and $\displaystyle\iint_S (\mathbf{E} \cdot \mathbf{n})\, dS = -\iint_{S_a} (\mathbf{E} \cdot \mathbf{n})\, dS$. On S_a,

$|\mathbf{r}| = a$, $\mathbf{n} = -(x\mathbf{i} + y\mathbf{j} + z\mathbf{k})/a = -\mathbf{r}/a$ and $\mathbf{E} \cdot \mathbf{n} = (q\mathbf{r}/a^3) \cdot (-\mathbf{r}/a) = -qa^2/a^4 = -q/a^2$. Thus

$$\iint_S (\mathbf{E} \cdot \mathbf{n})\, dS = -\iint_{S_a}\left(-\frac{q}{a^2}\right) dS = \frac{q}{a^2}\iint_{S_a} dS = \frac{q}{a^2} \times (\text{area of } S_a) = \frac{q}{a^2}(4\pi a^2) = 4\pi q.$$

18. By the Divergence Theorem and Problem 30 in Section 9.7,

$$\iint_S (\text{curl } \mathbf{F} \cdot \mathbf{n})\, dS = \iiint_D \text{div (curl } \mathbf{F})\, dV = \iiint_D 0\, dV = 0.$$

21. If $G(x, y, z)$ is a vector valued function then we define surface integrals and triple integrals of $\mathbf{G}$ component-wise. In this case, if $\mathbf{a}$ is a constant vector it is easily shown that

$$\iint_S \mathbf{a} \cdot \mathbf{G}\, dS = \mathbf{a} \cdot \iint_S \mathbf{G}\, dS \quad \text{and} \quad \iiint_D \mathbf{a} \cdot \mathbf{G}\, dV = \mathbf{a} \cdot \iiint_D \mathbf{G}\, dV.$$

Now let $\mathbf{F} = f\mathbf{a}$. Then

$$\iint_S \mathbf{F} \cdot \mathbf{n}\, dS = \iint_S (f\mathbf{a}) \cdot \mathbf{n}\, dS = \iint_S \mathbf{a} \cdot (f\mathbf{n})\, dS$$

and, using Problem 27 in Section 9.7 and the fact that $\nabla \cdot \mathbf{a} = 0$, we have

$$\iiint_D \text{div } \mathbf{F}\, dV = \iiint_D \nabla \cdot (f\mathbf{a})\, dV = \iiint_D [f(\nabla \cdot \mathbf{a}) + \mathbf{a} \cdot \nabla f]\, dV = \iiint_D \mathbf{a} \cdot \nabla f\, dV.$$

By the Divergence Theorem,

$$\iint_S \mathbf{a} \cdot (f\mathbf{n})\, dS = \iint_S \mathbf{F} \cdot \mathbf{n}\, dS = \iiint_D \text{div } \mathbf{F}\, dV = \iiint_D \mathbf{a} \cdot \nabla f\, dV$$

and

$$\mathbf{a} \cdot \left(\iint_S f\mathbf{n}\, dS\right) = \mathbf{a} \cdot \left(\iiint_D \nabla f\, dV\right) \quad \text{or} \quad \mathbf{a} \cdot \left(\iint_S f\mathbf{n}\, dS - \iiint_D \nabla f\, dV\right) = 0.$$

Since $\mathbf{a}$ is arbitrary,

$$\iint_S f\mathbf{n}\, dS - \iiint_D \nabla f\, dV = 0 \quad \text{and} \quad \iint_S f\mathbf{n}\, dS = \iiint_D \nabla f\, dV.$$

EXERCISES 9.17

Change of Variables in Multiple Integrals

3. The uv-corner points $(0,0)$, $(2,0)$, $(2,2)$ correspond to xy-points $(0,0)$, $(4,2)$, $(6,-4)$.

$v = 0$: $x = 2u$, $y = u \implies y = x/2$

$u = 2$: $x = 4 + v$, $y = 2 - 3v \implies$
$$y = 2 - 3(x-4) = -3x + 14$$

$v = u$: $x = 3u$, $y = -2u \implies y = -2x/3$

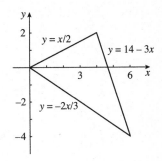

6. The uv-corner points $(1,1)$, $(2,1)$, $(2,2)$, $(1,2)$ correspond to the xy-points $(1,1)$, $(2,1)$, $(4,4)$, $(2,4)$.

$v = 1$: $x = u$, $y = 1 \implies y = 1$, $1 \le x \le 2$

$u = 2$: $x = 2v$, $y = v^2 \implies y = x^2/4$

$v = 2$: $x = 2u$, $y = 4 \implies y = 4$, $2 \le x \le 4$

$u = 1$: $x = v$, $y = v^2 \implies y = x^2$

9. $\dfrac{\partial(u,v)}{\partial(x,y)} = \begin{vmatrix} -2y/x^3 & 1/x^2 \\ -y^2/x^2 & 2y/x \end{vmatrix} = -\dfrac{3y^2}{x^4} = -3\left(\dfrac{y}{x^2}\right)^2 = -3u^2;$ $\quad \dfrac{\partial(x,y)}{\partial(u,v)} = \dfrac{1}{-3u^2} = -\dfrac{1}{3u^2}$

12. $\dfrac{\partial(x,y)}{\partial(u,v)} = \begin{vmatrix} 1-v & v \\ -u & u \end{vmatrix} = u.$ The transformation is 0 when u is 0, for $0 \le v \le 1$.

15. $R1$: $y = x^2 \implies u = 1$
$R2$: $x = y^2 \implies v = 1$
$R3$: $y = \frac{1}{2}x^2 \implies u = 2$
$R4$: $x = \frac{1}{2}y^2 \implies v = 2$

$$\dfrac{\partial(u,v)}{\partial(x,y)} = \begin{vmatrix} 2x/y & -x^2/y^2 \\ -y^2/x^2 & 2y/x \end{vmatrix} = 3 \implies \dfrac{\partial(x,y)}{\partial(u,v)} = \dfrac{1}{3}$$

$$\iint_R \frac{y^2}{x}\, dA = \iint_S v\left(\frac{1}{3}\right) dA' = \frac{1}{3}\int_1^2 \int_1^2 v\, du\, dv = \frac{1}{3}\int_1^2 v\, dv = \frac{1}{6}v^2 \Big|_1^2 = \frac{1}{2}$$

18. $R1$: $xy = -2 \implies v = -2$

$R2$: $x^2 - y^2 = 9 \implies u = 9$

$R3$: $xy = 2 \implies v = 2$

$R4$: $x^2 - y^2 = 1 \implies u = 1$

$$\frac{\partial(u,v)}{\partial(x,y)} = \begin{vmatrix} 2x & -2y \\ y & x \end{vmatrix} = 2(x^2 + y^2)$$

$$\implies \frac{\partial(x,y)}{\partial(u,v)} = \frac{1}{2(x^2 + y^2)}$$

$$\iint_R (x^2 + y^2) \sin xy \, dA = \iint_S (x^2 + y^2) \sin v \left(\frac{1}{2(x^2 + y^2)}\right) dA' = \frac{1}{2} \int_{-2}^{2} \int_{1}^{9} \sin v \, du \, dv = \frac{1}{2} \int_{-2}^{2} 8 \sin v \, dv = 0$$

21. $R1$: $y = 1/x \implies u = 1$

$R2$: $y = x \implies v = 1$

$R3$: $y = 4/x \implies u = 4$

$R4$: $y = 4x \implies v = 4$

$$\frac{\partial(u,v)}{\partial(x,y)} = \begin{vmatrix} y & x \\ -y/x^2 & 1/x \end{vmatrix} = \frac{2y}{x} \implies \frac{\partial(x,y)}{\partial(u,v)} = \frac{x}{2y}$$

$$8 \iint_R y^4 \, dA = \iint_S u^2 v^2 \left(\frac{1}{2v}\right) du \, dv = \frac{1}{2} \int_1^4 u^2 v \, du \, dv = \frac{1}{2} \int_1^4 \frac{1}{3} u^3 v \Big|_1^4 dv = \frac{1}{6} \int_1^4 63 v \, dv = \frac{21}{4} v^2 \Big|_1^4 = \frac{315}{4}$$

24. We let $u = y - x$ and $v = y$.

$R1$: $y = 0 \implies v = 0$, $u = -x \implies v = 0, 0 \le u \le 2$

$R2$: $x = 0 \implies v = u$

$R3$: $y = x + 2 \implies u = 2$

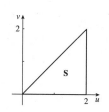

$$\frac{\partial(u,v)}{\partial(x,y)} = \begin{vmatrix} -1 & 1 \\ 0 & 1 \end{vmatrix} = -1 \implies \frac{\partial(x,y)}{\partial(u,v)} = -1$$

$$\iint_R e^{y^2 - 2xy + x^2} \, dA = \iint_S e^{u^2} |-1| \, dA' = \int_0^2 \int_0^u e^{u^2} \, dv \, du = \int_0^2 u e^{u^2} \, du = \frac{1}{2} e^{u^2} \Big|_0^2 = \frac{1}{2}(e^4 - 1)$$

27. Let $u = xy$ and $v = xy^{1.4}$. Then $xy^{1.4} = c \implies v = c$; $xy = b \implies u = b$; $xy^{1.4} = d \implies v = d$; $xy = a \implies u = a$.

$$\frac{\partial(u,v)}{\partial(x,y)} = \begin{vmatrix} y & x \\ y^{1.4} & 1.4xy^{0.4} \end{vmatrix} = 0.4xy^{1.4} = 0.4v \implies \frac{\partial(x,y)}{\partial(u,v)} = \frac{5}{2v}$$

$$\iint_R dA = \iint_S \frac{5}{2v} \, dA' = \int_c^d \int_a^b \frac{5}{2v} \, du \, dv = \frac{5}{2}(b - a) \int_c^d \frac{dv}{v} = \frac{5}{2}(b - a)(\ln d - \ln c)$$

30. $\dfrac{\partial(x,y,z)}{\partial(\rho,\phi,\theta)} = \begin{vmatrix} \sin\phi\cos\theta & \rho\cos\phi\cos\theta & -\rho\sin\phi\sin\theta \\ \sin\phi\sin\theta & \rho\cos\phi\sin\theta & \rho\sin\phi\cos\theta \\ \cos\phi & -\rho\sin\phi & 0 \end{vmatrix}$

$= \cos\phi(\rho^2\sin\phi\cos\phi\cos^2\theta + \rho^2\sin\phi\cos\phi\sin^2\theta) + \rho\sin\phi(\rho\sin^2\phi\cos^2\theta + \rho\sin^2\phi\sin^2\theta)$

$= \rho^2\sin\phi\cos^2\phi(\cos^2\theta + \sin^2\theta) + \rho^2\sin^3\phi(\cos^2\theta + \sin^2\theta) = \rho^2\sin\phi(\cos^2\phi + \sin^2\phi) = \rho^2\sin\phi$

CHAPTER 9 REVIEW EXERCISES

3. True

6. False; consider $f(x,y) = xy$ at $(0,0)$.

9. False; $\displaystyle\int_C x\,dx + x^2\,dy = 0$ from $(-1,0)$ to $(1,0)$ along the x-axis and along the semicircle $y = \sqrt{1-x^2}$, but since $x\,dx + x^2\,dy$ is not exact, the integral is not independent of path.

12. True

15. True   **18.** True

21. $\mathbf{v}(t) = 6\mathbf{i} + \mathbf{j} + 2t\mathbf{k}$; $\mathbf{a}(t) = 2\mathbf{k}$. To find when the particle passes through the plane, we solve $-6t + t + t^2 = -4$ or $t^2 - 5t + 4 = 0$. This gives $t = 1$ and $t = 4$. $\mathbf{v}(1) = 6\mathbf{i} + \mathbf{j} + 2\mathbf{k}$, $\mathbf{a}(1) = 2\mathbf{k}$; $\mathbf{v}(4) = 6\mathbf{i} + \mathbf{j} + 8\mathbf{k}$, $\mathbf{a}(4) = 2\mathbf{k}$

24. $\mathbf{v}(t) = t\mathbf{i} + t^2\mathbf{j} - t\mathbf{k}$; $|\mathbf{v}| = t\sqrt{t^2+2}$, $t > 0$; $\mathbf{a}(t) = \mathbf{i} + 2t\mathbf{j} - \mathbf{k}$; $\mathbf{v}\cdot\mathbf{a} = t + 2t^3 + t = 2t + 2t^3$;

$\mathbf{v}\times\mathbf{a} = t^2\mathbf{i} + t^2\mathbf{k}$, $|\mathbf{v}\times\mathbf{a}| = t^2\sqrt{2}$; $a_T = \dfrac{2t+2t^3}{t\sqrt{t^2+2}} = \dfrac{2+2t^2}{\sqrt{t^2+2}}$, $a_N = \dfrac{t^2\sqrt{2}}{t\sqrt{t^2+2}} = \dfrac{\sqrt{2}\,t}{\sqrt{t^2+2}}$;

$\kappa = \dfrac{t^2\sqrt{2}}{t^3(t^2+2)^{3/2}} = \dfrac{\sqrt{2}}{t(t^2+2)^{3/2}}$

27. $\nabla f = (2xy - y^2)\mathbf{i} + (x^2 - 2xy)\mathbf{j}$; $\mathbf{u} = \dfrac{2}{\sqrt{40}}\mathbf{i} + \dfrac{6}{\sqrt{40}}\mathbf{j} = \dfrac{1}{\sqrt{10}}(\mathbf{i} + 3\mathbf{j})$;

$D_{\mathbf{u}}f = \dfrac{1}{\sqrt{10}}(2xy - y^2 + 3x^2 - 6xy) = \dfrac{1}{\sqrt{10}}(3x^2 - 4xy - y^2)$

30. (a) $\dfrac{dw}{dt} = \dfrac{\partial w}{\partial x}\dfrac{dx}{dt} + \dfrac{\partial w}{\partial y}\dfrac{dy}{dt} + \dfrac{\partial w}{\partial z}\dfrac{dz}{dt}$

$= \dfrac{x}{\sqrt{x^2+y^2+z^2}}6\cos 2t + \dfrac{y}{\sqrt{x^2+y^2+z^2}}(-8\sin 2t) + \dfrac{z}{\sqrt{x^2+y^2+z^2}}15t^2$

$= \dfrac{(6x\cos 2t - 8y\sin 2t + 15zt^2)}{\sqrt{x^2+y^2+z^2}}$

(b) $\dfrac{\partial w}{\partial t} = \dfrac{\partial w}{\partial x}\dfrac{\partial x}{\partial t} + \dfrac{\partial w}{\partial y}\dfrac{\partial y}{\partial t} + \dfrac{\partial w}{\partial z}\dfrac{\partial z}{\partial t}$

$$= \frac{x}{\sqrt{x^2+y^2+z^2}}\frac{6}{r}\cos\frac{2t}{r} + \frac{y}{\sqrt{x^2+y^2+z^2}}\left(\frac{8r}{t^2}\sin\frac{2r}{t}\right) + \frac{z}{\sqrt{x^2+y^2+z^2}}15t^2r^3$$

$$= \frac{\left(\dfrac{6x}{r}\cos\dfrac{2t}{r} + \dfrac{8yr}{t^2}\sin\dfrac{2r}{t} + 15zt^2r^3\right)}{\sqrt{x^2+y^2+z^2}}$$

33. (a) $V = \displaystyle\int_0^1\int_x^{2x}\sqrt{1-x^2}\,dy\,dx = \int_0^1 y\sqrt{1-x^2}\,\Big|_x^{2x}\,dx = \int_0^1 x\sqrt{1-x^2}\,dx = -\frac{1}{3}(1-x^2)^{3/2}\,\Big|_0^1 = \frac{1}{3}$

(b) $V = \displaystyle\int_0^1\int_{y/2}^y\sqrt{1-x^2}\,dx\,dy + \int_1^2\int_{y/2}^1\sqrt{1-x^2}\,dx\,dy$

36. (a) Using symmetry,

$$V = 8\int_0^a\int_0^{\sqrt{a^2-x^2}}\int_0^{\sqrt{a^2-x^2-y^2}} dz\,dy\,dx = 8\int_0^a\int_0^{\sqrt{a^2-x^2}}\sqrt{a^2-x^2-y^2}\,dy\,dx$$

$\boxed{\text{Trig substitution}}$

$$= 8\int_0^a\left(\frac{y}{2}\sqrt{a^2-x^2-y^2} + \frac{a^2-x^2}{2}\sin^{-1}\frac{y}{\sqrt{a^2-x^2}}\right]\,\Big|_0^{\sqrt{a^2-x^2}}\,dx = 8\int_0^a\frac{\pi}{2}\frac{a^2-x^2}{2}\,dx$$

$$= 2\pi\left(a^2x - \frac{1}{3}x^3\right)\,\Big|_0^a = \frac{4}{3}\pi a^3$$

(b) Using symmetry,

$$V = 2\int_0^{2\pi}\int_0^a\int_0^{\sqrt{a^2-r^2}} r\,dz\,dr\,d\theta = 2\int_0^{2\pi}\int_0^a r\sqrt{a^2-r^2}\,dr\,d\theta$$

$$= 2\int_0^{2\pi}-\frac{1}{3}(a^2-r^2)^{3/2}\,\Big|_0^a\,d\theta = \frac{2}{3}\int_0^{2\pi}a^3\,d\theta = \frac{4}{3}\pi a^3$$

(c) $V = \displaystyle\int_0^{2\pi}\int_0^\pi\int_0^a \rho^2\sin\phi\,d\rho\,d\phi\,d\theta = \int_0^{2\pi}\int_0^\pi\frac{1}{3}\rho^3\sin\phi\,\Big|_0^a\,d\phi\,d\theta$

$$= \frac{1}{3}\int_0^{2\pi}\int_0^\pi a^3\sin\phi\,d\phi\,d\theta = \frac{1}{3}\int_0^{2\pi}-a^3\cos\phi\,\Big|_0^\pi\,d\theta = \frac{1}{3}\int_0^{2\pi}2a^3\,d\theta = \frac{4}{3}\pi a^3$$

39. $2xy + 2xy + 2xy = 6xy$

42. $\nabla(6xy) = 6y\mathbf{i} + 6x\mathbf{j}$

45. Since $P_y = 6x^2y = Q_x$, the integral is independent of path.

$\phi_x = 3x^2y^2$, $\phi = x^3y^2 + g(y)$, $\phi_y = 2x^3y + g'(y) = 2x^3y - 3y^2$; $g(y) = -y^3$; $\phi = x^3y^2 - y^3$;

$$\int_{(0,0)}^{(1,-2)} 3x^2y^2\,dx + (2x^3y - 3y^2)\,dy = (x^3y^2 - y^3)\,\Big|_{(0,0)}^{(1,-2)} = 12$$

48. Parameterize C by $x = \cos t$, $y = \sin t$; $0 \le t \le 2\pi$. Then

$$\oint_C \mathbf{F}\cdot d\mathbf{r} = \int_0^{2\pi}[4\sin t(-\sin t\,dt) + 6\cos t(\cos t)\,dt] = \int_0^{2\pi}(6\cos^2 t - 4\sin^2 t)\,dt$$

$$= \int_0^{2\pi}(10\cos^2 t - 4)\,dt = \left(5t + \frac{5}{2}\sin 2t - 4t\right)\,\Big|_0^{2\pi} = 2\pi.$$

Using Green's Theorem, $Q_x - P_y = 6 - 4 = 2$ and $\displaystyle\oint_C \mathbf{F}\cdot d\mathbf{r} = \iint_R 2\,dA = 2(\pi\cdot 1^2) = 2\pi.$

51. $z_x = 2x$, $z_y = 0$; $dS = \sqrt{1 + 4x^2}\, dA$

$$\iint_S \frac{z}{xy}\, dS = \int_1^3 \int_1^2 \frac{x^2}{xy} \sqrt{1 + 4x^2}\, dx\, dy = \int_1^3 \frac{1}{y} \left[\frac{1}{12}(1 + 4x^2)^{3/2} \right] \Big|_1^2 \, dy$$

$$= \frac{1}{12} \int_1^3 \frac{17^{3/2} - 5^{3/2}}{y}\, dy = \frac{17\sqrt{17} - 5\sqrt{5}}{12} \ln y \,\Big|_1^3$$

$$= \frac{17\sqrt{17} - 5\sqrt{5}}{12} \ln 3$$

54. In Problem 53, $\mathbf{F}$ is not continuous at $(0,0,0)$ which is in any acceptable region containing the sphere.

57. Identify $\mathbf{F} = -2y\mathbf{i} + 3x\mathbf{j} + 10z\mathbf{k}$. Then curl $\mathbf{F} = 5\mathbf{k}$. The curve C lies in the plane $z = 3$, so $\mathbf{n} = \mathbf{k}$ and $dS = dA$. Thus,

$$\oint_C \mathbf{F} \cdot d\mathbf{r} = \iint_S (\text{curl } \mathbf{F}) \cdot \mathbf{n}\, dS = \iint_R 5\, dA = 5 \times (\text{area of } R) = 5(25\pi) = 125\pi.$$

60. div $\mathbf{F} = x^2 + y^2 + z^2$. Using cylindrical coordinates,

$$\iint_S \mathbf{F} \cdot \mathbf{n}\, dS = \iiint_D \text{div } \mathbf{F}\, dV = \iiint_D (x^2 + y^2 + z^2)\, dV = \int_0^{2\pi} \int_0^1 \int_0^1 (r^2 + z^2) r\, dz\, dr\, d\theta$$

$$= \int_0^{2\pi} \int_0^1 \left(r^3 z + \frac{1}{3} r z^3 \right) \Big|_0^1 \, dr\, d\theta = \int_0^{2\pi} \int_0^1 \left(r^3 + \frac{1}{3} r \right) dr\, d\theta$$

$$= \int_0^{2\pi} \left(\frac{1}{4} r^4 + \frac{1}{6} r^2 \right) \Big|_0^1 \, d\theta = \int_0^{2\pi} \frac{5}{12}\, d\theta = \frac{5\pi}{6}.$$

63. $x = 0 \implies u = 0,\ v = -y^2 \implies u = 0,\ -1 \le v \le 0$

$x = 1 \implies u = 2y,\ v = 1 - y^2 = 1 - u^2/4$

$y = 0 \implies u = 0,\ v = x^2 \implies u = 0,\ 0 \le v \le 1$

$y = 1 \implies u = 2x,\ v = x^2 - 1 = u^2/4 - 1$

$$\frac{\partial(u,v)}{\partial(x,y)} = \begin{vmatrix} 2y & 2x \\ 2x & -2y \end{vmatrix} = -4(x^2 + y^2) \implies \frac{\partial(x,y)}{\partial(u,v)} = -\frac{1}{4(x^2 + y^2)}$$

$$\iint_R (x^2 + y^2) \sqrt[3]{x^2 - y^2}\, dA = \iint_S (x^2 + y^2) \sqrt[3]{v} \left| -\frac{1}{4(x^2 + y^2)} \right| dA' = \frac{1}{4} \int_0^2 \int_{u^2/4-1}^{1-u^2/4} v^{1/3}\, dv\, du$$

$$= \frac{1}{4} \int_0^2 \frac{3}{4} v^{4/3} \,\Big|_{u^2/4-1}^{1-u^2/4} \, du = \frac{3}{16} \int_0^2 \left[(1 - u^2/4)^{4/3} - (u^2/4 - 1)^{4/3} \right] du$$

$$= \frac{3}{16} \int_0^2 \left[(1 - u^2/4)^{4/3} - (1 - u^2/4)^{4/3} \right] du = 0$$

66. (a) Both states span 7 degrees of longitude and 4 degrees of latitude, but Colorado is larger because it lies to the south of Wyoming. Lines of longitude converge as they go north, so the east-west dimensions of Wyoming are shorter than those of Colorado.

(b) We use the function $f(x,y) = \sqrt{R^2 - x^2 - y^2}$ to describe the northern hemisphere, where $R \approx 3960$ miles is the radius of the Earth. We need to compute the surface area over a polar rectangle P of the form $\theta_1 \leq \theta \leq \theta_2$, $R\cos\phi_2 \leq r \leq R\cos\phi_1$. We have

$$f_x = \frac{-x}{\sqrt{R^2 - x^2 - y^2}} \qquad \text{and} \qquad f_y = \frac{-y}{\sqrt{R^2 - x^2 - y^2}}$$

so that

$$\sqrt{1 + f_x^2 + f_y^2} = \sqrt{1 + \frac{x^2 + y^2}{R^2 - x^2 - y^2}} = \frac{R}{\sqrt{R^2 - r^2}}.$$

Thus

$$A = \iint_P \sqrt{1 + f_x^2 + f_y^2}\, dA = \int_{\theta_1}^{\theta_2} \int_{R\cos\phi_2}^{R\cos\phi_1} \frac{R}{\sqrt{R^2 - r^2}}\, r\, dr\, d\theta$$

$$= (\theta_2 - \theta_1) R\sqrt{R^2 - r^2}\,\Big|_{R\cos\phi_1}^{R\cos\phi_2} = (\theta_2 - \theta_1) R^2(\sin\phi_2 - \sin\phi_1).$$

The ratio of Wyoming to Colorado is then $\dfrac{\sin 45° - \sin 41°}{\sin 41° - \sin 37°} \approx 0.941$. Thus Wyoming is about 6% smaller than Colorado.

(c) $97{,}914/104{,}247 \approx 0.939$, which is close to the theoretical value of 0.941. (Our formula for the area says that the area of Colorado is approximately 103,924 square miles, while the area of Wyoming is approximately 97,801 square miles.)

10 Systems of Linear Differential Equations

EXERCISES 10.1

Preliminary Theory

3. Let $\mathbf{X} = \begin{pmatrix} x \\ y \\ z \end{pmatrix}$. Then $\mathbf{X}' = \begin{pmatrix} -3 & 4 & -9 \\ 6 & -1 & 0 \\ 10 & 4 & 3 \end{pmatrix} \mathbf{X}$.

6. Let $\mathbf{X} = \begin{pmatrix} x \\ y \\ z \end{pmatrix}$. Then $\mathbf{X}' = \begin{pmatrix} -3 & 4 & 0 \\ 5 & 9 & 0 \\ 0 & 1 & 6 \end{pmatrix} \mathbf{X} + \begin{pmatrix} e^{-t} \sin 2t \\ 4e^{-t} \cos 2t \\ -e^{-t} \end{pmatrix}$.

9. $\dfrac{dx}{dt} = x - y + 2z + e^{-t} - 3t; \quad \dfrac{dy}{dt} = 3x - 4y + z + 2e^{-t} + t; \quad \dfrac{dz}{dt} = -2x + 5y + 6z + 2e^{-t} - t$

12. Since

$$\mathbf{X}' = \begin{pmatrix} 5\cos t - 5\sin t \\ 2\cos t - 4\sin t \end{pmatrix} e^t \quad \text{and} \quad \begin{pmatrix} -2 & 5 \\ -2 & 4 \end{pmatrix} \mathbf{X} = \begin{pmatrix} 5\cos t - 5\sin t \\ 2\cos t - 4\sin t \end{pmatrix} e^t$$

we see that

$$\mathbf{X}' = \begin{pmatrix} -2 & 5 \\ -2 & 4 \end{pmatrix} \mathbf{X}.$$

15. Since

$$\mathbf{X}' = \begin{pmatrix} 0 \\ 0 \\ 0 \end{pmatrix} \quad \text{and} \quad \begin{pmatrix} 1 & 2 & 1 \\ 6 & -1 & 0 \\ -1 & -2 & -1 \end{pmatrix} \mathbf{X} = \begin{pmatrix} 0 \\ 0 \\ 0 \end{pmatrix}$$

we see that

$$\mathbf{X}' = \begin{pmatrix} 1 & 2 & 1 \\ 6 & -1 & 0 \\ -1 & -2 & -1 \end{pmatrix} \mathbf{X}.$$

18. Yes, since $W(\mathbf{X}_1, \mathbf{X}_2) = 8e^{2t} \neq 0$ the set $\mathbf{X}_1$, $\mathbf{X}_2$ is linearly independent on $-\infty < t < \infty$.

21. Since

$$\mathbf{X}'_p = \begin{pmatrix} 2 \\ -1 \end{pmatrix} \quad \text{and} \quad \begin{pmatrix} 1 & 4 \\ 3 & 2 \end{pmatrix} \mathbf{X}_p + \begin{pmatrix} 2 \\ -4 \end{pmatrix} t + \begin{pmatrix} -7 \\ -18 \end{pmatrix} = \begin{pmatrix} 2 \\ -1 \end{pmatrix}$$

we see that

$$\mathbf{X}'_p = \begin{pmatrix} 1 & 4 \\ 3 & 2 \end{pmatrix} \mathbf{X}_p + \begin{pmatrix} 2 \\ -4 \end{pmatrix} t + \begin{pmatrix} -7 \\ -18 \end{pmatrix}.$$

24. Since

$$\mathbf{X}'_p = \begin{pmatrix} 3\cos 3t \\ 0 \\ -3\sin 3t \end{pmatrix} \quad \text{and} \quad \begin{pmatrix} 1 & 2 & 3 \\ -4 & 2 & 0 \\ -6 & 1 & 0 \end{pmatrix} \mathbf{X}_p + \begin{pmatrix} -1 \\ 4 \\ 3 \end{pmatrix} \sin 3t = \begin{pmatrix} 3\cos 3t \\ 0 \\ -3\sin 3t \end{pmatrix}$$

we see that

$$\mathbf{X}'_p = \begin{pmatrix} 1 & 2 & 3 \\ -4 & 2 & 0 \\ -6 & 1 & 0 \end{pmatrix} \mathbf{X}_p + \begin{pmatrix} -1 \\ 4 \\ 3 \end{pmatrix} \sin 3t.$$

EXERCISES 10.2

Homogeneous Linear Systems

3. The system is

$$\mathbf{X}' = \begin{pmatrix} -4 & 2 \\ -\frac{5}{2} & 2 \end{pmatrix} \mathbf{X}$$

and $\det(\mathbf{A} - \lambda \mathbf{I}) = (\lambda - 1)(\lambda + 3) = 0$. For $\lambda_1 = 1$ we obtain

$$\begin{pmatrix} -5 & 2 & | & 0 \\ -\frac{5}{2} & 1 & | & 0 \end{pmatrix} \Longrightarrow \begin{pmatrix} -5 & 2 & | & 0 \\ 0 & 0 & | & 0 \end{pmatrix} \quad \text{so that} \quad \mathbf{K}_1 = \begin{pmatrix} 2 \\ 5 \end{pmatrix}.$$

For $\lambda_2 = -3$ we obtain

$$\begin{pmatrix} -1 & 2 & | & 0 \\ -\frac{5}{2} & 5 & | & 0 \end{pmatrix} \Longrightarrow \begin{pmatrix} -1 & 2 & | & 0 \\ 0 & 0 & | & 0 \end{pmatrix} \quad \text{so that} \quad \mathbf{K}_2 = \begin{pmatrix} 2 \\ 1 \end{pmatrix}.$$

Then

$$\mathbf{X} = c_1 \begin{pmatrix} 2 \\ 5 \end{pmatrix} e^t + c_2 \begin{pmatrix} 2 \\ 1 \end{pmatrix} e^{-3t}.$$

6. The system is

$$\mathbf{X}' = \begin{pmatrix} -6 & 2 \\ -3 & 1 \end{pmatrix} \mathbf{X}$$

and $\det(\mathbf{A} - \lambda \mathbf{I}) = \lambda(\lambda + 5) = 0$. For $\lambda_1 = 0$ we obtain

$$\begin{pmatrix} -6 & 2 & | & 0 \\ -3 & 1 & | & 0 \end{pmatrix} \Longrightarrow \begin{pmatrix} 1 & -\frac{1}{3} & | & 0 \\ 0 & 0 & | & 0 \end{pmatrix} \quad \text{so that} \quad \mathbf{K}_1 = \begin{pmatrix} 1 \\ 3 \end{pmatrix}.$$

For $\lambda_2 = -5$ we obtain

$$\begin{pmatrix} -1 & 2 & | & 0 \\ -3 & 6 & | & 0 \end{pmatrix} \Longrightarrow \begin{pmatrix} 1 & -2 & | & 0 \\ 0 & 0 & | & 0 \end{pmatrix} \quad \text{so that} \quad \mathbf{K}_2 = \begin{pmatrix} 2 \\ 1 \end{pmatrix}.$$

Then

$$\mathbf{X} = c_1 \begin{pmatrix} 1 \\ 3 \end{pmatrix} + c_2 \begin{pmatrix} 2 \\ 1 \end{pmatrix} e^{-5t}.$$

9. We have $\det(\mathbf{A} - \lambda\mathbf{I}) = -(\lambda + 1)(\lambda - 3)(\lambda + 2) = 0$. For $\lambda_1 = -1$, $\lambda_2 = 3$, and $\lambda_3 = -2$ we obtain

$$\mathbf{K}_1 = \begin{pmatrix} -1 \\ 0 \\ 1 \end{pmatrix}, \quad \mathbf{K}_2 = \begin{pmatrix} 1 \\ 4 \\ 3 \end{pmatrix}, \quad \text{and} \quad \mathbf{K}_3 = \begin{pmatrix} 1 \\ -1 \\ 3 \end{pmatrix},$$

so that

$$\mathbf{X} = c_1 \begin{pmatrix} -1 \\ 0 \\ 1 \end{pmatrix} e^{-t} + c_2 \begin{pmatrix} 1 \\ 4 \\ 3 \end{pmatrix} e^{3t} + c_3 \begin{pmatrix} 1 \\ -1 \\ 3 \end{pmatrix} e^{-2t}.$$

12. We have $\det(\mathbf{A} - \lambda\mathbf{I}) = (\lambda - 3)(\lambda + 5)(6 - \lambda) = 0$. For $\lambda_1 = 3$, $\lambda_2 = -5$, and $\lambda_3 = 6$ we obtain

$$\mathbf{K}_1 = \begin{pmatrix} 1 \\ 1 \\ 0 \end{pmatrix}, \quad \mathbf{K}_2 = \begin{pmatrix} 1 \\ -1 \\ 0 \end{pmatrix}, \quad \text{and} \quad \mathbf{K}_3 = \begin{pmatrix} 2 \\ -2 \\ 11 \end{pmatrix},$$

so that

$$\mathbf{X} = c_1 \begin{pmatrix} 1 \\ 1 \\ 0 \end{pmatrix} e^{3t} + c_2 \begin{pmatrix} 1 \\ -1 \\ 0 \end{pmatrix} e^{-5t} + c_3 \begin{pmatrix} 2 \\ -2 \\ 11 \end{pmatrix} e^{6t}.$$

15. $\mathbf{X} = c_1 \begin{pmatrix} 0.382175 \\ 0.851161 \\ 0.359815 \end{pmatrix} e^{8.58979t} + c_2 \begin{pmatrix} 0.405188 \\ -0.676043 \\ 0.615458 \end{pmatrix} e^{2.25684t} + c_3 \begin{pmatrix} -0.923562 \\ -0.132174 \\ 0.35995 \end{pmatrix} e^{-0.0466321t}$

18. In Problem 2, letting $c_1 = 1$ and $c_2 = 0$ we get $x = -2e^t$, $y = e^t$. Eliminating the parameter we find $y = -\frac{1}{2}x$, $x < 0$. When $c_1 = -1$ and $c_2 = 0$ we find $y = -\frac{1}{2}x$, $x > 0$. Letting $c_1 = 0$ and $c_2 = 1$ we get $x = e^{4t}$, $y = e^{4t}$. Eliminating the parameter we find $y = x$, $x > 0$. When $c_1 = 0$ and $c_2 = -1$ we find $y = x$, $x < 0$.

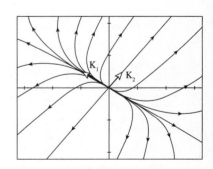

In Problem 4, letting $c_1 = 1$ and $c_2 = 0$ we get $x = -2e^{-7t/2}$, $y = e^{-7t/2}$. Eliminating the parameter we find $y = -\frac{1}{2}x$, $x < 0$. When $c_1 = -1$ and $c_2 = 0$ we find $y = -\frac{1}{2}x$, $x > 0$. Letting $c_1 = 0$ and $c_2 = 1$ we get $x = 4e^{-t}$, $y = 3e^{-t}$. Eliminating the parameter we find $y = \frac{3}{4}x$, $x > 0$. When $c_1 = 0$ and $c_2 = -1$ we find $y = \frac{3}{4}x$, $x < 0$.

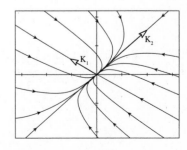

21. We have $\det(\mathbf{A} - \lambda\mathbf{I}) = (\lambda - 2)^2 = 0$. For $\lambda_1 = 2$ we obtain

$$\mathbf{K} = \begin{pmatrix} 1 \\ 1 \end{pmatrix}.$$

A solution of $(\mathbf{A} - \lambda_1\mathbf{I})\mathbf{P} = \mathbf{K}$ is

$$\mathbf{P} = \begin{pmatrix} -\frac{1}{3} \\ 0 \end{pmatrix}$$

so that

$$\mathbf{X} = c_1 \begin{pmatrix} 1 \\ 1 \end{pmatrix} e^{2t} + c_2 \left[\begin{pmatrix} 1 \\ 1 \end{pmatrix} t e^{2t} + \begin{pmatrix} -\frac{1}{3} \\ 0 \end{pmatrix} e^{2t} \right].$$

24. We have $\det(\mathbf{A} - \lambda\mathbf{I}) = (\lambda - 8)(\lambda + 1)^2 = 0$. For $\lambda_1 = 8$ we obtain

$$\mathbf{K}_1 = \begin{pmatrix} 2 \\ 1 \\ 2 \end{pmatrix}.$$

For $\lambda_2 = -1$ we obtain

$$\mathbf{K}_2 = \begin{pmatrix} 0 \\ -2 \\ 1 \end{pmatrix} \quad \text{and} \quad \mathbf{K}_3 = \begin{pmatrix} 1 \\ -2 \\ 0 \end{pmatrix}.$$

Then

$$\mathbf{X} = c_1 \begin{pmatrix} 2 \\ 1 \\ 2 \end{pmatrix} e^{8t} + c_2 \begin{pmatrix} 0 \\ -2 \\ 1 \end{pmatrix} e^{-t} + c_3 \begin{pmatrix} 1 \\ -2 \\ 0 \end{pmatrix} e^{-t}.$$

27. We have $\det(\mathbf{A} - \lambda\mathbf{I}) = -(\lambda - 1)^3 = 0$. For $\lambda_1 = 1$ we obtain

$$\mathbf{K} = \begin{pmatrix} 0 \\ 1 \\ 1 \end{pmatrix}.$$

Solutions of $(\mathbf{A} - \lambda_1\mathbf{I})\mathbf{P} = \mathbf{K}$ and $(\mathbf{A} - \lambda_1\mathbf{I})\mathbf{Q} = \mathbf{P}$ are

$$\mathbf{P} = \begin{pmatrix} 0 \\ 1 \\ 0 \end{pmatrix} \quad \text{and} \quad \mathbf{Q} = \begin{pmatrix} \frac{1}{2} \\ 0 \\ 0 \end{pmatrix}$$

so that

$$\mathbf{X} = c_1 \begin{pmatrix} 0 \\ 1 \\ 1 \end{pmatrix} e^t + c_2 \left[\begin{pmatrix} 0 \\ 1 \\ 1 \end{pmatrix} t e^t + \begin{pmatrix} 0 \\ 1 \\ 0 \end{pmatrix} e^t \right] + c_3 \left[\begin{pmatrix} 0 \\ 1 \\ 1 \end{pmatrix} \frac{t^2}{2} e^t + \begin{pmatrix} 0 \\ 1 \\ 0 \end{pmatrix} t e^t + \begin{pmatrix} \frac{1}{2} \\ 0 \\ 0 \end{pmatrix} e^t \right].$$

30. We have $\det(\mathbf{A} - \lambda\mathbf{I}) = -(\lambda + 1)(\lambda - 1)^2 = 0$. For $\lambda_1 = -1$ we obtain

$$\mathbf{K}_1 = \begin{pmatrix} -1 \\ 0 \\ 1 \end{pmatrix}.$$

For $\lambda_2 = 1$ we obtain

$$\mathbf{K}_2 = \begin{pmatrix} 1 \\ 0 \\ 1 \end{pmatrix} \quad \text{and} \quad \mathbf{K}_3 = \begin{pmatrix} 0 \\ 1 \\ 0 \end{pmatrix}.$$

so that

$$\mathbf{X} = c_1 \begin{pmatrix} -1 \\ 0 \\ 1 \end{pmatrix} e^{-t} + c_2 \begin{pmatrix} 1 \\ 0 \\ 1 \end{pmatrix} e^{t} + c_3 \begin{pmatrix} 0 \\ 1 \\ 0 \end{pmatrix} e^{t}.$$

If

$$\mathbf{X}(0) = \begin{pmatrix} 1 \\ 2 \\ 5 \end{pmatrix}$$

then $c_1 = 2$, $c_2 = 3$, and $c_3 = 2$.

33. We have $\det(\mathbf{A} - \lambda\mathbf{I}) = \lambda^2 - 8\lambda + 17 = 0$. For $\lambda_1 = 4 + i$ we obtain

$$\mathbf{K}_1 = \begin{pmatrix} 2 + i \\ 5 \end{pmatrix}$$

so that

$$\mathbf{X}_1 = \begin{pmatrix} 2 + i \\ 5 \end{pmatrix} e^{(4+i)t} = \begin{pmatrix} 2\cos t - \sin t \\ 5\cos t \end{pmatrix} e^{4t} + i \begin{pmatrix} \cos t + 2\sin t \\ 5\sin t \end{pmatrix} e^{4t}.$$

Then

$$\mathbf{X} = c_1 \begin{pmatrix} 2\cos t - \sin t \\ 5\cos t \end{pmatrix} e^{4t} + c_2 \begin{pmatrix} \cos t + 2\sin t \\ 5\sin t \end{pmatrix} e^{4t}.$$

36. We have $\det(\mathbf{A} - \lambda\mathbf{I}) = \lambda^2 - 10\lambda + 34 = 0$. For $\lambda_1 = 5 + 3i$ we obtain

$$\mathbf{K}_1 = \begin{pmatrix} 1 - 3i \\ 2 \end{pmatrix}$$

so that

$$\mathbf{X}_1 = \begin{pmatrix} 1 - 3i \\ 2 \end{pmatrix} e^{(5+3i)t} = \begin{pmatrix} \cos 3t + 3\sin 3t \\ 2\cos 3t \end{pmatrix} e^{5t} + i \begin{pmatrix} \sin 3t - 3\cos 3t \\ 2\sin 3t \end{pmatrix} e^{5t}.$$

Then

$$\mathbf{X} = c_1 \begin{pmatrix} \cos 3t + 3\sin 3t \\ 2\cos 3t \end{pmatrix} e^{5t} + c_2 \begin{pmatrix} \sin 3t - 3\cos 3t \\ 2\sin 3t \end{pmatrix} e^{5t}.$$

39. We have $\det(\mathbf{A} - \lambda\mathbf{I}) = -\lambda\left(\lambda^2 + 1\right) = 0$. For $\lambda_1 = 0$ we obtain

$$\mathbf{K}_1 = \begin{pmatrix} 1 \\ 0 \\ 0 \end{pmatrix}.$$

For $\lambda_2 = i$ we obtain

$$\mathbf{K}_2 = \begin{pmatrix} -i \\ i \\ 1 \end{pmatrix}$$

so that

$$\mathbf{X}_2 = \begin{pmatrix} -i \\ i \\ 1 \end{pmatrix} e^{it} = \begin{pmatrix} \sin t \\ -\sin t \\ \cos t \end{pmatrix} + i \begin{pmatrix} -\cos t \\ \cos t \\ \sin t \end{pmatrix}.$$

Then

$$\mathbf{X} = c_1 \begin{pmatrix} 1 \\ 0 \\ 0 \end{pmatrix} + c_2 \begin{pmatrix} \sin t \\ -\sin t \\ \cos t \end{pmatrix} + c_3 \begin{pmatrix} -\cos t \\ \cos t \\ \sin t \end{pmatrix}.$$

42. We have $\det(\mathbf{A} - \lambda\mathbf{I}) = -(\lambda - 6)(\lambda^2 - 8\lambda + 20) = 0$. For $\lambda_1 = 6$ we obtain

$$\mathbf{K}_1 = \begin{pmatrix} 0 \\ 1 \\ 0 \end{pmatrix}.$$

For $\lambda_2 = 4 + 2i$ we obtain

$$\mathbf{K}_2 = \begin{pmatrix} -i \\ 0 \\ 2 \end{pmatrix}$$

so that

$$\mathbf{X}_2 = \begin{pmatrix} -i \\ 0 \\ 2 \end{pmatrix} e^{(4+2i)t} = \begin{pmatrix} \sin 2t \\ 0 \\ 2\cos 2t \end{pmatrix} e^{4t} + i \begin{pmatrix} -\cos 2t \\ 0 \\ 2\sin 2t \end{pmatrix} e^{4t}.$$

Then

$$\mathbf{X} = c_1 \begin{pmatrix} 0 \\ 1 \\ 0 \end{pmatrix} e^{6t} + c_2 \begin{pmatrix} \sin 2t \\ 0 \\ 2\cos 2t \end{pmatrix} e^{4t} + c_3 \begin{pmatrix} -\cos 2t \\ 0 \\ 2\sin 2t \end{pmatrix} e^{4t}.$$

45. We have $\det(\mathbf{A} - \lambda\mathbf{I}) = (1 - \lambda)(\lambda^2 + 25) = 0$. For $\lambda_1 = 1$ we obtain

$$\mathbf{K}_1 = \begin{pmatrix} 25 \\ -7 \\ 6 \end{pmatrix}.$$

For $\lambda_2 = 5i$ we obtain

$$\mathbf{K}_2 = \begin{pmatrix} 1 + 5i \\ 1 \\ 1 \end{pmatrix}$$

so that

$$\mathbf{X}_2 = \begin{pmatrix} 1 + 5i \\ 1 \\ 1 \end{pmatrix} e^{5it} = \begin{pmatrix} \cos 5t - 5\sin 5t \\ \cos 5t \\ \cos 5t \end{pmatrix} + i \begin{pmatrix} \sin 5t + 5\cos 5t \\ \sin 5t \\ \sin 5t \end{pmatrix}.$$

Then

$$\mathbf{X} = c_1 \begin{pmatrix} 25 \\ -7 \\ 6 \end{pmatrix} e^t + c_2 \begin{pmatrix} \cos 5t - 5\sin 5t \\ \cos 5t \\ \cos 5t \end{pmatrix} + c_3 \begin{pmatrix} \sin 5t + 5\cos 5t \\ \sin 5t \\ \sin 5t \end{pmatrix}.$$

If

$$\mathbf{X}(0) = \begin{pmatrix} 4 \\ 6 \\ -7 \end{pmatrix}$$

then $c_1 = c_2 = -1$ and $c_3 = 6$.

48. (a) From $\det(\mathbf{A} - \lambda\mathbf{I}) = \lambda(\lambda - 2) = 0$ we get $\lambda_1 = 0$ and $\lambda_2 = 2$. For $\lambda_1 = 0$ we obtain

$$\left(\begin{array}{cc|c} 1 & 1 & 0 \\ 1 & 1 & 0 \end{array} \right) \Longrightarrow \left(\begin{array}{cc|c} 1 & 1 & 0 \\ 0 & 0 & 0 \end{array} \right) \quad \text{so that} \quad \mathbf{K}_1 = \begin{pmatrix} -1 \\ 1 \end{pmatrix}.$$

For $\lambda_2 = 2$ we obtain

$$\left(\begin{array}{cc|c} -1 & 1 & 0 \\ 1 & -1 & 0 \end{array} \right) \Longrightarrow \left(\begin{array}{cc|c} -1 & 1 & 0 \\ 0 & 0 & 0 \end{array} \right) \quad \text{so that} \quad \mathbf{K}_2 = \begin{pmatrix} 1 \\ 1 \end{pmatrix}.$$

Then

$$\mathbf{X} = c_1 \begin{pmatrix} -1 \\ 1 \end{pmatrix} + c_2 \begin{pmatrix} 1 \\ 1 \end{pmatrix} e^{2t}.$$

The line $y = -x$ is not a trajectory of the system. Trajectories are $x = -c_1 + c_2 e^{2t}$, $y = c_1 + c_2 e^{2t}$ or $y = x + 2c_1$. This is a family of lines perpendicular to the line $y = -x$. All of the constant solutions of the system do, however, lie on the line $y = -x$.

(b) From $\det(\mathbf{A} - \lambda\mathbf{I}) = \lambda^2 = 0$ we get $\lambda_1 = 0$ and

$$\mathbf{K} = \begin{pmatrix} -1 \\ 1 \end{pmatrix}.$$

A solution of $(\mathbf{A} - \lambda_1\mathbf{I})\mathbf{P} = \mathbf{K}$ is

$$\mathbf{P} = \begin{pmatrix} -1 \\ 0 \end{pmatrix}$$

so that

$$\mathbf{X} = c_1 \begin{pmatrix} -1 \\ 1 \end{pmatrix} + c_2 \left[\begin{pmatrix} -1 \\ 1 \end{pmatrix} t + \begin{pmatrix} -1 \\ 0 \end{pmatrix} \right].$$

All trajectories are parallel to $y = -x$, but $y = -x$ is not a trajectory. There are constant solutions of the system, however, that do lie on the line $y = -x$.

51. Suppose the eigenvalues are $\alpha \pm i\beta$, $\beta > 0$. In Problem 36 the eigenvalues are $5 \pm 3i$, in Problem 37 they are $\pm 3i$, and in Problem 38 they are $-1 \pm 2i$. From Problem 47 we deduce that the phase portrait will consist of a family of closed curves when $\alpha = 0$ and spirals when $\alpha \neq 0$. The origin will be a repellor when $\alpha > 0$, and an attractor when $\alpha < 0$.

EXERCISES 10.3
Solution by Diagonalization

3. $\lambda_1 = \frac{1}{2}$, $\lambda_2 = \frac{3}{2}$, $\mathbf{K}_1 = \begin{pmatrix} 1 \\ -2 \end{pmatrix}$, $\mathbf{K}_2 = \begin{pmatrix} 1 \\ 2 \end{pmatrix}$, $\mathbf{P} = \begin{pmatrix} 1 & 1 \\ -2 & 2 \end{pmatrix}$;

$$\mathbf{X} = \mathbf{PY} = \begin{pmatrix} 1 & 1 \\ -2 & 2 \end{pmatrix} \begin{pmatrix} c_1 e^{t/2} \\ c_2 e^{3t/2} \end{pmatrix} = \begin{pmatrix} c_1 e^{t/2} + c_2 e^{3t/2} \\ -2c_1 e^{t/2} + 2c_2 e^{3t/2} \end{pmatrix} = c_1 \begin{pmatrix} 1 \\ -2 \end{pmatrix} e^{t/2} + c_2 \begin{pmatrix} 1 \\ 2 \end{pmatrix} e^{3t/2}$$

6. $\lambda_1 = -1$, $\lambda_2 = 1$, $\lambda_3 = 4$, $\mathbf{K}_1 = \begin{pmatrix} -1 \\ 0 \\ 1 \end{pmatrix}$, $\mathbf{K}_2 = \begin{pmatrix} 1 \\ -2 \\ 1 \end{pmatrix}$, $\mathbf{K}_3 = \begin{pmatrix} 1 \\ 1 \\ 1 \end{pmatrix}$, $\mathbf{P} = \begin{pmatrix} -1 & 1 & 1 \\ 0 & -2 & 1 \\ 1 & 1 & 1 \end{pmatrix}$;

$$\mathbf{X} = \mathbf{PY} = \begin{pmatrix} -1 & 1 & 1 \\ 0 & -2 & 1 \\ 1 & 1 & 1 \end{pmatrix} \begin{pmatrix} c_1 e^{-t} \\ c_2 e^t \\ c_3 e^{4t} \end{pmatrix} = \begin{pmatrix} -c_1 + c_2 e^t + c_3 e^{4t} \\ -2c_2 e^t + c_3 e^{4t} \\ c_1 e^{-t} + c_2 e^t + c_3 e^{4t} \end{pmatrix} = c_1 \begin{pmatrix} -1 \\ 0 \\ 1 \end{pmatrix} e^{-t} + c_2 \begin{pmatrix} 1 \\ -2 \\ 1 \end{pmatrix} e^t + c_3 \begin{pmatrix} 1 \\ 1 \\ 1 \end{pmatrix} e^{4t}$$

9. $\lambda_1 = 1$, $\lambda_2 = 2$, $\lambda_3 = 3$, $\mathbf{K}_1 = \begin{pmatrix} 1 \\ 1 \\ 1 \end{pmatrix}$, $\mathbf{K}_2 = \begin{pmatrix} 2 \\ 2 \\ 3 \end{pmatrix}$, $\mathbf{K}_3 = \begin{pmatrix} 3 \\ 4 \\ 5 \end{pmatrix}$, $\mathbf{P} = \begin{pmatrix} 1 & 2 & 3 \\ 1 & 2 & 4 \\ 1 & 3 & 5 \end{pmatrix}$;

$$\mathbf{X} = \mathbf{PY} = \begin{pmatrix} 1 & 2 & 3 \\ 1 & 2 & 4 \\ 1 & 3 & 5 \end{pmatrix} \begin{pmatrix} c_1 e^t \\ c_2 e^{2t} \\ c_3 e^{3t} \end{pmatrix} = \begin{pmatrix} c_1 e^t + 2c_2 e^{2t} + 3c_3 e^{3t} \\ c_1 e^t + 2c_2 e^{2t} + 4c_3 e^{3t} \\ c_1 e^t + 3c_2 e^{2t} + 5c_3 e^{3t} \end{pmatrix} = c_1 \begin{pmatrix} 1 \\ 1 \\ 1 \end{pmatrix} e^t + c_2 \begin{pmatrix} 2 \\ 2 \\ 3 \end{pmatrix} e^{2t} + c_3 \begin{pmatrix} 3 \\ 4 \\ 5 \end{pmatrix} e^{3t}$$

EXERCISES 10.4
Nonhomogeneous Linear Systems

3. Solving

$$\det(\mathbf{A} - \lambda\mathbf{I}) = \begin{vmatrix} 1 - \lambda & 3 \\ 3 & 1 - \lambda \end{vmatrix} = \lambda^2 - 2\lambda - 8 = (\lambda - 4)(\lambda + 2) = 0$$

we obtain eigenvalues $\lambda_1 = -2$ and $\lambda_2 = 4$. Corresponding eigenvectors are

$$\mathbf{K}_1 = \begin{pmatrix} 1 \\ -1 \end{pmatrix} \quad \text{and} \quad \mathbf{K}_2 = \begin{pmatrix} 1 \\ 1 \end{pmatrix}.$$

Thus

$$\mathbf{X}_c = c_1 \begin{pmatrix} 1 \\ -1 \end{pmatrix} e^{-2t} + c_2 \begin{pmatrix} 1 \\ 1 \end{pmatrix} e^{4t}.$$

Substituting

$$\mathbf{X}_p = \begin{pmatrix} a_3 \\ b_3 \end{pmatrix} t^2 + \begin{pmatrix} a_2 \\ b_2 \end{pmatrix} t + \begin{pmatrix} a_1 \\ b_1 \end{pmatrix}$$

into the system yields

$$a_3 + 3b_3 = 2 \qquad a_2 + 3b_2 = 2a_3 \qquad a_1 + 3b_1 = a_2$$

$$3a_3 + b_3 = 0 \qquad 3a_2 + b_2 + 1 = 2b_3 \qquad 3a_1 + b_1 + 5 = b_2$$

from which we obtain $a_3 = -1/4$, $b_3 = 3/4$, $a_2 = 1/4$, $b_2 = -1/4$, $a_1 = -2$, and $b_1 = 3/4$. Then

$$\mathbf{X}(t) = c_1 \begin{pmatrix} 1 \\ -1 \end{pmatrix} e^{-2t} + c_2 \begin{pmatrix} 1 \\ 1 \end{pmatrix} e^{4t} + \begin{pmatrix} -\frac{1}{4} \\ \frac{3}{4} \end{pmatrix} t^2 + \begin{pmatrix} \frac{1}{4} \\ -\frac{1}{4} \end{pmatrix} t + \begin{pmatrix} -2 \\ \frac{3}{4} \end{pmatrix}.$$

6. Solving

$$\det(\mathbf{A} - \lambda\mathbf{I}) = \begin{vmatrix} -1 - \lambda & 5 \\ -1 & 1 - \lambda \end{vmatrix} = \lambda^2 + 4 = 0$$

we obtain the eigenvalues $\lambda_1 = 2i$ and $\lambda_2 = -2i$. Corresponding eigenvectors are

$$\mathbf{K}_1 = \begin{pmatrix} 5 \\ 1 + 2i \end{pmatrix} \quad \text{and} \quad \mathbf{K}_2 = \begin{pmatrix} 5 \\ 1 - 2i \end{pmatrix}.$$

Thus

$$\mathbf{X}_c = c_1 \begin{pmatrix} 5\cos 2t \\ \cos 2t - 2\sin 2t \end{pmatrix} + c_2 \begin{pmatrix} 5\sin 2t \\ 2\cos 2t + \sin 2t \end{pmatrix}.$$

Substituting

$$\mathbf{X}_p = \begin{pmatrix} a_2 \\ b_2 \end{pmatrix} \cos t + \begin{pmatrix} a_1 \\ b_1 \end{pmatrix} \sin t$$

into the system yields

$$-a_2 + 5b_2 - a_1 = 0$$

$$-a_2 + b_2 - b_1 - 2 = 0$$

$$-a_1 + 5b_1 + a_2 + 1 = 0$$

$$-a_1 + b_1 + b_2 = 0$$

from which we obtain $a_2 = -3$, $b_2 = -2/3$, $a_1 = -1/3$, and $b_1 = 1/3$. Then

$$\mathbf{X}(t) = c_1 \begin{pmatrix} 5\cos 2t \\ \cos 2t - 2\sin 2t \end{pmatrix} + c_2 \begin{pmatrix} 5\sin 2t \\ 2\cos 2t + \sin 2t \end{pmatrix} + \begin{pmatrix} -3 \\ -\frac{2}{3} \end{pmatrix} \cos t + \begin{pmatrix} -\frac{1}{3} \\ \frac{1}{3} \end{pmatrix} \sin t.$$

9. Solving

$$\det(\mathbf{A} - \lambda\mathbf{I}) = \begin{vmatrix} -1 - \lambda & -2 \\ 3 & 4 - \lambda \end{vmatrix} = \lambda^2 - 3\lambda + 2 = (\lambda - 1)(\lambda - 2) = 0$$

we obtain the eigenvalues $\lambda_1 = 1$ and $\lambda_2 = 2$. Corresponding eigenvectors are

$$\mathbf{K}_1 = \begin{pmatrix} 1 \\ -1 \end{pmatrix} \quad \text{and} \quad \mathbf{K}_2 = \begin{pmatrix} -4 \\ 6 \end{pmatrix}.$$

Thus

$$\mathbf{X}_c = c_1 \begin{pmatrix} 1 \\ -1 \end{pmatrix} e^t + c_2 \begin{pmatrix} -4 \\ 6 \end{pmatrix} e^{2t}.$$

Substituting

$$\mathbf{X}_p = \begin{pmatrix} a_1 \\ b_1 \end{pmatrix}$$

into the system yields

$$-a_1 - 2b_1 = -3$$

$$3a_1 + 4b_1 = -3$$

from which we obtain $a_1 = -9$ and $b_1 = 6$. Then

$$\mathbf{X}(t) = c_1 \begin{pmatrix} 1 \\ -1 \end{pmatrix} e^t + c_2 \begin{pmatrix} -4 \\ 6 \end{pmatrix} e^{2t} + \begin{pmatrix} -9 \\ 6 \end{pmatrix}.$$

Setting

$$\mathbf{X}(0) = \begin{pmatrix} -4 \\ 5 \end{pmatrix}$$

we obtain

$$c_1 - 4c_2 - 9 = -4$$
$$-c_1 + 6c_2 + 6 = 5.$$

Then $c_1 = 13$ and $c_2 = 2$ so

$$\mathbf{X}(t) = 13 \begin{pmatrix} 1 \\ -1 \end{pmatrix} e^t + 2 \begin{pmatrix} -4 \\ 6 \end{pmatrix} e^{2t} + \begin{pmatrix} -9 \\ 6 \end{pmatrix}.$$

12. From

$$\mathbf{X}' = \begin{pmatrix} 2 & -1 \\ 3 & -2 \end{pmatrix} \mathbf{X} + \begin{pmatrix} 0 \\ 4 \end{pmatrix} t$$

we obtain

$$\mathbf{X}_c = c_1 \begin{pmatrix} 1 \\ 1 \end{pmatrix} e^t + c_2 \begin{pmatrix} 1 \\ 3 \end{pmatrix} e^{-t}.$$

Then

$$\mathbf{\Phi} = \begin{pmatrix} e^t & e^{-t} \\ e^t & 3e^{-t} \end{pmatrix} \quad \text{and} \quad \mathbf{\Phi}^{-1} = \begin{pmatrix} \frac{3}{2}e^{-t} & -\frac{1}{2}e^{-t} \\ -\frac{1}{2}e^t & \frac{1}{2}e^t \end{pmatrix}$$

so that

$$\mathbf{U} = \int \mathbf{\Phi}^{-1} \mathbf{F} \, dt = \int \begin{pmatrix} -2te^{-t} \\ 2te^t \end{pmatrix} dt = \begin{pmatrix} 2te^{-t} + 2e^{-t} \\ 2te^t - 2e^t \end{pmatrix}$$

and

$$\mathbf{X}_p = \mathbf{\Phi}\mathbf{U} = \begin{pmatrix} 4 \\ 8 \end{pmatrix} t + \begin{pmatrix} 0 \\ -4 \end{pmatrix}.$$

15. From

$$\mathbf{X}' = \begin{pmatrix} 0 & 2 \\ -1 & 3 \end{pmatrix} \mathbf{X} + \begin{pmatrix} 1 \\ -1 \end{pmatrix} e^t$$

we obtain

$$\mathbf{X}_c = c_1 \begin{pmatrix} 2 \\ 1 \end{pmatrix} e^t + c_2 \begin{pmatrix} 1 \\ 1 \end{pmatrix} e^{2t}.$$

Then

$$\mathbf{\Phi} = \begin{pmatrix} 2e^t & e^{2t} \\ e^t & e^{2t} \end{pmatrix} \quad \text{and} \quad \mathbf{\Phi}^{-1} = \begin{pmatrix} e^{-t} & -e^{-t} \\ -e^{-2t} & 2e^{-2t} \end{pmatrix}$$

so that

$$\mathbf{U} = \int \mathbf{\Phi}^{-1} \mathbf{F} \, dt = \int \begin{pmatrix} 2 \\ -3e^{-t} \end{pmatrix} dt = \begin{pmatrix} 2t \\ 3e^{-t} \end{pmatrix}$$

and

$$\mathbf{X}_p = \mathbf{\Phi}\mathbf{U} = \begin{pmatrix} 4 \\ 2 \end{pmatrix} te^t + \begin{pmatrix} 3 \\ 3 \end{pmatrix} e^t.$$

18. From

$$\mathbf{X}' = \begin{pmatrix} 1 & 8 \\ 1 & -1 \end{pmatrix} \mathbf{X} + \begin{pmatrix} e^{-t} \\ te^t \end{pmatrix}$$

we obtain

$$\mathbf{X}_c = c_1 \begin{pmatrix} 4 \\ 1 \end{pmatrix} e^{3t} + c_2 \begin{pmatrix} -2 \\ 1 \end{pmatrix} e^{-3t}.$$

Then

$$\mathbf{\Phi} = \begin{pmatrix} 4e^{3t} & -2e^{3t} \\ e^{3t} & e^{-3t} \end{pmatrix} \quad \text{and} \quad \mathbf{\Phi}^{-1} = \begin{pmatrix} \frac{1}{6}e^{-3t} & \frac{1}{3}e^{-3t} \\ -\frac{1}{6}e^{3t} & \frac{2}{3}e^{3t} \end{pmatrix}$$

so that

$$\mathbf{U} = \int \mathbf{\Phi}^{-1}\mathbf{F}\, dt = \int \begin{pmatrix} \frac{1}{6}e^{-4t} + \frac{1}{3}te^{-2t} \\ -\frac{1}{6}e^{2t} + \frac{2}{3}te^{4t} \end{pmatrix} dt = \begin{pmatrix} -\frac{1}{24}e^{-4t} - \frac{1}{6}te^{-2t} - \frac{1}{12}e^{-2t} \\ -\frac{1}{12}e^{2t} + \frac{1}{6}te^{4t} - \frac{1}{24}e^{4t} \end{pmatrix}$$

and

$$\mathbf{X}_p = \mathbf{\Phi}\mathbf{U} = \begin{pmatrix} -te^{t} - \frac{1}{4}e^{t} \\ -\frac{1}{8}e^{-t} - \frac{1}{8}e^{t} \end{pmatrix}.$$

21. From

$$\mathbf{X}' = \begin{pmatrix} 0 & -1 \\ 1 & 0 \end{pmatrix} \mathbf{X} + \begin{pmatrix} \sec t \\ 0 \end{pmatrix}$$

we obtain

$$\mathbf{X}_c = c_1 \begin{pmatrix} \cos t \\ \sin t \end{pmatrix} + c_2 \begin{pmatrix} \sin t \\ -\cos t \end{pmatrix}.$$

Then

$$\mathbf{\Phi} = \begin{pmatrix} \cos t & \sin t \\ \sin t & -\cos t \end{pmatrix} \quad \text{and} \quad \mathbf{\Phi}^{-1} = \begin{pmatrix} \cos t & \sin t \\ \sin t & -\cos t \end{pmatrix}$$

so that

$$\mathbf{U} = \int \mathbf{\Phi}^{-1}\mathbf{F}\, dt = \int \begin{pmatrix} 1 \\ \tan t \end{pmatrix} dt = \begin{pmatrix} t \\ -\ln|\cos t| \end{pmatrix}$$

and

$$\mathbf{X}_p = \mathbf{\Phi}\mathbf{U} = \begin{pmatrix} t\cos t - \sin t \ln|\cos t| \\ t\sin t + \cos t \ln|\cos t| \end{pmatrix}.$$

24. From

$$\mathbf{X}' = \begin{pmatrix} 2 & -2 \\ 8 & -6 \end{pmatrix} \mathbf{X} + \begin{pmatrix} 1 \\ 3 \end{pmatrix} \frac{1}{t}e^{-2t}$$

we obtain

$$\mathbf{X}_c = c_1 \begin{pmatrix} 1 \\ 2 \end{pmatrix} e^{-2t} + c_2 \left[\begin{pmatrix} 1 \\ 2 \end{pmatrix} te^{-2t} + \begin{pmatrix} \frac{1}{2} \\ \frac{1}{2} \end{pmatrix} e^{-2t} \right].$$

Then

$$\mathbf{\Phi} = \begin{pmatrix} 1 & t + \frac{1}{2} \\ 2 & 2t + \frac{1}{2} \end{pmatrix} e^{-2t} \quad \text{and} \quad \mathbf{\Phi}^{-1} = \begin{pmatrix} -4t - 1 & 2t + 1 \\ 4 & -2 \end{pmatrix} e^{2t}$$

so that

$$\mathbf{U} = \int \mathbf{\Phi}^{-1}\mathbf{F}\, dt = \int \begin{pmatrix} 2 + 2/t \\ -2/t \end{pmatrix} dt = \begin{pmatrix} 2t + 2\ln t \\ -2\ln t \end{pmatrix}$$

and

$$\mathbf{X}_p = \mathbf{\Phi}\mathbf{U} = \begin{pmatrix} 2t + \ln t - 2t\ln t \\ 4t + 3\ln t - 4t\ln t \end{pmatrix} e^{-2t}.$$

27. From

$$\mathbf{X}' = \begin{pmatrix} 1 & 2 \\ -\frac{1}{2} & 1 \end{pmatrix} \mathbf{X} + \begin{pmatrix} \csc t \\ \sec t \end{pmatrix} e^{t}$$

221

we obtain

$$\mathbf{X}_c = c_1 \begin{pmatrix} 2\sin t \\ \cos t \end{pmatrix} e^t + c_2 \begin{pmatrix} 2\cos t \\ -\sin t \end{pmatrix} e^t.$$

Then

$$\mathbf{\Phi} = \begin{pmatrix} 2\sin t & 2\cos t \\ \cos t & -\sin t \end{pmatrix} e^t \quad \text{and} \quad \mathbf{\Phi}^{-1} = \begin{pmatrix} \frac{1}{2}\sin t & \cos t \\ \frac{1}{2}\cos t & -\sin t \end{pmatrix} e^{-t}$$

so that

$$\mathbf{U} = \int \mathbf{\Phi}^{-1}\mathbf{F}\,dt = \int \begin{pmatrix} \frac{3}{2} \\ \frac{1}{2}\cot t - \tan t \end{pmatrix} dt = \begin{pmatrix} \frac{3}{2}t \\ \frac{1}{2}\ln|\sin t| + \ln|\cos t| \end{pmatrix}$$

and

$$\mathbf{X}_p = \mathbf{\Phi}\mathbf{U} = \begin{pmatrix} 3\sin t \\ \frac{3}{2}\cos t \end{pmatrix} te^t + \begin{pmatrix} \cos t \\ -\frac{1}{2}\sin t \end{pmatrix} e^t \ln|\sin t| + \begin{pmatrix} 2\cos t \\ -\sin t \end{pmatrix} e^t \ln|\cos t|.$$

30. From

$$\mathbf{X}' = \begin{pmatrix} 3 & -1 & -1 \\ 1 & 1 & -1 \\ 1 & -1 & 1 \end{pmatrix}\mathbf{X} + \begin{pmatrix} 0 \\ t \\ 2e^t \end{pmatrix}$$

we obtain

$$\mathbf{X}_c = c_1 \begin{pmatrix} 1 \\ 1 \\ 1 \end{pmatrix} e^t + c_2 \begin{pmatrix} 1 \\ 1 \\ 0 \end{pmatrix} e^{2t} + c_3 \begin{pmatrix} 1 \\ 0 \\ 1 \end{pmatrix} e^{2t}.$$

Then

$$\mathbf{\Phi} = \begin{pmatrix} e^t & e^{2t} & e^{2t} \\ e^t & e^{2t} & 0 \\ e^t & 0 & e^{2t} \end{pmatrix} \quad \text{and} \quad \mathbf{\Phi}^{-1} = \begin{pmatrix} -e^{-t} & e^{-t} & e^{-t} \\ e^{-2t} & 0 & -e^{-2t} \\ e^{-2t} & -e^{-2t} & 0 \end{pmatrix}$$

so that

$$\mathbf{U} = \int \mathbf{\Phi}^{-1}\mathbf{F}\,dt = \int \begin{pmatrix} te^{-t} + 2 \\ -2e^{-t} \\ -te^{-2t} \end{pmatrix} dt = \begin{pmatrix} -te^{-t} - e^{-t} + 2t \\ 2e^{-t} \\ \frac{1}{2}te^{-2t} + \frac{1}{4}e^{-2t} \end{pmatrix}$$

and

$$\mathbf{X}_p = \mathbf{\Phi}\mathbf{U} = \begin{pmatrix} -\frac{1}{2} \\ -1 \\ -\frac{1}{2} \end{pmatrix} t + \begin{pmatrix} -\frac{3}{4} \\ -1 \\ -\frac{3}{4} \end{pmatrix} + \begin{pmatrix} 2 \\ 2 \\ 0 \end{pmatrix} e^t + \begin{pmatrix} 2 \\ 2 \\ 2 \end{pmatrix} te^t.$$

33. Let $\mathbf{I} = \begin{pmatrix} i_1 \\ i_2 \end{pmatrix}$ so that

$$\mathbf{I}' = \begin{pmatrix} -11 & 3 \\ 3 & -3 \end{pmatrix}\mathbf{I} + \begin{pmatrix} 100\sin t \\ 0 \end{pmatrix}.$$

and

$$\mathbf{I}_c = c_1 \begin{pmatrix} 1 \\ 3 \end{pmatrix} e^{-2t} + c_2 \begin{pmatrix} 3 \\ -1 \end{pmatrix} e^{-12t}.$$

Then

$$\mathbf{\Phi} = \begin{pmatrix} e^{-2t} & 3e^{-12t} \\ 3e^{-2t} & -e^{-12t} \end{pmatrix}, \quad \mathbf{\Phi}^{-1} = \begin{pmatrix} \frac{1}{10}e^{2t} & \frac{3}{10}e^{2t} \\ \frac{3}{10}e^{12t} & -\frac{1}{10}e^{12t} \end{pmatrix},$$

$$\mathbf{U} = \int \mathbf{\Phi}^{-1}\mathbf{F}\,dt = \int \begin{pmatrix} 10e^{2t}\sin t \\ 30e^{12t}\sin t \end{pmatrix} dt = \begin{pmatrix} 2e^{2t}(2\sin t - \cos t) \\ \frac{6}{29}e^{12t}(12\sin t - \cos t) \end{pmatrix},$$

and

$$\mathbf{I}_p = \mathbf{\Phi U} = \begin{pmatrix} \frac{332}{29} \sin t - \frac{76}{29} \cos t \\ \frac{276}{29} \sin t - \frac{168}{29} \cos t \end{pmatrix}$$

so that

$$\mathbf{I} = c_1 \begin{pmatrix} 1 \\ 3 \end{pmatrix} e^{-2t} + c_2 \begin{pmatrix} 3 \\ -1 \end{pmatrix} e^{-12t} + \mathbf{I}_p.$$

If $\mathbf{I}(0) = \begin{pmatrix} 0 \\ 0 \end{pmatrix}$ then $c_1 = 2$ and $c_2 = \frac{6}{29}$.

36. $\lambda_1 = -1$, $\lambda_2 = 4$, $\mathbf{K}_1 = \begin{pmatrix} 3 \\ -2 \end{pmatrix}$, $\mathbf{K}_2 = \begin{pmatrix} 1 \\ 1 \end{pmatrix}$, $\mathbf{P} = \begin{pmatrix} 3 & 1 \\ -2 & 1 \end{pmatrix}$, $\mathbf{P}^{-1} = \frac{1}{5} \begin{pmatrix} 1 & -1 \\ 2 & 3 \end{pmatrix}$, $\mathbf{P}^{-1}\mathbf{F} = \begin{pmatrix} 0 \\ e^t \end{pmatrix}$;

$$\mathbf{Y}' = \begin{pmatrix} -1 & 0 \\ 0 & 4 \end{pmatrix} \mathbf{Y} + \begin{pmatrix} 0 \\ e^t \end{pmatrix}$$

$$y_1 = c_1 e^{-t}, \qquad y_2 = -\frac{1}{3} e^t + c_2 e^{4t}$$

$$\mathbf{X} = \mathbf{PY} = \begin{pmatrix} 3 & 1 \\ -2 & 1 \end{pmatrix} \begin{pmatrix} c_1 e^{-t} \\ -\frac{1}{3} e^t + c_2 e^{4t} \end{pmatrix} = \begin{pmatrix} -\frac{1}{3} e^t + 3c_1 e^{-t} + c_2 e^{4t} \\ -\frac{1}{3} e^t - 2c_1 e^{-t} + c_2 e^{4t} \end{pmatrix} = c_1 \begin{pmatrix} 3 \\ -2 \end{pmatrix} e^{-t} + c_2 \begin{pmatrix} 1 \\ 1 \end{pmatrix} e^{4t} - \frac{1}{3} \begin{pmatrix} 1 \\ 1 \end{pmatrix} e^t$$

EXERCISES 10.5

Matrix Exponential

3. For

$$\mathbf{A} = \begin{pmatrix} 1 & 1 & 1 \\ 1 & 1 & 1 \\ -2 & -2 & -2 \end{pmatrix}$$

we have

$$\mathbf{A}^2 = \begin{pmatrix} 1 & 1 & 1 \\ 1 & 1 & 1 \\ -2 & -2 & -2 \end{pmatrix} \begin{pmatrix} 1 & 1 & 1 \\ 1 & 1 & 1 \\ -2 & -2 & -2 \end{pmatrix} = \begin{pmatrix} 0 & 0 & 0 \\ 0 & 0 & 0 \\ 0 & 0 & 0 \end{pmatrix}.$$

Thus, $\mathbf{A}^3 = \mathbf{A}^4 = \mathbf{A}^5 = \cdots = \mathbf{0}$ and

$$e^{\mathbf{A}t} = \mathbf{I} + \mathbf{A}t = \begin{pmatrix} 1 & 0 & 0 \\ 0 & 1 & 0 \\ 0 & 0 & 1 \end{pmatrix} + \begin{pmatrix} t & t & t \\ t & t & t \\ -2t & -2t & -2t \end{pmatrix} = \begin{pmatrix} t+1 & t & t \\ t & t+1 & t \\ -2t & -2t & -2t+1 \end{pmatrix}.$$

6. Using the result of Problem 2,

$$\mathbf{X} = \begin{pmatrix} \cosh t & \sinh t \\ \sinh t & \cosh t \end{pmatrix} \begin{pmatrix} c_1 \\ c_2 \end{pmatrix} = c_1 \begin{pmatrix} \cosh t \\ \sinh t \end{pmatrix} + c_2 \begin{pmatrix} \sinh t \\ \cosh t \end{pmatrix}.$$

9. To solve

$$\mathbf{X}' = \begin{pmatrix} 1 & 0 \\ 0 & 2 \end{pmatrix} \mathbf{X} + \begin{pmatrix} 3 \\ -1 \end{pmatrix}$$

we identify $t_0 = 0$, $\mathbf{F}(t) = \begin{pmatrix} 3 \\ -1 \end{pmatrix}$, and use the results of Problem 1 and equation (6) in the text.

$$\mathbf{X}(t) = e^{\mathbf{A}t}\mathbf{C} + e^{\mathbf{A}t}\int_{t_0}^{t} e^{-\mathbf{A}s}\mathbf{F}(s)\,ds$$

$$= \begin{pmatrix} e^t & 0 \\ 0 & e^{2t} \end{pmatrix}\begin{pmatrix} c_1 \\ c_2 \end{pmatrix} + \begin{pmatrix} e^t & 0 \\ 0 & e^{2t} \end{pmatrix}\int_0^t \begin{pmatrix} e^{-s} & 0 \\ 0 & e^{-2s} \end{pmatrix}\begin{pmatrix} 3 \\ -1 \end{pmatrix}ds$$

$$= \begin{pmatrix} c_1 e^t \\ c_2 e^{2t} \end{pmatrix} + \begin{pmatrix} e^t & 0 \\ 0 & e^{2t} \end{pmatrix}\int_0^t \begin{pmatrix} 3e^{-s} \\ -e^{-2s} \end{pmatrix}ds$$

$$= \begin{pmatrix} c_1 e^t \\ c_2 e^{2t} \end{pmatrix} + \begin{pmatrix} e^t & 0 \\ 0 & e^{2t} \end{pmatrix}\begin{pmatrix} -3e^{-s} \\ \frac{1}{2}e^{-2s} \end{pmatrix}\Big|_0^t$$

$$= \begin{pmatrix} c_1 e^t \\ c_2 e^{2t} \end{pmatrix} + \begin{pmatrix} e^t & 0 \\ 0 & e^{2t} \end{pmatrix}\begin{pmatrix} -3e^{-t} + 3 \\ \frac{1}{2}e^{-2t} - \frac{1}{2} \end{pmatrix}$$

$$= \begin{pmatrix} c_1 e^t \\ c_2 e^{2t} \end{pmatrix} + \begin{pmatrix} -3 + 3e^t \\ \frac{1}{2} - \frac{1}{2}e^{2t} \end{pmatrix} = c_3\begin{pmatrix} 1 \\ 0 \end{pmatrix}e^t + c_4\begin{pmatrix} 0 \\ 1 \end{pmatrix}e^{2t} + \begin{pmatrix} -3 \\ \frac{1}{2} \end{pmatrix}.$$

12. To solve

$$\mathbf{X}' = \begin{pmatrix} 0 & 1 \\ 1 & 0 \end{pmatrix}\mathbf{X} + \begin{pmatrix} \cosh t \\ \sinh t \end{pmatrix}$$

we identify $t_0 = 0$, $\mathbf{F}(t) = \begin{pmatrix} \cosh t \\ \sinh t \end{pmatrix}$, and use the results of Problem 2 and equation (6) in the text.

$$\mathbf{X}(t) = e^{\mathbf{A}t}\mathbf{C} + e^{\mathbf{A}t}\int_{t_0}^{t} e^{-\mathbf{A}s}\mathbf{F}(s)\,ds$$

$$= \begin{pmatrix} \cosh t & \sinh t \\ \sinh t & \cosh t \end{pmatrix}\begin{pmatrix} c_1 \\ c_2 \end{pmatrix} + \begin{pmatrix} \cosh t & \sinh t \\ \sinh t & \cosh t \end{pmatrix}\int_0^t \begin{pmatrix} \cosh s & -\sinh s \\ -\sinh s & \cosh s \end{pmatrix}\begin{pmatrix} \cosh s \\ \sinh s \end{pmatrix}ds$$

$$= \begin{pmatrix} c_1\cosh t + c_2\sinh t \\ c_1\sinh t + c_2\cosh t \end{pmatrix} + \begin{pmatrix} \cosh t & \sinh t \\ \sinh t & \cosh t \end{pmatrix}\int_0^t \begin{pmatrix} 1 \\ 0 \end{pmatrix}ds$$

$$= \begin{pmatrix} c_1\cosh t + c_2\sinh t \\ c_1\sinh t + c_2\cosh t \end{pmatrix} + \begin{pmatrix} \cosh t & \sinh t \\ \sinh t & \cosh t \end{pmatrix}\begin{pmatrix} s \\ 0 \end{pmatrix}\Big|_0^t$$

$$= \begin{pmatrix} c_1\cosh t + c_2\sinh t \\ c_1\sinh t + c_2\cosh t \end{pmatrix} + \begin{pmatrix} \cosh t & \sinh t \\ \sinh t & \cosh t \end{pmatrix}\begin{pmatrix} t \\ 0 \end{pmatrix}$$

$$= \begin{pmatrix} c_1\cosh t + c_2\sinh t \\ c_1\sinh t + c_2\cosh t \end{pmatrix} + \begin{pmatrix} t\cosh t \\ t\sinh t \end{pmatrix} = c_1\begin{pmatrix} \cosh t \\ \sinh t \end{pmatrix} + c_2\begin{pmatrix} \sinh t \\ \cosh t \end{pmatrix} + t\begin{pmatrix} \cosh t \\ \sinh t \end{pmatrix}.$$

15. From $s\mathbf{I} - \mathbf{A} = \begin{pmatrix} s - 4 & -3 \\ 4 & s + 4 \end{pmatrix}$ we find

$$(s\mathbf{I} - \mathbf{A})^{-1} = \begin{pmatrix} \dfrac{3/2}{s-2} - \dfrac{1/2}{s+2} & \dfrac{3/4}{s-2} - \dfrac{3/4}{s+2} \\[2ex] \dfrac{-1}{s-2} + \dfrac{1}{s+2} & \dfrac{-1/2}{s-2} + \dfrac{3/2}{s+2} \end{pmatrix}$$

and

$$e^{\mathbf{A}t} = \begin{pmatrix} \frac{3}{2}e^{2t} - \frac{1}{2}e^{-2t} & \frac{3}{4}e^{2t} - \frac{3}{4}e^{-2t} \\ -e^{2t} + e^{-2t} & -\frac{1}{2}e^{2t} + \frac{3}{2}e^{-2t} \end{pmatrix}.$$

The general solution of the system is then

$$\mathbf{X} = e^{\mathbf{A}t}\mathbf{C} = \begin{pmatrix} \frac{3}{2}e^{2t} - \frac{1}{2}e^{-2t} & \frac{3}{4}e^{2t} - \frac{3}{4}e^{-2t} \\ -e^{2t} + e^{-2t} & -\frac{1}{2}e^{2t} + \frac{3}{2}e^{-2t} \end{pmatrix} \begin{pmatrix} c_1 \\ c_2 \end{pmatrix}$$

$$= c_1 \begin{pmatrix} \frac{3}{2} \\ -1 \end{pmatrix} e^{2t} + c_1 \begin{pmatrix} -\frac{1}{2} \\ 1 \end{pmatrix} e^{-2t} + c_2 \begin{pmatrix} \frac{3}{4} \\ -\frac{1}{2} \end{pmatrix} e^{2t} + c_2 \begin{pmatrix} -\frac{3}{4} \\ \frac{3}{2} \end{pmatrix} e^{-2t}$$

$$= \left(\frac{1}{2}c_1 + \frac{1}{4}c_2\right) \begin{pmatrix} 3 \\ -2 \end{pmatrix} e^{2t} + \left(-\frac{1}{2}c_1 - \frac{3}{4}c_2\right) \begin{pmatrix} 1 \\ -2 \end{pmatrix} e^{-2t}$$

$$= c_3 \begin{pmatrix} 3 \\ -2 \end{pmatrix} e^{2t} + c_4 \begin{pmatrix} 1 \\ -2 \end{pmatrix} e^{-2t}.$$

18. From $s\mathbf{I} - \mathbf{A} = \begin{pmatrix} s & -1 \\ 2 & s+2 \end{pmatrix}$ we find

$$(s\mathbf{I} - \mathbf{A})^{-1} = \begin{pmatrix} \dfrac{s+1+1}{(s+1)^2 + 1} & \dfrac{1}{(s+1)^2 + 1} \\ \dfrac{-2}{(s+1)^2 + 1} & \dfrac{s+1-1}{(s+1)^2 + 1} \end{pmatrix}$$

and

$$e^{\mathbf{A}t} = \begin{pmatrix} e^{-t}\cos t + e^{-t}\sin t & e^{-t}\sin t \\ -2e^{-t}\sin t & e^{-t}\cos t - e^{-t}\sin t \end{pmatrix}.$$

The general solution of the system is then

$$\mathbf{X} = e^{\mathbf{A}t}\mathbf{C} = \begin{pmatrix} e^{-t}\cos t + e^{-t}\sin t & e^{-t}\sin t \\ -2e^{-t}\sin t & e^{-t}\cos t - e^{-t}\sin t \end{pmatrix} \begin{pmatrix} c_1 \\ c_2 \end{pmatrix}$$

$$= c_1 \begin{pmatrix} 1 \\ 0 \end{pmatrix} e^{-t}\cos t + c_1 \begin{pmatrix} 1 \\ -2 \end{pmatrix} e^{-t}\sin t + c_2 \begin{pmatrix} 0 \\ 1 \end{pmatrix} e^{-t}\cos t + c_2 \begin{pmatrix} 1 \\ -1 \end{pmatrix} e^{-t}\sin t \cdot$$

$$= c_1 \begin{pmatrix} \cos t + \sin t \\ -2\sin t \end{pmatrix} e^{-t} + c_2 \begin{pmatrix} \sin t \\ \cos t - \sin t \end{pmatrix} e^{-t}$$

21. The eigenvalues are $\lambda_1 = -1$ and $\lambda_2 = 3$. This leads to the system

$$e^{-t} = b_0 - b_1$$

$$e^{3t} = b_0 + 3b_1,$$

which has the solution $b_0 = \frac{3}{4}e^{-t} + \frac{1}{4}e^{3t}$ and $b_1 = -\frac{1}{4}e^{-t} + \frac{1}{4}e^{3t}$. Then

$$e^{\mathbf{A}t} = b_0\mathbf{I} + b_1\mathbf{A} = \begin{pmatrix} e^{3t} & -2e^{-t} + 2e^{3t} \\ 0 & e^{-t} \end{pmatrix}.$$

The general solution of the system is then

$$\mathbf{X} = e^{\mathbf{A}t}\mathbf{C} = \begin{pmatrix} e^{3t} & -2e^{-t} + 2e^{3t} \\ 0 & e^{-t} \end{pmatrix} \begin{pmatrix} c_1 \\ c_2 \end{pmatrix}$$

$$= c_1 \begin{pmatrix} 1 \\ 0 \end{pmatrix} e^{3t} + c_2 \begin{pmatrix} -2 \\ 1 \end{pmatrix} e^{-t} + c_2 \begin{pmatrix} 2 \\ 0 \end{pmatrix} e^{3t}$$

$$= c_2 \begin{pmatrix} -2 \\ 1 \end{pmatrix} e^{-t} + (c_1 + 2c_2) \begin{pmatrix} 1 \\ 0 \end{pmatrix} e^{3t}$$

$$= c_3 \begin{pmatrix} -2 \\ 1 \end{pmatrix} e^{-t} + c_4 \begin{pmatrix} 1 \\ 0 \end{pmatrix} e^{3t}.$$

24. From equation (3) in the text

$$e^{\mathbf{D}t} = \begin{pmatrix} 1 & 0 & \cdots & 0 \\ 0 & 1 & \cdots & 0 \\ \vdots & \vdots & \ddots & \vdots \\ 0 & 0 & \cdots & 1 \end{pmatrix} + \begin{pmatrix} \lambda_1 & 0 & \cdots & 0 \\ 0 & \lambda_2 & \cdots & 0 \\ \vdots & \vdots & \ddots & \vdots \\ 0 & 0 & \cdots & \lambda_n \end{pmatrix} + \frac{1}{2!}t^2 \begin{pmatrix} \lambda_1^2 & 0 & \cdots & 0 \\ 0 & \lambda_2^2 & \cdots & 0 \\ \vdots & \vdots & \ddots & \vdots \\ 0 & 0 & \cdots & \lambda_n^2 \end{pmatrix}$$

$$+ \frac{1}{3!}t^3 \begin{pmatrix} \lambda_1^3 & 0 & \cdots & 0 \\ 0 & \lambda_2^3 & \cdots & 0 \\ \vdots & \vdots & \ddots & \vdots \\ 0 & 0 & \cdots & \lambda_n^3 \end{pmatrix} + \cdots$$

$$= \begin{pmatrix} 1 + \lambda_1 t + \frac{1}{2!}(\lambda_1 t)^2 + \cdots & 0 & \cdots & 0 \\ 0 & 1 + \lambda_2 t + \frac{1}{2!}(\lambda_2 t)^2 + \cdots & \cdots & 0 \\ \vdots & \vdots & \ddots & \vdots \\ 0 & 0 & \cdots & 1 + \lambda_n t + \frac{1}{2!}(\lambda_n t)^2 + \cdots \end{pmatrix}$$

$$= \begin{pmatrix} e^{\lambda_1 t} & 0 & \cdots & 0 \\ 0 & e^{\lambda_2 t} & \cdots & 0 \\ \vdots & \vdots & \ddots & \vdots \\ 0 & 0 & \cdots & e^{\lambda_n t} \end{pmatrix}$$

27. (a) The following commands can be used in *Mathematica*:

```
A={{4, 2},{3, 3}};
c={c1, c2};
m=MatrixExp[A t];
sol=Expand[m.c]
Collect[sol, {c1, c2}]//MatrixForm
```

The output gives

$$x(t) = c_1 \left(\frac{2}{5}e^t + \frac{3}{5}e^{6t} \right) + c_2 \left(-\frac{2}{5}e^t + \frac{2}{5}e^{6t} \right)$$

$$y(t) = c_1 \left(-\frac{3}{5}e^t + \frac{3}{5}e^{6t} \right) + c_2 \left(\frac{3}{5}e^t + \frac{2}{5}e^{6t} \right).$$

The eigenvalues are 1 and 6 with corresponding eigenvectors

$$\begin{pmatrix} -2 \\ 3 \end{pmatrix} \quad \text{and} \quad \begin{pmatrix} 1 \\ 1 \end{pmatrix},$$

so the solution of the system is

$$\mathbf{X}(t) = b_1 \begin{pmatrix} -2 \\ 3 \end{pmatrix} e^t + b_2 \begin{pmatrix} 1 \\ 1 \end{pmatrix} e^{6t}$$

or

$$x(t) = -2b_1 e^t + b_2 e^{6t}$$
$$y(t) = 3b_1 e^t + b_2 e^{6t}.$$

If we replace b_1 with $-\frac{1}{5}c_1 + \frac{1}{5}c_2$ and b_2 with $\frac{3}{5}c_1 + \frac{2}{5}c_2$, we obtain the solution found using the matrix exponential.

(b) $x(t) = c_1 e^{-2t}\cos t - (c_1 + c_2)e^{-2t}\sin t$

$y(t) = c_2 e^{-2t}\cos t + (2c_1 + c_2)e^{-2t}\sin t$

30. Since $\mathbf{A}^3 = \mathbf{0}$, $\mathbf{A}$ is nilpotent. Since

$$e^{\mathbf{A}t} = \mathbf{I} + \mathbf{A}t + \mathbf{A}^2\frac{t^2}{2!} + \cdots + \mathbf{A}^k\frac{t^k}{k!} + \cdots,$$

if $\mathbf{A}$ is nilpotent and $\mathbf{A}^m = \mathbf{0}$, then $\mathbf{A}^k = \mathbf{0}$ for $k \geq m$ and

$$e^{\mathbf{A}t} = \mathbf{I} + \mathbf{A}t + \mathbf{A}^2\frac{t^2}{2!} + \cdots + \mathbf{A}^{m-1}\frac{t^{m-1}}{(m-1)!}.$$

In this problem $\mathbf{A}^3 = \mathbf{0}$, so

$$e^{\mathbf{A}t} = \mathbf{I} + \mathbf{A}t + \mathbf{A}^2\frac{t^2}{2} = \begin{pmatrix} 1 & 0 & 0 \\ 0 & 1 & 0 \\ 0 & 0 & 1 \end{pmatrix} + \begin{pmatrix} -1 & 1 & 1 \\ -1 & 0 & 1 \\ -1 & 1 & 1 \end{pmatrix}t + \begin{pmatrix} -1 & 0 & 1 \\ 0 & 0 & 0 \\ -1 & 0 & 1 \end{pmatrix}\frac{t^2}{2}$$

$$= \begin{pmatrix} 1 - t - t^2/2 & t & t + t^2/2 \\ -t & 1 & t \\ -t - t^2/2 & t & 1 + t + t^2/2 \end{pmatrix}$$

and the solution of $\mathbf{X}' = \mathbf{A}\mathbf{X}$ is

$$\mathbf{X}(t) = e^{\mathbf{A}t}\mathbf{C} = e^{\mathbf{A}t}\begin{pmatrix} c_1 \\ c_2 \\ c_3 \end{pmatrix} = \begin{pmatrix} c_1(1 - t - t^2/2) + c_2 t + c_3(t + t^2/2) \\ -c_1 t + c_2 + c_3 t \\ c_1(-t - t^2/2) + c_2 t + c_3(1 + t + t^2/2) \end{pmatrix}.$$

3. Since

$$\begin{pmatrix} 4 & 6 & 6 \\ 1 & 3 & 2 \\ -1 & -4 & -3 \end{pmatrix} \begin{pmatrix} 3 \\ 1 \\ -1 \end{pmatrix} = \begin{pmatrix} 12 \\ 4 \\ -4 \end{pmatrix} = 4 \begin{pmatrix} 3 \\ 1 \\ -1 \end{pmatrix},$$

we see that $\lambda = 4$ is an eigenvalue with eigenvector $\mathbf{K}_3$. The corresponding solution is $\mathbf{X}_3 = \mathbf{K}_3 e^{4t}$.

6. We have $\det(\mathbf{A} - \lambda\mathbf{I}) = (\lambda + 6)(\lambda + 2) = 0$ so that

$$\mathbf{X} = c_1 \begin{pmatrix} 1 \\ -1 \end{pmatrix} e^{-6t} + c_2 \begin{pmatrix} 1 \\ 1 \end{pmatrix} e^{-2t}.$$

9. We have $\det(\mathbf{A} - \lambda\mathbf{I}) = -(\lambda - 2)(\lambda - 4)(\lambda + 3) = 0$ so that

$$\mathbf{X} = c_1 \begin{pmatrix} -2 \\ 3 \\ 1 \end{pmatrix} e^{2t} + c_2 \begin{pmatrix} 0 \\ 1 \\ 1 \end{pmatrix} e^{4t} + c_3 \begin{pmatrix} 7 \\ 12 \\ -16 \end{pmatrix} e^{-3t}.$$

12. We have

$$\mathbf{X}_c = c_1 \begin{pmatrix} 2\cos t \\ -\sin t \end{pmatrix} e^t + c_2 \begin{pmatrix} 2\sin t \\ \cos t \end{pmatrix} e^t.$$

Then

$$\mathbf{\Phi} = \begin{pmatrix} 2\cos t & 2\sin t \\ -\sin t & \cos t \end{pmatrix} e^t, \quad \mathbf{\Phi}^{-1} = \begin{pmatrix} \frac{1}{2}\cos t & -\sin t \\ \frac{1}{2}\sin t & \cos t \end{pmatrix} e^{-t},$$

and

$$\mathbf{U} = \int \mathbf{\Phi}^{-1}\mathbf{F}\,dt = \int \begin{pmatrix} \cos t - \sec t \\ \sin t \end{pmatrix} dt = \begin{pmatrix} \sin t - \ln|\sec t + \tan t| \\ -\cos t \end{pmatrix},$$

so that

$$\mathbf{X}_p = \mathbf{\Phi}\mathbf{U} = \begin{pmatrix} -2\cos t \ln|\sec t + \tan t| \\ -1 + \sin t \ln|\sec t + \tan t| \end{pmatrix} e^t.$$

15. (a) Letting

$$\mathbf{K} = \begin{pmatrix} k_1 \\ k_2 \\ k_3 \end{pmatrix}$$

we note that $(\mathbf{A} - 2\mathbf{I})\mathbf{K} = \mathbf{0}$ implies that $3k_1 + 3k_2 + 3k_3 = 0$, so $k_1 = -(k_2 + k_3)$. Choosing $k_2 = 0$, $k_3 = 1$ and then $k_2 = 1$, $k_3 = 0$ we get

$$\mathbf{K}_1 = \begin{pmatrix} -1 \\ 0 \\ 1 \end{pmatrix} \quad \text{and} \quad \mathbf{K}_2 = \begin{pmatrix} -1 \\ 1 \\ 0 \end{pmatrix},$$

respectively. Thus,

$$\mathbf{X}_1 = \begin{pmatrix} -1 \\ 0 \\ 1 \end{pmatrix} e^{2t} \quad \text{and} \quad \mathbf{X}_2 = \begin{pmatrix} -1 \\ 1 \\ 0 \end{pmatrix} e^{2t}$$

are two solutions.

(b) From $\det(\mathbf{A} - \lambda \mathbf{I}) = \lambda^2(3 - \lambda) = 0$ we see that $\lambda_1 = 3$, and 0 is an eigenvalue of multiplicity two. Letting

$$\mathbf{K} = \begin{pmatrix} k_1 \\ k_2 \\ k_3 \end{pmatrix},$$

as in part **(a)**, we note that $(\mathbf{A} - 0\mathbf{I})\mathbf{K} = \mathbf{A}\mathbf{K} = \mathbf{0}$ implies that $k_1 + k_2 + k_3 = 0$, so $k_1 = -(k_2 + k_3)$. Choosing $k_2 = 0$, $k_3 = 1$, and then $k_2 = 1$, $k_3 = 0$ we get

$$\mathbf{K}_2 = \begin{pmatrix} -1 \\ 0 \\ 1 \end{pmatrix} \quad \text{and} \quad \mathbf{K}_3 = \begin{pmatrix} -1 \\ 1 \\ 0 \end{pmatrix},$$

respectively. Since the eigenvector corresponding to $\lambda_1 = 3$ is

$$\mathbf{K}_1 = \begin{pmatrix} 1 \\ 1 \\ 1 \end{pmatrix},$$

the general solution of the system is

$$\mathbf{X} = c_1 \begin{pmatrix} 1 \\ 1 \\ 1 \end{pmatrix} e^{3t} + c_2 \begin{pmatrix} -1 \\ 0 \\ 1 \end{pmatrix} + c_3 \begin{pmatrix} -1 \\ 1 \\ 0 \end{pmatrix}.$$

11 Systems of Nonlinear Differential Equations

3. The corresponding plane autonomous system is

$$x' = y, \quad y' = x^2 - y(1 - x^3).$$

If (x, y) is a critical point, $y = 0$ and so $x^2 - y(1 - x^3) = x^2 = 0$. Therefore $(0, 0)$ is the sole critical point.

6. The corresponding plane autonomous system is

$$x' = y, \quad y' = -x + \epsilon x|x|.$$

If (x, y) is a critical point, $y = 0$ and $-x + \epsilon x|x| = x(-1 + \epsilon|x|) = 0$. Hence $x = 0, 1/\epsilon, -1/\epsilon$. The critical points are $(0, 0)$, $(1/\epsilon, 0)$ and $(-1/\epsilon, 0)$.

9. From $x - y = 0$ we have $y = x$. Substituting into $3x^2 - 4y = 0$ we obtain $3x^2 - 4x = x(3x - 4) = 0$. It follows that $(0, 0)$ and $(4/3, 4/3)$ are the critical points of the system.

12. Adding the two equations we obtain $10 - 15y/(y + 5) = 0$. It follows that $y = 10$, and from $-2x + y + 10 = 0$ we can conclude that $x = 10$. Therefore $(10, 10)$ is the sole critical point of the system.

15. From $x(1 - x^2 - 3y^2) = 0$ we have $x = 0$ or $x^2 + 3y^2 = 1$. If $x = 0$, then substituting into $y(3 - x^2 - 3y^2)$ gives $y(3 - 3y^2) = 0$. Therefore $y = 0, 1, -1$. Likewise $x^2 = 1 - 3y^2$ yields $2y = 0$ so that $y = 0$ and $x^2 = 1 - 3(0)^2 = 1$. The critical points of the system are therefore $(0, 0)$, $(0, 1)$, $(0, -1)$, $(1, 0)$, and $(-1, 0)$.

18. **(a)** From Exercises 10.2, Problem 6, $x = c_1 + 2c_2 e^{-5t}$ and $y = 3c_1 + c_2 e^{-5t}$, which is not periodic.

(b) From $\mathbf{X}(0) = (3, 4)$ it follows that $c_1 = c_2 = 1$. Therefore $x = 1 + 2e^{-5t}$ and $y = 3 + e^{-5t}$ gives $y = \frac{1}{2}(x - 1) + 3$.

(c)

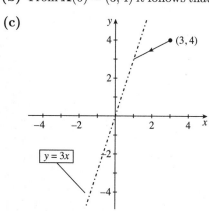

21. (a) From Exercises 10.2, Problem 35, $x = c_1(\sin t - \cos t)e^{4t} + c_2(-\sin t - \cos t)e^{4t}$ and $y = 2c_1(\cos t) e^{4t} + 2c_2(\sin t) e^{4t}$. Because of the presence of e^{4t}, there are no periodic solutions.

(b) From $\mathbf{X}(0) = (-1, 2)$ it follows that $c_1 = 1$ and $c_2 = 0$. Therefore $x = (\sin t - \cos t)e^{4t}$ and $y = 2(\cos t) e^{4t}$.

(c)

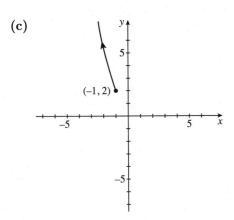

24. Switching to polar coordinates,

$$\frac{dr}{dt} = \frac{1}{r}\left(x\frac{dx}{dt} + y\frac{dy}{dt}\right) = \frac{1}{r}(xy - x^2r^2 - xy + y^2r^2) = r^3$$

$$\frac{d\theta}{dt} = \frac{1}{r^2}\left(-y\frac{dx}{dt} + x\frac{dy}{dt}\right) = \frac{1}{r^2}(-y^2 - xyr^2 - x^2 + xyr^2) = -1.$$

If we use separation of variables, it follows that

$$r = \frac{1}{\sqrt{-2t + c_1}} \quad \text{and} \quad \theta = -t + c_2.$$

Since $\mathbf{X}(0) = (4, 0)$, $r = 4$ and $\theta = 0$ when $t = 0$. It follows that $c_2 = 0$ and $c_1 = \frac{1}{16}$. The final solution can be written as

$$r = \frac{4}{\sqrt{1 - 32t}}, \quad \theta = -t.$$

Note that $r \to \infty$ as $t \to \left(\frac{1}{32}\right)$. Because $0 \le t \le \frac{1}{32}$, the curve is not a spiral.

27. The system has no critical points, so there are no periodic solutions.

30. The system has no critical points, so there are no periodic solutions.

EXERCISES 11.2

Stability of Linear Systems

3. (a) All solutions are unstable spirals which become unbounded as t increases.

(b)

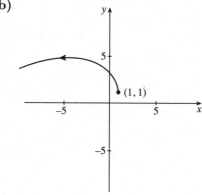

6. (a) All solutions become unbounded and $y = x/2$ serves as the asymptote.

(b)

9. Since $\Delta = -41 < 0$, we can conclude from Figure 11.18 that $(0,0)$ is a saddle point.

12. Since $\Delta = 1$ and $\tau = -1$, $\tau^2 - 4\Delta = -3$ and so from Figure 11.18, $(0,0)$ is a stable spiral point.

15. Since $\Delta = 0.01$ and $\tau = -0.03$, $\tau^2 - 4\Delta < 0$ and so from Figure 11.18, $(0,0)$ is a stable spiral point.

18. Note that $\Delta = 1$ and $\tau = \mu$. Therefore we need both $\tau = \mu < 0$ and $\tau^2 - 4\Delta = \mu^2 - 4 < 0$ for $(0,0)$ to be a stable spiral point. These two conditions can be written as $-2 < \mu < 0$.

21. $\mathbf{AX}_1 + \mathbf{F} = \mathbf{0}$ implies that $\mathbf{AX}_1 = -\mathbf{F}$ or $\mathbf{X}_1 = -\mathbf{A}^{-1}\mathbf{F}$. Since $\mathbf{X}_p(t) = -\mathbf{A}^{-1}\mathbf{F}$ is a particular solution, it follows from Theorem 8.6 that $\mathbf{X}(t) = \mathbf{X}_c(t) + \mathbf{X}_1$ is the general solution to $\mathbf{X}' = \mathbf{AX} + \mathbf{F}$. If $\tau < 0$ and $\Delta > 0$ then $\mathbf{X}_c(t)$ approaches $(0,0)$ by Theorem 11.1(a). It follows that $\mathbf{X}(t)$ approaches $\mathbf{X}_1$ as $t \to \infty$.

24. (a) The critical point is $\mathbf{X}_1 = (-1, -2)$.

(b) From the graph, $\mathbf{X}_1$ appears to be a stable node or a degenerate stable node.

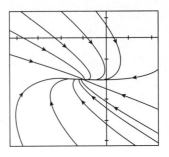

(c) Since $\tau = -16$, $\Delta = 64$, and $\tau^2 - 4\Delta = 0$, $(0, 0)$ is a degenerate stable node.

EXERCISES 11.3
Linearization and Local Stability

3. The critical points are $x = 0$ and $x = n + 1$. Since $g'(x) = k(n + 1) - 2kx$, $g'(0) = k(n + 1) > 0$ and $g'(n + 1) = -k(n + 1) < 0$. Therefore $x = 0$ is unstable while $x = n + 1$ is asymptotically stable. See Theorem 11.2.

6. The only critical point is $v = mg/k$. Now $g(v) = g - (k/m)v$ and so $g'(v) = -k/m < 0$. Therefore $v = mg/k$ is an asymptotically stable critical point by Theorem 11.2.

9. Critical points occur at $P = a/b$, c but not at $P = 0$. Since $g'(P) = (a - bP) + (P - c)(-b)$,

$$g'(a/b) = (a/b - c)(-b) = -a + bc \quad \text{and} \quad g'(c) = a - bc.$$

Since $a < bc$, $-a + bc > 0$ and $a - bc < 0$. Therefore $P = a/b$ is unstable while $P = c$ is asymptotically stable.

12. Critical points are $(1, 0)$ and $(-1, 0)$, and

$$\mathbf{g}'(\mathbf{X}) = \begin{pmatrix} 2x & -2y \\ 0 & 2 \end{pmatrix}.$$

At $\mathbf{X} = (1, 0)$, $\tau = 4$, $\Delta = 4$, and so $\tau^2 - 4\Delta = 0$. We can conclude that $(1, 0)$ is unstable but we are unable to classify this critical point any further. At $\mathbf{X} = (-1, 0)$, $\Delta = -4 < 0$ and so $(-1, 0)$ is a saddle point.

15. Since $x^2 - y^2 = 0$, $y^2 = x^2$ and so $x^2 - 3x + 2 = (x - 1)(x - 2) = 0$. It follows that the critical points are $(1, 1)$, $(1, -1)$, $(2, 2)$, and $(2, -2)$. We next use the Jacobian

$$\mathbf{g}'(\mathbf{X}) = \begin{pmatrix} -3 & 2y \\ 2x & -2y \end{pmatrix}$$

to classify these four critical points. For $\mathbf{X} = (1, 1)$, $\tau = -5$, $\Delta = 2$, and so $\tau^2 - 4\Delta = 17 > 0$. Therefore $(1, 1)$ is a stable node. For $\mathbf{X} = (1, -1)$, $\Delta = -2 < 0$ and so $(1, -1)$ is a saddle point. For $\mathbf{X} = (2, 2)$, $\Delta = -4 < 0$ and so we have another saddle point. Finally, if $\mathbf{X} = (2, -2)$, $\tau = 1$, $\Delta = 4$, and so $\tau^2 - 4\Delta = -15 < 0$. Therefore $(2, -2)$ is an unstable spiral point.

18. We found that $(0,0)$, $(0,1)$, $(0,-1)$, $(1,0)$ and $(-1,0)$ were the critical points in Problem 15, Section 11.1. The Jacobian is

$$\mathbf{g}'(\mathbf{X}) = \begin{pmatrix} 1 - 3x^2 - 3y^2 & -6xy \\ -2xy & 3 - x^2 - 9y^2 \end{pmatrix}.$$

For $\mathbf{X} = (0,0)$, $\tau = 4$, $\Delta = 3$ and so $\tau^2 - 4\Delta = 4 > 0$. Therefore $(0,0)$ is an unstable node. Both $(0,1)$ and $(0,-1)$ give $\tau = -8$, $\Delta = 12$, and $\tau^2 - 4\Delta = 16 > 0$. These two critical points are therefore stable nodes. For $\mathbf{X} = (1,0)$ or $(-1,0)$, $\Delta = -4 < 0$ and so saddle points occur.

21. The corresponding plane autonomous system is

$$\theta' = y, \quad y' = \left(\cos\theta - \frac{1}{2}\right)\sin\theta.$$

Since $|\theta| < \pi$, it follows that critical points are $(0,0)$, $(\pi/3, 0)$ and $(-\pi/3, 0)$. The Jacobian matrix is

$$\mathbf{g}'(\mathbf{X}) = \begin{pmatrix} 0 & 1 \\ \cos 2\theta - \frac{1}{2}\cos\theta & 0 \end{pmatrix}$$

and so at $(0,0)$, $\tau = 0$ and $\Delta = -1/2$. Therefore $(0,0)$ is a saddle point. For $\mathbf{X} = (\pm\pi/3, 0)$, $\tau = 0$ and $\Delta = 3/4$. It is not possible to classify either critical point in this borderline case.

24. The corresponding plane autonomous system is

$$x' = y, \quad y' = -\frac{4x}{1+x^2} - 2y$$

and the only critical point is $(0,0)$. Since the Jacobian matrix is

$$\mathbf{g}'(\mathbf{X}) = \begin{pmatrix} 0 & 1 \\ -4\dfrac{1-x^2}{(1+x^2)^2} & -2 \end{pmatrix},$$

$\tau = -2$, $\Delta = 4$, $\tau^2 - 4\Delta = -12$, and so $(0,0)$ is a stable spiral point.

27. The corresponding plane autonomous system is

$$x' = y, \quad y' = -\frac{(\beta + \alpha^2 y^2)x}{1 + \alpha^2 x^2}$$

and the Jacobian matrix is

$$\mathbf{g}'(\mathbf{X}) = \begin{pmatrix} 0 & 1 \\ \dfrac{(\beta + \alpha y^2)(\alpha^2 x^2 - 1)}{(1 + \alpha^2 x^2)^2} & \dfrac{-2\alpha^2 yx}{1 + \alpha^2 x^2} \end{pmatrix}.$$

For $\mathbf{X} = (0,0)$, $\tau = 0$ and $\Delta = \beta$. Since $\beta < 0$, we can conclude that $(0,0)$ is a saddle point.

30. (a) The corresponding plane autonomous system is

$$x' = y, \quad y' = \epsilon\left(y - \frac{1}{3}y^3\right) - x$$

and so the only critical point is $(0,0)$. Since the Jacobian matrix is

$$\mathbf{g}'(\mathbf{X}) = \begin{pmatrix} 0 & 1 \\ -1 & \epsilon(1 - y^2) \end{pmatrix},$$

$\tau = \epsilon$, $\Delta = 1$, and so $\tau^2 - 4\Delta = \epsilon^2 - 4$ at the critical point $(0,0)$.

(b) When $\tau = \epsilon > 0$, $(0,0)$ is an unstable critical point.

(c) When $\epsilon < 0$ and $\tau^2 - 4\Delta = \epsilon^2 - 4 < 0$, $(0,0)$ is a stable spiral point. These two requirements can be written as $-2 < \epsilon < 0$.

(d) When $\epsilon = 0$, $x'' + x = 0$ and so $x = c_1 \cos t + c_2 \sin t$. Therefore all solutions are periodic (with period 2π) and so $(0,0)$ is a center.

33. (a) $x' = 2xy = 0$ implies that either $x = 0$ or $y = 0$. If $x = 0$, then from $1 - x^2 + y^2 = 0$, $y^2 = -1$ and there are no real solutions. If $y = 0$, $1 - x^2 = 0$ and so $(1,0)$ and $(-1,0)$ are critical points. The Jacobian matrix is

$$\mathbf{g}'(\mathbf{X}) = \begin{pmatrix} 2y & 2x \\ -2x & 2y \end{pmatrix}$$

and so $\tau = 0$ and $\Delta = 4$ at either $\mathbf{X} = (1,0)$ or $(-1,0)$. We obtain no information about these critical points in this borderline case.

(b) The differential equation is

$$\frac{dy}{dx} = \frac{y'}{x'} = \frac{1 - x^2 + y^2}{2xy}$$

or

$$2xy \frac{dy}{dx} = 1 - x^2 + y^2.$$

Letting $\mu = y^2/x$, it follows that $d\mu/dx = (1/x^2) - 1$ and so $\mu = -(1/x) - x + 2c$. Therefore $y^2/x = -(1/x) - x + 2c$ which can be put in the form $(x - c)^2 + y^2 = c^2 - 1$. The solution curves are shown and so both $(1,0)$ and $(-1,0)$ are centers.

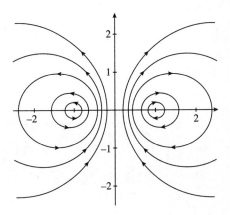

36. The corresponding plane autonomous system is

$$x' = y, \quad y' = \epsilon x^2 - x + 1$$

and so the critical points must satisfy $y = 0$ and

$$x = \frac{1 \pm \sqrt{1 - 4\epsilon}}{2\epsilon}.$$

Therefore we must require that $\epsilon \leq \frac{1}{4}$ for real solutions to exist. We will use the Jacobian matrix

$$\mathbf{g}'(\mathbf{X}) = \begin{pmatrix} 0 & 1 \\ 2\epsilon x - 1 & 0 \end{pmatrix}$$

to attempt to classify $((1 \pm \sqrt{1 - 4\epsilon})/2\epsilon, 0)$ when $\epsilon \leq 1/4$. Note that $\tau = 0$ and $\Delta = \mp\sqrt{1 - 4\epsilon}$. For $\mathbf{X} = ((1 + \sqrt{1 - 4\epsilon})/2\epsilon, 0)$ and $\epsilon < 1/4$, $\Delta < 0$ and so a saddle point occurs. For $\mathbf{X} = ((1 - \sqrt{1 - 4\epsilon})/2\epsilon, 0)$, $\Delta \geq 0$ and we are not able to classify this critical point using linearization.

39. (a) Letting $x = \theta$ and $y = x'$ we obtain the system $x' = y$ and $y' = 1/2 - \sin x$. Since $\sin \pi/6 = \sin 5\pi/6 = 1/2$ we see that $(\pi/6, 0)$ and $(5\pi/6, 0)$ are critical points of the system.

(b) The Jacobian matrix is

$$\mathbf{g}'(\mathbf{X}) = \begin{pmatrix} 0 & 1 \\ -\cos x & 0 \end{pmatrix}$$

and so

$$\mathbf{A}_1 = \mathbf{g}' = ((\pi/6, 0)) = \begin{pmatrix} 0 & 1 \\ -\sqrt{3}/2 & 0 \end{pmatrix} \quad \text{and} \quad \mathbf{A}_2 = \mathbf{g}' = ((5\pi/6, 0)) = \begin{pmatrix} 0 & 1 \\ \sqrt{3}/2 & 0 \end{pmatrix}.$$

Since $\det \mathbf{A}_1 > 0$ and the trace of $\mathbf{A}_1$ is 0, no conclusion can be drawn regarding the critical point $(\pi/6, 0)$. Since $\det \mathbf{A}_2 < 0$, we see that $(5\pi/6, 0)$ is a saddle point.

(c) From the system in part **(a)** we obtain the first-order differential equation

$$\frac{dy}{dx} = \frac{1/2 - \sin x}{y}.$$

Separating variables and integrating we obtain

$$\int y \, dy = \int \left(\frac{1}{2} - \sin x \right) dx$$

and

$$\frac{1}{2}y^2 = \frac{1}{2}x + \cos x + c_1$$

or

$$y^2 = x + 2\cos x + c_2.$$

For x_0 near $\pi/6$, if $\mathbf{X}(0) = (x_0, 0)$ then $c_2 = -x_0 - 2\cos x_0$ and $y^2 = x + 2\cos x - x_0 - 2\cos x_0$. Thus, there are two values of y for each x in a sufficiently small interval around $\pi/6$. Therefore $(\pi/6, 0)$ is a center.

EXERCISES 11.4

Autonomous Systems as Mathematical Models

3. The corresponding plane autonomous system is

$$x' = y, \quad y' = -g\frac{f'(x)}{1 + [f'(x)]^2} - \frac{\beta}{m}y$$

and

$$\frac{\partial}{\partial x}\left(-g\frac{f'(x)}{1 + [f'(x)]^2} - \frac{\beta}{m}y \right) = -g\frac{(1 + [f'(x)]^2)f''(x) - f'(x)2f'(x)f''(x)}{(1 + [f'(x)]^2)^2}.$$

If $\mathbf{X}_1 = (x_1, y_1)$ is a critical point, $y_1 = 0$ and $f'(x_1) = 0$. The Jacobian at this critical point is therefore

$$\mathbf{g}'(\mathbf{X}_1) = \begin{pmatrix} 0 & 1 \\ -gf''(x_1) & -\dfrac{\beta}{m} \end{pmatrix}.$$

6. (a) If $f(x) = \cosh x$, $f'(x) = \sinh x$ and $[f'(x)]^2 + 1 = \sinh^2 x + 1 = \cosh^2 x$. Therefore

$$\frac{dy}{dx} = \frac{y'}{x'} = -g\frac{\sinh x}{\cosh^2 x}\frac{1}{y}.$$

We can separate variables to show that $y^2 = 2g/\cosh x + c$. But $x(0) = x_0$ and $y(0) = x'(0) = v_0$. Therefore $c = v_0^2 - (2g/\cosh x_0)$ and so

$$y^2 = \frac{2g}{\cosh x} - \frac{2g}{\cosh x_0} + v_0^2.$$

Now

$$\frac{2g}{\cosh x} - \frac{2g}{\cosh x_0} + v_0^2 \geq 0 \quad \text{if and only if} \quad \cosh x \leq \frac{2g\cosh x_0}{2g - v_0^2 \cosh x_0}$$

and the solution to this inequality is an interval $[-a, a]$. Therefore each x in $(-a, a)$ has two corresponding values of y and so the solution is periodic.

(b) Since $z = \cosh x$, the maximum height occurs at the largest value of x on the cycle. From (a), $x_{\max} = a$ where $\cosh a = 2g \cosh x_0/(2g - v_0^2 \cosh x_0)$. Therefore

$$z_{\max} = \frac{2g \cosh x_0}{2g - v_0^2 \cosh x_0}.$$

9. (a) In the Lotka-Volterra Model the average number of predators is d/c and the average number of prey is a/b. But

$$x' = -ax + bxy - \epsilon_1 x = -(a + \epsilon_1)x + bxy$$
$$y' = -cxy + dy - \epsilon_2 y = -cxy + (d - \epsilon_2)y$$

and so the new critical point in the first quadrant is $(d/c - \epsilon_2/c, a/b + \epsilon_1/b)$.

(b) The average number of predators $d/c - \epsilon_2/c$ has decreased while the average number of prey $a/b + \epsilon_1/b$ has increased. The fishery science model is consistent with Volterra's principle.

12. $\Delta = r_1 r_2$, $\tau = r_1 + r_2$ and $\tau^2 - 4\Delta = (r_1 + r_2)^2 - 4r_1 r_2 = (r_1 - r_2)^2$. Therefore when $r_1 \neq r_2$, $(0,0)$ is an unstable node.

15. $K_1/\alpha_{12} < K_2 < K_1 \alpha_{21}$ and so $\alpha_{12}\alpha_{21} > 1$. Therefore $\Delta = (1 - \alpha_{12}\alpha_{21})\hat{x}\hat{y}\, r_1 r_2/K_1 K_2 < 0$ and so $(\hat{x}, \hat{y})$ is a saddle point.

18. (a) The magnitude of the frictional force between the bead and the wire is $\mu(mg \cos \theta)$ for some $\mu > 0$. The component of this frictional force in the x-direction is

$$(\mu mg \cos \theta) \cos \theta = \mu mg \cos^2 \theta.$$

But

$$\cos \theta = \frac{1}{\sqrt{1 + [f'(x)]^2}} \quad \text{and so} \quad \mu mg \cos^2 \theta = \frac{\mu mg}{1 + [f'(x)]^2}.$$

It follows from Newton's Second Law that

$$mx'' = -mg \frac{f'(x)}{1 + [f'(x)]^2} - \beta x' + mg \frac{\mu}{1 + [f'(x)]^2}$$

and so

$$x'' = g \frac{\mu - f'(x)}{1 + [f'(x)]^2} - \frac{\beta}{m} x'.$$

(b) A critical point (x, y) must satisfy $y = 0$ and $f'(x) = \mu$. Therefore critical points occur at $(x_1, 0)$ where $f'(x_1) = \mu$. The Jacobian matrix of the plane autonomous system is

$$g'(\mathbf{X}) = \begin{pmatrix} 0 & 1 \\ g \dfrac{(1 + [f'(x)]^2)(-f''(x)) - (\mu - f'(x))2f'(x)f''(x)}{(1 + [f'(x)]^2)^2} & -\dfrac{\beta}{m} \end{pmatrix}$$

and so at a critical point $\mathbf{X}_1$,

$$g'(\mathbf{X}) = \begin{pmatrix} 0 & 1 \\ \dfrac{-gf''(x_1)}{1 + \mu^2} & -\dfrac{\beta}{m} \end{pmatrix}.$$

Therefore $\tau = -\beta/m < 0$ and $\Delta = gf''(x_1)/(1 + \mu^2)$. When $f''(x_1) < 0$, $\Delta < 0$ and so a saddle point occurs. When $f''(x_1) > 0$ and

$$\tau^2 - 4\Delta = \frac{\beta^2}{m^2} - 4g \frac{f''(x_1)}{1 + \mu^2} < 0,$$

$(x_1, 0)$ is a stable spiral point. This condition can also be written as

$$\beta^2 < 4gm^2 \frac{f''(x_1)}{1 + \mu^2}.$$

21. The equation

$$x' = \alpha \frac{y}{1 + y} x - x = x \left(\frac{\alpha y}{1 + y} - 1 \right) = 0$$

implies that $x = 0$ or $y = 1/(\alpha - 1)$. When $\alpha > 0$, $\hat{y} = 1/(\alpha - 1) > 0$. If $x = 0$, then from the differential equation for y', $y = \beta$. On the other hand, if $\hat{y} = 1/(\alpha - 1)$, $\hat{y}/(1 + \hat{y}) = 1/\alpha$ and so $\hat{x}/\alpha - 1/(\alpha - 1) + \beta = 0$. It follows that

$$\hat{x} = \alpha \left(\beta - \frac{1}{\alpha - 1} \right) = \frac{\alpha}{\alpha - 1} [(\alpha - 1)\beta - 1]$$

and if $\beta(\alpha - 1) > 1$, $\hat{x} > 0$. Therefore $(\hat{x}, \hat{y})$ is the unique critical point in the first quadrant. The Jacobian matrix is

$$\mathbf{g}'(\mathbf{X}) = \begin{pmatrix} \alpha \dfrac{y}{y + 1} - 1 & \dfrac{\alpha x}{(1 + y)^2} \\ -\dfrac{y}{1 + y} & \dfrac{-x}{(1 + y)^2} - 1 \end{pmatrix}$$

and for $\mathbf{X} = (\hat{x}, \hat{y})$, the Jacobian can be written in the form

$$\mathbf{g}'((\hat{x}, \hat{y})) = \begin{pmatrix} 0 & \dfrac{(\alpha - 1)^2}{\alpha} \hat{x} \\ -\dfrac{1}{\alpha} & -\dfrac{(\alpha - 1)^2}{\alpha^2} - 1 \end{pmatrix}.$$

It follows that

$$\tau = -\left[\frac{(\alpha - 1)^2}{\alpha^2} \hat{x} + 1 \right] < 0, \quad \Delta = \frac{(\alpha - 1)^2}{\alpha^2} \hat{x}$$

and so $\tau = -(\Delta + 1)$. Therefore $\tau^2 - 4\Delta = (\Delta + 1)^2 - 4\Delta = (\Delta - 1)^2 > 0$. Therefore $(\hat{x}, \hat{y})$ is a stable node.

EXERCISES 11.5
Periodic Solutions, Limit Cycles, and Global Stability

3. For $P = -x + y^2$ and $Q = x - y$, $\dfrac{\partial P}{\partial x} + \dfrac{\partial Q}{\partial y} = -2 < 0$. Therefore there are no periodic solutions by Theorem 11.5.

6. From $y' = xy - y = y(x - 1) = 0$ either $y = 0$ or $x = 1$. If $y = 0$, then from $2x + y^2 = 0$, $x = 0$. Likewise $x = 1$ implies that $2 + y^2 = 0$, which has no real solutions. Therefore $(0, 0)$ is the only critical point. But $\mathbf{g}'((0, 0))$ has determinant $\Delta = -2$. The single critical point is a saddle point and so, by the corollary to Theorem 11.4, there are no periodic solutions.

9. For $\delta(x, y) = e^{ax + by}$, $\dfrac{\partial}{\partial x}(\delta P) + \dfrac{\partial}{\partial y}(\delta Q)$ can be simplified to

$$e^{ax + by}[-bx^2 - 2ax + axy + (2b + 1)y].$$

Setting $a = 0$ and $b = -1/2$,

$$\frac{\partial}{\partial}(\delta P) + \frac{\partial}{\partial y}(\delta Q) = \frac{1}{2}x^2 e^{-\frac{1}{2}y}$$

which does not change signs. Therefore by Theorem 11.6 there are no periodic solutions.

12. The corresponding plane autonomous system is $x' = y$, $y' = g(x, y)$ and so

$$\frac{\partial P}{\partial x} + \frac{\partial Q}{\partial y} = \frac{\partial g}{\partial y} = \frac{\partial g}{\partial x'} \neq 0$$

in the region R. Therefore $\dfrac{\partial P}{\partial x} + \dfrac{\partial Q}{\partial y}$ cannot change signs and so there are no periodic solutions by Theorem 11.5.

15. If $\mathbf{n} = (-2x, -2y)$,

$$\mathbf{V} \cdot \mathbf{n} = 2x^2 - 4xy + 2y^2 + 2y^4 = 2(x - y)^2 + 2y^4 \geq 0.$$

Therefore $x^2 + y^2 \leq r$ serves as an invariant region for any $r > 0$ by Theorem 11.7.

18. The corresponding plane autonomous system is

$$x' = y, \quad y' = y(1 - 3x^2 - 2y^2) - x$$

and it is easy to see that $(0, 0)$ is the only critical point. If $\mathbf{n} = (-2x, -2y)$ then

$$\mathbf{V} \cdot \mathbf{n} = -2xy - 2y^2(1 - 3x^2 - 2y^2) + 2xy = -2y^2(1 - 2r^2 - x^2).$$

If $r = \frac{1}{2}\sqrt{2}$, $2r^2 = 1$ and so $\mathbf{V} \cdot \mathbf{n} = 2x^2y^2 \geq 0$. Therefore $\frac{1}{4} \leq x^2 + y^2 \leq \frac{1}{2}$ serves as an invariant region. By Theorem 11.8(ii) there is at least one periodic solution.

21. (a) $\dfrac{\partial P}{\partial x} + \dfrac{\partial Q}{\partial y} = 2xy - 1 - x^2 \leq 2x - 1 - x^2 = -(x - 1)^2 \leq 0$. Therefore there are no periodic solutions.

(b) If (x, y) is a critical point, $x^2 y = \frac{1}{2}$ and so from $x' = x^2 y - x + 1$, $\frac{1}{2} - x + 1 = 0$. Therefore $x = 3/2$ and so $y = 2/9$. For this critical point, $\tau = -31/12 < 0$, $\Delta = 9/4 > 0$, and $\tau^2 - 4\Delta < 0$. Therefore $(3/2, 2/9)$ is a stable spiral point and so, from Theorem 11.9(ii), $\lim_{x \to \infty} \mathbf{X}(t) = (3/2, 2/9)$.

CHAPTER 11 REVIEW EXERCISES

3. A center or a saddle point

6. True

9. True

12. (a) If $\mathbf{X}(0) = \mathbf{X}_0$ lies on the line $y = -2x$, then $\mathbf{X}(t)$ approaches $(0, 0)$ along this line. For all other initial conditions, $\mathbf{X}(t)$ approaches $(0, 0)$ from the direction determined by the line $y = x$.

(b) If $\mathbf{X}(0) = \mathbf{X}_0$ lies on the line $y = -x$, then $\mathbf{X}(t)$ approaches $(0, 0)$ along this line. For all other initial conditions, $\mathbf{X}(t)$ becomes unbounded and $y = 2x$ serves as an asymptote.

15. The corresponding plane autonomous system is $x' = y$, $y' = \mu(1 - x^2) - x$ and so the Jacobian at the critical point $(0,0)$ is

$$\mathbf{g}'((0,0)) = \begin{pmatrix} 0 & 1 \\ -1 & \mu \end{pmatrix}.$$

Therefore $\tau = \mu$, $\Delta = 1$ and $\tau^2 - 4\Delta = \mu^2 - 4$. Now $\mu^2 - 4 < 0$ if and only if $-2 < \mu < 2$. We may therefore conclude that $(0,0)$ is a stable node for $\mu < -2$, a stable spiral point for $-2 < \mu < 0$, an unstable spiral point for $0 < \mu < 2$, and an unstable node for $\mu > 2$.

18. The corresponding plane autonomous system is

$$x' = y, \quad y' = -\frac{\beta}{m}y - \frac{k}{m}(s+x)^3 + g$$

and so the Jacobian is

$$\mathbf{g}'(\mathbf{X}) = \begin{pmatrix} 0 & 1 \\ -\frac{3k}{m}(s+x)^2 & -\frac{\beta}{m} \end{pmatrix}.$$

For $\mathbf{X} = (0,0)$, $\tau = -\dfrac{\beta}{m} < 0$, $\Delta = \dfrac{3k}{m}s^2 > 0$. Therefore

$$\tau^2 - 4\Delta = \frac{\beta^2}{m^2} - \frac{12k}{m}s^2 = \frac{1}{m^2}(\beta^2 - 12kms^2).$$

Therefore $(0,0)$ is a stable node if $\beta^2 > 12kms^2$ and a stable spiral point provided $\beta^2 < 12kms^2$, where $ks^3 = mg$.

21. (a) If $x = \theta$ and $y = x' = \theta'$, the corresponding plane autonomous system is

$$x' = y, \quad y' = \omega^2 \sin x \cos x - \frac{g}{l}\sin x - \frac{\beta}{ml}y.$$

Therefore $\dfrac{\partial P}{\partial x} + \dfrac{\partial Q}{\partial y} = -\dfrac{\beta}{ml} < 0$ and so there are no periodic solutions.

(b) If (x, y) is a critical point, $y = 0$ and so $\sin x(\omega^2 \cos x - g/l) = 0$. Either $\sin x = 0$ (in which case $x = 0$) of $\cos x = g/\omega^2 l$. But if $\omega^2 < g/l$, $g/\omega^2 l > 1$ and so the latter equation has no real solutions. Therefore $(0,0)$ is the only critical point if $\omega^2 < g/l$. The Jacobian matrix is

$$\mathbf{g}'(\mathbf{X}) = \begin{pmatrix} 0 & 1 \\ \omega^2 \cos 2x - \frac{g}{l}\cos x & -\frac{\beta}{ml} \end{pmatrix}$$

and so $\tau = -\beta/ml < 0$ and $\Delta = g/l - \omega^2 > 0$ for $\mathbf{X} = (0,0)$. It follows that $(0,0)$ is asymptotically stable and so after a small displacement, the pendulum will return to $\theta = 0$, $\theta' = 0$.

(c) If $\omega^2 > g/l$, $\cos x = g/\omega^2 l$ will have two solutions $x = \pm\hat{x}$ that satisfy $-\pi < x < \pi$. Therefore $(\pm\hat{x}, 0)$ are two additional critical points. If $\mathbf{X}_1 = (0, 0)$, $\Delta = g/l - \omega^2 < 0$ and so $(0, 0)$ is a saddle point. If $\mathbf{X}_1 = (\pm\hat{x}, 0)$, $\tau = -\beta/ml < 0$ and

$$\Delta = \frac{g}{l} \cos\hat{x} - \omega^2 \cos 2\hat{x} = \frac{g^2}{\omega^2 l^2} - \omega^2 \left(2\frac{g^2}{\omega^4 l^2} - 1\right) = \omega^2 - \frac{g^2}{\omega^2 l^2} > 0.$$

Therefore $(\hat{x}, 0)$ and $(-\hat{x}, 0)$ are each stable. When $\theta(0) = \theta_0$, $\theta'(0) = 0$ and θ_0 is small we expect the pendulum to reach one of these two stable equilibrium positions.

(d) In (b), $(0, 0)$ is a stable spiral point provided

$$\tau^2 - 4\Delta = \frac{\beta^2}{m^2 l^2} - 4\left(\frac{g}{l} - \omega^2\right) < 0.$$

This condition is equivalent to $\beta < 2ml\sqrt{g/l - \omega^2}$. In (c), $(\pm\hat{x}, 0)$ are stable spiral points provided that

$$\tau^2 - 4\Delta = \frac{\beta^2}{m^2 l^2} - 4\left(\omega^2 - \frac{g^2}{\omega^2 l^2}\right) < 0.$$

This condition is equivalent to $\beta < 2ml\sqrt{\omega^2 - g^2/(\omega^2 l^2)}$.

12 Orthogonal Functions and Fourier Series

EXERCISES 12.1

Orthogonal Functions

3. $\displaystyle\int_0^2 e^x(xe^{-x} - e^{-x})dx = \int_0^2 (x-1)dx = \left(\frac{1}{2}x^2 - x\right)\bigg|_0^2 = 0$

6. $\displaystyle\int_{\pi/4}^{5\pi/4} e^x \sin x\, dx = \left(\frac{1}{2}e^x \sin x - \frac{1}{2}e^x \cos x\right)\bigg|_{\pi/4}^{5\pi/4} = 0$

9. For $m \neq n$

$$\int_0^\pi \sin nx \sin mx\, dx = \frac{1}{2}\int_0^\pi \Big(\cos(n-m)x - \cos(n+m)x\Big)\, dx$$

$$= \frac{1}{2(n-m)}\sin(n-m)x\bigg|_0^\pi - \frac{1}{2(n+m)}\sin(n+m)x\bigg|_0^\pi = 0.$$

For $m = n$

$$\int_0^\pi \sin^2 nx\, dx = \int_0^\pi \left(\frac{1}{2} - \frac{1}{2}\cos 2nx\right)dx = \frac{1}{2}x\bigg|_0^\pi - \frac{1}{4n}\sin 2nx\bigg|_0^\pi = \frac{\pi}{2}$$

so that

$$\|\sin nx\| = \sqrt{\frac{\pi}{2}}\,.$$

12. For $m \neq n$, we use Problems 11 and 10:

$$\int_{-p}^{p} \cos\frac{n\pi}{p}x\, \cos\frac{m\pi}{p}x\, dx = 2\int_0^p \cos\frac{n\pi}{p}x\, \cos\frac{m\pi}{p}x\, dx = 0$$

$$\int_{-p}^{p} \sin\frac{n\pi}{p}x\, \sin\frac{m\pi}{p}x\, dx = 2\int_0^p \sin\frac{n\pi}{p}x\, \sin\frac{m\pi}{p}x\, dx = 0.$$

Also

$$\int_{-p}^{p} \sin\frac{n\pi}{p}x\, \cos\frac{m\pi}{p}x\, dx = \frac{1}{2}\int_{-p}^{p}\left(\sin\frac{(n-m)\pi}{p}x + \sin\frac{(n+m)\pi}{p}x\right)dx = 0,$$

$$\int_{-p}^{p} 1\cdot\cos\frac{n\pi}{p}x\, dx = \frac{p}{n\pi}\sin\frac{n\pi}{p}x\bigg|_{-p}^{p} = 0,$$

$$\int_{-p}^{p} 1\cdot\sin\frac{n\pi}{p}x\, dx = -\frac{p}{n\pi}\cos\frac{n\pi}{p}x\bigg|_{-p}^{p} = 0,$$

and

$$\int_{-p}^{p} \sin \frac{n\pi}{p} x \cos \frac{n\pi}{p} x \, dx = \int_{-p}^{p} \frac{1}{2} \sin \frac{2n\pi}{p} x \, dx = -\frac{p}{4n\pi} \cos \frac{2n\pi}{p} x \Big|_{-p}^{p} = 0.$$

For $m = n$

$$\int_{-p}^{p} \cos^2 \frac{n\pi}{p} x \, dx = \int_{-p}^{p} \left(\frac{1}{2} + \frac{1}{2} \cos \frac{2n\pi}{p} x \right) dx = p,$$

$$\int_{-p}^{p} \sin^2 \frac{n\pi}{p} x \, dx = \int_{-p}^{p} \left(\frac{1}{2} - \frac{1}{2} \cos \frac{2n\pi}{p} x \right) dx = p,$$

and

$$\int_{-p}^{p} 1^2 dx = 2p$$

so that

$$\|1\| = \sqrt{2p}, \quad \left\| \cos \frac{n\pi}{p} x \right\| = \sqrt{p}, \quad \text{and} \quad \left\| \sin \frac{n\pi}{p} x \right\| = \sqrt{p}.$$

15. By orthogonality $\int_a^b \phi_0(x)\phi_n(x)dx = 0$ for $n = 1, 2, 3, \ldots$; that is, $\int_a^b \phi_n(x)dx = 0$ for $n = 1, 2, 3, \ldots$.

18. Setting

$$0 = \int_{-2}^{2} f_3(x)f_1(x) \, dx = \int_{-2}^{2} \left(x^2 + c_1 x^3 + c_2 x^4 \right) dx = \frac{16}{3} + \frac{64}{5} c_2$$

and

$$0 = \int_{-2}^{2} f_3(x)f_2(x) \, dx = \int_{-2}^{2} \left(x^3 + c_1 x^4 + c_2 x^5 \right) dx = \frac{64}{5} c_1$$

we obtain $c_1 = 0$ and $c_2 = -5/12$.

21. (a) The fundamental period is $2\pi/2\pi = 1$.

(b) The fundamental period is $2\pi/(4/L) = \frac{1}{2}\pi L$.

(c) The fundamental period of $\sin x + \sin 2x$ is 2π.

(d) The fundamental period of $\sin 2x + \cos 4x$ is $2\pi/2 = \pi$.

(e) The fundamental period of $\sin 3x + \cos 4x$ is 2π since the smallest integer multiples of $2\pi/3$ and $2\pi/4 = \pi/2$ that are equal are 3 and 4, respectively.

(f) The fundamental period of $f(x)$ is $2\pi/(n\pi/p) = 2p/n$.

24. (i): $(f_1, f_2) = \int_a^b f_1(x)f_2(x)dx = \int_a^b f_2(x)f_1(x)dx = (f_2, f_1)$.

(ii): $(kf_1, f_2) = \int_a^b kf_1(x)f_2(x)dx = k\int_a^b f_1(x)f_2(x)dx = k(f_1, f_2)$.

(iii): If $f_1(x) = 0$ then $(f_1, f_1) = \int_a^b 0 \, dx = 0$; if $f_1(x) \neq 0$ then $(f_1, f_1) = \int_a^b [f_1(x)]^2 dx > 0$ since $[f_1(x)]^2 > 0$.

(iv): $(f_1 + f_2, f_3) = \int_a^b [f_1(x) + f_2(x)]f_3(x)dx = \int_a^b [f_1(x)f_3(x) + f_2(x)f_3(x)]dx$

$$= \int_a^b f_1(x)f_3(x)dx + \int_a^b f_2(x)f_3(x)dx = (f_1, f_3) + (f_2, f_3).$$

EXERCISES 12.2

Fourier Series

3. $a_0 = \displaystyle\int_{-1}^{1} f(x)\,dx = \int_{-1}^{0} 1\,dx + \int_{0}^{1} x\,dx = \dfrac{3}{2}$

$a_n = \displaystyle\int_{-1}^{1} f(x)\cos n\pi x\,dx = \int_{-1}^{0} \cos n\pi x\,dx + \int_{0}^{1} x\cos n\pi x\,dx = \dfrac{1}{n^2\pi^2}[(-1)^n - 1]$

$b_n = \displaystyle\int_{-1}^{1} f(x)\sin n\pi x\,dx = \int_{-1}^{0} \sin n\pi x\,dx + \int_{0}^{1} x\sin n\pi x\,dx = -\dfrac{1}{n\pi}$

$f(x) = \dfrac{3}{4} + \displaystyle\sum_{n=1}^{\infty}\left[\dfrac{(-1)^n - 1}{n^2\pi^2}\cos n\pi x - \dfrac{1}{n\pi}\sin n\pi x\right]$

6. $a_0 = \dfrac{1}{\pi}\displaystyle\int_{-\pi}^{\pi} f(x)\,dx = \dfrac{1}{\pi}\int_{-\pi}^{0} \pi^2\,dx + \dfrac{1}{\pi}\int_{0}^{\pi} \left(\pi^2 - x^2\right)\,dx = \dfrac{5}{3}\pi^2$

$a_n = \dfrac{1}{\pi}\displaystyle\int_{-\pi}^{\pi} f(x)\cos nx\,dx = \dfrac{1}{\pi}\int_{-\pi}^{0} \pi^2\cos nx\,dx + \dfrac{1}{\pi}\int_{0}^{\pi}\left(\pi^2 - x^2\right)\cos nx\,dx$

$\quad = \dfrac{1}{\pi}\left(\dfrac{\pi^2 - x^2}{n}\sin nx\,\Big|_{0}^{\pi} + \dfrac{2}{n}\int_{0}^{\pi} x\sin nx\,dx\right) = \dfrac{2}{n^2}(-1)^{n+1}$

$b_n = \dfrac{1}{\pi}\displaystyle\int_{-\pi}^{\pi} f(x)\sin nx\,dx = \dfrac{1}{\pi}\int_{-\pi}^{0} \pi^2\sin nx\,dx + \dfrac{1}{\pi}\int_{0}^{\pi}\left(\pi^2 - x^2\right)\sin nx\,dx$

$\quad = \dfrac{\pi}{n}[(-1)^n - 1] + \dfrac{1}{\pi}\left(\dfrac{x^2 - \pi^2}{n}\cos nx\,\Big|_{0}^{\pi} - \dfrac{2}{n}\int_{0}^{\pi} x\cos nx\,dx\right) = \dfrac{\pi}{n}(-1)^n + \dfrac{2}{n^3\pi}[1 - (-1)^n]$

$f(x) = \dfrac{5\pi^2}{6} + \displaystyle\sum_{n=1}^{\infty}\left[\dfrac{2}{n^2}(-1)^{n+1}\cos nx + \left(\dfrac{\pi}{n}(-1)^n + \dfrac{2[1-(-1)^n]}{n^3\pi}\right)\sin nx\right]$

9. $a_0 = \dfrac{1}{\pi}\displaystyle\int_{-\pi}^{\pi} f(x)\,dx = \dfrac{1}{\pi}\int_{0}^{\pi} \sin x\,dx = \dfrac{2}{\pi}$

$a_n = \dfrac{1}{\pi}\displaystyle\int_{-\pi}^{\pi} f(x)\cos nx\,dx = \dfrac{1}{\pi}\int_{0}^{\pi} \sin x\,\cos nx\,dx = \dfrac{1}{2\pi}\int_{0}^{\pi}\left(\sin(1+n)x + \sin(1-n)x\right)dx$

$\quad = \dfrac{1 + (-1)^n}{\pi(1 - n^2)}\quad \text{for } n = 2, 3, 4, \ldots$

$a_1 = \dfrac{1}{2\pi}\displaystyle\int_{0}^{\pi} \sin 2x\,dx = 0$

$b_n = \dfrac{1}{\pi}\displaystyle\int_{-\pi}^{\pi} f(x)\sin nx\,dx = \dfrac{1}{\pi}\int_{0}^{\pi} \sin x\,\sin nx\,dx$

$\quad = \dfrac{1}{2\pi}\displaystyle\int_{0}^{\pi}\left(\cos(1-n)x - \cos(1+n)x\right)dx = 0\quad \text{for } n = 2, 3, 4, \ldots$

$b_1 = \dfrac{1}{2\pi}\displaystyle\int_{0}^{\pi}(1 - \cos 2x)\,dx = \dfrac{1}{2}$

$f(x) = \dfrac{1}{\pi} + \dfrac{1}{2}\sin x + \displaystyle\sum_{n=2}^{\infty}\dfrac{1 + (-1)^n}{\pi(1 - n^2)}\cos nx$

12. $a_0 = \dfrac{1}{2} \displaystyle\int_{-2}^{2} f(x)\,dx = \dfrac{1}{2}\left(\int_{0}^{1} x\,dx + \int_{1}^{2} 1\,dx \right) = \dfrac{3}{4}$

$a_n = \dfrac{1}{2} \displaystyle\int_{-2}^{2} f(x) \cos\dfrac{n\pi}{2} x\,dx = \dfrac{1}{2}\left(\int_{0}^{1} x\cos\dfrac{n\pi}{2} x\,dx + \int_{1}^{2} \cos\dfrac{n\pi}{2} x\,dx \right) = \dfrac{2}{n^2\pi^2}\left(\cos\dfrac{n\pi}{2} - 1 \right)$

$b_n = \dfrac{1}{2} \displaystyle\int_{-2}^{2} f(x) \sin\dfrac{n\pi}{2} x\,dx = \dfrac{1}{2}\left(\int_{0}^{1} x\sin\dfrac{n\pi}{2} x\,dx + \int_{1}^{2} \sin\dfrac{n\pi}{2} x\,dx \right)$

$\qquad = \dfrac{2}{n^2\pi^2}\left(\sin\dfrac{n\pi}{2} + \dfrac{n\pi}{2}(-1)^{n+1} \right)$

$f(x) = \dfrac{3}{8} + \displaystyle\sum_{n=1}^{\infty}\left[\dfrac{2}{n^2\pi^2}\left(\cos\dfrac{n\pi}{2} - 1 \right)\cos\dfrac{n\pi}{2}x + \dfrac{2}{n^2\pi^2}\left(\sin\dfrac{n\pi}{2} + \dfrac{n\pi}{2}(-1)^{n+1} \right)\sin\dfrac{n\pi}{2}x \right]$

15. $a_0 = \dfrac{1}{\pi} \displaystyle\int_{-\pi}^{\pi} f(x)\,dx = \dfrac{1}{\pi}\int_{-\pi}^{\pi} e^x\,dx = \dfrac{1}{\pi}(e^{\pi} - e^{-\pi})$

$a_n = \dfrac{1}{\pi} \displaystyle\int_{-\pi}^{\pi} f(x) \cos nx\,dx = \dfrac{(-1)^n(e^{\pi} - e^{-\pi})}{\pi(1 + n^2)}$

$b_n = \dfrac{1}{\pi} \displaystyle\int_{-\pi}^{\pi} f(x) \sin nx\,dx = \dfrac{1}{\pi}\int_{-\pi}^{\pi} e^x\sin nx\,dx = \dfrac{(-1)^n n(e^{-\pi} - e^{\pi})}{\pi(1 + n^2)}$

$f(x) = \dfrac{e^{\pi} - e^{-\pi}}{2\pi} + \displaystyle\sum_{n=1}^{\infty}\left[\dfrac{(-1)^n(e^{\pi} - e^{-\pi})}{\pi(1 + n^2)}\cos nx + \dfrac{(-1)^n n(e^{-\pi} - e^{\pi})}{\pi(1 + n^2)}\sin nx \right]$

18. From Problem 17

$$\dfrac{\pi^2}{8} = \dfrac{1}{2}\left(\dfrac{\pi^2}{6} + \dfrac{\pi^2}{12} \right) = \dfrac{1}{2}\left(2 + \dfrac{2}{3^2} + \dfrac{2}{5^2} + \cdots \right) = 1 + \dfrac{1}{3^2} + \dfrac{1}{5^2} + \cdots.$$

21. Writing

$$f(x) = \dfrac{a_0}{2} + a_1\cos\dfrac{\pi}{p}x + \cdots + a_n\cos\dfrac{n\pi}{p}x + \cdots + b_1\sin\dfrac{\pi}{p}x + \cdots + b_n\sin\dfrac{n\pi}{p}x + \cdots$$

we see that $f^2(x)$ consists exclusively of squared terms of the form

$$\dfrac{a_0^2}{4}, \qquad a_n^2\cos^2\dfrac{n\pi}{p}x, \qquad b_n^2\sin^2\dfrac{n\pi}{p}x$$

and cross-product terms, with $m \neq n$, of the form

$$a_0 a_n\cos\dfrac{n\pi}{p}x, \qquad a_0 b_n\sin\dfrac{n\pi}{p}x, \qquad 2a_m a_n\cos\dfrac{m\pi}{p}x\cos\dfrac{n\pi}{p}x,$$

$$2a_m b_n\cos\dfrac{m\pi}{p}x\sin\dfrac{n\pi}{p}x, \qquad 2b_m b_n\sin\dfrac{m\pi}{p}x\sin\dfrac{n\pi}{p}x.$$

The integral of each cross-product term taken over the interval $(-p, p)$ is zero by orthogonality. For the squared terms we have

$$\dfrac{a_0^2}{4}\int_{-p}^{p} dx = \dfrac{a_0^2 p}{2}, \qquad a_n^2\int_{-p}^{p}\cos^2\dfrac{n\pi}{p}x\,dx = a_n^2 p, \qquad b_n^2\int_{-p}^{p}\sin^2\dfrac{n\pi}{p}x\,dx = b_n^2 p.$$

Thus

$$RMS(f) = \sqrt{\dfrac{1}{4}a_0^2 + \dfrac{1}{2}\sum_{n=1}^{\infty}(a_n^2 + b_n^2)}.$$

EXERCISES 12.3

Fourier Cosine and Sine Series

3. Since $f(-x) = (-x)^2 - x = x^2 - x$, $f(x)$ is neither even nor odd.

6. Since $f(-x) = e^{-x} - e^x = -f(x)$, $f(x)$ is an odd function.

9. Since $f(x)$ is not defined for $x < 0$, it is neither even nor odd.

12. Since $f(x)$ is an even function, we expand in a cosine series:

$$a_0 = \int_1^2 1\,dx = 1$$

$$a_n = \int_1^2 \cos\frac{n\pi}{2}x\,dx = -\frac{2}{n\pi}\sin\frac{n\pi}{2}.$$

Thus

$$f(x) = \frac{1}{2} + \sum_{n=1}^{\infty} \frac{-2}{n\pi}\sin\frac{n\pi}{2}\cos\frac{n\pi}{2}x.$$

15. Since $f(x)$ is an even function, we expand in a cosine series:

$$a_0 = 2\int_0^1 x^2\,dx = \frac{2}{3}$$

$$a_n = 2\int_0^1 x^2 \cos n\pi x\,dx = 2\left(\frac{x^2}{n\pi}\sin n\pi x\,\Big|_0^1 - \frac{2}{n\pi}\int_0^1 x\sin n\pi x\,dx\right) = \frac{4}{n^2\pi^2}(-1)^n.$$

Thus

$$f(x) = \frac{1}{3} + \sum_{n=1}^{\infty} \frac{4}{n^2\pi^2}(-1)^n \cos n\pi x.$$

18. Since $f(x)$ is an odd function, we expand in a sine series:

$$b_n = \frac{2}{\pi}\int_0^\pi x^3 \sin nx\,dx = \frac{2}{\pi}\left(-\frac{x^3}{n}\cos nx\,\Big|_0^\pi + \frac{3}{n}\int_0^\pi x^2\cos nx\,dx\right) = \frac{2\pi^2}{n}(-1)^{n+1} - \frac{12}{n^2\pi}\int_0^\pi x\sin nx\,dx$$

$$= \frac{2\pi^2}{n}(-1)^{n+1} - \frac{12}{n^2\pi}\left(-\frac{x}{n}\cos nx\,\Big|_0^\pi + \frac{1}{n}\int_0^\pi \cos nx\,dx\right) = \frac{2\pi^2}{n}(-1)^{n+1} + \frac{12}{n^3}(-1)^n.$$

Thus

$$f(x) = \sum_{n=1}^{\infty}\left(\frac{2\pi^2}{n}(-1)^{n+1} + \frac{12}{n^3}(-1)^n\right)\sin nx.$$

21. Since $f(x)$ is an even function, we expand in a cosine series:

$$a_0 = \int_0^1 x\,dx + \int_1^2 1\,dx = \frac{3}{2}$$

$$a_n = \int_0^1 x \cos \frac{n\pi}{2} x\,dx + \int_1^2 \cos \frac{n\pi}{2} x\,dx = \frac{4}{n^2\pi^2}\left(\cos\frac{n\pi}{2} - 1\right).$$

Thus

$$f(x) = \frac{3}{4} + \sum_{n=1}^{\infty} \frac{4}{n^2\pi^2}\left(\cos\frac{n\pi}{2} - 1\right)\cos\frac{n\pi}{2}x.$$

24. Since $f(x)$ is an even function, we expand in a cosine series. [See the solution of Problem 10 in Exercises 12.2 for the computation of the integrals.]

$$a_0 = \frac{2}{\pi/2}\int_0^{\pi/2} \cos x\,dx = \frac{4}{\pi}$$

$$a_n = \frac{2}{\pi/2}\int_0^{\pi/2} \cos x \cos\frac{n\pi}{\pi/2} x\,dx = \frac{4(-1)^{n+1}}{\pi\left(4n^2 - 1\right)}$$

Thus

$$f(x) = \frac{2}{\pi} + \sum_{n=1}^{\infty} \frac{4(-1)^{n+1}}{\pi\left(4n^2 - 1\right)} \cos 2nx.$$

27. $a_0 = \dfrac{4}{\pi}\displaystyle\int_0^{\pi/2} \cos x\,dx = \dfrac{4}{\pi}$

$$a_n = \frac{4}{\pi}\int_0^{\pi/2} \cos x \cos 2nx\,dx = \frac{2}{\pi}\int_0^{\pi/2} [\cos(2n+1)x + \cos(2n-1)x]\,dx = \frac{4(-1)^n}{\pi(1-4n^2)}$$

$$b_n = \frac{4}{\pi}\int_0^{\pi/2} \cos x \sin 2nx\,dx = \frac{2}{\pi}\int_0^{\pi/2} [\sin(2n+1)x + \sin(2n-1)x]\,dx = \frac{8n}{\pi(4n^2-1)}$$

$$f(x) = \frac{2}{\pi} + \sum_{n=1}^{\infty} \frac{4(-1)^n}{\pi(1-4n^2)} \cos 2nx$$

$$f(x) = \sum_{n=1}^{\infty} \frac{8n}{\pi(4n^2-1)} \sin 2nx$$

30. $a_0 = \dfrac{1}{\pi}\displaystyle\int_\pi^{2\pi} (x-\pi)\,dx = \dfrac{\pi}{2}$

$$a_n = \frac{1}{\pi}\int_\pi^{2\pi} (x-\pi)\cos\frac{n}{2}x\,dx = \frac{4}{n^2\pi}\left((-1)^n - \cos\frac{n\pi}{2}\right)$$

$$b_n = \frac{1}{\pi}\int_\pi^{2\pi} (x-\pi)\sin\frac{n}{2}x\,dx = \frac{2}{n}(-1)^{n+1} - \frac{4}{n^2\pi}\sin\frac{n\pi}{2}$$

$$f(x) = \frac{\pi}{4} + \sum_{n=1}^{\infty} \frac{4}{n^2\pi}\left((-1)^n - \cos\frac{n\pi}{2}\right)\cos\frac{n}{2}x$$

$$f(x) = \sum_{n=1}^{\infty} \left(\frac{2}{n}(-1)^{n+1} - \frac{4}{n^2\pi}\sin\frac{n\pi}{2}\right)\sin\frac{n}{2}x$$

33. $a_0 = 2 \int_0^1 (x^2 + x) \, dx = \dfrac{5}{3}$

$a_n = 2 \int_0^1 (x^2 + x) \cos n\pi x \, dx = \dfrac{2(x^2 + x)}{n\pi} \sin n\pi x \Big|_0^1 - \dfrac{2}{n\pi} \int_0^1 (2x + 1) \sin n\pi x \, dx = \dfrac{2}{n^2\pi^2}[3(-1)^n - 1]$

$b_n = 2 \int_0^1 (x^2 + x) \sin n\pi x \, dx = -\dfrac{2(x^2 + x)}{n\pi} \cos n\pi x \Big|_0^1 + \dfrac{2}{n\pi} \int_0^1 (2x + 1) \cos n\pi x \, dx$

$\qquad = \dfrac{4}{n\pi}(-1)^{n+1} + \dfrac{4}{n^3\pi^3}[(-1)^n - 1]$

$f(x) = \dfrac{5}{6} + \displaystyle\sum_{n=1}^{\infty} \dfrac{2}{n^2\pi^2}[3(-1)^n - 1] \cos n\pi x$

$f(x) = \displaystyle\sum_{n=1}^{\infty} \left(\dfrac{4}{n\pi}(-1)^{n+1} + \dfrac{4}{n^3\pi^3}[(-1)^n - 1] \right) \sin n\pi x$

36. $a_0 = \dfrac{2}{\pi} \int_0^\pi x \, dx = \pi$

$a_n = \dfrac{2}{\pi} \int_0^\pi x \cos 2nx \, dx = 0$

$b_n = \dfrac{2}{\pi} \int_0^\pi x \sin 2nx \, dx = -\dfrac{1}{n}$

$f(x) = \dfrac{\pi}{2} - \displaystyle\sum_{n=1}^{\infty} \dfrac{1}{n} \sin 2nx$

39. We have

$$b_n = \frac{2}{\pi} \int_0^\pi 5 \sin nt \, dt = \frac{10}{n\pi}[1 - (-1)^n]$$

so that

$$f(t) = \sum_{n=1}^{\infty} \frac{10[1 - (-1)^n]}{n\pi} \sin nt.$$

Substituting the assumption $x_p(t) = \sum_{n=1}^{\infty} B_n \sin nt$ into the differential equation then gives

$$x_p'' + 10x_p = \sum_{n=1}^{\infty} B_n (10 - n^2) \sin nt = \sum_{n=1}^{\infty} \frac{10[1 - (-1)^n]}{n\pi} \sin nt$$

and so $B_n = 10[1 - (-1)^n]/n\pi(10 - n^2)$. Thus

$$x_p(t) = \frac{10}{\pi} \sum_{n=1}^{\infty} \frac{1 - (-1)^n}{n(10 - n^2)} \sin nt.$$

42. We have

$$a_0 = \frac{2}{1/2} \int_0^{1/2} t \, dt = \frac{1}{2}$$

$$a_n = \frac{2}{1/2} \int_0^{1/2} t \cos 2n\pi t \, dt = \frac{1}{n^2\pi^2} \left[(-1)^n - 1\right]$$

so that

$$f(t) = \frac{1}{4} + \sum_{n=1}^{\infty} \frac{(-1)^n - 1}{n^2\pi^2} \cos 2n\pi t.$$

Substituting the assumption

$$x_p(t) = \frac{A_0}{2} + \sum_{n=1}^{\infty} A_n \cos 2n\pi t$$

into the differential equation then gives

$$\frac{1}{4} x_p'' + 12 x_p = 6A_0 + \sum_{n=1}^{\infty} A_n (12 - n^2\pi^2) \cos 2n\pi t = \frac{1}{4} + \sum_{n=1}^{\infty} \frac{(-1)^n - 1}{n^2\pi^2} \cos 2n\pi t$$

and $A_0 = 1/24$, $A_n = [(-1)^n - 1]/n^2\pi^2(12 - n^2\pi^2)$. Thus

$$x_p(t) = \frac{1}{48} + \frac{1}{\pi^2} \sum_{n=1}^{\infty} \frac{(-1)^n - 1}{n^2(12 - n^2\pi^2)} \cos 2n\pi t.$$

45. (a) We have

$$b_n = \frac{2}{L} \int_0^L \frac{w_0 x}{L} \sin \frac{n\pi}{L} x \, dx = \frac{2w_0}{n\pi} (-1)^{n+1}$$

so that

$$w(x) = \sum_{n=1}^{\infty} \frac{2w_0}{n\pi} (-1)^{n+1} \sin \frac{n\pi}{L} x.$$

(b) If we assume $y_p(x) = \sum_{n=1}^{\infty} B_n \sin(n\pi x/L)$ then

$$y_p^{(4)} = \sum_{n=1}^{\infty} \frac{n^4\pi^4}{L^4} B_n \sin \frac{n\pi}{L} x$$

and so the differential equation $EI y_p^{(4)} = w(x)$ gives

$$B_n = \frac{2w_0(-1)^{n+1} L^4}{EI n^5 \pi^5}.$$

Thus

$$y_p(x) = \frac{2w_0 L^4}{EI\pi^5} \sum_{n=1}^{\infty} \frac{(-1)^{n+1}}{n^5} \sin \frac{n\pi}{L} x.$$

48. The graph is obtained by summing the series from $n = 1$ to 10. It appears that

$$f(x) = \begin{cases} 1 - x, & 0 < x < 1 \\ 0, & 1 < x < 2. \end{cases}$$

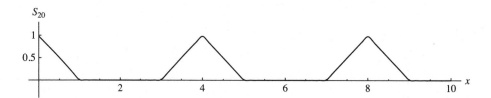

51. No, it is not a full Fourier series. A full Fourier series of $f(x) = e^x, 0 < x < \pi$, would converge to the π-periodic extension of f. The cosine and sine series converge to a 2π-periodic extension (even and odd, respectively). The average of the two series converges to a 2π-periodic extension of

$$f(x) = \begin{cases} e^x, & 0 < x < \pi \\ 0, & -\pi < x < 0. \end{cases}$$

EXERCISES 12.4

Complex Fourier Series

In this section we make use of the following identities due to Euler's formula:

$$e^{in\pi} = e^{-in\pi} = (-1)^n, \qquad e^{-2in\pi} = 1, \qquad e^{-in\pi/2} = (-i)^n.$$

3. Identifying $p = 1/2$ we have

$$c_n = \int_{-1/2}^{1/2} f(x)e^{-2in\pi x}\,dx = \int_{0}^{1/4} e^{-2in\pi x}\,dx = -\frac{1}{2in\pi}e^{-2in\pi x}\Big|_{0}^{1/4}$$

$$= -\frac{1}{2in\pi}\left[e^{-in\pi/2} - 1\right] = -\frac{1}{2in\pi}[(-i)^n - 1] = \frac{i}{2n\pi}[(-i)^n - 1]$$

and

$$c_0 = \int_{0}^{1/4} dx = \frac{1}{4}.$$

Thus

$$f(x) = \frac{1}{4} + \frac{i}{2\pi}\sum_{\substack{n=-\infty \\ n \neq 0}}^{\infty} \frac{(-i)^n - 1}{n}e^{2in\pi x}.$$

6. Identifying $p = 1$ we have

$$c_n = \frac{1}{2} \int_{-1}^{1} f(x)e^{-in\pi x}dx = \frac{1}{2}\left[\int_{-1}^{0} e^x e^{-in\pi x}dx + \int_{0}^{1} e^{-x}e^{-in\pi x}dx\right]$$

$$= \frac{1}{2}\left[-\frac{1}{1-in\pi}e^{(1-in\pi)x}\Big|_{-1}^{0} - \frac{1}{1+in\pi}e^{-(1+in\pi)x}\Big|_{0}^{1}\right]$$

$$= \frac{e-(-1)^n}{e(1-in\pi)} + \frac{1-e^{-1}(-1)^n}{1+in\pi} = \frac{2[e-(-1)^n]}{e(1+n^2\pi^2)}.$$

Thus

$$f(x) = \sum_{n=-\infty}^{\infty} \frac{2[e-(-1)^n]}{e(1+n^2\pi^2)}e^{in\pi x}.$$

9. Identifying $2p = \pi$ or $p = \pi/2$, and using $\sin x = (e^{ix} - e^{-ix})/2i$, we have

$$c_n = \frac{1}{\pi}\int_{0}^{\pi} f(x)e^{-2inx/\pi}dx = \frac{1}{\pi}\int_{0}^{\pi}(\sin x)e^{-2inx/\pi}dx$$

$$= \frac{1}{\pi}\int_{0}^{\pi}\frac{1}{2i}(e^{ix}-e^{-ix})e^{-2inx/\pi}dx$$

$$= \frac{1}{2\pi i}\int_{0}^{\pi}\left(e^{(1-2n/\pi)ix} - e^{-(1+2n/\pi)ix}\right)dx$$

$$= \frac{1}{2\pi i}\left[\frac{1}{i(1-2n/\pi)}e^{(1-2n/\pi)ix}\right.$$

$$\left. + \frac{1}{i(1+2n/\pi)}e^{-(1+2n/\pi)ix}\right]_{0}^{\pi}$$

$$= \frac{\pi(1+e^{-2in})}{\pi^2 - 4n^2}.$$

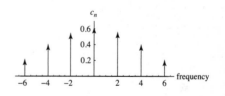

The fundamental period is $T = \pi$, so $\omega = 2\pi/\pi = 2$ and the values of $n\omega$ are $0, \pm 2, \pm 4, \pm 6, \ldots$. Values of $|c_n|$ for $n = 0, \pm 1, \pm 2, \pm 3, \pm 4$, and ± 5 are shown in the table. The bottom graph is a portion of the frequency spectrum.

n	-5	-4	-3	-2	-1	0	1	2	3	4	5
c_n	0.0198	0.0759	0.2380	0.4265	0.5784	0.6366	0.5784	0.4285	0.2380	0.0759	0.0198

12. From Problem 11 and the fact that f is odd, $c_n + c_{-n} = a_n = 0$, so $c_{-n} = -c_n$. Then $b_n = i(c_n - c_{-n}) = 2ic_n$. From Problem 1, $b_n = 2i[1-(-1)^n]/n\pi i = 2[1-(-1)^n]/n\pi$, and the Fourier sine series of f is

$$f(x) = \sum_{i=1}^{\infty} \frac{2[1-(-1)^n]}{n\pi}\sin\frac{n\pi x}{2}.$$

EXERCISES 12.5

Sturm-Liouville Problem

3. For $\lambda = 0$ the solution of $y'' = 0$ is $y = c_1 x + c_2$. The condition $y'(0) = 0$ implies $c_1 = 0$, so $\lambda = 0$ is an eigenvalue with corresponding eigenfunction 1.

For $\lambda = -\alpha^2 < 0$ we have $y = c_1 \cosh \alpha x + c_2 \sinh \alpha x$ and $y' = c_1 \alpha \sinh \alpha x + c_2 \alpha \cosh \alpha x$. The condition $y'(0) = 0$ implies $c_2 = 0$ and so $y = c_1 \cosh \alpha x$. Now the condition $y'(L) = 0$ implies $c_1 = 0$. Thus $y = 0$ and there are no negative eigenvalues.

For $\lambda = \alpha^2 > 0$ we have $y = c_1 \cos \alpha x + c_2 \sin \alpha x$ and $y' = -c_1 \alpha \sin \alpha x + c_2 \alpha \cos \alpha x$. The condition $y'(0) = 0$ implies $c_2 = 0$ and so $y = c_1 \cos \alpha x$. Now the condition $y'(L) = 0$ implies $-c_1 \alpha \sin \alpha L = 0$. For $c_1 \neq 0$ this condition will hold when $\alpha L = n\pi$ or $\lambda = \alpha^2 = n^2 \pi^2 / L^2$, where $n = 1, 2, 3, \ldots$. These are the positive eigenvalues with corresponding eigenfunctions $\cos(n\pi x / L)$, $n = 1, 2, 3, \ldots$.

6. The eigenfunctions are $\sin \alpha_n x$ where $\tan \alpha_n = -\alpha_n$. Thus

$$\| \sin \alpha_n x \|^2 = \int_0^1 \sin^2 \alpha_n x \, dx = \frac{1}{2} \int_0^1 (1 - \cos 2\alpha_n x) \, dx$$

$$= \frac{1}{2} \left(x - \frac{1}{2\alpha_n} \sin 2\alpha_n x \right) \Big|_0^1 = \frac{1}{2} \left(1 - \frac{1}{2\alpha_n} \sin 2\alpha_n \right)$$

$$= \frac{1}{2} \left[1 - \frac{1}{2\alpha_n} (2 \sin \alpha_n \cos \alpha_n) \right]$$

$$= \frac{1}{2} \left[1 - \frac{1}{\alpha_n} \tan \alpha_n \cos \alpha_n \cos \alpha_n \right]$$

$$= \frac{1}{2} \left[1 - \frac{1}{\alpha_n} \left(-\alpha_n \cos^2 \alpha_n \right) \right] = \frac{1}{2} \left(1 + \cos^2 \alpha_n \right).$$

9. To obtain the self-adjoint form we note that an integrating factor is $(1/x)e^{\int (1-x)dx/x} = e^{-x}$. Thus, the differential equation is

$$x e^{-x} y'' + (1 - x)e^{-x} y' + n e^{-x} y = 0$$

and the self-adjoint form is

$$\frac{d}{dx} \left[x e^{-x} y' \right] + n e^{-x} y = 0.$$

Identifying the weight function $p(x) = e^{-x}$ and noting that since $r(x) = x e^{-x}$, $r(0) = 0$ and $\lim_{x \to \infty} r(x) = 0$, we have the orthogonality relation

$$\int_0^\infty e^{-x} L_m(x) L_n(x) \, dx = 0, \ m \neq n.$$

12. (a) Letting $\lambda = \alpha^2$ the differential equation becomes $x^2 y'' + x y' + (\alpha^2 x^2 - 1)y = 0$. This is the parametric Bessel equation with $\nu = 1$. The general solution is

$$y = c_1 J_1(\alpha x) + c_2 Y_1(\alpha x).$$

Since Y is unbounded at 0 we must have $c_2 = 0$, so that $y = c_1 J_1(\alpha x)$. The condition $J_1(3\alpha) = 0$ defines the eigenvalues $\lambda_n = \alpha_n^2$ for $n = 1, 2, 3, \ldots$. The corresponding eigenfunctions are $J_1(\alpha_n x)$.

(b) Using a CAS or Table 5.1 in the text to solve $J_1(3\alpha) = 0$ we find $3\alpha_1 = 3.8317$, $3\alpha_2 = 7.0156$, $3\alpha_3 = 10.1735$, and $3\alpha_4 = 13.3237$. The corresponding eigenvalues are $\lambda_1 = \alpha_1^2 = 1.6313$, $\lambda_2 = \alpha_2^2 = 5.4687$, $\lambda_3 = \alpha_3^2 = 11.4999$, and $\lambda_4 = \alpha_4^2 = 19.7245$.

15. (a) An orthogonality relation is

$$\int_0^1 (x_m \cos x_m x - \sin x_m x)(x_n \cos x_n x - \sin x_n x)\, dx = 0$$

where $x_m \neq x_n$ are positive solutions of $\tan x = x$.

(b) Referring to Problem 2 we use a CAS to compute

$$\int_0^1 (4.4934 \cos 4.4934 x - \sin 4.4934 x)(7.7253 \cos 7.7253 x - \sin 7.7253 x)\, dx = -2.5650 \times 10^{-4} \approx 0.$$

EXERCISES 12.6

Bessel and Legendre Series

3. The boundary condition indicates that we use (15) and (16) in the text. With $b = 2$ we obtain

$$c_i = \frac{2}{4J_1^2(2\alpha_i)} \int_0^2 x J_0(\alpha_i x)\, dx \qquad \boxed{t = \alpha_i x \qquad dt = \alpha_i\, dx}$$

$$= \frac{1}{2J_1^2(2\alpha_i)} \cdot \frac{1}{\alpha_i^2} \int_0^{2\alpha_i} t J_0(t)\, dt$$

$$= \frac{1}{2\alpha_i^2 J_1^2(2\alpha_i)} \int_0^{2\alpha_i} \frac{d}{dt}[t J_1(t)]\, dt \qquad \text{[From (5) in the text]}$$

$$= \frac{1}{2\alpha_i^2 J_1^2(2\alpha_i)} t J_1(t) \Big|_0^{2\alpha_i}$$

$$= \frac{1}{\alpha_i J_1(2\alpha_i)}.$$

Thus

$$f(x) = \sum_{i=1}^{\infty} \frac{1}{\alpha_i J_1(2\alpha_i)} J_0(\alpha_i x).$$

6. Writing the boundary condition in the form

$$2 J_0(2\alpha) + 2\alpha J_0'(2\alpha) = 0$$

we identify $b = 2$ and $h = 2$. Using (17) and (18) in the text we obtain

$$c_i = \frac{2\alpha_i^2}{(4\alpha_i^2 + 4)J_0^2(2\alpha_i)} \int_0^2 x J_0(\alpha_i x)\, dx \qquad \boxed{t = \alpha_i x \qquad dt = \alpha_i\, dx}$$

$$= \frac{\alpha_i^2}{2(\alpha_i^2+1)J_0^2(2\alpha_i)} \cdot \frac{1}{\alpha_i^2} \int_0^{2\alpha_i} t J_0(t)\, dt$$

$$= \frac{1}{2(\alpha_i^2+1)J_0^2(2\alpha_i)} \int_0^{2\alpha_i} \frac{d}{dt}[t J_1(t)]\, dt \qquad \text{[From (5) in the text]}$$

$$= \frac{1}{2(\alpha_i^2+1)J_0^2(2\alpha_i)} t J_1(t) \Big|_0^{2\alpha_i}$$

$$= \frac{\alpha_i J_1(2\alpha_i)}{(\alpha_i^2+1)J_0^2(2\alpha_i)} \cdot$$

Thus

$$f(x) = \sum_{i=1}^{\infty} \frac{\alpha_i J_1(2\alpha_i)}{(\alpha_i^2+1)J_0^2(2\alpha_i)} J_0(\alpha_i x).$$

9. The boundary condition indicates that we use (19) and (20) in the text. With $b=3$ we obtain

$$c_1 = \frac{2}{9} \int_0^3 x x^2\, dx = \frac{2}{9} \frac{x^4}{4} \Big|_0^3 = \frac{9}{2},$$

$$c_i = \frac{2}{9 J_0^2(3\alpha_i)} \int_0^3 x J_0(\alpha_i x) x^2\, dx$$

$$\boxed{t = \alpha_i x \qquad dt = \alpha_i\, dx}$$

$$= \frac{2}{9 J_0^2(3\alpha_i)} \cdot \frac{1}{\alpha_i^4} \int_0^{3\alpha_i} t^3 J_0(t)\, dt$$

$$= \frac{2}{9\alpha_i^4 J_0^2(3\alpha_i)} \int_0^{3\alpha_i} t^2 \frac{d}{dt}[t J_1(t)]\, dt$$

$$\boxed{\begin{array}{ll} u = t^2 & dv = \frac{d}{dt}[t J_1(t)]\, dt \\ du = 2t\, dt & v = t J_1(t) \end{array}}$$

$$= \frac{2}{9\alpha_i^4 J_0^2(3\alpha_i)} \left(t^3 J_1(t) \Big|_0^{3\alpha_i} - 2 \int_0^{3\alpha_i} t^2 J_1(t)\, dt \right).$$

With $n=0$ in equation (6) in the text we have $J_0'(x) = -J_1(x)$, so the boundary condition $J_0'(3\alpha_i) = 0$ implies $J_1(3\alpha_i) = 0$. Then

$$c_i = \frac{2}{9\alpha_i^4 J_0^2(3\alpha_i)} \left(-2 \int_0^{3\alpha_i} \frac{d}{dt}[t^2 J_2(t)]\, dt \right) = \frac{2}{9\alpha_i^4 J_0^2(3\alpha_i)} \left(-2 t^2 J_2(t) \Big|_0^{3\alpha_i} \right)$$

$$= \frac{2}{9\alpha_i^4 J_0^2(3\alpha_i)} \left[-18\alpha_i^2 J_2(3\alpha_i) \right] = \frac{-4 J_2(3\alpha_i)}{\alpha_i^2 J_0^2(3\alpha_i)} \cdot$$

Thus

$$f(x) = \frac{9}{2} - 4 \sum_{i=1}^{\infty} \frac{J_2(3\alpha_i)}{\alpha_i^2 J_0^2(3\alpha_i)} J_0(\alpha_i x).$$

12. (a) From Problem 7, the coefficients of the Fourier-Bessel series are

$$c_i = \frac{20\alpha_i J_2(4\alpha_i)}{(2\alpha_i^2+1)J_1^2(4\alpha_i)} \cdot$$

254

Using a CAS we find $c_1 = 26.7896$, $c_2 = -12.4624$, $c_3 = 7.1404$, $c_4 = -4.68705$, and $c_5 = 3.35619$.

(b)

(c)

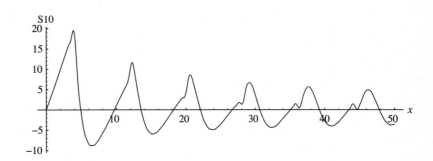

15. We compute

$$c_0 = \frac{1}{2}\int_0^1 xP_0(x)\,dx = \frac{1}{2}\int_0^1 x\,dx = \frac{1}{4}$$

$$c_1 = \frac{3}{2}\int_0^1 xP_1(x)\,dx = \frac{3}{2}\int_0^1 x^2\,dx = \frac{1}{2}$$

$$c_2 = \frac{5}{2}\int_0^1 xP_2(x)\,dx = \frac{5}{2}\int_0^1 \frac{1}{2}(3x^3 - x)dx = \frac{5}{16}$$

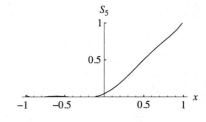

$$c_3 = \frac{7}{2}\int_0^1 xP_3(x)\,dx = \frac{7}{2}\int_0^1 \frac{1}{2}(5x^4 - 3x^2)dx = 0$$

$$c_4 = \frac{9}{2}\int_0^1 xP_4(x)\,dx = \frac{9}{2}\int_0^1 \frac{1}{8}(35x^5 - 30x^3 + 3x)dx = -\frac{3}{32}$$

$$c_5 = \frac{11}{2}\int_0^1 xP_5(x)\,dx = \frac{11}{2}\int_0^1 \frac{1}{8}(63x^6 - 70x^4 + 15x^2)dx = 0$$

$$c_6 = \frac{13}{2}\int_0^1 xP_6(x)\,dx = \frac{13}{2}\int_0^1 \frac{1}{16}(231x^7 - 315x^5 + 105x^3 - 5x)dx = \frac{13}{256}.$$

Thus

$$f(x) = \frac{1}{4}P_0(x) + \frac{1}{2}P_1(x) + \frac{5}{16}P_2(x) - \frac{3}{32}P_4(x) + \frac{13}{256}P_6(x) + \cdots.$$

The figure above is the graph of $S_5(x) = \frac{1}{4}P_0(x) + \frac{1}{2}P_1(x) + \frac{5}{16}P_2(x) - \frac{3}{32}P_4(x) + \frac{13}{256}P_6(x)$.

18. From Problem 17 we have

$$P_2(\cos\theta) = \frac{1}{4}(3\cos 2\theta + 1) \qquad \text{or} \qquad \cos 2\theta = \frac{4}{3}P_2(\cos\theta) - \frac{1}{3}.$$

Then, using $P_0(\cos\theta) = 1$,

$$F(\theta) = 1 - \cos 2\theta = 1 - \left[\frac{4}{3}P_2(\cos\theta) - \frac{1}{3}\right]$$

$$= \frac{4}{3} - \frac{4}{3}P_2(\cos\theta) = \frac{4}{3}P_0(\cos\theta) - \frac{4}{3}P_2(\cos\theta).$$

21. From (26) in Problem 19 in the text we find

$$c_0 = \int_0^1 xP_0(x)\,dx = \int_0^1 x\,dx = \frac{1}{2},$$

$$c_2 = 5\int_0^1 xP_2(x)\,dx = 5\int_0^1 \frac{1}{2}(3x^3 - x)dx = \frac{5}{8},$$

$$c_4 = 9\int_0^1 xP_4(x)\,dx = 9\int_0^1 \frac{1}{8}(35x^5 - 30x^3 + 3x)dx = -\frac{3}{16},$$

and

$$c_6 = 13\int_0^1 xP_6(x)\,dx = 13\int_0^1 \frac{1}{16}(231x^7 - 315x^5 + 105x^3 - 5x)dx = \frac{13}{128}.$$

Hence, from (25) in the text,

$$f(x) = \frac{1}{2}P_0(x) + \frac{5}{8}P_2(x) - \frac{3}{16}P_4(x) + \frac{13}{128}P_6 + \cdots.$$

On the interval $-1 < x < 1$ this series represents the function $f(x) = |x|$.

24. We want to express both x^2 and x^3 as linear combinations of $P_0(x) = 1, P_1(x) = x, P_2(x) = \frac{1}{2}(3x^2 - 1)$, and $P_3(x) = \frac{1}{2}(5x^3 - 3x)$. Setting

$$x^2 = c_0P_0(x) + c_1P_1(x) + c_2P_2(x) = c_0 + c_1x + c_2\left[\frac{1}{2}(3x^2 - 1)\right] = \left(c_0 - \frac{3}{2}c_2\right) + c_1x + \frac{3}{2}c_2x^2,$$

we obtain the system

$$c_0 - \frac{1}{2}c_2 = 0$$

$$c_1 = 0$$

$$\frac{3}{2}c_2 = 1.$$

The solution is $c_0 = \frac{1}{3}, c_1 = 0, c_2 = \frac{2}{3}$. Thus, $x^2 = \frac{1}{3}P_0(x) + \frac{2}{3}P_2(x)$. Setting

$$x^3 = c_0P_0(x) + c_1P_1(x) + c_2P_2(x) + c_3P_3(x) = c_0 + c_1x + c_2\left[\frac{1}{2}(3x^2 - 1)\right] + c_3\left[\frac{1}{2}(5x^3 - 3x)\right]$$

$$= \left(c_0 - \frac{1}{2}c_2\right) + \left(c_1 - \frac{3}{2}c_3\right)x + \frac{3}{2}c_2x^2 + \frac{5}{2}c_3x^3,$$

we obtain the system

$$c_0 - \frac{1}{2}c_2 = 0$$

$$c_1 - \frac{3}{2}c_3 = 0$$

$$\frac{3}{2}c_2 = 0$$

$$\frac{5}{2}c_3 = 1.$$

The solution is $c_0 = 0, c_1 = \frac{3}{5}, c_2 = 0, c_3 = \frac{2}{5}$. Thus $x^3 = \frac{3}{5}P_1(x) + \frac{2}{5}P_3(x)$.

CHAPTER 12 REVIEW EXERCISES

3. cosine, since f is even

6. Periodically extending the function we see that at $x = -1$ the function converges to $\frac{1}{2}(-1+0) = -\frac{1}{2}$; at $x = 0$ it converges to $\frac{1}{2}(0+1) = \frac{1}{2}$, and at $x = 1$ it converges to $\frac{1}{2}(-1+0) = -\frac{1}{2}$.

9. True, since $\int_0^1 P_{2m}(x)P_{2n}(x)\,dx = \frac{1}{2}\int_{-1}^1 P_{2m}(x)P_{2n}(x)\,dx = 0$ when $m \neq n$.

12. **(a)** For $m \neq n$

$$\int_0^L \sin\frac{(2n+1)\pi}{2L}x \, \sin\frac{(2m+1)\pi}{2L}x\,dx = \frac{1}{2}\int_0^L \left(\cos\frac{n-m}{L}\pi x - \cos\frac{n+m+1}{L}\pi x \right) dx = 0.$$

(b) From

$$\int_0^L \sin^2\frac{(2n+1)\pi}{2L}x\,dx = \int_0^L \left(\frac{1}{2} - \frac{1}{2}\cos\frac{(2n+1)\pi}{L}x \right) dx = \frac{L}{2}$$

we see that

$$\left\| \sin\frac{(2n+1)\pi}{2L}x \right\| = \sqrt{\frac{L}{2}}.$$

15. Since

$$a_0 = \frac{2}{1}\int_0^1 e^x dx = 2(e-1)$$

and

$$a_n = \frac{2}{1}\int_0^1 e^x \cos n\pi x\,dx = \frac{2}{1+n^2\pi^2}[e(-1)^n - 1],$$

for $n = 1, 2, 3, \ldots$, we have the cosine series

$$f(x) = e - 1 + 2\sum_{n=1}^{\infty} \frac{e(-1)^n - 1}{1+n^2\pi^2}\cos n\pi x.$$

Since

$$b_n = \frac{2}{1} \int_0^1 e^x \sin n\pi x \, dx = \frac{2n\pi}{1 + n^2\pi^2}[1 - e(-1)^n],$$

for $n = 1, 2, 3, \ldots$, we have the sine series

$$f(x) = 2\pi \sum_{n=1}^{\infty} \frac{n[1 - e(-1)^n]}{1 + n^2\pi^2} \sin n\pi x.$$

18. To obtain the self-adjoint form of the differential equation in Problem 17 we note that an integrating factor is $(1/x^2)e^{\int dx/x} = 1/x$. Thus the weight function is $1/x$ and an orthogonality relation is

$$\int_1^e \frac{1}{x} \cos\left(\frac{2n-1}{2}\pi \ln x\right) \cos\left(\frac{2m-1}{2}\pi \ln x\right) dx = 0, \ m \neq n.$$

21. The boundary condition indicates that we use (15) and (16) of Section 12.6 in the text. With $b = 4$ we obtain

$$c_i = \frac{2}{16 J_1^2(4\alpha_i)} \int_0^4 x J_0(\alpha_i x) f(x) \, dx$$

$$= \frac{1}{8 J_1^2(4\alpha_i)} \int_0^2 x J_0(\alpha_i x) \, dx \qquad \boxed{t = \alpha_i x \qquad dt = \alpha_i \, dx}$$

$$= \frac{1}{8 J_1^2(4\alpha_i)} \cdot \frac{1}{\alpha_i^2} \int_0^{2\alpha_i} t J_0(t) \, dt$$

$$= \frac{1}{8 J_1^2(4\alpha_i)} \int_0^{2\alpha_i} \frac{d}{dt}[t J_1(t)] \, dt \qquad \text{[From (5) in 12.6 in the text]}$$

$$= \frac{1}{8 J_1^2(4\alpha_i)} t J_1(t) \Big|_0^{2\alpha_i} = \frac{J_1(2\alpha_i)}{4\alpha_i J_1^2(4\alpha_i)}.$$

Thus

$$f(x) = \frac{1}{4} \sum_{i=1}^{\infty} \frac{J_1(2\alpha_i)}{\alpha_i J_1^2(4\alpha_i)} J_0(\alpha_i x).$$

13 Boundary-Value Problems in Rectangular Coordinates

EXERCISES 13.1
Separable Partial Differential Equations

3. Substituting $u(x, y) = X(x)Y(y)$ into the partial differential equation yields $X'Y + XY' = XY$. Separating variables and using the separation constant $-\lambda$ we obtain

$$\frac{X'}{X} = \frac{Y - Y'}{Y} = -\lambda.$$

Then

$$X' + \lambda X = 0 \quad \text{and} \quad Y' - (1 + \lambda)Y = 0$$

so that

$$X = c_1 e^{-\lambda x} \quad \text{and} \quad Y = c_2 e^{(1+\lambda)y}.$$

A particular product solution of the partial differential equation is

$$u = XY = c_3 e^{y + \lambda(y - x)}.$$

6. Substituting $u(x, y) = X(x)Y(y)$ into the partial differential equation yields $yX'Y + xXY' = 0$. Separating variables and using the separation constant $-\lambda$ we obtain

$$\frac{X'}{xX} = -\frac{Y'}{yY} = -\lambda.$$

When $\lambda \neq 0$

$$X' + \lambda x X = 0 \quad \text{and} \quad Y' - \lambda y Y = 0$$

so that

$$X = c_1 e^{\lambda x^2/2} \quad \text{and} \quad Y = c_2 e^{-\lambda y^2/2}.$$

A particular product solution of the partial differential equation is

$$u = XY = c_3 e^{\lambda(x^2 - y^2)/2}.$$

When $\lambda = 0$ the differential equations become $X' = 0$ and $Y' = 0$, so in this case $X = c_4$, $Y = c_5$, and $u = XY = c_6$.

9. Substituting $u(x, t) = X(x)T(t)$ into the partial differential equation yields $kX''T - XT = XT'$. Separating variables and using the separation constant $-\lambda$ we obtain

$$\frac{kX'' - X}{X} = \frac{T'}{T} = -\lambda.$$

Then

$$X'' + \frac{\lambda - 1}{k}X = 0 \quad \text{and} \quad T' + \lambda T = 0.$$

The second differential equation implies $T(t) = c_1 e^{-\lambda t}$. For the first differential equation we consider three cases:

I. If $(\lambda - 1)/k = 0$ then $\lambda = 1$, $X'' = 0$, and $X(x) = c_2 x + c_3$, so

$$u = XT = e^{-t}(A_1 x + A_2).$$

II. If $(\lambda - 1)/k = -\alpha^2 < 0$, then $\lambda = 1 - k\alpha^2$, $X'' - \alpha^2 X = 0$, and $X(x) = c_4 \cosh \alpha x + c_5 \sinh \alpha x$, so

$$u = XT = (A_3 \cosh \alpha x + A_4 \sinh \alpha x)e^{-(1-k\alpha^2)t}.$$

III. If $(\lambda - 1)/k = \alpha^2 > 0$, then $\lambda = 1 + \lambda\alpha^2$, $X'' + \alpha^2 X = 0$, and $X(x) = c_6 \cos \alpha x + c_7 \sin \alpha x$, so

$$u = XT = (A_5 \cos \alpha x + A_6 \sin \alpha x)e^{-(1+\lambda\alpha^2)t}.$$

12. Substituting $u(x,t) = X(x)T(t)$ into the partial differential equation yields $a^2 X''T = XT'' + 2kXT'$. Separating variables and using the separation constant $-\lambda$ we obtain

$$\frac{X''}{X} = \frac{T'' + 2kT'}{a^2 T} = -\lambda.$$

Then

$$X'' + \lambda X = 0 \quad \text{and} \quad T'' + 2kT' + a^2 \lambda T = 0.$$

We consider three cases:

I. If $\lambda = 0$ then $X'' = 0$ and $X(x) = c_1 x + c_2$. Also, $T'' + 2kT' = 0$ and $T(t) = c_3 + c_4 e^{-2kt}$, so

$$u = XT = (c_1 x + c_2)(c_3 + c_4 e^{-2kt}).$$

II. If $\lambda = -\alpha^2 < 0$, then $X'' - \alpha^2 X = 0$, and $X(x) = c_5 \cosh \alpha x + c_6 \sinh \alpha x$. The auxiliary equation of $T'' + 2kT' - \alpha^2 a^2 T = 0$ is $m^2 + 2km - \alpha^2 a^2 = 0$. Solving for m we obtain $m = -k \pm \sqrt{k^2 + \alpha^2 a^2}$, so $T(t) = c_7 e^{(-k+\sqrt{k^2+\alpha^2 a^2})t} + c_8 e^{(-k-\sqrt{k^2+\alpha^2 a^2})t}$. Then

$$u = XT = (c_5 \cosh \alpha x + c_6 \sinh \alpha x)\left(c_7 e^{(-k+\sqrt{k^2+\alpha^2 a^2})t} + c_8 e^{(-k-\sqrt{k^2+\alpha^2 a^2})t}\right).$$

III. If $\lambda = \alpha^2 > 0$, then $X'' + \alpha^2 X = 0$, and $X(x) = c_9 \cos \alpha x + c_{10} \sin \alpha x$. The auxiliary equation of $T'' + 2kT' + \alpha^2 a^2 T = 0$ is $m^2 + 2km + \alpha^2 a^2 = 0$. Solving for m we obtain $m = -k \pm \sqrt{k^2 - \alpha^2 a^2}$. We consider three possibilities for the discriminant $k^2 - \alpha^2 a^2$:

(i) If $k^2 - \alpha^2 a^2 = 0$ then $T(t) = c_{11}e^{-kt} + c_{12}te^{-kt}$ and

$$u = XT = (c_9 \cos \alpha x + c_{10} \sin \alpha x)(c_{11}e^{-kt} + c_{12}te^{-kt}).$$

From $k^2 - \alpha^2 a^2 = 0$ we have $\alpha = k/a$ so the solution can be written

$$u = XT = (c_9 \cos kx/a + c_{10} \sin kx/a)(c_{11}e^{-kt} + c_{12}te^{-kt}).$$

(ii) If $k^2 - \alpha^2 a^2 < 0$ then $T(t) = e^{-kt}\left(c_{13} \cos \sqrt{\alpha^2 a^2 - k^2}\, t + c_{14} \sin \sqrt{\alpha^2 a^2 - k^2}\, t\right)$ and

$$u = XT = (c_9 \cos \alpha x + c_{10} \sin \alpha x)e^{-kt}\left(c_{13} \cos \sqrt{\alpha^2 a^2 - k^2}\, t + c_{14} \sin \sqrt{\alpha^2 a^2 - k^2}\, t\right).$$

(iii) If $k^2 - \alpha^2 a^2 > 0$ then $T(t) = c_{15}e^{(-k+\sqrt{k^2-\alpha^2 a^2})t} + c_{16}e^{(-k-\sqrt{k^2-\alpha^2 a^2})t}$ and

$$u = XT = (c_9 \cos \alpha x + c_{10} \sin \alpha x)\left(c_{15}e^{(-k+\sqrt{k^2-\alpha^2 a^2})t} + c_{16}e^{(-k-\sqrt{k^2-\alpha^2 a^2})t}\right).$$

15. Substituting $u(x, y) = X(x)Y(y)$ into the partial differential equation yields $X''Y + XY'' = XY$. Separating variables and using the separation constant $-\lambda$ we obtain

$$\frac{X''}{X} = \frac{Y - Y''}{Y} = -\lambda.$$

Then

$$X'' + \lambda X = 0 \quad \text{and} \quad Y'' - (1 + \lambda)Y = 0.$$

We consider three cases:

I. If $\lambda = 0$ then $X'' = 0$ and $X(x) = c_1 x + c_2$. Also $Y'' - Y = 0$ and $Y(y) = c_3 \cosh y + c_4 \sinh y$ so

$$u = XY = (c_1 x + c_2)(c_3 \cosh y + c_4 \sinh y).$$

II. If $\lambda = -\alpha^2 < 0$ then $X'' - \alpha^2 X = 0$ and $Y'' + (\alpha^2 - 1)Y = 0$. The solution of the first differential equation is $X(x) = c_5 \cosh \alpha x + c_6 \sinh \alpha x$. The solution of the second differential equation depends on the nature of $\alpha^2 - 1$. We consider three cases:

(i) If $\alpha^2 - 1 = 0$, or $\alpha^2 = 1$, then $Y(y) = c_7 y + c_8$ nad

$$u = XY = (c_5 \cosh \alpha x + c_6 \sinh \alpha x)(c_7 y + c_8).$$

(ii) If $\alpha^2 - 1 < 0$, or $0 < \alpha^2 < 1$, then $Y(y) = c_9 \cosh \sqrt{1 - \alpha^2}\, y + c_{10} \sinh \sqrt{1 - \alpha^2}\, y$ and

$$u = XY = (c_5 \cosh \alpha x + c_6 \sinh \alpha x)\left(c_9 \cosh \sqrt{1 - \alpha^2}\, y + c_{10} \sinh \sqrt{1 - \alpha^2}\, y\right).$$

(iii) If $\alpha^2 - 1 > 0$, or $\alpha^2 > 1$, then $Y(y) = c_{11} \cos \sqrt{\alpha^2 - 1}\, y + c_{12} \sin \sqrt{\alpha^2 - 1}\, y$ and

$$u = XY = (c_5 \cosh \alpha x + c_6 \sinh \alpha x)\left(c_{11} \cos \sqrt{\alpha^2 - 1}\, y + c_{12} \sin \sqrt{\alpha^2 - 1}\, y\right).$$

III. If $\lambda = \alpha^2 > 0$, then $X'' + \alpha^2 X = 0$ and $X(x) = c_{13} \cos \alpha x + c_{14} \sin \alpha x$. Also,

$Y'' - (1 + \alpha^2)Y = 0$ and $Y(y) = c_{15} \cosh \sqrt{1 + \alpha^2}\, y + c_{16} \sinh \sqrt{1 + \alpha^2}\, y$ so

$$u = XY = (c_{13} \cos \alpha x + c_{14} \sin \alpha x)\left(c_{15} \cosh \sqrt{1 + \alpha^2}\, y + c_{16} \sinh \sqrt{1 + \alpha^2}\, y\right).$$

18. Identifying $A = 3$, $B = 5$, and $C = 1$, we compute $B^2 - 4AC = 13 > 0$. The equation is hyperbolic.

21. Identifying $A = 1$, $B = -9$, and $C = 0$, we compute $B^2 - 4AC = 81 > 0$. The equation is hyperbolic.

24. Identifying $A = 1$, $B = 0$, and $C = 1$, we compute $B^2 - 4AC = -4 < 0$. The equation is elliptic.

27. Substituting $u(r, t) = R(r)T(t)$ into the partial differential equation yields

$$k\left(R''T + \frac{1}{r}R'T\right) = RT'.$$

Separating variables and using the separation constant $-\lambda$ we obtain

$$\frac{rR'' + R'}{rR} = \frac{T'}{kT} = -\lambda.$$

Then

$$rR'' + R' + \lambda rR = 0 \quad \text{and} \quad T' + \lambda kT = 0.$$

Letting $\lambda = \alpha^2$ and writing the first equation as $r^2 R'' + rR' = \alpha^2 r^2 R = 0$ we see that it is a parametric Bessel equation of order 0. As discussed in Chapter 5 of the text, it has solution $R(r) = c_1 J_0(\alpha r) + c_2 Y_0(\alpha r)$. Since a solution of $T' + \alpha^2 kT$ is $T(t) = e^{-k\alpha^2 t}$, we see that a solution of the partial differential equation is

$$u = RT = e^{-k\alpha^2 t}[c_1 J_0(\alpha r) + c_2 Y_0(\alpha r)].$$

30. We identify $A = xy + 1$, $B = x + 2y$, and $C = 1$. Then $B^2 - 4AC = x^2 + 4y^2 - 4$. The equation $x^2 + 4y^2 = 4$ defines an ellipse. The partial differential equation is hyperbolic outside the ellipse, parabolic on the ellipse, and elliptic inside the ellipse.

EXERCISES 13.2

Classical Equations and Boundary-Value Problems

3. $k\dfrac{\partial^2 u}{\partial x^2} = \dfrac{\partial u}{\partial t}, \quad 0 < x < L,\ t > 0$

$u(0,t) = 100, \quad \dfrac{\partial u}{\partial x}\bigg|_{x=L} = -hu(L,t), \quad t > 0$

$u(x,0) = f(x), \quad 0 < x < L$

6. $k\dfrac{\partial^2 u}{\partial x^2} + h(u - 50) = \dfrac{\partial u}{\partial t}, \quad 0 < x < L,\ t > 0$

$\dfrac{\partial u}{\partial x}\bigg|_{x=0} = 0, \quad \dfrac{\partial u}{\partial x}\bigg|_{x=L} = 0, \quad t > 0$

$u(x,0) = 100, \quad 0 < x < L$

9. $a^2\dfrac{\partial^2 u}{\partial x^2} - 2\beta\dfrac{\partial u}{\partial t} = \dfrac{\partial^2 u}{\partial t^2}, \quad 0 < x < L,\ t > 0$

$u(0,t) = 0, \quad u(L,t) = \sin \pi t, \quad t > 0$

$u(x,0) = f(x), \quad \dfrac{\partial u}{\partial t}\bigg|_{t=0} = 0, \quad 0 < x < L$

12. $\dfrac{\partial^2 u}{\partial x^2} + \dfrac{\partial^2 u}{\partial y^2} = 0, \quad 0 < x < \pi,\ y > 0$

$u(0,y) = e^{-y}, \quad u(\pi,y) = \begin{cases} 100, & 0 < y \le 1 \\ 0, & y > 1 \end{cases}$

$u(x,0) = f(x), \quad 0 < x < \pi$

EXERCISES 13.3

Heat Equation

3. Using $u = XT$ and $-\lambda$ as a separation constant we obtain

$$X'' + \lambda X = 0,$$
$$X'(0) = 0,$$
$$X'(L) = 0,$$

and

$$T' + k\lambda T = 0.$$

This leads to

$$X = c_1 \cos \frac{n\pi}{L}x \quad \text{and} \quad T = c_2 e^{-kn^2\pi^2 t/L^2}$$

for $n = 0, 1, 2, \ldots$ ($\lambda = 0$ is an eigenvalue in this case) so that

$$u = \sum_{n=0}^{\infty} A_n e^{-kn^2\pi^2 t/L^2} \cos \frac{n\pi}{L} x.$$

Imposing

$$u(x, 0) = f(x) = A_0 + \sum_{n=1}^{\infty} A_n \cos \frac{n\pi}{L} x$$

gives

$$u(x, t) = \frac{1}{L} \int_0^L f(x)\, dx + \frac{2}{L} \sum_{n=1}^{\infty} \left(\int_0^L f(x) \cos \frac{n\pi}{L} x\, dx \right) e^{-kn^2\pi^2 t/L^2} \cos \frac{n\pi}{L} x.$$

6. In Problem 5 we instead find that $X(0) = 0$ and $X(L) = 0$ so that

$$X = c_1 \sin \frac{n\pi}{L} x$$

and

$$u = \frac{2e^{-ht}}{L} \sum_{n=1}^{\infty} \left(\int_0^L f(x) \sin \frac{n\pi}{L} x\, dx \right) e^{-kn^2\pi^2 t/L^2} \sin \frac{n\pi}{L} x.$$

9. (a) The solution is

$$u(x, t) = \sum_{n=1}^{\infty} A_n e^{-kn^2\pi^2 t/100^2} \sin \frac{n\pi}{100} x,$$

where

$$A_n = \frac{2}{100} \left[\int_0^{50} 0.8x \sin \frac{n\pi}{100} x\, dx + \int_{50}^{100} 0.8(100 - x) \sin \frac{n\pi}{100} x\, dx \right] = \frac{320}{n^2\pi^2} \sin \frac{n\pi}{2}.$$

Thus,

$$u(x, t) = \frac{320}{\pi^2} \sum_{n=1}^{\infty} \frac{1}{n^2} \left(\sin \frac{n\pi}{2} \right) e^{-kn^2\pi^2 t/100^2} \sin \frac{n\pi}{100} x.$$

(b) Since $A_n = 0$ for n even, the first five nonzero terms correspond to $n = 1, 3, 5, 7, 9$. In this case $\sin(n\pi/2) = \sin(2p - 1)/2 = (-1)^{p+1}$ for $p = 1, 2, 3, 4, 5$, and

$$u(x, t) = \frac{320}{\pi^2} \sum_{p=1}^{\infty} \frac{(-1)^{p+1}}{(2p - 1)^2} e^{(-1.6352(2p-1)^2\pi^2/100^2)t} \sin \frac{(2p - 1)\pi}{100} x.$$

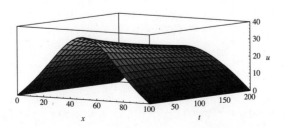

3. Using $u = XT$ and $-\lambda$ as a separation constant we obtain

$$X'' + \lambda X = 0,$$
$$X(0) = 0,$$
$$X(L) = 0,$$

and

$$T'' + \lambda a^2 T = 0,$$
$$T'(0) = 0.$$

Solving the differential equations we get

$$X = c_1 \sin \frac{n\pi}{L} x + c_2 \cos \frac{n\pi}{L} x \quad \text{and} \quad T = c_3 \cos \frac{n\pi a}{L} t + c_4 \sin \frac{n\pi a}{L} t$$

for $n = 1, 2, 3, \ldots$. The boundary and initial conditions give

$$u = \sum_{n=1}^{\infty} A_n \cos \frac{n\pi a}{L} t \sin \frac{n\pi}{L} x.$$

Imposing

$$u(x, 0) = \sum_{n=1}^{\infty} A_n \sin \frac{n\pi}{L} x$$

gives

$$A_n = \frac{2}{L} \left(\int_0^{L/3} \frac{3}{L} x \sin \frac{n\pi}{L} x \, dx + \int_{L/3}^{2L/3} \sin \frac{n\pi}{L} x \, dx + \int_{2L/3}^{L} \left(3 - \frac{3}{L} x \right) \sin \frac{n\pi}{L} x \, dx \right)$$

so that

$$A_1 = \frac{6\sqrt{3}}{\pi^2},$$
$$A_2 = A_3 = A_4 = 0,$$
$$A_5 = -\frac{6\sqrt{3}}{5^2 \pi^2},$$
$$A_6 = 0,$$
$$A_7 = \frac{6\sqrt{3}}{7^2 \pi^2}$$
$$\vdots$$

and

$$u(x, t) = \frac{6\sqrt{3}}{\pi^2} \left(\cos \frac{\pi a}{L} t \sin \frac{\pi}{L} x - \frac{1}{5^2} \cos \frac{5\pi a}{L} t \sin \frac{5\pi}{L} x + \frac{1}{7^2} \cos \frac{7\pi a}{L} t \sin \frac{7\pi}{L} x - \cdots \right).$$

6. Using $u = XT$ and $-\lambda$ as a separation constant we obtain

$$X'' + \lambda X = 0,$$
$$X(0) = 0,$$
$$X(1) = 0,$$

and

$$T'' + \lambda a^2 T = 0,$$
$$T'(0) = 0.$$

Solving the differential equations we get

$$X = c_1 \sin n\pi x + c_2 \cos n\pi x \qquad \text{and} \qquad T = c_3 \cos n\pi a t + c_4 \sin n\pi a t$$

for $n = 1, 2, 3, \ldots$. The boundary and initial conditions give

$$u = \sum_{n=1}^{\infty} A_n \cos nt \, \sin nx.$$

Imposing

$$u(x, 0) = 0.01 \sin 3\pi x = \sum_{n=1}^{\infty} A_n \sin n\pi x$$

gives $A_3 = 0.01$, and $A_n = 0$ for $n = 1, 2, 4, 5, 6, \ldots$ so that

$$u(x, t) = 0.01 \sin 3\pi x \, \cos 3\pi a t.$$

9. Using $u = XT$ and $-\lambda$ as a separation constant we obtain

$$X'' + \lambda X = 0,$$
$$X(0) = 0,$$
$$X(\pi) = 0,$$

and

$$T'' + 2\beta T' + \lambda T = 0,$$
$$T'(0) = 0.$$

Solving the differential equations we get

$$X = c_1 \sin nx + c_2 \cos nx \qquad \text{and} \qquad T = e^{-\beta t}\left(c_3 \cos \sqrt{n^2 - \beta^2}\, t + c_4 \sin \sqrt{n^2 - \beta^2}\, t \right)$$

The boundary conditions on X imply $c_2 = 0$ so

$$X = c_1 \sin nx \qquad \text{and} \qquad T = e^{-\beta t}\left(c_3 \cos \sqrt{n^2 - \beta^2}\, t + c_4 \sin \sqrt{n^2 - \beta^2}\, t \right)$$

and

$$u = \sum_{n=1}^{\infty} e^{-\beta t}\left(A_n \cos \sqrt{n^2 - \beta^2}\, t + B_n \sin \sqrt{n^2 - \beta^2}\, t \right) \sin nx.$$

Imposing

$$u(x, 0) = f(x) = \sum_{n=1}^{\infty} A_n \sin nx$$

and

$$u_t(x, 0) = 0 = \sum_{n=1}^{\infty} \left(B_n \sqrt{n^2 - \beta^2} - \beta A_n \right) \sin nx$$

gives

$$u(x,t) = e^{-\beta t} \sum_{n=1}^{\infty} A_n \left(\cos \sqrt{n^2 - \beta^2}\, t + \frac{\beta}{\sqrt{n^2 - \beta^2}} \sin \sqrt{n^2 - \beta^2}\, t \right) \sin nx,$$

where

$$A_n = \frac{2}{\pi} \int_0^{\pi} f(x) \sin nx \, dx.$$

12. (a) We note that $\xi_x = \eta_x = 1$, $\xi_t = a$, and $\eta_t = -a$. Then

$$\frac{\partial u}{\partial x} = \frac{\partial u}{\partial \xi} \frac{\partial \xi}{\partial x} + \frac{\partial u}{\partial \eta} \frac{\partial \eta}{\partial x} = u_\xi + u_\eta$$

and

$$\frac{\partial^2 u}{\partial x^2} = \frac{\partial}{\partial x}(u_\xi + u_\eta) = \frac{\partial u_\xi}{\partial \xi} \frac{\partial \xi}{\partial x} + \frac{\partial u_\xi}{\partial \eta} \frac{\partial \eta}{\partial x} + \frac{\partial u_\eta}{\partial \xi} \frac{\partial \xi}{\partial x} + \frac{\partial u_\eta}{\partial \eta} \frac{\partial \eta}{\partial x}$$

$$= u_{\xi\xi} + 2u_{\xi\eta} + u_{\eta\eta}.$$

Similarly

$$\frac{\partial^2 u}{\partial t^2} = a^2 (u_{\xi\xi} - 2u_{\xi\eta} + u_{\eta\eta}).$$

Thus

$$a^2 \frac{\partial^2 u}{\partial x^2} = \frac{\partial^2 u}{\partial t^2} \qquad \text{becomes} \qquad \frac{\partial^2 u}{\partial \xi \partial \eta} = 0.$$

(b) Integrating

$$\frac{\partial^2 u}{\partial \xi \partial \eta} = \frac{\partial}{\partial \eta} u_\xi = 0$$

we obtain

$$\int \frac{\partial}{\partial \eta} u_\xi \, d\eta = \int 0 \, d\eta$$

$$u_\xi = f(\xi).$$

Integrating this result with respect to ξ we obtain

$$\int \frac{\partial u}{\partial \xi} \, d\xi = \int f(\xi) \, d\xi$$

$$u = F(\xi) + G(\eta).$$

Since $\xi = x + at$ and $\eta = x - at$, we then have

$$u = F(\xi) + G(\eta) = F(x + at) + G(x - at).$$

Next, we have

$$u(x,t) = F(x + at) + G(x - at)$$

$$u(x,0) = F(x) + G(x) = f(x)$$

$$u_t(x,0) = aF'(x) - aG'(x) = g(x)$$

Integrating the last equation with respect to x gives

$$F(x) - G(x) = \frac{1}{a} \int_{x_0}^{x} g(s) \, ds + c_1.$$

Substituting $G(x) = f(x) - F(x)$ we obtain

$$F(x) = \frac{1}{2} f(x) + \frac{1}{2a} \int_{x_0}^{x} g(s) \, ds + c$$

where $c = c_1/2$. Thus

$$G(x) = \frac{1}{2}f(x) - \frac{1}{2a}\int_{x_0}^{x} g(s)\,ds - c.$$

(c) From the expressions for F and G,

$$F(x+at) = \frac{1}{2}f(x+at) + \frac{1}{2a}\int_{x_0}^{x+at} g(s)\,ds + c$$

$$G(x-at) = \frac{1}{2}f(x-at) - \frac{1}{2a}\int_{x_0}^{x-at} g(s)\,ds - c.$$

Thus,

$$u(x,t) = F(x+at) + G(x-at) = \frac{1}{2}[f(x+at) + f(x-at)] + \frac{1}{2a}\int_{x-at}^{x+at} g(s)\,ds.$$

Here we have used $-\int_{x_0}^{x-at} g(s)\,ds = \int_{x-at}^{x_0} g(s)\,ds.$

15. $u(x,t) = 0 + \dfrac{1}{2a}\displaystyle\int_{x-at}^{x+at}\sin 2s\,ds = \dfrac{1}{2a}\left[\dfrac{-\cos(2x+2at)+\cos(2x-2at)}{2}\right]$

$$= \frac{1}{4a}[-\cos 2x \cos 2at + \sin 2x \sin 2at + \cos 2x \cos 2at + \sin 2x \sin 2at] = \frac{1}{2a}\sin 2x \sin 2at$$

18. (a) Write the differential equation in X from Problem 17 as $X^{(4)} - \alpha^4 X = 0$ where the eigenvalues are $\lambda = \alpha^2$. Then

$$X = c_1 \cosh \alpha x + c_2 \sinh \alpha x + c_3 \cos \alpha x + c_4 \sin \alpha x$$

and using $X(0) = 0$ and $X'(0) = 0$ we find $c_3 = -c_1$ and $c_4 = -c_2$. The conditions $X(L) = 0$ and $X'(L) = 0$ yield the system of equations

$$c_1(\cosh \alpha L - \cos \alpha L) + c_2(\sinh \alpha L - \sin \alpha L) = 0$$

$$c_1(\alpha \sinh \alpha L + \alpha \sin \alpha L) + c_2(\alpha \cosh \alpha L - \alpha \cos \alpha L) = 0.$$

In order for this system to have nontrivial solutions the determinant of the coefficients must be zero. That is,

$$\alpha(\cosh \alpha L - \cos \alpha L)^2 - \alpha(\sinh^2 \alpha L - \sin^2 \alpha L) = 0.$$

Since $\alpha = 0$ leads to $X = 0$, $\lambda = \alpha^2 = 0^2 = 0$ is not an eigenvalue. Then, dividing the above equation by α, we have

$$(\cosh \alpha L - \cos \alpha L)^2 - (\sinh^2 \alpha L - \sin^2 \alpha L)$$

$$= \cosh^2 \alpha L - 2\cosh \alpha L \cos \alpha L + \cos^2 \alpha L - \sinh^2 \alpha L + \sin^2 \alpha L$$

$$= -2\cosh \alpha L \cos \alpha L + 2 = 0$$

or $\cosh \alpha L \cos \alpha L = 1$. Letting $x = \alpha L$ we see that the eigenvalues are $\lambda_n = \alpha_n^2 = x_n^2/L^2$ where x_n, $n = 1$, 2, 3, $\ldots$, are the positive roots of the equation $\cosh x \cos x = 1$.

(b) The equation $\cosh x \cos x = 1$ is the same as $\cos x = \operatorname{sech} x$. The figure indicates that the equation has an infinite number of roots.

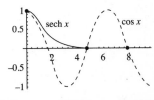

(c) Using a CAS we find the first four positive roots of $\cosh x \cos x = 1$ to be $x_1 = 4.7300$, $x_2 = 7.8532$, $x_3 = 10.9956$, and $x_4 = 14.1372$. Thus the first four eigenvalues are $\lambda_1 = x_1^2/L = 22.3733/L$, $\lambda_2 = x_2^2/L = 61.6728/L$, $\lambda_3 = x_3^2/L = 120.9034/L$, and $\lambda_4 = x_4^2/L = 199.8594/L$.

21. (a) and **(b)** With the given parameters, the solution is

$$u(x,t) = \frac{8}{\pi^2} \sum_{n=1}^{\infty} \frac{1}{n^2} \sin\frac{n\pi}{2} \cos nt \sin nx.$$

For n even, $\sin(n\pi/2) = 0$, so the first six nonzero terms correspond to $n = 1, 3, 5, 7, 9, 11$. In this case $\sin(n\pi/2) = \sin(2p-1)/2 = (-1)^{p+1}$ for $p = 1, 2, 3, 4, 5, 6$, and

$$u(x,t) = \frac{8}{\pi^2} \sum_{p=1}^{\infty} \frac{(-1)^{p+1}}{(2p-1)^2} \cos(2p-1)t \sin(2p-1)x.$$

Frames of the movie corresponding to $t = 0.5, 1, 1.5$, and 2 are shown.

EXERCISES 13.5

Laplace's Equation

3. Using $u = XY$ and $-\lambda$ as a separation constant we obtain

$$X'' + \lambda X = 0,$$
$$X(0) = 0,$$
$$X(a) = 0,$$

and

$$Y'' - \lambda Y = 0,$$
$$Y(b) = 0.$$

With $\lambda = \alpha^2 > 0$ the solutions of the differential equations are

$$X = c_1 \cos\alpha x + c_2 \sin\alpha x \quad \text{and} \quad Y = c_3 \cosh\alpha y + c_4 \sinh\alpha y$$

The boundary and initial conditions imply

$$X = c_2 \sin\frac{n\pi}{a}x \quad \text{and} \quad Y = c_2 \cosh\frac{n\pi}{a}y - c_2 \frac{\cosh\frac{n\pi b}{a}}{\sinh\frac{n\pi b}{a}} \sinh\frac{n\pi}{a}y$$

for $n = 1, 2, 3, \ldots$ so that

$$u = \sum_{n=1}^{\infty} A_n \left(\cosh \frac{n\pi}{a} y - \frac{\cosh \frac{n\pi b}{a}}{\sinh \frac{n\pi b}{a}} \sinh \frac{n\pi}{a} y \right) \sin \frac{n\pi}{a} x.$$

Imposing

$$u(x, 0) = f(x) = \sum_{n=1}^{\infty} A_n \sin \frac{n\pi}{a} x$$

gives

$$A_n = \frac{2}{a} \int_0^a f(x) \sin \frac{n\pi}{a} x \, dx$$

so that

$$u(x, y) = \frac{2}{a} \sum_{n=1}^{\infty} \left(\int_0^a f(x) \sin \frac{n\pi}{a} x \, dx \right) \left(\cosh \frac{n\pi}{a} y - \frac{\cosh \frac{n\pi b}{a}}{\sinh \frac{n\pi b}{a}} \sinh \frac{n\pi}{a} y \right) \sin \frac{n\pi}{a} x.$$

6. Using $u = XY$ and $-\lambda$ as a separation constant we obtain

$$X'' + \lambda X = 0,$$
$$X'(1) = 0$$

and

$$Y'' - \lambda Y = 0,$$
$$Y'(0) = 0,$$
$$Y'(\pi) = 0.$$

With $\lambda = \alpha^2 < 0$ the solutions of the differential equations are

$$X = c_1 \cosh \alpha x + c_2 \sinh \alpha x \quad \text{and} \quad Y = c_3 \cos \alpha y + c_4 \sin \alpha y$$

The boundary and initial conditions imply

$$X = c_1 \cosh nx - c_1 \frac{\sinh n}{\cosh n} \sinh nx \quad \text{and} \quad Y = c_3 \cos ny$$

for $n = 1, 2, 3, \ldots$. Since $\lambda = 0$ is an eigenvalue for both differential equations with corresponding eigenfunctions 1 and 1 we have

$$u = A_0 + \sum_{n=1}^{\infty} A_n \left(\cosh nx - \frac{\sinh n}{\cosh n} \sinh nx \right) \cos ny.$$

Imposing

$$u(0, y) = g(y) = A_0 + \sum_{n=1}^{\infty} A_n \cos ny$$

gives

$$A_0 = \frac{1}{\pi} \int_0^\pi g(y) \, dy \quad \text{and} \quad A_n = \frac{2}{\pi} \int_0^\pi g(y) \cos ny \, dy$$

for $n = 1, 2, 3, \ldots$ so that

$$u(x, y) = \frac{1}{\pi} \int_0^\pi g(y) \, dy + \sum_{n=1}^{\infty} \left(\frac{2}{\pi} \int_0^\pi g(y) \cos ny \, dy \right) \left(\cosh nx - \frac{\sinh n}{\cosh n} \sinh nx \right) \cos ny.$$

9. This boundary-value problem has the form of Problem 1 in this section, with $a = b = 1$, $f(x) = 100$, and $g(x) = 200$. The solution, then, is

$$u(x, y) = \sum_{n=1}^{\infty} (A_n \cosh n\pi y + B_n \sinh n\pi y) \sin n\pi x,$$

where

$$A_n = 2 \int_0^1 100 \sin n\pi x \, dx = 200 \left(\frac{1 - (-1)^n}{n\pi} \right)$$

and

$$B_n = \frac{1}{\sinh n\pi} \left[2 \int_0^1 200 \sin n\pi x \, dx - A_n \cosh n\pi \right]$$

$$= \frac{1}{\sinh n\pi} \left[400 \left(\frac{1 - (-1)^n}{n\pi} \right) - 200 \left(\frac{1 - (-1)^n}{n\pi} \right) \cosh n\pi \right]$$

$$= 200 \left[\frac{1 - (-1)^n}{n\pi} \right] [2 \operatorname{csch} n\pi - \coth n\pi].$$

12. Using $u = XY$ and $-\lambda$ as a separation constant we obtain

$$X'' + \lambda X = 0,$$
$$X'(0) = 0,$$
$$X'(\pi) = 0,$$

and

$$Y'' - \lambda Y = 0.$$

With $\lambda = \alpha^2 > 0$ the solutions of the differential equations are

$$X = c_1 \cos \alpha x + c_2 \sin \alpha x \qquad \text{and} \qquad Y = c_3 e^{\alpha y} + c_4 e^{-\alpha y}$$

The boundary conditions at $x = 0$ and $x = \pi$ imply $c_2 = 0$ so $X = c_1 \cos nx$ for $n = 1, 2, 3, \ldots$. Now the boundedness of u as $y \to \infty$ implies $c_3 = 0$, so $Y = c_4 e^{-ny}$. In this problem $\lambda = 0$ is also an eigenvalue with corresponding eigenfunction 1 so that

$$u = A_0 + \sum_{n=1}^{\infty} A_n e^{-ny} \cos nx.$$

Imposing

$$u(x, 0) = f(x) = A_0 + \sum_{n=1}^{\infty} A_n \cos nx$$

gives

$$A_0 = \frac{1}{\pi} \int_0^{\pi} f(x) \, dx \qquad \text{and} \qquad A_n = \frac{2}{\pi} \int_0^{\pi} f(x) \cos nx \, dx$$

so that

$$u(x, y) = \frac{1}{\pi} \int_0^{\pi} f(x) \, dx + \sum_{n=1}^{\infty} \left(\frac{2}{\pi} \int_0^{\pi} f(x) \cos nx \, dx \right) e^{-ny} \cos nx.$$

15. Referring to the discussion in this section of the text we identify $a = b = \pi$, $f(x) = 0$, $g(x) = 1$, $F(y) = 1$, and $G(y) = 1$. Then $A_n = 0$ and

$$u_1(x, y) = \sum_{n=1}^{\infty} B_n \sinh ny \sin nx$$

where

$$B_n = \frac{2}{\pi \sinh n\pi} \int_0^{\pi} \sin nx \, dx = \frac{2[1 - (-1)^n]}{n\pi \sinh n\pi}.$$

Next

$$u_2(x, y) = \sum_{n=1}^{\infty} (A_n \cosh nx + B_n \sinh nx) \sin ny$$

where

$$A_n = \frac{2}{\pi} \int_0^{\pi} \sin ny \, dy = \frac{2[1 - (-1)^n]}{n\pi}$$

and

$$B_n = \frac{1}{\sinh n\pi} \left(\frac{2}{\pi} \int_0^{\pi} \sin ny \, dy - A_n \cosh n\pi \right)$$

$$= \frac{1}{\sinh n\pi} \left(\frac{2[1 - (-1)^n]}{n\pi} - \frac{2[1 - (-1)^n]}{n\pi} \cosh n\pi \right)$$

$$= \frac{2[1 - (-1)^n]}{n\pi \sinh n\pi} (1 - \cosh n\pi).$$

Now

$$A_n \cosh nx + B_n \sinh nx = \frac{2[1 - (-1)^n]}{n\pi} \left[\cosh nx + \frac{\sinh nx}{\sinh n\pi} (1 - \cosh n\pi) \right]$$

$$= \frac{2[1 - (-1)^n]}{n\pi \sinh n\pi} [\cosh nx \sinh n\pi + \sinh nx - \sinh nx \cosh n\pi]$$

$$= \frac{2[1 - (-1)^n]}{n\pi \sinh n\pi} [\sinh nx + \sinh n(\pi - x)]$$

and

$$u(x, y) = u_1 + u_2 = \frac{2}{\pi} \sum_{n=1}^{\infty} \frac{1 - (-1)^n}{n \sinh n\pi} \sinh ny \sin nx$$

$$+ \frac{2}{\pi} \sum_{n=1}^{\infty} \frac{[1 - (-1)^n][\sinh nx + \sinh n(\pi - x)]}{n \sinh n\pi} \sin ny.$$

18. (a)

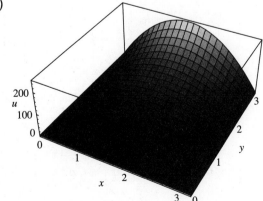

(b) The maximum value occurs at $(\pi/2, \pi)$ and is $f(\pi/2) = 25\pi^2$.

(c) The coefficients are

$$A_n = \frac{2}{\pi} \operatorname{csch} n\pi \int_0^\pi 100x(\pi - x) \sin nx \, dx$$

$$= \frac{200 \operatorname{csch} n\pi}{\pi} \left[\frac{200}{n^3} \left(1 - (-1)^n \right) \right] = \frac{400}{n^3 \pi} \left[1 - (-1)^n \right] \operatorname{csch} n\pi.$$

See part **(a)** for the graph.

21. (a) Assuming $u(x, y) = X(x)Y(y)$ and substituting into the partial differential equationwe get $X''Y + XY'' = 0$. Separating variables and using $\lambda = \alpha^2$ we get

$$X'' - \alpha^2 X = 0, \quad X'(0) = 0,$$

which implies $X(x) = c_3 \cosh \alpha x$. From

$$Y'' + \alpha^2 Y = 0, \quad Y'(0) = 0, \quad Y'(b) = 0$$

we get $Y(y) = c_1 \cos \alpha y + c_2 \sin \alpha y$ and eigenvalues $\lambda_n = n^2 \pi^2 / b^2$, $n = 1, 2, 3, \ldots$. The corresponding eigenfunctions are $Y(y) = c_1 \cos(n\pi y / b)$. For $\lambda = 0$ the boundary conditions applied to $X(x) = c_3 + c_4 x$ and $Y(y) = c_1 + c_2 y$ imply $X = c_3$ and $Y = c_1$. Forming products and using the superposition principle then gives

$$u(x, y) = A_0 + \sum_{n=1}^\infty A_n \cosh \frac{n\pi}{b} x \cos \frac{n\pi}{b} y.$$

The remaining boundary condition, $u_x(a) = g(y)$ implies

$$g(y) = \left. \frac{\partial u}{\partial x} \right|_{x=a} = \sum_{n=1}^\infty A_n \frac{n\pi}{b} \sinh \frac{n\pi}{b} x \cos \frac{n\pi}{b} y,$$

and so A_0 remains arbitrary. In order that the series expression for $g(y)$ be a cosine series, the constant term in the series, $a_0/2$, must be 0. Thus, from Section 12.3 in the text,

$$a_0 = \frac{2}{b} \int_0^b g(y) dy = 0 \quad \text{so} \quad \int_0^b g(y) dy = 0.$$

Also,

$$A_n \frac{n\pi}{b} \sinh \frac{n\pi}{b} a = \frac{2}{b} \int_0^b g(y) \cos \frac{n\pi}{b} y \, dy$$

and

$$A_n = \frac{2}{n\pi \sinh n\pi a/b} \int_0^b g(y) \cos \frac{n\pi}{b} y \, dy.$$

The solution is then

$$u(x, y) = A_0 + \sum_{n=1}^\infty A_n \cosh \frac{n\pi}{b} x \cos \frac{n\pi}{b} y,$$

where the A_n are defined above and A_0 is arbitrary. In general, Neumann problems do not have unique solutions.

For a physical interpretation of the compatibility condition $\int_0^b g(y) dy = 0$ see the texts *Elementary Partial Differential Equations* by Paul Berg and James McGregor (Holden-Day) and *Partial Differential Equations of Mathematical Physics* by Tyn Myint-U (North Holland).

(b) Since the derivative of a constant is 0, $v(x, y) = u(x, y) + c$ satisfies the five boundary conditions in the Neumann problem as given in part **(a)** of this problem in the text.

EXERCISES 13.6

Nonhomogeneous Equations and Boundary Conditions

3. If we let $u(x,t) = v(x,t) + \psi(x)$, then we obtain as in Example 1 in the text

$$k\psi'' + r = 0$$

or

$$\psi(x) = -\frac{r}{2k}x^2 + c_1 x + c_2.$$

The boundary conditions become

$$u(0,t) = v(0,t) + \psi(0) = u_0$$
$$u(1,t) = v(1,t) + \psi(1) = u_0.$$

Letting $\psi(0) = \psi(1) = u_0$ we obtain homogeneous boundary conditions in v:

$$v(0,t) = 0 \quad \text{and} \quad v(1,t) = 0.$$

Now $\psi(0) = \psi(1) = u_0$ implies $c_2 = u_0$ and $c_1 = r/2k$. Thus

$$\psi(x) = -\frac{r}{2k}x^2 + \frac{r}{2k}x + u_0 = u_0 - \frac{r}{2k}x(x-1).$$

To determine $v(x,t)$ we solve

$$k\frac{\partial^2 v}{\partial x^2} = \frac{\partial v}{dt}, \quad 0 < x < 1, \ t > 0$$

$$v(0,t) = 0, \quad v(1,t) = 0,$$

$$v(x,0) = \frac{r}{2k}x(x-1) - u_0.$$

Separating variables, we find

$$v(x,t) = \sum_{n=1}^{\infty} A_n e^{-kn^2\pi^2 t} \sin n\pi x,$$

where

$$A_n = 2\int_0^1 \left[\frac{r}{2k}x(x-1) - u_0\right]\sin n\pi x\, dx = 2\left[\frac{u_0}{n\pi} + \frac{r}{kn^3\pi^3}\right][(-1)^n - 1]. \tag{1}$$

Hence, a solution of the original problem is

$$u(x,t) = \psi(x) + v(x,t) = u_0 - \frac{r}{2k}x(x-1) + \sum_{n=1}^{\infty} A_n e^{-kn^2\pi^2 t}\sin n\pi x,$$

where A_n is defined in (1).

6. Substituting $u(x,t) = v(x,t) + \psi(x)$ into the partial differential equation gives

$$k\frac{\partial^2 v}{\partial x^2} + k\psi'' - hv - h\psi = \frac{\partial v}{\partial t}.$$

This equation will be homogeneous provided ψ satisfies

$$k\psi'' - h\psi = 0.$$

Since k and h are positive, the general solution of this latter linear second-order equation is

$$\psi(x) = c_1 \cosh\sqrt{\frac{h}{k}}\,x + c_2 \sinh\sqrt{\frac{h}{k}}\,x.$$

273

From $\psi(0) = 0$ and $\psi(\pi) = u_0$ we find $c_1 = 0$ and $c_2 = u_0 / \sinh \sqrt{h/k}\,\pi$. Hence

$$\psi(x) = u_0 \frac{\sinh \sqrt{h/k}\,x}{\sinh \sqrt{h/k}\,\pi}.$$

Now the new problem is

$$k \frac{\partial^2 v}{\partial x^2} - hv = \frac{\partial v}{\partial t}, \quad 0 < x < \pi, \ t > 0$$

$$v(0, t) = 0, \quad v(\pi, t) = 0, \quad t > 0$$

$$v(x, 0) = -\psi(x), \quad 0 < x < \pi.$$

If we let $v = XT$ then

$$\frac{X''}{X} = \frac{T' + hT}{kT} = -\lambda.$$

With $\lambda = \alpha^2 > 0$, the separated differential equations

$$X'' + \alpha^2 X = 0 \quad \text{and} \quad T' + \left(h + k\alpha^2\right) T = 0.$$

have the respective solutions

$$X(x) = c_3 \cos \alpha x + c_4 \sin \alpha x$$

$$T(t) = c_5 e^{-\left(h + k\alpha^2\right)t}.$$

From $X(0) = 0$ we get $c_3 = 0$ and from $X(\pi) = 0$ we find $\alpha = n$ for $n = 1, 2, 3, \ldots$. Consequently, it follows that

$$v(x, t) = \sum_{n=1}^{\infty} A_n e^{-\left(h + kn^2\right)t} \sin nx$$

where

$$A_n = -\frac{2}{\pi} \int_0^\pi \psi(x) \sin nx \, dx.$$

Hence a solution of the original problem is

$$u(x, t) = u_0 \frac{\sinh \sqrt{h/k}\,x}{\sinh \sqrt{h/k}\,\pi} + e^{-ht} \sum_{n=1}^{\infty} A_n e^{-kn^2 t} \sin nx$$

where

$$A_n = -\frac{2}{\pi} \int_0^\pi u_0 \frac{\sinh \sqrt{h/k}\,x}{\sinh \sqrt{h/k}\,\pi} \sin nx \, dx.$$

Using the exponential definition of the hyperbolic sine and integration by parts we find

$$A_n = \frac{2u_0 nk(-1)^n}{\pi \left(h + kn^2\right)}.$$

9. Substituting $u(x, t) = v(x, t) + \psi(x)$ into the partial differential equation gives

$$a^2 \frac{\partial^2 v}{\partial x^2} + a^2 \psi'' + Ax = \frac{\partial^2 v}{\partial t^2}.$$

This equation will be homogeneous provided ψ satisfies

$$a^2 \psi'' + Ax = 0.$$

The solution of this differential equation is

$$\psi(x) = -\frac{A}{6a^2} x^3 + c_1 x + c_2.$$

274

From $\psi(0) = 0$ we obtain $c_2 = 0$, and from $\psi(1) = 0$ we obtain $c_1 = A/6a^2$. Hence

$$\psi(x) = \frac{A}{6a^2}(x - x^3).$$

Now the new problem is

$$a^2 \frac{\partial^2 v}{\partial x^2} = \frac{\partial^2 v}{\partial t^2}$$

$$v(0, t) = 0, \quad v(1, t) = 0, \quad t > 0,$$

$$v(x, 0) = -\psi(x), \quad v_t(x, 0) = 0, \quad 0 < x < 1.$$

Identifying this as the wave equation solved in Section 13.4 in the text with $L = 1$, $f(x) = -\psi(x)$, and $g(x) = 0$ we obtain

$$v(x, t) = \sum_{n=1}^{\infty} A_n \cos n\pi a t \sin n\pi x$$

where

$$A_n = 2 \int_0^1 [-\psi(x)] \sin n\pi x \, dx = \frac{A}{3a^2} \int_0^1 (x^3 - x) \sin n\pi x \, dx = \frac{2A(-1)^n}{a^2 \pi^3 n^3}.$$

Thus

$$u(x, t) = \frac{A}{6a^2}(x - x^3) + \frac{2A}{a^2 \pi^3} \sum_{n=1}^{\infty} \frac{(-1)^n}{n^3} \cos n\pi a t \sin n\pi x.$$

12. Substituting $u(x, y) = v(x, y) + \psi(x)$ into Poisson's equation we obtain

$$\frac{\partial^2 v}{\partial x^2} + \psi''(x) + h + \frac{\partial^2 v}{\partial y^2} = 0.$$

The equation will be homogeneous provided ψ satisfies $\psi''(x) + h = 0$ or $\psi(x) = -\frac{h}{2}x^2 + c_1 x + c_2$. From $\psi(0) = 0$ we obtain $c_2 = 0$. From $\psi(\pi) = 1$ we obtain

$$c_1 = \frac{1}{\pi} + \frac{h\pi}{2}.$$

Then

$$\psi(x) = \left(\frac{1}{\pi} + \frac{h\pi}{2}\right)x - \frac{h}{2}x^2.$$

The new boundary-value problem is

$$\frac{\partial^2 v}{\partial x^2} + \frac{\partial^2 v}{\partial y^2} = 0$$

$$v(0, y) = 0, \quad v(\pi, y) = 0,$$

$$v(x, 0) = -\psi(x), \quad 0 < x < \pi.$$

This is Problem 11 in Section 13.5. The solution is

$$v(x, y) = \sum_{n=1}^{\infty} A_n e^{-ny} \sin nx$$

where

$$A_n = \frac{2}{\pi} \int_0^{\pi} [-\psi(x) \sin nx] \, dx$$

$$= \frac{2(-1)^n}{m}\left(\frac{1}{\pi} + \frac{h\pi}{2}\right) - h(-1)^n\left(\frac{\pi}{n} + \frac{2}{n^2}\right).$$

Thus

$$u(x,y) = v(x,y) + \psi(x) = \left(\frac{1}{\pi} + \frac{h\pi}{2}\right)x - \frac{h}{2}x^2 + \sum_{n=1}^{\infty} A_n e^{-ny}\sin nx.$$

15. Identifying $k = 1$ and $L = 1$ we see that the eigenfunctions of $X'' + \lambda X = 0$, $X(0) = 0$, $X(1) = 0$ are $\sin n\pi x$, $n = 1, 2, 3, \ldots$. Assuming that $u(x,t) = \sum_{n=1}^{\infty} u_n(t)\sin n\pi x$, the formal partial derivatives of u are

$$\frac{\partial^2 u}{\partial x^2} = \sum_{n=1}^{\infty} u_n(t)(-n^2\pi^2)\sin n\pi x \qquad \text{and} \qquad \frac{\partial u}{\partial t} = \sum_{n=1}^{\infty} u_n'(t)\sin n\pi x.$$

Assuming that $-1 + x - x\cos t = \sum_{n=1}^{\infty} F_n(t)\sin n\pi x$ we have

$$F_n(t) = \frac{2}{1}\int_0^1 (-1 + x - x\cos t)\sin n\pi x\, dx = \frac{2[-1 + (-1)^n\cos t]}{n\pi}.$$

Then

$$-1 + x - x\cos t = \frac{2}{\pi}\sum_{n=1}^{\infty} \frac{-1 + (-1)^n\cos t}{n}\sin n\pi x$$

and

$$u_t - u_{xx} = \sum_{n=1}^{\infty}[u_n'(t) + n^2\pi^2 u_n(t)]\sin n\pi x$$

$$= -1 + x - x\cos t = \frac{2}{\pi}\sum_{n=1}^{\infty}\frac{-1 + (-1)^n\cos t}{n}\sin n\pi x.$$

Equating coefficients we obtain

$$u_n'(t) + n^2\pi^2 u_n(t) = \frac{2[-1 + (-1)^n\cos t]}{n\pi}.$$

This is a linear first-order differential equation whose solution is

$$u_n(t) = \frac{2}{n\pi}\left[-\frac{1}{n^2\pi^2} + (-1)^n\frac{n^2\pi^2\cos t + \sin t}{n^4\pi^4 + 1}\right] + C_n e^{-n^2\pi^2 t}.$$

Thus

$$u(x,t) = \sum_{n=1}^{\infty}\frac{2}{n\pi}\left[-\frac{1}{n^2\pi^2} + (-1)^n\frac{n^2\pi^2\cos t + \sin t}{n^4\pi^4 + 1}\right]\sin n\pi x + \sum_{n=1}^{\infty} C_n e^{-n^2\pi^2 t}\sin n\pi x$$

and $u(x,0) = x(1-x)$ implies

$$\sum_{n=1}^{\infty}\frac{2}{n\pi}\left[-\frac{1}{n^2\pi^2} + (-1)^n\frac{n^2\pi^2}{n^4\pi^4 + 1} + C_n\right]\sin n\pi x = x(1-x).$$

Hence

$$\frac{2}{n\pi}\left[-\frac{1}{n^2\pi^2} + (-1)^n\frac{n^2\pi^2}{n^4\pi^4 + 1} + C_n\right] = \frac{2}{1}\int_0^1 x(1-x)\sin n\pi x\, dx = 2\left[\frac{1 - (-1)^n}{n^3\pi^3}\right]$$

and

$$C_n = \frac{4 - 2(-1)^n}{n^3\pi^3} - (-1)^n\frac{2n\pi}{n^4\pi^4 + 1}.$$

Therefore

$$u(x,t) = \sum_{n=1}^{\infty}\frac{2}{n\pi}\left[-\frac{1}{n^2\pi^2} + (-1)^n\frac{n^2\pi^2\cos t + \sin t}{n^4\pi^4 + 1}\right]\sin n\pi x$$

$$+ \sum_{n=1}^{\infty}\left[\frac{4 - 2(-1)^n}{n^3\pi^3} - (-1)^n\frac{2n\pi}{n^4\pi^4 + 1}\right]e^{-n^2\pi^2 t}\sin n\pi x.$$

18. To match this problem to (1) in the text we identify $k = 1$, $L = 1$, $F(x,t) = 2t + 3tx$, $u_0(t) = t^2$, $u_1(t) = 1$, and $f(x) = x^2$. To construct $\psi(x,t)$ we use

$$\psi(x,t) = u_0(t) + \frac{x}{L}[u_1(t) - u_0(t)] = x + (1-x)t^2,$$

so $G(x,t) = F(x,t) - \psi_t(x,t) = 2t + 3tx - 2(1-x)t = 5tx$. Then the substitution

$$u(x,t) = v(x,t) + \psi(x,t) = v(x,t) + x + (1-x)t^2$$

leads to the boundary-value problem

$$\frac{\partial^2 v}{\partial x^2} + 5tx = \frac{\partial v}{\partial t}, \quad 0 < x < 1, \quad t > 0$$
$$v(0,t) = 0, \quad v(1,t) = 0, \quad t > 0$$
$$v(x,0) = x^2 - x, \quad 0 < x < 1.$$

The eigenvalues and eigenfunctions of the Sturm-Liouville problem

$$X'' + \lambda X = 0, \quad X(0) = 0, \quad X(1) = 0$$

are $\lambda_n = \alpha_n^2 = n^2\pi^2$ and $\sin n\pi x$, $n = 1, 2, 3, \ldots$. With $G(x,t) = 5tx$ we assume for fixed t that v and G can be written as Fourier sine series:

$$v(x,t) = \sum_{n=1}^{\infty} v_n(t)\sin n\pi x$$

and

$$G(x,t) = \sum_{n=1}^{\infty} G_n(t)\sin n\pi x.$$

By treating t as a parameter, the coefficients G_n can be computed:

$$G_n(t) = \frac{2}{1}\int_0^1 5tx\sin n\pi x\,dx = 10t\int_0^1 x\sin n\pi x\,dx = \frac{10t}{n\pi}(-1)^{n+1}.$$

Hence

$$5tx = \sum_{n=1}^{\infty}(-1)^{n+1}\frac{10t}{n\pi}\sin n\pi x.$$

Now, using the series representation for $v(x,t)$, we have

$$\frac{\partial^2 v}{\partial x^2} = \sum_{n=1}^{\infty} v_n(t)(-n^2\pi^2)\sin n\pi x \quad \text{and} \quad \frac{\partial v}{\partial t} = \sum_{n=1}^{\infty} v_n'(t)\sin n\pi x.$$

Writing the partial differential equation as $v_t - v_{xx} = 5tx$ and using the above results we have

$$\sum_{n=1}^{\infty}[v_n'(t) + n^2\pi^2 v_n(t)]\sin n\pi x = \sum_{n=1}^{\infty}(-1)^{n+1}\frac{10t}{n\pi}\sin n\pi x.$$

Equating coefficients we get

$$v_n'(t) + n^2\pi^2 v_n(t) = (-1)^{n+1}\frac{10t}{n\pi}.$$

For each n this is a linear first-order differential equation whose general solution is

$$v_n(t) = 10(-1)^{n+1}\frac{n^2\pi^2 t - 1}{n^5\pi^5} + C_n e^{-n^2\pi^2 t}.$$

Thus

$$v(x,t) = \sum_{n=1}^{\infty}\left[10(-1)^{n+1}\frac{n^2\pi^2 t - 1}{n^5\pi^5} + C_n e^{-n^2\pi^2 t}\right]\sin n\pi x.$$

The initial condition $v(x,0) = x^2 - x$ implies

$$\sum_{n=1}^{\infty} \left[10(-1)^{n+1} \frac{-1}{n^5 \pi^5} + C_n \right] \sin n\pi x = x^2 - x.$$

Thinking of $x^2 - x$ as a Fourier sine series with coefficients $2\int_0^1 (x^2 - x) \sin n\pi x\, dx = [4(-1)^n - 4]/n^3\pi^3$ we equate coefficients to obtain

$$\frac{10(-1)^n}{n^5 \pi^5} + C_n = \frac{4(-1)^n - 4}{n^3 \pi^3}$$

so

$$C_n = \frac{4(-1)^n - 4}{n^3 \pi^3} - \frac{10(-1)^n}{n^5 \pi^5}.$$

Therefore

$$v(x,t) = \sum_{n=1}^{\infty} \left[10(-1)^{n+1} \frac{n^2\pi^2 t - 1}{n^5 \pi^5} + \left(\frac{4(-1)^n - 4}{n^3 \pi^3} - \frac{10(-1)^n}{n^5 \pi^5} \right) e^{-n^2\pi^2 t} \right] \sin n\pi x$$

and

$$u(x,t) = v(x,t) + \psi(x,t)$$

$$= x + (1-x)t^2 + \sum_{n=1}^{\infty} \left[10(-1)^{n+1} \frac{n^2\pi^2 t - 1}{n^5 \pi^5} + \left(\frac{4(-1)^n - 4}{n^3 \pi^3} - \frac{10(-1)^n}{n^5 \pi^5} \right) e^{-n^2\pi^2 t} \right] \sin n\pi x.$$

EXERCISES 13.7

Orthogonal Series Expansions

3. Separating variables in Laplace's equation gives

$$X'' + \alpha^2 X = 0$$
$$Y'' - \alpha^2 Y = 0$$

and

$$X(x) = c_1 \cos \alpha x + c_2 \sin \alpha x$$
$$Y(y) = c_3 \cosh \alpha y + c_4 \sinh \alpha y.$$

From $u(0,y) = 0$ we obtain $X(0) = 0$ and $c_1 = 0$. From $u_x(a,y) = -hu(a,y)$ we obtain $X'(a) = -hX(a)$ and

$$\alpha \cos \alpha a = -h \sin \alpha a \qquad \text{or} \qquad \tan \alpha a = -\frac{\alpha}{h}.$$

Let α_n, where $n = 1, 2, 3, \ldots$, be the consecutive positive roots of this equation. From $u(x,0) = 0$ we obtain $Y(0) = 0$ and $c_3 = 0$. Thus

$$u(x,y) = \sum_{n=1}^{\infty} A_n \sinh \alpha_n y \sin \alpha_n x.$$

Now

$$f(x) = \sum_{n=1}^{\infty} A_n \sinh \alpha_n b \sin \alpha_n x$$

and

$$A_n \sinh \alpha_n b = \frac{\int_0^a f(x) \sin \alpha_n x\, dx}{\int_0^a \sin^2 \alpha_n x\, dx}.$$

Since

$$\int_0^a \sin^2 \alpha_n x \, dx = \frac{1}{2} \left[a - \frac{1}{2\alpha_n} \sin 2\alpha_n a \right] = \frac{1}{2} \left[a - \frac{1}{\alpha_n} \sin \alpha_n a \cos \alpha_n a \right]$$

$$= \frac{1}{2} \left[a - \frac{1}{h\alpha_n} (h \sin \alpha_n a) \cos \alpha_n a \right]$$

$$= \frac{1}{2} \left[a - \frac{1}{h\alpha_n} (-\alpha_n \cos \alpha_n a) \cos \alpha_n a \right] = \frac{1}{2h} \left[ah + \cos^2 \alpha_n a \right],$$

we have

$$A_n = \frac{2h}{\sinh \alpha_n b [ah + \cos^2 \alpha_n a]} \int_0^a f(x) \sin \alpha_n x \, dx.$$

6. Substituting $u(x,t) = v(x,t) + \psi(x)$ into the partial differential equation gives

$$a^2 \frac{\partial^2 v}{\partial x^2} + \psi''(x) = \frac{\partial^2 v}{\partial t^2}.$$

This equation will be homogeneous if $\psi''(x) = 0$ or $\psi(x) = c_1 x + c_2$. The boundary condition $u(0,t) = 0$ implies $\psi(0) = 0$ which implies $c_2 = 0$. Thus $\psi(x) = c_1 x$. Using the second boundary condition, we obtain

$$E \left(\frac{\partial v}{\partial x} + \psi' \right) \bigg|_{x=L} = F_0,$$

which will be homogeneous when

$$E\psi'(L) = F_0.$$

Since $\psi'(x) = c_1$ we conclude that $c_1 = F_0/E$ and

$$\psi(x) = \frac{F_0}{E} x.$$

The new boundary-value problem is

$$a^2 \frac{\partial^2 v}{\partial x^2} = \frac{\partial^2 v}{\partial t^2}, \quad 0 < x < L, \quad t > 0$$

$$v(0,t) = 0, \quad \frac{\partial v}{\partial x} \bigg|_{x=L} = 0, \quad t > 0,$$

$$v(x,0) = -\frac{F_0}{E} x, \quad \frac{\partial v}{\partial t} \bigg|_{t=0} = 0, \quad 0 < x < L.$$

Referring to Example 2 in the text we see that

$$v(x,t) = \sum_{n=1}^{\infty} A_n \cos a \left(\frac{2n-1}{2L} \right) \pi t \sin \left(\frac{2n-1}{2L} \right) \pi x$$

where

$$-\frac{F_0}{E} x = \sum_{n=1}^{\infty} A_n \sin \left(\frac{2n-1}{2L} \right) \pi x$$

and

$$A_n = \frac{-F_0 \int_0^L x \sin \left(\frac{2n-1}{2L} \right) \pi x \, dx}{E \int_0^L \sin^2 \left(\frac{2n-1}{2L} \right) \pi x \, dx} = \frac{8 F_0 L (-1)^n}{E \pi^2 (2n-1)^2}.$$

Thus

$$u(x,t) = v(x,t) + \psi(x) = \frac{F_0}{E} x + \frac{8 F_0 L}{E \pi^2} \sum_{n=1}^{\infty} \frac{(-1)^n}{(2n-1)^2} \cos a \left(\frac{2n-1}{2L} \right) \pi t \sin \left(\frac{2n-1}{2L} \right) \pi x.$$

9. The eigenfunctions of the associated homogeneous boundary-value problem are $\sin \alpha_n x$, $n = 1, 2, 3, \ldots$, where the α_n are the consecutive positive roots of $\tan \alpha = -\alpha$. We assume that

$$u(x, t) = \sum_{n=1}^{\infty} u_n(t) \sin \alpha_n x \qquad \text{and} \qquad x e^{-2t} = \sum_{n=1}^{\infty} F_n(t) \sin \alpha_n x.$$

Then

$$F_n(t) = \frac{e^{-2t} \int_0^1 x \sin \alpha_n x \, dx}{\int_0^1 \sin^2 \alpha_n x \, dx}.$$

Since $\alpha_n \cos \alpha_n = -\sin \alpha_n$ and

$$\int_0^1 \sin^2 \alpha_n x \, dx = \frac{1}{2} \left[1 - \frac{1}{2\alpha_n} \sin 2\alpha_n \right],$$

we have

$$e^{-2t} \int_0^1 x \sin \alpha_n x \, dx = e^{-2t} \left(\frac{\sin \alpha_n - \alpha_n \cos \alpha_n}{\alpha_n^2} \right) = \frac{2 \sin \alpha_n}{\alpha_n^2} e^{-2t}$$

$$\int_0^1 \sin^2 \alpha_n x \, dx = \frac{1}{2} [1 + \cos^2 \alpha_n]$$

and so

$$F_n(t) = \frac{4 \sin \alpha_n}{\alpha_n^2 (1 + \cos^2 \alpha_n)} e^{-2t}.$$

Substituting the assumptions into $u_t - k u_{xx} = x e^{-2t}$ and equating coefficients leads to the linear first-order differential equation

$$u_n'(t) + k \alpha_n^2 u(t) = \frac{4 \sin \alpha_n}{\alpha_n^2 (1 + \cos^2 \alpha_n)} e^{-2t}$$

whose solution is

$$u_n(t) = \frac{4 \sin \alpha_n}{\alpha_n^2 (1 + \cos^2 \alpha_n)(k \alpha_n^2 - 2)} e^{-2t} + C_n e^{-k \alpha_n^2 t}.$$

From

$$u(x, t) = \sum_{n=1}^{\infty} \left[\frac{4 \sin \alpha_n}{\alpha_n^2 (1 + \cos^2 \alpha_n)(k \alpha_n^2 - 2)} e^{-2t} + C_n e^{-k \alpha_n^2 t} \right] \sin \alpha_n x$$

and the initial condition $u(x, 0) = 0$ we see

$$C_n = -\frac{4 \sin \alpha_n}{\alpha_n^2 (1 + \cos^2 \alpha_n)(k \alpha_n^2 - 2)}.$$

The formal solution of the original problem is then

$$u(x, t) = \sum_{n=1}^{\infty} \frac{4 \sin \alpha_n}{\alpha_n^2 (1 + \cos^2 \alpha_n)(k \alpha_n^2 - 2)} \left(e^{-2t} - e^{-k \alpha_n^2 t} \right) \sin \alpha_n x.$$

3. In this problem we need to solve the partial differential equation

$$a^2 \left(\frac{\partial^2 u}{\partial x^2} + \frac{\partial^2 u}{\partial y^2} \right) = \frac{\partial^2 u}{\partial t^2} \,.$$

To separate this equation we try $u(x, y, t) = X(x)Y(y)T(t)$:

$$a^2 (X''YT + XY''T) = XYT''$$

$$\frac{X''}{X} = -\frac{Y''}{Y} + \frac{T''}{a^2 T} = -\alpha^2.$$

Then

$$X'' + \alpha^2 X = 0 \tag{1}$$

$$\frac{Y''}{Y} = \frac{T''}{a^2 T} + \alpha^2 = -\beta^2$$

$$Y'' + \beta^2 Y = 0 \tag{2}$$

$$T'' + a^2 \left(\alpha^2 + \beta^2 \right) T = 0. \tag{3}$$

The general solutions of equations **(1)**, **(2)**, and **(3)** are, respectively,

$$X(x) = c_1 \cos \alpha x + c_2 \sin \alpha x$$

$$Y(y) = c_3 \cos \beta y + c_4 \sin \beta y$$

$$T(t) = c_5 \cos a\sqrt{\alpha^2 + \beta^2}\, t + c_6 \sin a\sqrt{\alpha^2 + \beta^2}\, t.$$

The conditions $X(0) = 0$ and $Y(0) = 0$ give $c_1 = 0$ and $c_3 = 0$. The conditions $X(\pi) = 0$ and $Y(\pi) = 0$ yield two sets of eigenvalues:

$$\alpha = m, \ m = 1, 2, 3, \ldots \quad \text{and} \quad \beta = n, \ n = 1, 2, 3, \ldots.$$

A product solution of the partial differential equation that satisfies the boundary conditions is

$$u_{mn}(x, y, t) = \left(A_{mn} \cos a\sqrt{m^2 + n^2}\, t + B_{mn} \sin a\sqrt{m^2 + n^2}\, t \right) \sin mx \sin ny.$$

To satisfy the initial conditions we use the superposition principle:

$$u(x, y, t) = \sum_{m=1}^{\infty} \sum_{n=1}^{\infty} \left(A_{mn} \cos a\sqrt{m^2 + n^2}\, t + B_{mn} \sin a\sqrt{m^2 + n^2}\, t \right) \sin mx \sin ny.$$

The initial condition $u_t(x, y, 0) = 0$ implies $B_{mn} = 0$ and

$$u(x, y, t) = \sum_{m=1}^{\infty} \sum_{n=1}^{\infty} A_{mn} \cos a\sqrt{m^2 + n^2}\, t \sin mx \sin ny.$$

At $t = 0$ we have

$$xy(x - \pi)(y - \pi) = \sum_{m=1}^{\infty} \sum_{n=1}^{\infty} A_{mn} \sin mx \sin ny.$$

Using (12) and (13) in the text, it follows that

$$A_{mn} = \frac{4}{\pi^2} \int_0^\pi \int_0^\pi xy(x - \pi)(y - \pi)\sin mx \sin ny\, dx\, dy$$

$$= \frac{4}{\pi^2} \int_0^\pi x(x - \pi)\sin mx\, dx \int_0^\pi y(y - \pi)\sin ny\, dy$$

$$= \frac{16}{m^3 n^3 \pi^2}[(-1)^m - 1][(-1)^n - 1].$$

6. In this problem we try $u(x, y, z) = X(x)Y(y)Z(z)$ to separate Laplace's equation in three dimensions:

$$X''YZ + XY''Z + XYZ'' = 0$$
$$\frac{X''}{X} = -\frac{Y''}{Y} - \frac{Z''}{Z} = -\alpha^2.$$

Then

$$X'' + \alpha^2 X = 0 \tag{4}$$

$$\frac{Y''}{Y} = -\frac{Z''}{Z} + \alpha^2 = -\beta^2$$

$$Y'' + \beta^2 Y = 0 \tag{5}$$

$$Z'' - (\alpha^2 + \beta^2)Z = 0. \tag{6}$$

The general solutions of equations (**4**), (**5**), and (**6**) are, respectively

$$X(x) = c_1 \cos \alpha x + c_2 \sin \alpha x$$

$$Y(y) = c_3 \cos \beta y + c_4 \sin \beta y$$

$$Z(z) = c_5 \cosh \sqrt{\alpha^2 + \beta^2}\, z + c_6 \sinh \sqrt{\alpha^2 + \beta^2}\, z.$$

The boundary and initial conditions are

$$u(0, y, z) = 0, \qquad u(a, y, z) = 0,$$
$$u(x, 0, z) = 0, \qquad u(x, b, z) = 0,$$
$$u(x, y, 0) = f(x, y), \qquad u(x, y, c) = 0.$$

The conditions $X(0) = Y(0) = 0$ give $c_1 = c_3 = 0$. The conditions $X(a) = Y(b) = 0$ yield two sets of eigenvalues:

$$\alpha = \frac{m\pi}{a},\ m = 1, 2, 3, \ldots \qquad \text{and} \qquad \beta = \frac{n\pi}{b},\ n = 1, 2, 3, \ldots.$$

Let

$$\omega_{mn}^2 = \frac{m^2 \pi^2}{a^2} + \frac{n^2 \pi^2}{b^2}.$$

Then the boundary condition $Z(c) = 0$ gives

$$c_5 \cosh c\omega_{mn} + c_6 \sinh c\omega_{mn} = 0$$

from which we obtain

$$Z(z) = c_5 \left(\cosh w_{mn} z - \frac{\cosh c\omega_{mn}}{\sinh c\omega_{mn}} \sinh \omega z \right)$$

$$= \frac{c_5}{\sinh c\omega_{mn}} (\sinh c\omega_{mn} \cosh \omega_{mn} z - \cosh c\omega_{mn} \sinh \omega_{mn} z) = c_{mn} \sinh \omega_{mn}(c - z).$$

By the superposition principle

$$u(x,y,t) = \sum_{m=1}^{\infty}\sum_{n=1}^{\infty} A_{mn}\sinh\omega_{mn}(c-z)\sin\frac{m\pi}{a}x\sin\frac{n\pi}{b}y$$

where

$$A_{mn} = \frac{4}{ab\sinh c\omega_{mn}}\int_0^b\int_0^a f(x,y)\sin\frac{m\pi}{a}x\sin\frac{n\pi}{b}y\,dx\,dy.$$

CHAPTER 13 REVIEW EXERCISES

3. Substituting $u(x,t) = v(x,t) + \psi(x)$ into the partial differential equation we obtain

$$k\frac{\partial^2 v}{\partial x^2} + k\psi''(x) = \frac{\partial v}{\partial t}.$$

This equation will be homogeneous provided ψ satisfies

$$k\psi'' = 0 \qquad\text{or}\qquad \psi = c_1 x + c_2.$$

Considering

$$u(0,t) = v(0,t) + \psi(0) = u_0$$

we set $\psi(0) = u_0$ so that $\psi(x) = c_1 x + u_0$. Now

$$-\frac{\partial u}{\partial x}\bigg|_{x=\pi} = -\frac{\partial v}{\partial x}\bigg|_{x=\pi} - \psi'(x) = v(\pi,t) + \psi(\pi) - u_1$$

is equivalent to

$$\frac{\partial v}{\partial x}\bigg|_{x=\pi} + v(\pi,t) = u_1 - \psi'(x) - \psi(\pi) = u_1 - c_1 - (c_1\pi + u_0),$$

which will be homogeneous when

$$u_1 - c_1 - c_1\pi - u_0 = 0 \qquad\text{or}\qquad c_1 = \frac{u_1 - u_0}{1+\pi}.$$

The steady-state solution is

$$\psi(x) = \left(\frac{u_1 - u_0}{1+\pi}\right)x + u_0.$$

6. The boundary-value problem is

$$\frac{\partial^2 u}{\partial x^2} + x^2 = \frac{\partial^2 u}{\partial t^2}, \quad 0 < x < 1, \quad t > 0,$$
$$u(0,t) = 1, \quad u(1,t) = 0, \quad t > 0,$$
$$u(x,0) = f(x), \quad u_t(x,0) = 0, \quad 0 < x < 1.$$

Substituting $u(x,t) = v(x,t) + \psi(x)$ into the partial differential equation gives

$$\frac{\partial^2 v}{\partial x^2} + \psi''(x) + x^2 = \frac{\partial^2 v}{\partial t^2}.$$

This equation will be homogeneous provided $\psi''(x) + x^2 = 0$ or

$$\psi(x) = -\frac{1}{12}x^4 + c_1 x + c_2.$$

From $\psi(0) = 1$ and $\psi(1) = 0$ we obtain $c_1 = -11/12$ and $c_2 = 1$. The new problem is

$$\frac{\partial^2 v}{\partial x^2} = \frac{\partial^2 v}{\partial t^2}, \quad 0 < x < 1, \quad t > 0,$$

$$v(0,t) = 0, \quad v(1,t) = 0, \quad t > 0,$$

$$v(x,0) = f(x) - \psi(x), \quad v_t(x,0) = 0, \quad 0 < x < 1.$$

From Section 13.4 in the text we see that $B_n = 0$,

$$A_n = 2\int_0^1 [f(x) - \psi(x)] \sin n\pi x \, dx = 2\int_0^1 \left[f(x) + \frac{1}{12}x^4 + \frac{11}{12}x - 1 \right] \sin n\pi x \, dx,$$

and

$$v(x,t) = \sum_{n=1}^{\infty} A_n \cos n\pi t \sin n\pi x.$$

Thus

$$u(x,t) = v(x,t) + \psi(x) = -\frac{1}{12}x^4 - \frac{11}{12}x + 1 + \sum_{n=1}^{\infty} A_n \cos n\pi t \sin n\pi x.$$

9. Using $u = XY$ and $-\lambda$ as a separation constant leads to

$$X'' - \lambda X = 0,$$

and

$$Y'' + \lambda Y = 0,$$

$$Y(0) = 0,$$

$$Y(\pi) = 0.$$

Then

$$X = c_1 e^{nx} + c_2 e^{-nx} \quad \text{and} \quad Y = c_3 \cos ny + c_4 \sin ny$$

for $n = 1, 2, 3, \ldots$. Since u must be bounded as $x \to \infty$, we define $c_1 = 0$. Also $Y(0) = 0$ implies $c_3 = 0$ so

$$u = \sum_{n=1}^{\infty} A_n e^{-nx} \sin ny.$$

Imposing

$$u(0,y) = 50 = \sum_{n=1}^{\infty} A_n \sin ny$$

gives

$$A_n = \frac{2}{\pi} \int_0^{\pi} 50 \sin ny \, dy = \frac{100}{n\pi}[1 - (-1)^n]$$

so that

$$u(x,y) = \sum_{n=1}^{\infty} \frac{100}{n\pi}[1 - (-1)^n] e^{-nx} \sin ny.$$

12. Substituting $u(x,t) = v(x,t) + \psi(x)$ into the partial differential equation results in $\psi'' = -\sin x$ and $\psi(x) = c_1 x + c_2 + \sin x$. The boundary conditions $\psi(0) = 400$ and $\psi(\pi) = 200$ imply $c_1 = -200/\pi$ and $c_2 = 400$ so

$$\psi(x) = -\frac{200}{\pi}x + 400 + \sin x.$$

Solving

$$\frac{\partial^2 v}{\partial x^2} = \frac{\partial v}{\partial t}, \qquad 0 < x < \pi, \quad t > 0$$

$$v(0,t) = 0, \qquad v(\pi, t) = 0, \quad t > 0$$

$$u(x,0) = 400 + \sin x - \left(-\frac{200}{\pi}x + 400 + \sin x\right) = \frac{200}{\pi}x, \qquad 0 < x < \pi$$

using separation of variables with separation constant $-\lambda$, where $\lambda = \alpha^2$, gives

$$X'' + \alpha^2 X = 0 \qquad \text{and} \qquad T' + \alpha^2 T = 0.$$

Using $X(0) = 0$ and $X(\pi) = 0$ we determine $\alpha^2 = n^2$, $X(x) = c_2 \sin nx$, and $T(t) = c_3 e^{-n^2 t}$. Then

$$v(x,t) = \sum_{n=1}^{\infty} A_n e^{-n^2 t} \sin nx$$

and

$$v(x,0) = \frac{200}{\pi}x = \sum_{n=1}^{\infty} A_n \sin nx$$

so

$$A_n = \frac{400}{\pi^2}\int_0^{\pi} x \sin nx \, dx = \frac{400}{n\pi}(-1)^{n+1}.$$

Thus

$$u(x,t) = -\frac{200}{\pi}x + 400 + \sin x + \frac{400}{\pi}\sum_{n=1}^{\infty}\frac{(-1)^{n+1}}{n}e^{-n^2 t}\sin nx.$$

14 Boundary-Value Problems in Other Coordinate Systems

EXERCISES 14.1

Problems in Polar Coordinates

3. We have

$$A_0 = \frac{1}{2\pi} \int_0^{2\pi} (2\pi\theta - \theta^2)\, d\theta = \frac{2\pi^2}{3}$$

$$A_n = \frac{1}{\pi} \int_0^{2\pi} (2\pi\theta - \theta^2) \cos n\theta\, d\theta = -\frac{4}{n^2}$$

$$B_n = \frac{1}{\pi} \int_0^{2\pi} (2\pi\theta - \theta^2) \sin n\theta\, d\theta = 0$$

and so

$$u(r, \theta) = \frac{2\pi^2}{3} - 4 \sum_{n=1}^{\infty} \frac{r^n}{n^2} \cos n\theta.$$

6. Using the same reasoning as in Example 1 in the text we obtain

$$u(r, \theta) = A_0 + \sum_{n=1}^{\infty} r^n (A_n \cos n\theta + B_n \sin n\theta).$$

The boundary condition at $r = c$ implies

$$f(\theta) = \sum_{n=1}^{\infty} nc^{n-1} (A_n \cos n\theta + B_n \sin n\theta).$$

Since this condition does not determine A_0, it is an arbitrary constant. However, to be a full Fourier series on $[0, 2\pi]$ we must require that $f(\theta)$ satisfy the condition $A_0 = a_0/2 = 0$ or $\int_0^{2\pi} f(\theta)\, d\theta = 0$. If this integral were not 0, then the series for $f(\theta)$ would contain a nonzero constant, which it obviously does not. With this as a necessary compatibility condition we can then make the identifications

$$nc^{n-1} A_n = a_n \quad \text{and} \quad nc^{n-1} B_n = b_n$$

or

$$A_n = \frac{1}{nc^{n-1}\pi} \int_0^{2\pi} f(\theta) \cos n\theta\, d\theta \quad \text{and} \quad B_n = \frac{1}{nc^{n-1}\pi} \int_0^{2\pi} f(\theta) \sin n\theta\, d\theta.$$

9. This is similar to the solution to Problem 7 above. When $n = 0$, $\Theta(\theta) = c_5\theta + c_6$ and $R(r) = c_7 + c_8 \ln r$. Periodicity implies $c_5 = 0$ and the insulation condition at $r = a$ implies $c_8 = 0$. Thus, we take $u_0 = A_0 = c_6 c_7$. Then, for $n = 1, 2, 3, \ldots$, $\Theta(\theta) = c_1 \cos n\theta + c_2 \sin n\theta$ and $R(r) = c_3 r^n + c_4 r^{-n}$. From $R'(a) = 0$ we get $c_3 n a^{n-1} - c_4 n a^{-n-1} = 0$, which implies $c_4 = c_3 a^{2n}$. Then

$$R(r) = c_3 (r^n + a^{2n} r^{-n}) = c_3 \frac{r^{2n} + a^{2n}}{r^n}$$

and

$$u(r,\theta) = A_0 + \sum_{n=1}^{\infty} \frac{r^{2n} + a^{2n}}{r^n} \left(A_n \cos n\theta + B_n \sin n\theta\right).$$

Taking $r = b$ we have

$$f(\theta) = A_0 + \sum_{n=1}^{\infty} \frac{b^{2n} + a^{2n}}{b^n} \left(A_n \cos n\theta + B_n \sin n\theta\right),$$

which implies

$$A_0 = \frac{a_0}{2} = \frac{1}{2\pi} \int_0^{2\pi} f(\theta)\, d\theta$$

and

$$\frac{b^{2n} + a^{2n}}{b^n} A_n = \frac{1}{\pi} \int_0^{2\pi} f(\theta) \cos n\theta\, d\theta \quad \text{and} \quad \frac{b^{2n} + a^{2n}}{b^n} B_n = \frac{1}{\pi} \int_0^{2\pi} f(\theta) \sin 2n\theta\, d\theta.$$

Hence

$$A_n = \frac{b^n}{\pi(a^{2n} + b^{2n})} \int_0^{2\pi} f(\theta) \cos n\theta\, d\theta \quad \text{and} \quad B_n = \frac{b^n}{\pi(a^{2n} + b^{2n})} \int_0^{2\pi} f(\theta) \sin n\theta\, d\theta.$$

12. We solve

$$\frac{\partial^2 u}{\partial r^2} + \frac{1}{r} \frac{\partial u}{\partial r} + \frac{1}{r^2} \frac{\partial^2 u}{\partial \theta^2} = 0, \quad 0 < \theta < \pi/4, \quad r > 0$$

$$u(r, 0) = 0, \quad r > 0$$

$$u(r, \pi/4) = 30, \quad r > 0.$$

Proceeding as in Example 1 in the text we find the separated ordinary differential equations to be

$$r^2 R'' + r R' - \lambda R = 0$$

$$\Theta'' + \lambda \Theta = 0.$$

With $\lambda = \alpha^2 > 0$ the corresponding general solutions are

$$R(r) = c_1 r^\alpha + c_2 r^{-\alpha}$$

$$\Theta(\theta) = c_3 \cos \alpha\theta + c_4 \sin \alpha\theta.$$

The condition $\Theta(0) = 0$ implies $c_3 = 0$ so that $\Theta = c_4 \sin \alpha\theta$. Now, in order that the temperature be bounded as $r \to \infty$ we define $c_1 = 0$. Similarly, in order that the temperature be bounded as $r \to 0$ we are forced to define $c_2 = 0$. Thus $R(r) = 0$ and so no nontrivial solution exists for $\lambda > 0$. For $\lambda = 0$ the separated differential equations are

$$r^2 R'' + r R' = 0 \quad \text{and} \quad \Theta'' = 0.$$

Solutions of these latter equations are

$$R(r) = c_1 + c_2 \ln r \quad \text{and} \quad \Theta(\theta) = c_3 \theta + c_4.$$

$\Theta(0) = 0$ still implies $c_4 = 0$, whereas boundedness as $r \to 0$ demands $c_2 = 0$. Thus, a product solution is

$$u = c_1 c_3 \theta = A\theta.$$

From $u(r, \pi/4) = 0$ we obtain $A = 120/\pi$. Thus, a solution to the problem is

$$u(r, \theta) = \frac{120}{\pi} \theta.$$

287

15. We solve

$$\frac{\partial^2 u}{\partial r^2} + \frac{1}{r}\frac{\partial u}{\partial r} + \frac{1}{r^2}\frac{\partial^2 u}{\partial \theta^2} = 0, \quad 0 < \theta < \pi, \quad 0 < r < 2,$$

$$u(2,\theta) = \begin{cases} u_0, & 0 < \theta < \pi/2 \\ 0, & \pi/2 < \theta < \pi \end{cases}$$

$$\frac{\partial u}{\partial \theta}\bigg|_{\theta=0} = 0, \quad \frac{\partial u}{\partial \theta}\bigg|_{\theta=\pi} = 0, \quad 0 < r < 2.$$

Proceeding as in Example 1 in the text we obtain the separated differential equations

$$r^2 R'' + r R' - \lambda R = 0$$

$$\Theta'' + \lambda \Theta = 0.$$

Taking $\lambda = \alpha^2$ the solutions are

$$\Theta(\theta) = c_1 \cos \alpha\theta + c_2 \sin \alpha\theta$$

$$R(r) = c_3 r^\alpha + c_4 r^{-\alpha}.$$

Applying the boundary conditions $\Theta'(0) = 0$ and $\Theta'(\pi) = 0$ we find that $c_2 = 0$ and $\alpha = n$ for $n = 0, 1, 2, \ldots$. Since we want $R(r)$ to be bounded as $r \to 0$ we require $c_4 = 0$. Thus

$$u(r,\theta) = A_0 + \sum_{n=1}^{\infty} A_n r^n \cos n\theta.$$

From

$$u(2,\theta) = \begin{cases} u_0, & 0 < \theta < \pi/2 \\ 0, & \pi/2 < \theta < \pi \end{cases} = A_0 + \sum_{n=1}^{\infty} A_n 2^n \cos n\theta$$

we find

$$A_0 = \frac{1}{2}\frac{2}{\pi}\int_0^{\pi/2} u_0 \, d\theta = \frac{u_0}{2}$$

and

$$2^n A_n = \frac{2u_0}{\pi}\int_0^{\pi/2} \cos n\theta \, d\theta = \frac{2u_0}{\pi}\frac{\sin n\pi/2}{n}.$$

Therefore

$$u(r,\theta) = \frac{u_0}{2} + \frac{2u_0}{\pi}\sum_{n=1}^{\infty}\frac{1}{n}\left(\sin\frac{n\pi}{2}\right)\left(\frac{r}{2}\right)^n \cos n\theta.$$

EXERCISES 14.2

Problems in Cylindrical Coordinates

3. Referring to Example 2 in the text we have

$$R(r) = c_1 J_0(\alpha r) + c_2 Y_0(\alpha r)$$

$$Z(z) = c_3 \cosh \alpha z + c_4 \sinh \alpha z$$

where $c_2 = 0$ and $J_0(2\alpha) = 0$ defines the positive eigenvalues $\lambda_n = \alpha_n^2$. From $Z(4) = 0$ we obtain

$$c_3 \cosh 4\alpha_n + c_4 \sinh 4\alpha_n = 0 \quad \text{or} \quad c_4 = -c_3 \frac{\cosh 4\alpha_n}{\sinh 4\alpha_n}.$$

Then

$$Z(z) = c_3 \left[\cosh \alpha_n z - \frac{\cosh 4\alpha_n}{\sinh 4\alpha_n} \sinh \alpha_n z \right] = c_3 \frac{\sinh 4\alpha_n \cosh \alpha_n z - \cosh 4\alpha_n \sinh \alpha_n z}{\sinh 4\alpha_n}$$

$$= c_3 \frac{\sinh \alpha_n (4 - z)}{\sinh 4\alpha_n}$$

and

$$u(r, z) = \sum_{n=1}^{\infty} A_n \frac{\sinh \alpha_n (4 - z)}{\sinh 4\alpha_n} J_0(\alpha_n r).$$

From

$$u(r, 0) = u_0 = \sum_{n=1}^{\infty} A_n J_0(\alpha_n r)$$

we obtain

$$A_n = \frac{2u_0}{4 J_1^2(2\alpha_n)} \int_0^2 r J_0(\alpha_n r) \, dr = \frac{u_0}{\alpha_n J_1(2\alpha_n)}.$$

Thus the temperature in the cylinder is

$$u(r, z) = u_0 \sum_{n=1}^{\infty} \frac{\sinh \alpha_n (4 - z) J_0(\alpha_n r)}{\alpha_n \sinh 4\alpha_n J_1(2\alpha_n)}.$$

6. Using λ as the separation constant the separated equations are

$$r R'' + R' - r\lambda R = 0 \quad \text{and} \quad Z'' + \lambda Z = 0.$$

The boundary conditions are $Z'(0) = 0$ and $Z'(1) = 0$.

If $\lambda = 0$ the solutions of the ordinary differential equations are

$$R = c_1 + c_2 \ln r \quad \text{and} \quad Z = c_3 + c_4 z.$$

Since $Z'(0) = 0$, $c_4 = 0$. Therefore $Z = c_3$ and $Z(1) = 0$ so $Z(1) = c_3 = 0$. The product solution is $u = R(r)Z(z) = 0$. Thus $\lambda = 0$ is not an eigenvalue.

If $\lambda = -\alpha^2 < 0$, the solutions of the ordinary differential equations are

$$R = c_1 J_0(\alpha r) + c_2 Y_0(\alpha r) \quad \text{and} \quad Z = c_3 \cosh \alpha z + c_4 \sinh \alpha z.$$

Since $Z'(0) = 0$, $c_4 = 0$. Therefore $Z = c_3 \cosh \alpha z$ and $Z(1) = 0$ so $c_4 \cosh \alpha = 0$, which implies $c_4 = 0$. Thus $Z = 0$ and therefore $u = 0$.

If $\lambda = \alpha^2 > 0$, the solutions of the ordinary differential equations are

$$R = c_1 I_0(\alpha r) + c_2 K_0(\alpha r) \quad \text{and} \quad Z = c_3 \cos \alpha z + c_4 \sin \alpha z.$$

Since $Z'(0) = 0$, $c_4 = 0$ and so $Z = c_3 \cos \alpha z$. Now $Z(1) = 0$ so $c_4 \cos \alpha = 0$ and $\alpha = (2n-1)\pi/2$, $n = 1, 2, 3, \ldots$. The eigenvalues are $\lambda_n = (2n-1)^2 \pi^2/4$ and corresponding eigenfunctions are $Z = c_3 \cos(2n-1)\pi z/2$. Now, the usual implicit requirement that u be bounded at $r = 0$ implies $c_2 = 0$. (See Figure 5.3.4 in the text.) Therefore $R = c_1 I_0(\alpha r)$ or $R = c_1 I_0\big((2n-1)\pi r/2\big)$. The superposition principle then yields

$$u(r, z) = \sum_{n=1}^{\infty} A_n I_0\left(\frac{2n-1}{2}\pi r\right) \cos \frac{2n-1}{2}\pi z.$$

At $r = 1$

$$u(1, z) = z = \sum_{n=1}^{\infty} A_n I_0\left(\frac{2n-1}{2}\pi\right) \cos \frac{2n-1}{2}\pi z,$$

which is not a Fourier series. Thus

$$A_n I_0\left(\frac{2n-1}{2}\pi\right) = \frac{\int_0^1 z \cos \frac{2n-1}{2}\pi z\, dz}{\int_0^1 \cos^2 \frac{2n-1}{2}\pi z\, dz} = \frac{-8 - 4(2n-1)\pi(-1)^n}{(2n-1)^2 \pi^2}.$$

Therefore

$$u(r, z) = 4 \sum_{n=1}^{\infty} \frac{(2n-1)\pi(-1)^{n+1} - 2}{\frac{(2n-1)^2 \pi^2 I_0(2n-1)}{2\pi)}} I_0\left(\frac{2n-1}{2}\pi r\right) \cos \frac{2n-1}{2}\pi z.$$

9. Letting $u(r, t) = R(r)T(t)$ and separating variables we obtain

$$\frac{R'' + \frac{1}{r}R'}{R} = \frac{T'}{kT} = -\lambda \quad \text{and} \quad R'' + \frac{1}{r}R' + \lambda R = 0, \quad T' + \lambda k T = 0.$$

From the last equation we find $T(t) = e^{-\lambda k t}$. If $\lambda < 0$, $T(t)$ increases without bound as $t \to \infty$. Thus we assume $\lambda = \alpha^2 > 0$. Now

$$R'' + \frac{1}{r}R' + \alpha^2 R = 0$$

is a parametric Bessel equation with solution

$$R(r) = c_1 J_0(\alpha r) + c_2 Y_0(\alpha r).$$

Since Y_0 is unbounded as $r \to 0$ we take $c_2 = 0$. Then $R(r) = c_1 J_0(\alpha r)$ and the boundary condition $u(c, t) = R(c)T(t) = 0$ implies $J_0(\alpha c) = 0$. This latter equation defines the positive eigenvalues $\lambda_n = \alpha_n^2$. Thus

$$u(r, t) = \sum_{n=1}^{\infty} A_n J_0(\alpha_n r) e^{-\alpha_n^2 k t}.$$

From

$$u(r, 0) = f(r) = \sum_{n=1}^{\infty} A_n J_0(\alpha_n r)$$

we find

$$A_n = \frac{2}{c^2 J_1^2(\alpha_n c)} \int_0^c r J_0(\alpha_n r) f(r)\, dr, \quad n = 1, 2, 3, \ldots .$$

12. We solve

$$\frac{\partial^2 u}{\partial r^2} + \frac{1}{r}\frac{\partial u}{\partial r} + \frac{\partial^2 u}{\partial z^2} = 0, \quad 0 < r < 1, \quad z > 0$$

$$\left.\frac{\partial u}{\partial r}\right|_{r=1} = -hu(1, z), \quad z > 0$$

$$u(r, 0) = u_0, \quad 0 < r < 1.$$

assuming $u = RZ$ we get

$$\frac{R'' + \frac{1}{r}R'}{R} = -\frac{Z''}{Z} = -\lambda$$

and so

$$rR'' + R' + \lambda^2 rR = 0 \qquad \text{and} \qquad Z'' - \lambda Z = 0.$$

Letting $\lambda = \alpha^2$ we then have

$$R(r) = c_1 J_0(\alpha r) + c_2 Y_0(\alpha r) \qquad \text{and} \qquad Z(z) = c_3 e^{-\alpha z} + c_4 e^{\alpha z}.$$

We use the exponential form of the solution of $Z'' - \alpha^2 Z = 0$ since the domain of the variable z is a semi-infinite interval. As usual we define $c_2 = 0$ since the temperature is surely bounded as $r \to 0$. Hence $R(r) = c_1 J_0(\alpha r)$. Now the boundary-condition $u_r(1, z) + hu(1, z) = 0$ is equivalent to

$$\alpha J_0'(\alpha) + h J_0(\alpha) = 0. \tag{4}$$

The eigenvalues α_n are the positive roots of (**4**) above. Finally, we must now define $c_4 = 0$ since the temperature is also expected to be bounded as $z \to \infty$. A product solution of the partial differential equation that satisfies the first boundary condition is given by

$$u_n(r, z) = A_n e^{-\alpha_n z} J_0(\alpha_n r).$$

Therefore

$$u(r, z) = \sum_{n=1}^{\infty} A_n e^{-\alpha_n z} J_0(\alpha_n r)$$

is another formal solution. At $z = 0$ we have $u_0 = A_n J_0(\alpha_n r)$. In view of (**4**) above we use equations (**17**) and (**18**) of Section 12.6 in the text with the identification $b = 1$:

$$A_n = \frac{2\alpha_n^2}{(\alpha_n^2 + h^2) J_0^2(\alpha_n)} \int_0^1 r J_0(\alpha_n r) u_0 \, dr$$

$$= \frac{2\alpha_n^2 u_0}{(\alpha_n^2 + h^2) J_0^2(\alpha_n) \alpha_n^2} t J_1(t) \Big|_0^{\alpha_n} = \frac{2\alpha_n u_0 J_1(\alpha_n)}{(\alpha_n^2 + h^2) J_0^2(\alpha_n)}. \tag{5}$$

Since $J_0' = -J_1$ [see equation (**6**) of Section 11.5 in the text] it follows from (**4**) above that $\alpha_n J_1(\alpha_n) = h J_0(\alpha_n)$. Thus (**5**) above simplifies to

$$A_n = \frac{2u_0 h}{(\alpha_n^2 + h^2) J_0(\alpha_n)}.$$

A solution to the boundary-value problem is then

$$u(r, z) = 2u_0 h \sum_{n=1}^{\infty} \frac{e^{-\alpha_n z}}{(\alpha_n^2 + h^2) J_0(\alpha_n)} J_0(\alpha_n r).$$

15. **(a)** Writing the partial differential equation in the form

$$g\left(x\frac{\partial^2 u}{\partial x^2} + \frac{\partial u}{\partial x}\right) = \frac{\partial^2 u}{\partial t^2}$$

and separating variables we obtain

$$\frac{xX'' + X'}{X} = \frac{T''}{gT} = -\lambda.$$

Letting $\lambda = \alpha^2$ we obtain

$$xX'' + X' + \alpha^2 X = 0 \qquad \text{and} \qquad T'' + g\alpha^2 T = 0.$$

Letting $x = \tau^2/4$ in the first equation we obtain $dx/d\tau = \tau/2$ or $d\tau/dx = 2\tau$. Then

$$\frac{dX}{dx} = \frac{dX}{d\tau}\frac{d\tau}{dx} = \frac{2}{\tau}\frac{dX}{d\tau}$$

and

$$\frac{d^2X}{dx^2} = \frac{d}{dx}\left(\frac{2}{\tau}\frac{dX}{d\tau}\right) = \frac{2}{\tau}\frac{d}{dx}\left(\frac{dX}{d\tau}\right) + \frac{dX}{d\tau}\frac{d}{dx}\left(\frac{2}{\tau}\right)$$

$$= \frac{2}{\tau}\frac{d}{d\tau}\left(\frac{dX}{d\tau}\right)\frac{d\tau}{dx} + \frac{dX}{d\tau}\frac{d}{d\tau}\left(\frac{2}{\tau}\right)\frac{d\tau}{dx} = \frac{4}{\tau^2}\frac{d^2X}{d\tau^2} - \frac{4}{\tau^3}\frac{dX}{d\tau}.$$

Thus

$$xX'' + X' + \alpha^2 X = \frac{\tau^2}{4}\left(\frac{4}{\tau^2}\frac{d^2X}{d\tau^2} - \frac{4}{\tau^3}\frac{dX}{d\tau}\right) + \frac{2}{\tau}\frac{dX}{d\tau} + \alpha^2 X = \frac{d^2X}{d\tau^2} + \frac{1}{\tau}\frac{dX}{d\tau} + \alpha^2 X = 0.$$

This is a parametric Bessel equation with solution

$$X(\tau) = c_1 J_0(\alpha\tau) + c_2 Y_0(\alpha\tau).$$

(b) To insure a finite solution at $x = 0$ (and thus $\tau = 0$) we set $c_2 = 0$. The condition $u(L, t) = X(L)T(t) = 0$ implies $X\big|_{x=L} = X\big|_{\tau=2\sqrt{L}} = c_1 J_0(2\alpha\sqrt{L}) = 0$, which defines positive eigenvalues $\lambda_n = \alpha_n^2$. The solution of $T'' + g\alpha^2 T = 0$ is

$$T(t) = c_3 \cos(\alpha_n\sqrt{g}\,t) + c_4 \sin(\alpha_n\sqrt{g}\,t).$$

The boundary condition $u_t(x, 0) = X(x)T'(0) = 0$ implies $c_4 = 0$. Thus

$$u(\tau, t) = \sum_{n=1}^{\infty} A_n \cos(\alpha_n\sqrt{g}\,t) J_0(\alpha_n\tau).$$

From

$$u(\tau, 0) = f(\tau^2/4) = \sum_{n=1}^{\infty} A_n J_0(\alpha_n\tau)$$

we find

$$A_n = \frac{2}{(2\sqrt{L})^2 J_1^2(2\alpha_n\sqrt{L})} \int_0^{2\sqrt{L}} \tau J_0(\alpha_n\tau) f(\tau^2/4)\,d\tau \qquad \boxed{v = \tau/2,\ dv = d\tau/2}$$

$$= \frac{1}{2L J_1^2(2\alpha_n\sqrt{L})} \int_0^{\sqrt{L}} 2v J_0(2\alpha_n v) f(v^2) 2\,dv$$

$$= \frac{2}{L J_1^2(2\alpha_n\sqrt{L})} \int_0^{\sqrt{L}} v J_0(2\alpha_n v) f(v^2)\,dv.$$

The solution of the boundary-value problem is

$$u(x, t) = \sum_{n=1}^{\infty} A_n \cos(\alpha_n\sqrt{g}\,t) J_0(2\alpha_n\sqrt{x}).$$

18. Because of the nonhomogeneous boundary condition $u(c, t) = 200$ we use the substitution $u(r, t) = v(r, t) + \psi(r)$. This gives

$$k\left(\frac{\partial^2 v}{\partial r^2} + \frac{1}{r}\frac{\partial v}{\partial r} + \psi'' + \frac{1}{r}\psi'\right) = \frac{\partial v}{\partial t}.$$

292

This equation will be homogeneous provided $\psi'' + (1/r)\psi' = 0$ or $\psi(r) = c_1 \ln r + c_2$. Since $\ln r$ is unbounded as $r \to 0$ we take $c_1 = 0$. Then $\psi(r) = c_2$ and using $u(c,t) = v(c,t) + c_2 = 200$ we set $c_2 = 200$, giving $v(c,t) = 0$. Referring to Problem 5 in this section, the solution of the boundary-value problem

$$k\left(\frac{\partial^2 v}{\partial r^2} + \frac{1}{r}\frac{\partial v}{\partial r}\right) = \frac{\partial v}{\partial t}, \quad 0 < r < c,\ t > 0$$

$$v(c,t) = 0, \quad t > 0$$

$$v(r,0) = -200, \quad 0 < r < c$$

is

$$v(r,t) = \sum_{n=1}^{\infty} A_n J_0(\alpha_n r)e^{-\alpha_n^2 kt},$$

where the separation constant is $-\lambda = -\alpha^2$. The eigenvalues are $\lambda_n = \alpha^2 = x_n^2/c^2$ where x_n is a positive root of $J_0(x) = 0$ and the coefficients A_n are

$$A_n = \frac{2}{c^2 J_1^2(\alpha_n c)}\int_0^c rJ_0(\alpha_n r)(-200)dr = -\frac{400}{c^2 J_1^2(\alpha_n c)}\int_0^c rJ_0(\alpha_n r)dr.$$

Taking $c = 10$ and $k = 0.1$ we have

$$u(r,t) = v(r,t) + 200 = 200 + \sum_{n=1}^{\infty} A_n J_0(x_n r/10)e^{-0.01x_n^2 t/100}$$

where

$$A_n = -\frac{4}{J_1^2(x_n)}\int_0^{10} rJ_0(x_n r/10)\, dr.$$

Using a CAS we find that the first five values of x_n are

$x_1 = 2.4048$, $x_2 = 5.5201$, $x_3 = 8.6537$, $x_4 = 11.7915$,

and $x_5 = 14.9309$, with corresponding eigenvalues

$\lambda_1 = 0.0578$, $\lambda_2 = 0.3047$, $\lambda_3 = 0.7489$, $\lambda_4 = 1.3904$,

and $\lambda_5 = 2.2293$. The first five values of A_n are

$A_1 = -320.4$, $A_2 = 213.0$, $A_3 = -170.3$, $A_4 = 145.9$,

and $A_5 = -129.7$. Using a root finding application in

a CAS we find that $u(5,t) = 100$ when $t \approx 1331$ and

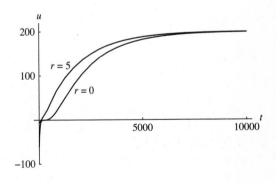

$u(0,t) = 100$ when $t \approx 2005$. Since $u = 200$ is an asymptote for the graphs of $u(0,t)$ and $u(5,t)$ we solve $u(5,t) = 199.9$ and $u(0,t) = 199.9$. This gives $t \approx 13{,}265$ and $t \approx 13{,}958$, respectively.

3. The coefficients are given by

$$A_n = \frac{2n+1}{2c^n} \int_0^\pi \cos\theta P_n(\cos\theta) \sin\theta \, d\theta = \frac{2n+1}{2c^n} \int_0^\pi P_1(\cos\theta) P_n(\cos\theta) \sin\theta \, d\theta$$

$$\boxed{x = \cos\theta, \ dx = -\sin\theta \, d\theta}$$

$$= \frac{2n+1}{2c^n} \int_{-1}^1 P_1(x) P_n(x) \, dx.$$

Since $P_n(x)$ and $P_m(x)$ are orthogonal for $m \neq n$, $A_n = 0$ for $n \neq 1$ and

$$A_1 = \frac{2(1)+1}{2c^1} \int_{-1}^1 P_1(x) P_1(x) \, dx = \frac{3}{2c} \int_{-1}^1 x^2 dx = \frac{1}{c}.$$

Thus

$$u(r,\theta) = \frac{r}{c} P_1(\cos\theta) = \frac{r}{c} \cos\theta.$$

6. Referring to Example 1 in the text we have

$$R(r) = c_1 r^n \qquad \text{and} \qquad \Theta(\theta) = P_n(\cos\theta).$$

Now $\Theta(\pi/2) = 0$ implies that n is odd, so

$$u(r,\theta) = \sum_{n=0}^\infty A_{2n+1} r^{2n+1} P_{2n+1}(\cos\theta).$$

From

$$u(c,\theta) = f(\theta) = \sum_{n=0}^\infty A_{2n+1} c^{2n+1} P_{2n+1}(\cos\theta)$$

we see that

$$A_{2n+1} c^{2n+1} = (4n+3) \int_0^{\pi/2} f(\theta) \sin\theta \, P_{2n+1}(\cos\theta) \, d\theta.$$

Thus

$$u(r,\theta) = \sum_{n=0}^\infty A_{2n+1} r^{2n+1} P_{2n+1}(\cos\theta)$$

where

$$A_{2n+1} = \frac{4n+3}{c^{2n+1}} \int_0^{\pi/2} f(\theta) \sin\theta \, P_{2n+1}(\cos\theta) \, d\theta.$$

9. Checking the hint, we find

$$\frac{1}{r} \frac{\partial^2}{\partial r^2}(ru) = \frac{1}{r} \frac{\partial}{\partial r}\left[r \frac{\partial u}{\partial r} + u \right] = \frac{1}{r}\left[r \frac{\partial^2 u}{\partial r^2} + \frac{\partial u}{\partial r} + \frac{\partial u}{\partial r} \right] = \frac{\partial^2 u}{\partial r^2} + \frac{2}{r} \frac{\partial u}{\partial r}.$$

The partial differential equation then becomes

$$\frac{\partial^2}{\partial r^2}(ru) = r \frac{\partial u}{\partial t}.$$

Now, letting $ru(r,t) = v(r,t) + \psi(r)$, since the boundary condition is nonhomogeneous, we obtain

$$\frac{\partial^2}{\partial r^2}\left[v(r,t) + \psi(r)\right] = r\frac{\partial}{\partial t}\left[\frac{1}{r}v(r,t) + \psi(r)\right]$$

or

$$\frac{\partial^2 v}{\partial r^2} + \psi''(r) = \frac{\partial v}{\partial t}.$$

This differential equation will be homogeneous if $\psi''(r) = 0$ or $\psi(r) = c_1 r + c_2$. Now

$$u(r,t) = \frac{1}{r}v(r,t) + \frac{1}{r}\psi(r) \qquad \text{and} \qquad \frac{1}{r}\psi(r) = c_1 + \frac{c_2}{r}.$$

Since we want $u(r,t)$ to be bounded as r approaches 0, we require $c_2 = 0$. Then $\psi(r) = c_1 r$. When $r = 1$

$$u(1,t) = v(1,t) + \psi(1) = v(1,t) + c_1 = 100,$$

and we will have the homogeneous boundary condition $v(1,t) = 0$ when $c_1 = 100$. Consequently, $\psi(r) = 100r$. The initial condition

$$u(r,0) = \frac{1}{r}v(r,0) + \frac{1}{r}\psi(r) = \frac{1}{r}v(r,0) + 100 = 0$$

implies $v(r,0) = -100r$. We are thus led to solve the new boundary-value problem

$$\frac{\partial^2 v}{\partial r^2} = \frac{\partial v}{\partial t}, \qquad 0 < r < 1, \quad t > 0,$$

$$v(1,t) = 0, \qquad \lim_{r\to 0}\frac{1}{r}v(r,t) < \infty,$$

$$v(r,0) = -100r.$$

Letting $v(r,t) = R(r)T(t)$ and using the separation constant $-\lambda$ we obtain

$$R'' + \lambda R = 0 \qquad \text{and} \qquad T' + \lambda T = 0.$$

Using $\lambda = \alpha^2 > 0$ we then have

$$R(r) = c_3\cos\alpha r + c_4\sin\alpha r \qquad \text{and} \qquad T(t) = c_5 e^{-\alpha^2 t}.$$

The boundary conditions are equivalent to $R(1) = 0$ and $\lim_{r\to 0} R(r)/r < \infty$. Since

$$\lim_{r\to 0}\frac{\cos\alpha r}{r}$$

does not exist we must have $c_3 = 0$. Then $R(r) = c_4\sin\alpha r$, and $R(1) = 0$ implies $\alpha = n\pi$ for $n = 1, 2, 3, \ldots$. Thus

$$v_n(r,t) = A_n e^{-n^2\pi^2 t}\sin n\pi r$$

for $n = 1, 2, 3, \ldots$. Using the condition $\lim_{r\to 0} R(r)/r < \infty$ it is easily shown that there are no eigenvalues for $\lambda = 0$, nor does setting the common constant to $-\lambda = \alpha^2$ when separating variables lead to any solutions. Now, by the superposition principle,

$$v(r,t) = \sum_{n=1}^{\infty} A_n e^{-n^2\pi^2 t}\sin n\pi r.$$

The initial condition $v(r,0) = -100r$ implies

$$-100r = \sum_{n=1}^{\infty} A_n\sin n\pi r.$$

This is a Fourier sine series and so

$$A_n = 2 \int_0^1 (-100r \sin n\pi r) \, dr = -200 \left[-\frac{r}{n\pi} \cos n\pi r \Big|_0^1 + \int_0^1 \frac{1}{n\pi} \cos n\pi r \, dr \right]$$

$$= -200 \left[-\frac{\cos n\pi}{n\pi} + \frac{1}{n^2\pi^2} \sin n\pi r \Big|_0^1 \right] = -200 \left[-\frac{(-1)^n}{n\pi} \right] = \frac{(-1)^n 200}{n\pi}.$$

A solution of the problem is thus

$$u(r,t) = \frac{1}{r}v(r,t) + \frac{1}{r}\psi(r) = \frac{1}{r}\sum_{n=1}^{\infty}(-1)^n \frac{20}{n\pi} e^{-n^2\pi^2 t} \sin n\pi r + \frac{1}{r}(100r)$$

$$= \frac{200}{\pi r}\sum_{n=1}^{\infty}\frac{(-1)^n}{n} e^{-n^2\pi^2 t} \sin n\pi r + 100.$$

12. Proceeding as in Example 1 we obtain

$$\Theta(\theta) = P_n(\cos\theta) \qquad \text{and} \qquad R(r) = c_1 r^n + c_2 r^{-(n+1)}$$

so that

$$u(r,\theta) = \sum_{n=0}^{\infty}(A_n r^n + B_n r^{-(n+1)})P_n(\cos\theta).$$

To satisfy $\lim_{r\to\infty} u(r,\theta) = -Er\cos\theta$ we must have $A_n = 0$ for $n = 2, 3, 4, \ldots$. Then

$$\lim_{r\to\infty} u(r,\theta) = -Er\cos\theta = A_0 \cdot 1 + A_1 r \cos\theta,$$

so $A_0 = 0$ and $A_1 = -E$. Thus

$$u(r,\theta) = -Er\cos\theta + \sum_{n=0}^{\infty} B_n r^{-(n+1)} P_n(\cos\theta).$$

Now

$$u(c,\theta) = 0 = -Ec\cos\theta + \sum_{n=0}^{\infty} B_n c^{-(n+1)} P_n(\cos\theta)$$

so

$$\sum_{n=0}^{\infty} B_n c^{-(n+1)} P_n(\cos\theta) = Ec\cos\theta$$

and

$$B_n c^{-(n+1)} = \frac{2n+1}{2} \int_0^{\pi} Ec\cos\theta \, P_n(\cos\theta) \sin\theta \, d\theta.$$

Now $\cos\theta = P_1(\cos\theta)$ so, for $n \neq 1$,

$$\int_0^{\pi} \cos\theta \, P_n(\cos\theta) \sin\theta \, d\theta = 0$$

by orthogonality. Thus $B_n = 0$ for $n \neq 1$ and

$$B_1 = \frac{3}{2} Ec^3 \int_0^{\pi} \cos^2\theta \sin\theta \, d\theta = Ec^3.$$

Therefore,

$$u(r,\theta) = -Er\cos\theta + Ec^3 r^{-2} \cos\theta.$$

CHAPTER 14 REVIEW EXERCISES

3. The conditions $\Theta(0) = 0$ and $\Theta(\pi) = 0$ applied to $\Theta = c_1 \cos \alpha\theta + c_2 \sin \alpha\theta$ give $c_1 = 0$ and $\alpha = n$, $n = 1, 2, 3, \ldots$, respectively. Thus we have the Fourier sine-series coefficients

$$A_n = \frac{2}{\pi} \int_0^\pi u_0 (\pi\theta - \theta^2) \sin n\theta \, d\theta = \frac{4u_0}{n^3 \pi} [1 - (-1)^n].$$

Thus

$$u(r, \theta) = \frac{4u_0}{\pi} \sum_{n=1}^\infty \frac{1 - (-1)^n}{n^3} r^n \sin n\theta.$$

6. Two of the boundary conditions are $u(r, 0) = 0$ and $u_\theta(r, \pi) = 0$ which imply $\Theta(0) = 0$ and $\Theta'(\pi) = 0$. For $\lambda = [(2n-1)/2]^2$, $n = 1, 2, 3, \ldots$, we have $\Theta(\theta) = c_2 \sin[(2n-1)/2]\theta$. Assuming a bounded solution at $r = 0$ we have $R(r) = c_3 r^n$, so

$$u(r, \theta) = \sum_{n=1}^\infty A_n r^n \sin\left(\frac{2n-1}{2}\theta\right) \quad \text{and} \quad f(\theta) = \sum_{n=1}^\infty A_n c^n \sin\left(\frac{2n-1}{2}\theta\right).$$

This is not a Fourier series but it is an orthogonal series expansion of f, so

$$A_n = c^{-n} \left(\frac{\displaystyle\int_0^\pi f(\theta) \sin\left(\frac{2n-1}{2}\theta\right) d\theta}{\displaystyle\int_0^\pi \sin^2\left(\frac{2n-1}{2}\theta\right) d\theta} \right).$$

9. The boundary-value problem in polar coordinates is

$$\frac{\partial^2 u}{\partial r^2} + \frac{1}{r}\frac{\partial u}{\partial r} + \frac{1}{r^2}\frac{\partial^2 u}{\partial \theta^2} = 0, \quad 0 < \theta < 2\pi, \quad 1 < r < 2,$$

$$u(1, \theta) = \sin^2 \theta, \quad \left.\frac{\partial u}{\partial r}\right|_{r=2} = 0, \quad 0 < \theta < 2\pi.$$

Letting $u(r, \theta) = R(r)\Theta(\theta)$ and separating variables we obtain

$$r^2 R'' + rR' - \lambda R = 0 \quad \text{and} \quad \Theta'' + \lambda \Theta = 0.$$

For $\lambda = 0$ we have $\Theta'' = 0$ and $r^2 R'' + rR' = 0$. This gives $\Theta = c_1 + c_2\theta$ and $R = c_3 + c_4 \ln r$. The periodicity assumption (as mentioned in Example 1 of Section 14.1 in the text) implies $c_2 = 0$, while the boundary condition $R'(2) = 0$ implies $c_4 = 0$. Thus, for $\lambda = 0$, $u = c_1 c_3 = A_0$. Now, for $\lambda = \alpha^2$, the differential equations become $\Theta'' + \alpha^2 \Theta = 0$ and $r^2 R'' + rR' - \alpha^2 R = 0$. The corresponding solutions are $\Theta = c_5 \cos \alpha\theta + c_6 \sin \alpha\theta$ and $R = c_7 r^\alpha + c_8 r^{-\alpha}$. In this case the periodicity assumption implies $\alpha = n$, $n = 1, 2, 3, \ldots$, while the boundary condition $R'(2) = 0$ implies $R = c_7(r^n + 4^n r^{-n})$. The product of the solutions is $u_n = (r^n + 4^n r^{-n})(A_n \cos n\theta + B_n \sin n\theta)$ and the superposition principle implies

$$u(r, \theta) = A_0 + \sum_{n=1}^\infty (r^n + 4^n r^{-n})(A_n \cos n\theta + B_n \sin n\theta).$$

Using the boundary condition at $r = 1$ we have

$$\sin^2 \theta = \frac{1}{2}(1 - \cos 2\theta) = A_0 + \sum_{n=1}^{\infty}(1 + 4^n)(A_n \cos n\theta + B_n \sin n\theta).$$

From this we conclude that $B_n = 0$ for all integers n, $A_0 = \frac{1}{2}$, $A_1 = 0$, $A_2 = -\frac{1}{34}$, and $A_m = 0$ for $m = 3, 4, 5, \dots$. Therefore

$$u(r, \theta) = A_0 + A_2 \cos 2\theta = \frac{1}{2} - \frac{1}{34}(r^2 + 16r^{-2}) \cos 2\theta = \frac{1}{2} - \left(\frac{1}{34}r^2 + \frac{8}{17}r^{-2}\right) \cos 2\theta.$$

12. Letting $\lambda = \alpha^2 > 0$ and proceeding in the usual manner we find

$$u(r, t) = \sum_{n=1}^{\infty} A_n \cos a\alpha_n t J_0(\alpha_n r)$$

where the eigenvalues $\lambda_n = \alpha_n^2$ are determined by $J_0(\alpha) = 0$. Then the initial condition gives

$$u_0 J_0(x_k r) = \sum_{n=1}^{\infty} A_n J_0(\alpha_n r)$$

and so

$$A_n = \frac{2}{J_1^2(\alpha_n)} \int_0^1 r\left(u_0 J_0(x_k r)\right) J_0(\alpha_n r)\, dr.$$

But $J_0(\alpha) = 0$ implies that the eigenvalues are the positive zeros of J_0, that is, $\alpha_n = x_n$ for $n = 1, 2, 3, \dots$. Therefore

$$A_n = \frac{2u_0}{J_1^2(\alpha_n)} \int_0^1 r J_0(\alpha_k r) J_0(\alpha_n r)\, dr = 0, \quad n \neq k$$

by orthogonality. For $n = k$,

$$A_k = \frac{2u_0}{J_1^2(\alpha_k)} \int_0^1 r J_0^2(\alpha_k)\, dr = u_0$$

by (7) of Section 12.6. Thus the solution $u(r, t)$ reduces to one term when $n = k$, and

$$u(r, t) = u_0 \cos a\alpha_k t J_0(\alpha_k r) = u_0 \cos a x_k t J_0(x_k r).$$

15. Referring to Example 1 in Section 14.3 of the text we have

$$u(r, \theta) = \sum_{n=0}^{\infty} A_n r^n P_n(\cos \theta).$$

For $x = \cos \theta$

$$u(1, \theta) = \begin{cases} 100 & 0 < \theta < \pi/2 \\ -100 & \pi/2 < \theta < \pi \end{cases} = 100 \begin{cases} -1, & -1 < x < 0 \\ 1, & 0 < x < 1 \end{cases} = g(x).$$

From Problem 22 in Exercise 12.6 we have

$$u(r, \theta) = 100 \left[\frac{3}{2}r P_1(\cos \theta) - \frac{7}{8}r^3 P_3(\cos \theta) + \frac{11}{16}r^5 P_5(\cos \theta) + \cdots\right].$$

18. Letting $u(r, t) = R(r)T(t)$ and the separation constant be $-\lambda = -\alpha^2$ we obtain

$$rR'' + R' + \alpha^2 rR = 0$$

$$T' + \alpha^2 T = 0,$$

with solutions

$$R(r) = c_1 J_0(\alpha r) + c_2 Y_0(\alpha r)$$

$$T(t) = c_3 e^{-\alpha^2 t}.$$

Now the boundary conditions imply

$$R(a) = 0 = c_1 J_0(\alpha a) + c_2 Y_0(\alpha a)$$
$$R(b) = 0 = c_1 J_0(\alpha b) + c_2 Y_0(\alpha b)$$

so that

$$c_2 = -\frac{c_1 J_0(\alpha a)}{Y_0(\alpha a)}$$

and

$$c_1 J_0(\alpha b) - \frac{c_1 J_0(\alpha a)}{Y_0(\alpha a)} Y_0(\alpha b) = 0$$

or

$$Y_0(\alpha a) J_0(\alpha b) - J_0(\alpha a) Y_0(\alpha b) = 0.$$

This equation defines α_n for $n = 1, 2, 3, \ldots$. Now

$$R(r) = c_1 J_0(\alpha r) - c_1 \frac{J_0(\alpha a)}{Y_0(\alpha a)} Y_0(\alpha r) = \frac{c_1}{Y_0(\alpha a)} \left[Y_0(\alpha a) J_0(\alpha r) - J_0(\alpha a) Y_0(\alpha r) \right]$$

and

$$u_n(r, t) = A_n \left[Y_0(\alpha_n a) J_0(\alpha_n r) - J_0(\alpha_n a) Y_0(\alpha_n r) \right] e^{-\alpha_n^2 t} = A_n u_n(r) e^{-\alpha_n^2 t}.$$

Thus

$$u(r, t) = \sum_{n=1}^{\infty} A_n u_n(r) e^{-\alpha_n^2 t}.$$

From the initial condition

$$u(r, 0) = f(r) = \sum_{n=1}^{\infty} A_n u_n(r)$$

we obtain

$$A_n = \frac{\int_a^b r f(r) u_n(r) \, dr}{\int_a^b r u_n^2(r) \, dr}.$$

21. In the boundary-value problem

$$\frac{\partial^2 u}{\partial r^2} + \frac{1}{r} \frac{\partial u}{\partial r} + \frac{\partial^2 u}{\partial z^2} = 0, \quad 0 < r < 1, \; 0 < z < 1$$
$$u|z = 100, \quad 0 < z < 1$$
$$\left. \frac{\partial u}{\partial z} \right|_{z=0} = 0, \quad 0 < r < 1$$
$$u|r1 = 200, \quad 0 < r < 1,$$

we substitute $u(r, z) = \psi(r) + v(r, z)$. The new problem is

$$\frac{\partial^2 v}{\partial r^2} + \frac{1}{r} \frac{\partial v}{\partial r} + \frac{\partial^2 v}{\partial z^2} = 0, \quad 0 < r < 1, \; 0 < z < 1$$
$$v(1, z) = 0, \quad 0 < z < 1$$
$$\left. \frac{\partial v}{\partial z} \right|_{z=0} = 0, \quad 0 < r < 1$$
$$v(r, 1) = 100, \quad 0 < r < 1.$$

The function $\psi(r)$ is determined by the problem

$$\frac{d^2\psi}{dr^2} + \frac{1}{r}\frac{d\psi}{dr} = 0, \quad \psi(1) = 100.$$

From the solution $\psi(r) = c_1 + c_2 \ln r$ we define $c_1 = 0$ for boundedness at $r = 0$. To match the condition at $r = 1$ we chose $c_1 = 100$. Thus $\psi(r) = 100$. Now, using $-\lambda$ as the separation constant in the second boundary-value problemyields

$$rR'' + R' + \lambda r R = 0 \quad \text{and} \quad Z'' - \lambda Z = 0.$$

The only choice of λ that results in nontrivial solutions is $\lambda = \alpha^2 > 0$ and the solutions are

$$R(r) = c_1 J_0(\alpha r) \quad \text{and} \quad Z(z) = c_3 \cosh \alpha z + c_4 \sinh \alpha z.$$

The usual boundedness argument implies $c_1 = 0$. The boundary condition $Z'(0) = 0$ implies $c_4 = 0$. The condition $R(1) = 0$ gives the defining equation for the eigenvalues $J_0(\alpha) = 0$. Forming the superposition of the product of the eigenfunctions and the Z solutions gives

$$v(r,z) = \sum_{n=1}^{\infty} A_n \cosh \alpha_n z J_0(\alpha_n r),$$

where the eigenvalues are $\lambda_n = \alpha_n^2$. When $z = 1$ we have

$$v(r,1) = 100 = \sum_{n=1}^{\infty} A_n \cosh \alpha_n J_0(\alpha_n r),$$

and so from (16) of Section 12.6 in the text the coefficients are

$$A_n \cosh \alpha_n = \frac{2}{J_1^2(\alpha_n)} \int_0^1 100 r J_0(\alpha_n r) dr \quad \longleftarrow t = \alpha_n r$$

$$= \frac{200}{\alpha_n^2 J_1^2(\alpha_n)} \int_0^{\alpha_n} t J_0(t) dt$$

$$= \frac{200}{\alpha_n^2 J_1^2(\alpha_n)} \int_0^{\alpha_n} \frac{d}{dt}[t J_1(t)] dt = \frac{200}{\alpha_n J_1(\alpha_n)}.$$

Thus

$$A_n \cosh \alpha_n = \frac{200}{\alpha_n J_1(\alpha_n)},$$

so

$$A_n = \frac{200}{\alpha_n \cosh \alpha_n J_1(\alpha_n)},$$

and

$$v(r,z) = 200 \sum_{n=1}^{\infty} \frac{\cosh \alpha_n z}{\alpha_n \cosh \alpha_n J_1(\alpha_n)} J_0(\alpha_n r).$$

Therefore $u(r,z) = \psi(r) + v(r,z)$ is given by

$$u(r,z) = 100 + 200 \sum_{n=1}^{\infty} \frac{\cosh \alpha_n z}{\cosh \alpha_n J_1(\alpha_n)} J_0(\alpha_n r).$$

300

15 Integral Transform Method

EXERCISES 15.1
Error Function

3. By the first translation theorem,

$$\mathscr{L}\left\{e^t \operatorname{erf}(\sqrt{t})\right\} = \mathscr{L}\left\{\operatorname{erf}(\sqrt{t})\right\}\bigg|_{s \to s-1} = \frac{1}{s\sqrt{s+1}}\bigg|_{s \to s-1} = \frac{1}{\sqrt{s}\,(s-1)}.$$

6. To find

$$\mathscr{L}^{-1}\left\{\frac{1}{1+\sqrt{s+1}}\right\}$$

we begin by rationalizing first the denominator and then the numerator:

$$\frac{1}{1+\sqrt{s+1}} = \frac{1}{1+\sqrt{s+1}} \cdot \frac{1-\sqrt{s+1}}{1-\sqrt{s+1}} = \frac{1-\sqrt{s+1}}{-s}$$

$$= -\frac{1}{s} + \frac{\sqrt{s+1}}{s} = -\frac{1}{s} + \frac{\sqrt{s+1}}{s} \cdot \frac{\sqrt{s+1}}{\sqrt{s+1}}$$

$$= -\frac{1}{s} + \frac{s+1}{s\sqrt{s+1}} \qquad \longleftarrow \text{ partial fractions}$$

$$= -\frac{1}{s} + \frac{1}{\sqrt{s+1}} + \frac{1}{s\sqrt{s+1}}.$$

Therefore, from the first translation theorem and Problem 1(b) in this section,

$$\mathscr{L}^{-1}\left\{-\frac{1}{s} + \frac{1}{\sqrt{s+1}} + \frac{1}{s\sqrt{s+1}}\right\} = \mathscr{L}\left\{\frac{1}{s}\right\} + \mathscr{L}^{-1}\left\{\frac{1}{\sqrt{s+1}}\right\} + \mathscr{L}^{-1}\left\{\frac{1}{s\sqrt{s+1}}\right\}$$

$$= -1 + \mathscr{L}^{-1}\left\{\frac{1}{s}\bigg|_{s \to s+1}\right\} + \mathscr{L}^{-1}\left\{\frac{1}{s\sqrt{s+1}}\right\}$$

$$= -1 + \frac{1}{\sqrt{\pi}}\mathscr{L}^{-1}\left\{\frac{\sqrt{\pi}}{s}\bigg|_{s \to s+1}\right\} + \mathscr{L}^{-1}\left\{\frac{1}{s\sqrt{s+1}}\right\}$$

$$= -1 + \frac{1}{\sqrt{\pi}}\frac{e^{-t}}{\sqrt{t}} + \operatorname{erf}(\sqrt{t}) = \frac{e^{-t}}{\sqrt{\pi t}} - (1 - \operatorname{erf}(\sqrt{t}))$$

$$= \frac{e^{-t}}{\sqrt{\pi t}} - \operatorname{erfc}(\sqrt{t}).$$

9. Taking the Laplace transform of both sides of the equation we obtain

$$\mathcal{L}\{y(t)\} = \mathcal{L}\{1\} - \mathcal{L}\left\{\int_0^t \frac{y(\tau)}{\sqrt{t-\tau}}\,d\tau\right\}$$

$$Y(s) = \frac{1}{s} - Y(s)\frac{\sqrt{\pi}}{\sqrt{s}}$$

$$\frac{\sqrt{s}+\sqrt{\pi}}{\sqrt{s}}Y(s) = \frac{1}{s}$$

$$Y(s) = \frac{1}{\sqrt{s}\left(\sqrt{s}+\sqrt{\pi}\right)}.$$

Thus

$$y(t) = \mathcal{L}^{-1}\left\{\frac{1}{\sqrt{s}\left(\sqrt{s}+\sqrt{\pi}\right)}\right\} = e^{\pi t}\,\mathrm{erfc}(\sqrt{\pi t}). \qquad \boxed{\text{By entry 5 in Table 15.1}}$$

12. Since $f(x) = e^{-x^2}$ is an even function,

$$\int_{-a}^a e^{-u^2}\,du = 2\int_0^a e^{-u^2}\,du.$$

Therefore,

$$\int_{-a}^a e^{-u^2}\,du = \sqrt{\pi}\,\mathrm{erf}(a).$$

EXERCISES 15.2

Applications of the Laplace Transform

3. The solution of

$$a^2\frac{d^2U}{dx^2} - s^2U = 0$$

is in this case

$$U(x,s) = c_1 e^{-(x/a)s} + c_2 e^{(x/a)s}.$$

Since $\lim_{x\to\infty} u(x,t) = 0$ we have $\lim_{x\to\infty} U(x,s) = 0$. Thus $c_2 = 0$ and

$$U(x,s) = c_1 e^{-(x/a)s}.$$

If $\mathcal{L}\{u(0,t)\} = \mathcal{L}\{f(t)\} = F(s)$ then $U(0,s) = F(s)$. From this we have $c_1 = F(s)$ and

$$U(x,s) = F(s)e^{-(x/a)s}.$$

Hence, by the second translation theorem,

$$u(x,t) = f\left(t - \frac{x}{a}\right)\mathcal{U}\left(t - \frac{x}{a}\right).$$

6. Transforming the partial differential equation gives

$$\frac{d^2U}{dx^2} - s^2U = -\frac{\omega}{s^2+\omega^2}\sin\pi x.$$

Using undetermined coefficients we obtain

$$U(x, s) = c_1 \cosh sx + c_2 \sinh sx + \frac{\omega}{(s^2 + \pi^2)(s^2 + \omega^2)} \sin \pi x.$$

The transformed boundary conditions $U(0, s) = 0$ and $U(1, s) = 0$ give, in turn, $c_1 = 0$ and $c_2 = 0$. Therefore

$$U(x, s) = \frac{\omega}{(s^2 + \pi^2)(s^2 + \omega^2)} \sin \pi x$$

and

$$u(x, t) = \omega \sin \pi x \, \mathscr{L}^{-1} \left\{ \frac{1}{(s^2 + \pi^2)(s^2 + \omega^2)} \right\}$$

$$= \frac{\omega}{\omega^2 - \pi^2} \sin \pi x \, \mathscr{L}^{-1} \left\{ \frac{1}{\pi} \frac{\pi}{s^2 + \pi^2} - \frac{1}{\omega} \frac{\omega}{s^2 + \omega^2} \right\}$$

$$= \frac{\omega}{\pi(\omega^2 - \pi^2)} \sin \pi t \sin \pi x - \frac{1}{\omega^2 - \pi^2} \sin \omega t \sin \pi x.$$

9. Transforming the partial differential equation gives

$$\frac{d^2 U}{dx^2} - s^2 U = -sx e^{-x}.$$

Using undetermined coefficients we obtain

$$U(x, s) = c_1 e^{-sx} + c_2 e^{sx} - \frac{2s}{(s^2 - 1)^2} e^{-x} + \frac{s}{s^2 - 1} x e^{-x}.$$

The transformed boundary conditions $\lim_{x \to \infty} U(x, s) = 0$ and $U(0, s) = 0$ give, in turn, $c_2 = 0$ and $c_1 = 2s/(s^2 - 1)^2$. Therefore

$$U(x, s) = \frac{2s}{(s^2 - 1)^2} e^{-sx} - \frac{2s}{(s^2 - 1)^2} e^{-x} + \frac{s}{s^2 - 1} x e^{-x}.$$

From entries (13) and (26) in the Table of Laplace transforms we obtain

$$u(x, t) = \mathscr{L}^{-1} \left\{ \frac{2s}{(s^2 - 1)^2} e^{-sx} - \frac{2s}{(s^2 - 1)^2} e^{-x} + \frac{s}{s^2 - 1} x e^{-x} \right\}$$

$$= 2(t - x) \sinh(t - x) \, \mathscr{U}(t - x) - t e^{-x} \sinh t + x e^{-x} \cosh t.$$

12. We use

$$U(x, s) = c_1 e^{-\sqrt{s}\, x} + c_2 e^{\sqrt{s}\, x} + \frac{u_1 x}{s}.$$

The condition $\lim_{x \to \infty} u(x, t)/x = u_1$ implies $\lim_{x \to \infty} U(x, s)/x = u_1/s$, so we define $c_2 = 0$. Then

$$U(x, s) = c_1 e^{-\sqrt{s}\, x} + \frac{u_1 x}{s}.$$

From $U(0, s) = u_0/s$ we obtain $c_1 = u_0/s$. Hence

$$U(x, s) = u_0 \frac{e^{-\sqrt{s}\, x}}{s} + \frac{u_1 x}{s}$$

and

$$u(x, t) = u_0 \, \mathscr{L}^{-1} \left\{ \frac{e^{-x\sqrt{s}}}{s} \right\} + u_1 x \, \mathscr{L}^{-1} \left\{ \frac{1}{s} \right\} = u_0 \, \mathrm{erfc}\left(\frac{x}{2\sqrt{t}} \right) + u_1 x.$$

15. We use

$$U(x, s) = c_1 e^{-\sqrt{s}\, x} + c_2 e^{\sqrt{s}\, x}.$$

The condition $\lim_{x \to \infty} u(x, t) = 0$ implies $\lim_{x \to \infty} U(x, s) = 0$, so we define $c_2 = 0$. Hence

$$U(x, s) = c_1 e^{-\sqrt{s}\, x}.$$

The transform of $u(0, t) = f(t)$ is $U(0, s) = F(s)$. Therefore

$$U(x, s) = F(s) e^{-\sqrt{s}\, x}$$

and

$$u(x, t) = \mathscr{L}^{-1}\left\{ F(s) e^{-x\sqrt{s}} \right\} = \frac{x}{2\sqrt{\pi}} \int_0^t \frac{f(t - \tau) e^{-x^2/4\tau}}{\tau^{3/2}} \, d\tau.$$

18. The solution of the transformed equation

$$\frac{d^2 U}{dx^2} - sU = -100$$

by undetermined coefficients is

$$U(x, s) = c_1 e^{\sqrt{s}\, x} + c_2 e^{-\sqrt{s}\, x} + \frac{100}{s}.$$

From the fact that $\lim_{x \to \infty} U(x, s) = 100/s$ we see that $c_1 = 0$. Thus

$$U(x, s) = c_2 e^{-\sqrt{s}\, x} + \frac{100}{s}. \tag{1}$$

Now the transform of the boundary condition at $x = 0$ is

$$U(0, s) = 20 \left[\frac{1}{s} - \frac{1}{s} e^{-s} \right].$$

It follows from (1) that

$$\frac{20}{s} - \frac{20}{s} e^{-s} = c_2 + \frac{100}{s} \qquad \text{or} \qquad c_2 = -\frac{80}{s} - \frac{20}{s} e^{-s}$$

and so

$$U(x, s) = \left(-\frac{80}{s} - \frac{20}{s} e^{-s} \right) e^{-\sqrt{s}\, x} + \frac{100}{s}$$

$$= \frac{100}{s} - \frac{80}{s} e^{-\sqrt{s}\, x} - \frac{20}{s} e^{-\sqrt{s}\, x} e^{-s}.$$

Thus

$$u(x, t) = 100 \, \mathscr{L}^{-1}\left\{ \frac{1}{s} \right\} - 80 \, \mathscr{L}^{-1}\left\{ \frac{e^{-\sqrt{s}\, x}}{s} \right\} - 20 \, \mathscr{L}^{-1}\left\{ \frac{e^{-\sqrt{s}\, x}}{s} e^{-s} \right\}$$

$$= 100 - 80 \operatorname{erfc}\left(x/2\sqrt{t} \right) - 20 \operatorname{erfc}\left(x/2\sqrt{t - 1} \right) \mathscr{U}(t - 1).$$

21. The solution of

$$\frac{d^2 U}{dx^2} - sU = -u_0 - u_0 \sin \frac{\pi}{L} x$$

is

$$U(x, s) = c_1 \cosh(\sqrt{s}\, x) + c_2 \sinh(\sqrt{s}\, x) + \frac{u_0}{s} + \frac{u_0}{s + \pi^2/L^2} \sin \frac{\pi}{L} x.$$

The transformed boundary conditions $U(0, s) = u_0/s$ and $U(L, s) = u_0/s$ give, in turn, $c_1 = 0$ and $c_2 = 0$. Therefore

$$U(x, s) = \frac{u_0}{s} + \frac{u_0}{s + \pi^2/L^2} \sin \frac{\pi}{L} x$$

and

$$u(x, t) = u_0 \, \mathscr{L}^{-1} - 1pt\left\{ \frac{1}{s} \right\} + u_0 \, \mathscr{L}^{-1}\left\{ \frac{1}{s + \pi^2/L^2} \right\} \sin \frac{\pi}{L} x = u_0 + u_0 e^{-\pi^2 t/L^2} \sin \frac{\pi}{L} x.$$

24. We use

$$c(x, s) = c_1 \cosh \sqrt{\frac{s}{D}}\, x + c_2 \sinh \sqrt{\frac{s}{D}}\, x.$$

The transform of the two boundary conditions are $c(0, s) = c_0/s$ and $c(1, s) = c_0/s$. From these conditions we obtain $c_1 = c_0/s$ and

$$c_2 = c_0(1 - \cosh \sqrt{s/D}\,)/s \sinh \sqrt{s/D}\,.$$

Therefore

$$c(x, s) = c_0 \left[\frac{\cosh \sqrt{s/D}\, x}{s} + \frac{(1 - \cosh \sqrt{s/D}\,)}{s \sinh \sqrt{s/D}} \sinh \sqrt{s/D}\, x \right]$$

$$= c_0 \left[\frac{\sinh \sqrt{s/D}\, (1 - x)}{s \sinh \sqrt{s/D}} + \frac{\sin \sqrt{s/D}\, x}{s \sinh \sqrt{s/D}} \right]$$

$$= c_0 \left[\frac{e^{\sqrt{s/D}\,(1-x)} - e^{-\sqrt{s/D}\,(1-x)}}{s(e^{\sqrt{s/D}} - e^{-\sqrt{s/D}})} + \frac{e^{\sqrt{s/D}\, x} - e^{-\sqrt{s/D}\, x}}{s(e^{\sqrt{s/D}} - e^{-\sqrt{s/D}})} \right]$$

$$= c_0 \left[\frac{e^{-\sqrt{s/D}\, x} - e^{-\sqrt{s/D}\,(2-x)}}{s(1 - e^{-2\sqrt{s/D}})} + \frac{e^{\sqrt{s/D}\,(x-1)} - e^{-\sqrt{s/D}\,(x+1)}}{s(1 - e^{-2\sqrt{s/D}})} \right]$$

$$= c_0 \frac{(e^{-\sqrt{s/D}\, x} - e^{-\sqrt{s/D}\,(2-x)})}{s} \left(1 + e^{-2\sqrt{s/D}} + e^{-4\sqrt{s/D}} + \cdots \right)$$

$$+ c_0 \frac{(e^{\sqrt{s/D}\,(x-1)} - e^{-\sqrt{s/D}\,(x+1)})}{s} \left(1 + e^{-2\sqrt{s/D}} + e^{-4\sqrt{s/D}} + \cdots \right)$$

$$= c_0 \sum_{n=0}^{\infty} \left[\frac{e^{-(2n+x)\sqrt{s/D}}}{s} - \frac{e^{-(2n+2-x)\sqrt{s/D}}}{s} \right]$$

$$+ c_0 \sum_{n=0}^{\infty} \left[\frac{e^{-(2n+1-x)\sqrt{s/D}}}{s} - \frac{e^{-(2n+1+x)\sqrt{s/D}}}{s} \right]$$

and so

$$c(x, t) = c_0 \sum_{n=0}^{\infty} \left[\mathcal{L}^{-1} \left\{ \frac{e^{-\frac{(2n+x)}{\sqrt{D}}\sqrt{s}}}{s} \right\} - \mathcal{L}^{-1} \left\{ \frac{e^{-\frac{(2n+2-x)}{\sqrt{D}}\sqrt{s}}}{s} \right\} \right]$$

$$+ c_0 \sum_{n=0}^{\infty} \left[\mathcal{L}^{-1} \left\{ \frac{e^{-\frac{(2n+1-x)}{\sqrt{D}}\sqrt{s}}}{s} \right\} - \mathcal{L}^{-1} \left\{ \frac{e^{-\frac{(2n+1+x)}{\sqrt{D}}\sqrt{s}}}{s} \right\} \right]$$

$$= c_0 \sum_{n=0}^{\infty} \left[\operatorname{erfc}\left(\frac{2n+x}{2\sqrt{Dt}} \right) - \operatorname{erfc}\left(\frac{2n+2-x}{2\sqrt{Dt}} \right) \right]$$

$$+ c_0 \sum_{n=0}^{\infty} \left[\operatorname{erfc}\left(\frac{2n+1-x}{2\sqrt{Dt}} \right) - \operatorname{erfc}\left(\frac{2n+1+x}{2\sqrt{Dt}} \right) \right].$$

Now using $\mathrm{erfc}(x) = 1 - \mathrm{erf}(x)$ we get

$$c(x,t) = c_0 \sum_{n=0}^{\infty} \left[\mathrm{erf}\left(\frac{2n+2-x}{2\sqrt{Dt}}\right) - \mathrm{erf}\left(\frac{2n+x}{2\sqrt{Dt}}\right) \right]$$

$$+ c_0 \sum_{n=0}^{\infty} \left[\mathrm{erf}\left(\frac{2n+1+x}{2\sqrt{Dt}}\right) - \mathrm{erf}\left(\frac{2n+1-x}{2\sqrt{Dt}}\right) \right].$$

27. Using the Laplace transform with respect to t of the partial differential equation gives

$$\frac{d^2U}{dr^2} + \frac{2}{r}\frac{dU}{dr} = sU - \overbrace{u(r,0)}^{0},$$

where $\mathscr{L}u(r,t) = U(r,s)$. The ordinary differential equation is equivalent to

$$rU'' + 2U' - srU = 0.$$

If we let $v(r,s = rU(r,s)$ then differentiation with respect to r gives $v'' = rU'' + 2U'$. The transformed ordinary differential equation becomes

$$v'' - sv = 0$$

$$v = c_1 e^{-\sqrt{s}\,r} + c_2 e^{\sqrt{s}\,r}$$

$$U(r,s) = c_1 \frac{e^{-\sqrt{s}\,r}}{r} + c_2 \frac{e^{\sqrt{s}\,r}}{r}.$$

The usual argument about boundedness as $r \to \infty$ implies $c_2 = 0$. Therefore $U(r,s) = c_1 e^{-\sqrt{s}\,r}/r$. Now the Laplace transform of $u(1,t) = 100$ is

$$U(1,s) = \frac{100}{s} = c_1 e^{-\sqrt{s}}$$

so

$$c_1 = \frac{e^{\sqrt{s}}}{s}$$

and

$$U(r,s) = \frac{100}{r}\frac{e^{-\sqrt{s}(r-1)}}{s}.$$

Thus

$$u(r,t) = \frac{100}{r} \mathscr{L}^{-1}\left\{ \frac{e^{-(r-1)\sqrt{s}}}{s} \right\}.$$

From the third entry in Table 15.1.1 we get

$$u(r,t) = \frac{100}{r}\, \mathrm{erfc}\left(\frac{r-1}{2\sqrt{t}}\right).$$

30. (a) Transforming the partial differential equation and using the initial condition gives

$$k\frac{d^2U}{dx^2} - sU = 0.$$

Since the domain of the variable x is an infinite interval we write the general solution of this differential equation as

$$U(x,s) = c_1 e^{-\sqrt{s/k}\,x} + c_2 e^{\sqrt{s/k}\,x}.$$

Transforming the boundary conditions gives $U'(0, s) = -A/s$ and $\lim\limits_{x \to \infty} U(x, s) = 0$. Hence we find $c_2 = 0$ and $c_1 = A\sqrt{k}/s\sqrt{s}$. From

$$U(x, s) = A\sqrt{k} \, \frac{e^{-\sqrt{s/k}\, x}}{s\sqrt{s}}$$

we see that

$$u(x, t) = A\sqrt{k} \, \mathscr{L}^{-1} \left\{ \frac{e^{-\sqrt{s/k}\, x}}{s\sqrt{s}} \right\}.$$

With the identification $a = x/\sqrt{k}$ it follows from the Table of Laplace transforms that

$$u(x, t) = A\sqrt{k} \left\{ 2\sqrt{\frac{t}{\pi}} \, e^{-x^2/4kt} - \frac{x}{\sqrt{k}} \, \text{erfc}\left(\frac{x}{2\sqrt{kt}} \right) \right\}$$

$$= 2A\sqrt{\frac{kt}{\pi}} \, e^{-x^2/4kt} - Ax \, \text{erfc}\left(x/2\sqrt{kt} \right).$$

Since $\text{erfc}(0) = 1$,

$$\lim_{t \to \infty} u(x, t) = \lim_{t \to \infty} \left(2A\sqrt{\frac{kt}{\pi}} e^{-x^2/4kt} - Ax \, \text{erfc}\left(\frac{x}{2\sqrt{kt}} \right) \right) = \infty.$$

(b)

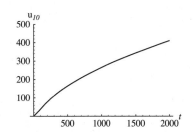

EXERCISES 15.3
Fourier Integral

3. From formulas (5) and (6) in the text,

$$A(\alpha) = \int_0^3 x \cos \alpha x \, dx = \left. \frac{x \sin \alpha x}{\alpha} \right|_0^3 - \frac{1}{\alpha} \int_0^3 \sin \alpha x \, dx$$

$$= \frac{3 \sin 3\alpha}{\alpha} + \left. \frac{\cos \alpha x}{\alpha^2} \right|_0^3 = \frac{3\alpha \sin 3\alpha + \cos 3\alpha - 1}{\alpha^2}$$

and

$$B(\alpha) = \int_0^3 x \sin \alpha x \, dx = \left. -\frac{x \cos \alpha x}{\alpha} \right|_0^3 + \frac{1}{\alpha} \int_0^3 \cos \alpha x \, dx$$

$$= -\frac{3 \cos 3\alpha}{\alpha} + \left. \frac{\sin \alpha x}{\alpha^2} \right|_0^3 = \frac{\sin 3\alpha - 3\alpha \cos 3\alpha}{\alpha^2}.$$

Hence

$$f(x) = \frac{1}{\pi} \int_0^\infty \frac{(3\alpha \sin 3\alpha + \cos 3\alpha - 1)\cos \alpha x + (\sin 3\alpha - 3\alpha \cos 3\alpha)\sin \alpha x}{\alpha^2} \, d\alpha$$

$$= \frac{1}{\pi} \int_0^\infty \frac{3\alpha(\sin 3\alpha \cos \alpha x - \cos 3\alpha \sin \alpha x) + \cos 3\alpha \cos \alpha x + \sin 3\alpha \sin \alpha x - \cos \alpha x}{\alpha^2} \, d\alpha$$

$$= \frac{1}{\pi} \int_0^\infty \frac{3\alpha \sin \alpha(3 - x) + \cos \alpha(3 - x) - \cos \alpha x}{\alpha^2} \, d\alpha.$$

6. From formulas (5) and (6) in the text,

$$A(\alpha) = \int_{-1}^1 e^x \cos \alpha x \, dx$$

$$= \frac{e(\cos \alpha + \alpha \sin \alpha) - e^{-1}(\cos \alpha - \alpha \sin \alpha)}{1 + \alpha^2}$$

$$= \frac{2(\sinh 1)\cos \alpha - 2\alpha(\cosh 1)\sin \alpha}{1 + \alpha^2}$$

and

$$B(\alpha) = \int_{-1}^1 e^x \sin \alpha x \, dx$$

$$= \frac{e(\sin \alpha - \alpha \cos \alpha) - e^{-1}(-\sin \alpha - \alpha \cos \alpha)}{1 + \alpha^2}$$

$$= \frac{2(\cosh 1)\sin \alpha - 2\alpha(\sinh 1)\cos \alpha}{1 + \alpha^2}.$$

Hence

$$f(x) = \frac{1}{\pi} \int_0^\infty [A(\alpha)\cos \alpha x + B(\alpha)\sin \alpha x] \, d\alpha.$$

9. The function is even. Thus from formula (9) in the text

$$A(\alpha) = \int_0^\pi x \cos \alpha x \, dx = \frac{x \sin \alpha x}{\alpha} \bigg|_0^\pi - \frac{1}{\alpha} \int_0^\pi \sin \alpha x \, dx$$

$$= \frac{\pi \alpha \sin \pi \alpha}{\alpha} + \frac{1}{\alpha^2} \cos \alpha x \bigg|_0^\pi = \frac{\pi \alpha \sin \pi \alpha + \cos \pi \alpha - 1}{\alpha^2}.$$

Hence from formula (8) in the text

$$f(x) = \frac{2}{\pi} \int_0^\infty \frac{(\pi \alpha \sin \pi \alpha + \cos \pi \alpha - 1)\cos \alpha x}{\alpha^2} \, d\alpha.$$

12. The function is odd. Thus from formula (11) in the text

$$B(\alpha) = \int_0^\infty x e^{-x} \sin \alpha x \, dx.$$

Now recall

$$\mathcal{L}\{t \sin kt\} = -\frac{d}{ds}\mathcal{L}\{\sin kt\} = 2ks/(s^2 + k^2)^2.$$

If we set $s = 1$ and $k = \alpha$ we obtain

$$B(\alpha) = \frac{2\alpha}{(1 + \alpha^2)^2}.$$

Hence from formula (10) in the text

$$f(x) = \frac{4}{\pi} \int_0^\infty \frac{\alpha \sin \alpha x}{(1 + \alpha^2)^2} \, d\alpha.$$

15. For the cosine integral,

$$A(\alpha) = \int_0^\infty xe^{-2x} \cos \alpha x \, dx.$$

But we know

$$\mathscr{L}\{t \cos kt\} = -\frac{d}{ds} \frac{s}{(s^2 + k^2)} = \frac{(s^2 - k^2)}{(s^2 + k^2)^2}.$$

If we set $s = 2$ and $k = \alpha$ we obtain

$$A(\alpha) = \frac{4 - \alpha^2}{(4 + \alpha^2)^2}.$$

Hence

$$f(x) = \frac{2}{\pi} \int_0^\infty \frac{(4 - \alpha^2) \cos \alpha x}{(4 + \alpha^2)^2} \, d\alpha.$$

For the sine integral,

$$B(\alpha) = \int_0^\infty xe^{-2x} \sin \alpha x \, dx.$$

From Problem 12, we know

$$\mathscr{L}\{t \sin kt\} = \frac{2ks}{(s^2 + k^2)^2}.$$

If we set $s = 2$ and $k = \alpha$ we obtain

$$B(\alpha) = \frac{4\alpha}{(4 + \alpha^2)^2}.$$

Hence

$$f(x) = \frac{8}{\pi} \int_0^\infty \frac{\alpha \sin \alpha x}{(4 + \alpha^2)^2} \, d\alpha.$$

18. From the formula for sine integral of $f(x)$ we have

$$f(x) = \frac{2}{\pi} \int_0^\infty \left(\int_0^\infty f(x) \sin \alpha x \, dx \right) \sin \alpha x \, dx$$

$$= \frac{2}{\pi} \left[\int_0^1 1 \cdot \sin \alpha x \, d\alpha + \int_1^\infty 0 \cdot \sin \alpha x \, d\alpha \right]$$

$$= \frac{2}{\pi} \frac{(-\cos \alpha x)}{x} \Big|_0^1 = \frac{2}{\pi} \frac{1 - \cos x}{x}.$$

21. (a) From the identity

$$\sin A \cos B = \frac{1}{2}[\sin(A + B) + \sin(A - B)]$$

we have

$$\sin \alpha \cos \alpha x = \frac{1}{2}[\sin(\alpha + \alpha x) + \sin(\alpha - \alpha x)]$$

$$= \frac{1}{2}[\sin \alpha(1 + x) + \sin \alpha(1 - x)]$$

$$= \frac{1}{2}[\sin \alpha(x + 1) - \sin \alpha(x - 1)].$$

Then

$$\frac{2}{\pi} \int_0^\infty \frac{\sin \alpha \cos \alpha x}{\alpha} \, d\alpha = \frac{1}{\pi} \int_0^\infty \frac{\sin \alpha(x + 1) - \sin \alpha(x - 1)}{\alpha} \, d\alpha.$$

(b) Noting that

$$F_b = \frac{1}{\pi} \int_0^b \frac{\sin\alpha(x+1) - \sin\alpha(x-1)}{\alpha} \, d\alpha$$

$$= \frac{1}{\pi} \left[\int_0^b \frac{\sin\alpha(x+1)}{\alpha} \, d\alpha - \int_0^b \frac{\sin\alpha(x-1)}{\alpha} \, d\alpha \right]$$

and letting $t = \alpha(x+1)$ so that $dt = (x+1)\,d\alpha$ in the first integral and $t = \alpha(x-1)$ so that $dt = (x-1)\,d\alpha$ in the second integral we have

$$F_b = \frac{1}{\pi} \left[\int_0^{b(x+1)} \frac{\sin t}{t} \, dt - \int_0^{b(x-1)} \frac{\sin t}{t} \, dt \right].$$

Since $\mathrm{Si}\,(x) = \int_0^x [(\sin t)/t]\, dt$, this becomes

$$F_b = \frac{1}{\pi} [\mathrm{Si}\,(b(x+1)) - \mathrm{Si}\,(b(x-1))].$$

(c) In *Mathematica* we define $\mathbf{f[b\_]} := (\mathbf{1/Pi})(\mathbf{SinIntegral}[\mathbf{b(x+1)}] - \mathbf{SinIntegral}[\mathbf{b(x-1)}])$. Graphs of $F_b(x)$ for $b = 4, 6, 15,$ and 75 are shown below.

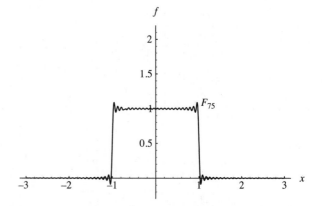

Fourier Transforms

For the boundary-value problems in this section it is sometimes useful to note that the identities

$$e^{i\alpha} = \cos\alpha + i\sin\alpha \qquad and \qquad e^{-i\alpha} = \cos\alpha - i\sin\alpha$$

imply

$$e^{i\alpha} + e^{-i\alpha} = 2\cos\alpha \qquad and \qquad e^{i\alpha} - e^{-i\alpha} = 2i\sin\alpha.$$

3. Using the Fourier sine transform, the partial differential equation becomes

$$\frac{dU}{dt} + k\alpha^2 U = k\alpha u_0.$$

The general solution of this linear equation is

$$U(\alpha, t) = ce^{-k\alpha^2 t} + \frac{u_0}{\alpha}.$$

But $U(\alpha, 0) = 0$ implies $c = -u_0/\alpha$ and so

$$U(\alpha, t) = u_0 \frac{1 - e^{-k\alpha^2 t}}{\alpha}$$

and

$$u(x, t) = \frac{2u_0}{\pi} \int_0^\infty \frac{1 - e^{-k\alpha^2 t}}{\alpha} \sin\alpha x \, d\alpha.$$

6. Since the domain of x is $(0, \infty)$ and the condition at $x = 0$ involves $\partial u/\partial x$ we use the Fourier cosine transform:

$$-k\alpha^2 U(\alpha, t) - ku_x(0, t) = \frac{dU}{dt}$$

$$\frac{dU}{dt} + k\alpha^2 U = kA$$

$$U(\alpha, t) = ce^{-k\alpha^2 t} + \frac{A}{\alpha^2}.$$

Since

$$\mathscr{F}\{u(x, 0)\} = U(\alpha, 0) = 0$$

we find $c = -A/\alpha^2$, so that

$$U(\alpha, t) = A\frac{1 - e^{-k\alpha^2 t}}{\alpha^2}.$$

Applying the inverse Fourier cosine transform we obtain

$$u(x, t) \overset{\mathscr{F}_c^{-1}}{=} \{U(\alpha, t)\} = \frac{2A}{\pi} \int_0^\infty \frac{1 - e^{-k\alpha^2 t}}{\alpha^2} \cos\alpha x \, d\alpha.$$

9. (a) Using the Fourier transform we obtain

$$U(\alpha, t) = c_1 \cos\alpha at + c_2 \sin\alpha at.$$

If we write

$$\mathscr{F}\{u(x,0)\} = \mathscr{F}\{f(x)\} = F(\alpha)$$

and

$$\mathscr{F}\{u_t(x,0)\} = \mathscr{F}\{g(x)\} = G(\alpha)$$

we first obtain $c_1 = F(\alpha)$ from $U(\alpha,0) = F(\alpha)$ and then $c_2 = G(\alpha)/\alpha a$ from $dU/dt\big|_{t=0} = G(\alpha)$. Thus

$$U(\alpha,t) = F(\alpha)\cos\alpha at + \frac{G(\alpha)}{\alpha a}\sin\alpha at$$

and

$$u(x,t) = \frac{1}{2\pi}\int_{-\infty}^{\infty}\left(F(\alpha)\cos\alpha at + \frac{G(\alpha)}{\alpha a}\sin\alpha at\right)e^{-i\alpha x}d\alpha.$$

(b) If $g(x) = 0$ then $c_2 = 0$ and

$$u(x,t) = \frac{1}{2\pi}\int_{-\infty}^{\infty}F(\alpha)\cos\alpha at\, e^{-i\alpha x}d\alpha$$

$$= \frac{1}{2\pi}\int_{-\infty}^{\infty}F(\alpha)\left(\frac{e^{\alpha ati}+e^{-\alpha ati}}{2}\right)e^{-i\alpha x}d\alpha$$

$$= \frac{1}{2}\left[\frac{1}{2\pi}\int_{-\infty}^{\infty}F(\alpha)e^{-i(x-at)\alpha}d\alpha + \frac{1}{2\pi}\int_{-\infty}^{\infty}F(\alpha)e^{-i(x+at)\alpha}d\alpha\right]$$

$$= \frac{1}{2}\left[f(x-at)+f(x+at)\right].$$

12. Since the boundary condition at $y = 0$ now involves $u(x,0)$ rather than $u'(x,0)$, we use the Fourier sine transform. The transform of the partial differential equation is then

$$\frac{d^2U}{dx^2} - \alpha^2 U + \alpha u(x,0) = 0 \qquad \text{or} \qquad \frac{d^2U}{dx^2} - \alpha^2 U = -\alpha.$$

The solution of this differential equation is

$$U(x,\alpha) = c_1\cosh\alpha x + c_2\sinh\alpha x + \frac{1}{\alpha}.$$

The transforms of the boundary conditions at $x = 0$ and $x = \pi$ in turn imply that $c_1 = 1/\alpha$ and

$$c_2 = \frac{\cosh\alpha\pi}{\alpha\sinh\alpha\pi} - \frac{1}{\alpha\sinh\alpha\pi} + \frac{\alpha}{(1+\alpha^2)\sinh\alpha\pi}.$$

Hence

$$U(x,\alpha) = \frac{1}{\alpha} - \frac{\cosh\alpha x}{\alpha} + \frac{\cosh\alpha\pi}{\alpha\sinh\alpha\pi}\sinh\alpha x - \frac{\sinh\alpha x}{\alpha\sinh\alpha\pi} + \frac{\alpha\sinh\alpha x}{(1+\alpha^2)\sinh\alpha\pi}$$

$$= \frac{1}{\alpha} - \frac{\sinh\alpha(\pi-x)}{\alpha\sinh\alpha\pi} - \frac{\sinh\alpha x}{\alpha(1+\alpha^2)\sinh\alpha\pi}.$$

Taking the inverse transform it follows that

$$u(x,y) = \frac{2}{\pi}\int_0^{\infty}\left(\frac{1}{\alpha} - \frac{\sinh\alpha(\pi-x)}{\alpha\sinh\alpha\pi} - \frac{\sinh\alpha x}{\alpha(1+\alpha^2)\sinh\alpha\pi}\right)\sin\alpha y\, d\alpha.$$

15. We use the Fourier sine transform with respect to x to obtain

$$U(\alpha,y) = c_1\cosh\alpha y + c_2\sinh\alpha y.$$

The transforms of $u(x,0) = f(x)$ and $u(x,2) = 0$ give, in turn, $U(\alpha,0) = F(\alpha)$ and $U(\alpha,2) = 0$. The first condition gives $c_1 = F(\alpha)$ and the second condition then yields

$$c_2 = -\frac{F(\alpha)\cosh 2\alpha}{\sinh 2\alpha}.$$

Hence

$$U(\alpha,y) = F(\alpha)\cosh \alpha y - \frac{F(\alpha)\cosh 2\alpha \sinh \alpha y}{\sinh 2\alpha}$$

$$= F(\alpha)\frac{\sinh 2\alpha \cosh \alpha y - \cosh 2\alpha \sinh \alpha y}{\sinh 2\alpha}$$

$$= F(\alpha)\frac{\sinh \alpha(2-y)}{\sinh 2\alpha}$$

and

$$u(x,y) = \frac{2}{\pi}\int_0^\infty F(\alpha)\frac{\sinh \alpha(2-y)}{\sinh 2\alpha}\sin \alpha x \, d\alpha.$$

18. We solve the three boundary-value problems:

Using separation of variables we find the solution of the first problem is

$$u_1(x,y) = \sum_{n=1}^{\infty} A_n e^{-ny}\sin nx \quad \text{where} \quad A_n = \frac{2}{\pi}\int_0^\pi f(x)\sin nx \, dx.$$

Using the Fourier sine transform with respect to y gives the solution of the second problem:

$$u_2(x,y) = \frac{200}{\pi}\int_0^\infty \frac{(1-\cos \alpha)\sinh \alpha(\pi-x)}{\alpha \sinh \alpha\pi}\sin \alpha y \, d\alpha.$$

Also, the Fourier sine transform with respect to y gives the solution of the third problem:

$$u_3(x,y) = \frac{2}{\pi}\int_0^\infty \frac{\alpha \sinh \alpha x}{(1+\alpha^2)\sinh \alpha\pi}\sin \alpha y \, d\alpha.$$

The solution of the original problem is

$$u(x,y) = u_1(x,y) + u_2(x,y) + u_3(x,y).$$

21. Using the Fourier transform with respect to x gives

$$U(\alpha, y) = c_1 \cosh \alpha y + c_2 \sinh \alpha y.$$

The transform of the boundary condition $\partial u/\partial y \,\big|_{y=0} = 0$ is $dU/dy \,\big|_{y=0} = 0$. This condition gives $c_2 = 0$. Hence

$$U(\alpha, y) = c_1 \cosh \alpha y.$$

Now by the given information the transform of the boundary condition $u(x, 1) = e^{-x^2}$ is $U(\alpha, 1) = \sqrt{\pi}\, e^{-\alpha^2/4}$. This condition then gives $c_1 = \sqrt{\pi}\, e^{-\alpha^2/4} \cosh \alpha$. Therefore

$$U(\alpha, y) = \sqrt{\pi}\, \frac{e^{-\alpha^2/4} \cosh \alpha y}{\cosh \alpha}$$

and

$$U(x, y) = \frac{1}{2\sqrt{\pi}} \int_{-\infty}^{\infty} \frac{e^{-\alpha^2/4} \cosh \alpha y}{\cosh \alpha} e^{-i\alpha x}\, d\alpha = \frac{1}{2\sqrt{\pi}} \int_{-\infty}^{\infty} \frac{e^{-\alpha^2/4} \cosh \alpha y}{\cosh \alpha} \cos \alpha x\, d\alpha$$

$$= \frac{1}{\sqrt{\pi}} \int_0^{\infty} \frac{e^{-\alpha^2/4} \cosh \alpha y}{\cosh \alpha} \cos \alpha x\, d\alpha.$$

24. We use the Fourier sine transform with respect to the variable z. The transform of the partial differential equation is

$$\frac{d^2 U}{dr^2} + \frac{1}{r}\frac{dU}{dr} - \alpha^2 U + \alpha U(r, 0) = 0 \quad \text{or} \quad \frac{d^2 U}{dr^2} + \frac{1}{r}\frac{dU}{dr} - \alpha^2 U = -\alpha u_0.$$

Thus

$$U(r, z) = c_1 I_0(\alpha r) + c_2 K_0(\alpha r) + \frac{u_0}{\alpha},$$

where $U(r, \alpha) = F_s\{u(r, z)\}$. The usual argument about u being bounded at $r = 0$ implies $c_2 = 0$ so $U(r, \alpha) = c_1 I_0(\alpha r) + u_0/\alpha$. Now the transform of the boundary condition $u(1, z) = 0$ is $U(1, \alpha) = 0$ and so $c_1 I_0(\alpha) + u_0/\alpha = 0$ or $c_1 = -u_0/\alpha I_0(\alpha)$. Thus

$$U(r, \alpha) = \frac{-u_0 I_0(\alpha r)}{\alpha I_0(\alpha)} + \frac{u_0}{\alpha}.$$

The inverse transform is

$$u(r, z) = \frac{2}{\pi} \int_0^{\infty} \left(-\frac{u_0 I_0(\alpha r)}{\alpha I_0(\alpha)} + \frac{u_0}{\alpha} \right) \sin \alpha z\, d\alpha$$

$$= \frac{2u_0}{\pi} \int_0^{\infty} \frac{\sin \alpha z}{\alpha}\, d\alpha - \frac{2u_0}{\pi} \int_0^{\infty} \frac{I_0(\alpha r)}{\alpha I_0(\alpha)} \sin \alpha z\, d\alpha. \qquad \longleftarrow \text{Problem 4, Exercises 15.4}$$

Therefore

$$u(r, z) = u_0 - \frac{2u_0}{\pi} \int_0^{\infty} \frac{I_0(\alpha r)}{\alpha I_0(\alpha)} \sin \alpha z\, d\alpha.$$

EXERCISES 15.5

Fast Fourier Transform

3. By the sifting property,

$$\mathscr{F}\{\delta(x)\} = \int_{-\infty}^{\infty} \delta(x)e^{i\alpha x}dx = e^{i\alpha 0} = 1.$$

6. Using a CAS we find

$$\mathscr{F}\{g(x)\} = \frac{1}{2}[\text{sign}(A - \alpha) + \text{sign}(A + \alpha)]$$

where $\text{sign}(t) = 1$ if $t > 0$ and $\text{sign}\,t = -1$ if $t < 0$. Thus

$$\mathscr{F}\{g(x)\} = \begin{cases} 1, & -A < \alpha < A \\ 0, & \text{elsewhere.} \end{cases}$$

9. The Fourier transform of $g(x) = (\sin 2x)/\pi x$ is

$$G(\alpha) = \begin{cases} 1, & -2 < \alpha < 2 \\ 0, & \text{elsewhere.} \end{cases}$$

This implies that $(f * g)(x) = \mathscr{F}^{-1}\{F(\alpha)G(\alpha)\}$ is band-limited.
The graph of $F(\alpha)G(\alpha)$, which is identical to the graph of $\mathscr{F}(f * g)$, is shown.

CHAPTER 15 REVIEW EXERCISES

3. The Laplace transform gives

$$U(x, s) = c_1 e^{-\sqrt{s+h}\,x} + c_2 e^{\sqrt{s+h}\,x} + \frac{u_0}{s + h}.$$

The condition $\lim_{x \to \infty} \partial u/\partial x = 0$ implies $\lim_{x \to \infty} dU/dx = 0$ and so we define $c_2 = 0$. Thus

$$U(x, s) = c_1 e^{-\sqrt{s+h}\,x} + \frac{u_0}{s + h}.$$

The condition $U(0, s) = 0$ then gives $c_1 = -u_0/(s + h)$ and so

$$U(x, s) = \frac{u_0}{s + h} - u_0 \frac{e^{-\sqrt{s+h}\,x}}{s + h}.$$

With the help of the first translation theorem we then obtain

$$u(x, t) = u_0 \mathscr{L}^{-1}\left\{\frac{1}{s + h}\right\} - u_0 \mathscr{L}^{-1}\left\{\frac{e^{-\sqrt{s+h}\,x}}{s + h}\right\} = u_0 e^{-ht} - u_0 e^{-ht} \text{erfc}\left(\frac{x}{2\sqrt{t}}\right)$$

$$= u_0 e^{-ht}\left[1 - \text{erfc}\left(\frac{x}{2\sqrt{t}}\right)\right] = u_0 e^{-ht}\text{erf}\left(\frac{x}{2\sqrt{t}}\right).$$

6. The Laplace transform and undetermined coefficients give

$$U(x, s) = c_1 \cosh sx + c_2 \sinh sx + \frac{s - 1}{s^2 + \pi^2} \sin \pi x.$$

The conditions $U(0, s) = 0$ and $U(1, s) = 0$ give, in turn, $c_1 = 0$ and $c_2 = 0$. Thus

$$U(x, s) = \frac{s - 1}{s^2 + \pi^2} \sin \pi x$$

and

$$u(x, t) = \sin \pi x \, \mathscr{L}^{-1} \left\{ \frac{s}{s^2 + \pi^2} \right\} - \frac{1}{\pi} \sin \pi x \, \mathscr{L}^{-1} \left\{ \frac{\pi}{s^2 + \pi^2} \right\}$$

$$= (\sin \pi x) \cos \pi t - \frac{1}{\pi} (\sin \pi x) \sin \pi t.$$

9. We solve the two problems

$$\frac{\partial^2 u_1}{\partial x^2} + \frac{\partial^2 u_1}{\partial y^2} = 0, \quad x > 0, \quad y > 0,$$

$$u_1(0, y) = 0, \quad y > 0,$$

$$u_1(x, 0) = \begin{cases} 100, & 0 < x < 1 \\ 0, & x > 1 \end{cases}$$

and

$$\frac{\partial^2 u_2}{\partial x^2} + \frac{\partial^2 u_2}{\partial y^2} = 0, \quad x > 0, \quad y > 0,$$

$$u_2(0, y) = \begin{cases} 50, & 0 < y < 1 \\ 0, & y > 1 \end{cases}$$

$$u_2(x, 0) = 0.$$

Using the Fourier sine transform with respect to x we find

$$u_1(x, y) = \frac{200}{\pi} \int_0^\infty \left(\frac{1 - \cos \alpha}{\alpha} \right) e^{-\alpha y} \sin \alpha x \, d\alpha.$$

Using the Fourier sine transform with respect to y we find

$$u_2(x, y) = \frac{100}{\pi} \int_0^\infty \left(\frac{1 - \cos \alpha}{\alpha} \right) e^{-\alpha x} \sin \alpha y \, d\alpha.$$

The solution of the problem is then

$$u(x, y) = u_1(x, y) + u_2(x, y).$$

12. Using the Laplace transform gives

$$U(x, s) = c_1 \cosh \sqrt{s} \, x + c_2 \sinh \sqrt{s} \, x.$$

The condition $u(0, t) = u_0$ transforms into $U(0, s) = u_0/s$. This gives $c_1 = u_0/s$. The condition $u(1, t) = u_0$ transforms into $U(1, s) = u_0/s$. This implies that $c_2 = u_0(1 - \cosh \sqrt{s})/s \sinh \sqrt{s}$. Hence

$$U(x, s) = \frac{u_0}{s} \cosh \sqrt{s} \, x + u_0 \left[\frac{1 - \cosh \sqrt{s}}{s \sinh \sqrt{s}} \right] \sinh \sqrt{s} \, x$$

$$= u_0 \left[\frac{\sinh \sqrt{s} \cosh \sqrt{s} \, x - \cosh \sqrt{s} \sinh \sqrt{s} \, x + \sinh \sqrt{s} \, x}{s \sinh \sqrt{s}} \right]$$

$$= u_0 \left[\frac{\sinh \sqrt{s} \, (1 - x) + \sinh \sqrt{s} \, x}{s \sinh \sqrt{s}} \right]$$

$$= u_0 \left[\frac{\sinh \sqrt{s} \, (1 - x)}{s \sinh \sqrt{s}} + \frac{\sinh \sqrt{s} \, x}{s \sinh \sqrt{s}} \right]$$

and

$$u(x,t) = u_0 \left[\mathscr{L}^{-1} \left\{ \frac{\sinh\sqrt{s}\,(1-x)}{s\sinh\sqrt{s}} \right\} + \mathscr{L}^{-1} \left\{ \frac{\sinh\sqrt{s}\,x}{s\sinh\sqrt{s}} \right\} \right]$$

$$= u_0 \sum_{n=0}^{\infty} \left[\operatorname{erf}\left(\frac{2n+2-x}{2\sqrt{t}} \right) - \operatorname{erf}\left(\frac{2n+x}{2\sqrt{t}} \right) \right]$$

$$+ u_0 \sum_{n=0}^{\infty} \left[\operatorname{erf}\left(\frac{2n+1+x}{2\sqrt{t}} \right) - \operatorname{erf}\left(\frac{2n+1-x}{2\sqrt{t}} \right) \right].$$

15. Using the Fourier transform with respect to x we obtain

$$\frac{d^2 U}{dy^2} - \alpha^2 U = 0.$$

Since $0 < y < 1$ is a finite interval we use the general solution

$$U(\alpha, y) = c_1 \cosh\alpha y + c_2 \sinh\alpha y.$$

The boundary condition at $y = 0$ transforms into $U'(\alpha, 0) = 0$, so $c_2 = 0$ and $U(\alpha, y) = c_1 \cosh\alpha y$. Now denote the Fourier transform of f as $F(\alpha)$. Then $U(\alpha, 1) = F(\alpha)$ so $F(\alpha) = c_1 \cosh\alpha$ and

$$U(\alpha, y) = F(\alpha) \frac{\cosh\alpha y}{\cosh\alpha}.$$

Taking the inverse Fourier transform we obtain

$$u(x, y) = \frac{1}{2\pi} \int_{-\infty}^{\infty} F(\alpha) \frac{\cosh\alpha y}{\cosh\alpha} e^{-i\alpha x}\, d\alpha.$$

But

$$F(\alpha) = \int_{-\infty}^{\infty} f(t) e^{i\alpha t}\, dt,$$

and so

$$u(x, y) = \frac{1}{2\pi} \int_{-\infty}^{\infty} \left(\int_{-\infty}^{\infty} f(t) e^{i\alpha t}\, dt \right) \frac{\cosh\alpha y}{\cosh\alpha} e^{-i\alpha x}\, d\alpha$$

$$= \frac{1}{2\pi} \int_{-\infty}^{\infty} \int_{-\infty}^{\infty} f(t) e^{i\alpha(t-x)} \frac{\cosh\alpha y}{\cosh\alpha}\, dt\, d\alpha$$

$$= \frac{1}{2\pi} \int_{-\infty}^{\infty} \int_{-\infty}^{\infty} f(t)(\cos\alpha(t-x) + i\sin\alpha(t-x)) \frac{\cosh\alpha y}{\cosh\alpha}\, dt\, d\alpha$$

$$= \frac{1}{2\pi} \int_{-\infty}^{\infty} \int_{-\infty}^{\infty} f(t) \cos\alpha(t-x) \frac{\cosh\alpha y}{\cosh\alpha}\, dt\, d\alpha$$

$$= \frac{1}{\pi} \int_{0}^{\infty} \int_{-\infty}^{\infty} f(t) \cos\alpha(t-x) \frac{\cosh\alpha y}{\cosh\alpha}\, dt\, d\alpha,$$

since the imaginary part of the integrand is an odd function of α followed by the fact that the remaining integrand is an even function of α.

18. The Laplace transform with respect to t of the partial differential equation gives

$$\frac{d^2 U}{dx^2} - sU = -50 \quad \text{so} \quad U(x, s) = c_1 e^{-\sqrt{s}\,x} + c_2 e^{\sqrt{s}\,x} + \frac{50}{s}.$$

The boundary condition

$$\lim_{x\to\infty} u(x, t) = 50 \quad \text{implies} \quad \lim_{x\to\infty} u(x, s) = \frac{50}{s}$$

so we take $c_2 = 0$. Thus

$$U(x, s) = c_1 e^{-\sqrt{s}\, x} + \frac{50}{s}.$$

The transform of the boundary condition at $x = 0$ is

Since

$$U(0, s) = \frac{100}{s} e^{-5s} - \frac{100}{s} e^{-10s}.$$

$$\frac{100}{s} e^{-5s} - \frac{100}{s} e^{-10s} = c_1 + \frac{50}{s}$$

we have

$$c_1 = -\frac{50}{s} + \frac{100}{s} e^{-5s} - \frac{100}{s} e^{-10s}$$

and

$$U(x, s) = \left(-\frac{50}{s} + \frac{100}{s} e^{-5s} - \frac{100}{s} e^{-10s} \right) e^{-\sqrt{s}\, x} + \frac{50}{s}.$$

Using the inverse form of the second translation theorem along with entry 3 of Table 15.1.1 yields

$$u(x, t) = 50 - 50 \operatorname{erfc}\left(\frac{x}{2\sqrt{t}} \right) + 100 \operatorname{erfc}\left(\frac{x}{2\sqrt{t-5}} \right) \mathcal{U}_{t-5} - 100 \operatorname{erfc}\left(\frac{x}{2\sqrt{t-10}} \right) \mathcal{U}_{t-10}$$

or

$$u(x, t) = \begin{cases} 50 - 50 \operatorname{erfc}\left(\dfrac{x}{2\sqrt{t}} \right), & 0 < t < 5 \\[2ex] 50 - 50 \operatorname{erfc}\left(\dfrac{x}{2\sqrt{t}} \right) + 100 \operatorname{erfc}\left(\dfrac{x}{2\sqrt{t-5}} \right), & 5 < t < 10 \\[2ex] 50 - 50 \operatorname{erfc}\left(\dfrac{x}{2\sqrt{t}} \right) + 100 \operatorname{erfc}\left(\dfrac{x}{2\sqrt{t-5}} \right) - 100 \operatorname{erfc}\left(\dfrac{x}{2\sqrt{t-10}} \right), & t > 10. \end{cases}$$

16 Numerical Solutions of Partial Differential Equations

Laplace's Equation

3. The figure shows the values of $u(x, y)$ along the boundary. We need to determine u_{11}, u_{21}, u_{12}, and u_{22}. By symmetry $u_{11} = u_{21}$ and $u_{12} = u_{22}$. The system is

$$u_{21} + u_{12} + 0 + 0 - 4u_{11} = 0$$

$$0 + u_{22} + u_{11} + 0 - 4u_{21} = 0$$

$$u_{22} + \sqrt{3}/2 + 0 + u_{11} - 4u_{12} = 0 \qquad \text{or}$$

$$0 + \sqrt{3}/2 + u_{12} + u_{21} - 4u_{22} = 0$$

$$3u_{11} + u_{12} = 0$$

$$u_{11} - 3u_{12} = -\frac{\sqrt{3}}{2}.$$

Solving we obtain $u_{11} = u_{21} = \sqrt{3}/16$ and $u_{12} = u_{22} = 3\sqrt{3}/16$.

6. The coefficients of the unknowns are the same as shown above in Problem 5. The constant terms are 7.5, 5, 20, 10, 0, 15, 17.5, 5, 27.5. We use 32.5 as the initial guess for each variable. Then $u_{11} = 21.92$, $u_{21} = 28.30$, $u_{31} = 38.17$, $u_{12} = 29.38$, $u_{22} = 33.13$, $u_{32} = 44.38$, $u_{13} = 22.46$, $u_{23} = 30.45$, and $u_{33} = 46.21$.

3. We identify $c = 1$, $a = 2$, $T = 1$, $n = 8$, and $m = 40$. Then $h = 2/8 = 0.25$, $k = 1/40 = 0.025$, and $\lambda = 2/5 = 0.4$.

TIME	X = 0.25	X = 0.50	X = 0.75	X = 1.00	X = 1.25	X = 1.50	X = 1.75
0.000	1.0000	1.0000	1.0000	1.0000	0.0000	0.0000	0.0000
0.025	0.7074	0.9520	0.9566	0.7444	0.2545	0.0371	0.0053
0.050	0.5606	0.8499	0.8685	0.6633	0.3303	0.1034	0.0223
0.075	0.4684	0.7473	0.7836	0.6191	0.3614	0.1529	0.0462
0.100	0.4015	0.6577	0.7084	0.5837	0.3753	0.1871	0.0684
0.125	0.3492	0.5821	0.6428	0.5510	0.3797	0.2101	0.0861
0.150	0.3069	0.5187	0.5857	0.5199	0.3778	0.2247	0.0990
0.175	0.2721	0.4652	0.5359	0.4901	0.3716	0.2329	0.1078
0.200	0.2430	0.4198	0.4921	0.4617	0.3622	0.2362	0.1132
0.225	0.2186	0.3809	0.4533	0.4348	0.3507	0.2358	0.1160
0.250	0.1977	0.3473	0.4189	0.4093	0.3378	0.2327	0.1166
0.275	0.1798	0.3181	0.3881	0.3853	0.3240	0.2275	0.1157
0.300	0.1643	0.2924	0.3604	0.3626	0.3097	0.2208	0.1136
0.325	0.1507	0.2697	0.3353	0.3412	0.2953	0.2131	0.1107
0.350	0.1387	0.2495	0.3125	0.3211	0.2808	0.2047	0.1071
0.375	0.1281	0.2313	0.2916	0.3021	0.2666	0.1960	0.1032
0.400	0.1187	0.2150	0.2725	0.2843	0.2528	0.1871	0.0989
0.425	0.1102	0.2002	0.2549	0.2675	0.2393	0.1781	0.0946
0.450	0.1025	0.1867	0.2387	0.2517	0.2263	0.1692	0.0902
0.475	0.0955	0.1743	0.2236	0.2368	0.2139	0.1606	0.0858
0.500	0.0891	0.1630	0.2097	0.2228	0.2020	0.1521	0.0814
0.525	0.0833	0.1525	0.1967	0.2096	0.1906	0.1439	0.0772
0.550	0.0779	0.1429	0.1846	0.1973	0.1798	0.1361	0.0731
0.575	0.0729	0.1339	0.1734	0.1856	0.1696	0.1285	0.0691
0.600	0.0683	0.1256	0.1628	0.1746	0.1598	0.1214	0.0653
0.625	0.0641	0.1179	0.1530	0.1643	0.1506	0.1145	0.0617
0.650	0.0601	0.1106	0.1438	0.1546	0.1419	0.1080	0.0582
0.675	0.0564	0.1039	0.1351	0.1455	0.1336	0.1018	0.0549
0.700	0.0530	0.0976	0.1270	0.1369	0.1259	0.0959	0.0518
0.725	0.0497	0.0917	0.1194	0.1288	0.1185	0.0904	0.0488
0.750	0.0467	0.0862	0.1123	0.1212	0.1116	0.0852	0.0460
0.775	0.0439	0.0810	0.1056	0.1140	0.1050	0.0802	0.0433
0.800	0.0413	0.0762	0.0993	0.1073	0.0989	0.0755	0.0408
0.825	0.0388	0.0716	0.0934	0.1009	0.0931	0.0711	0.0384
0.850	0.0365	0.0674	0.0879	0.0950	0.0876	0.0669	0.0362
0.875	0.0343	0.0633	0.0827	0.0894	0.0824	0.0630	0.0341
0.900	0.0323	0.0596	0.0778	0.0841	0.0776	0.0593	0.0321
0.925	0.0303	0.0560	0.0732	0.0791	0.0730	0.0558	0.0302
0.950	0.0285	0.0527	0.0688	0.0744	0.0687	0.0526	0.0284
0.975	0.0268	0.0496	0.0647	0.0700	0.0647	0.0495	0.0268
1.000	0.0253	0.0466	0.0609	0.0659	0.0608	0.0465	0.0252

(x, y)	exact	approx	abs error
(0.25, 0.1)	0.3794	0.4015	0.0221
(1, 0.5)	0.1854	0.2228	0.0374
(1.5, 0.8)	0.0623	0.0755	0.0132

6. (a) We identify $c = 15/88 \approx 0.1705$, $a = 20$, $T = 10$, $n = 10$, and $m = 10$. Then $h = 2$, $k = 1$, and $\lambda = 15/352 \approx 0.0426$.

TIME	X = 2	X = 4	X = 6	X = 8	X = 10	X = 12	X = 14	X = 16	X = 18
0	30.0000	30.0000	30.0000	30.0000	30.0000	30.0000	30.0000	30.0000	30.0000
1	28.7216	30.0000	30.0000	30.0000	30.0000	30.0000	30.0000	30.0000	28.7216
2	27.5521	29.9455	30.0000	30.0000	30.0000	30.0000	30.0000	29.9455	27.5521
3	26.4800	29.8459	29.9977	30.0000	30.0000	30.0000	29.9977	29.8459	26.4800
4	25.4951	29.7089	29.9913	29.9999	30.0000	29.9999	29.9913	29.7089	25.4951
5	24.5882	29.5414	29.9796	29.9995	30.0000	29.9995	29.9796	29.5414	24.5882
6	23.7515	29.3490	29.9618	29.9987	30.0000	29.9987	29.9618	29.3490	23.7515
7	22.9779	29.1365	29.9373	29.9972	29.9998	29.9972	29.9373	29.1365	22.9779
8	22.2611	28.9082	29.9057	29.9948	29.9996	29.9948	29.9057	28.9082	22.2611
9	21.5958	28.6675	29.8670	29.9912	29.9992	29.9912	29.8670	28.6675	21.5958
10	20.9768	28.4172	29.8212	29.9862	29.9985	29.9862	29.8212	28.4172	20.9768

(b) We identify $c = 15/88 \approx 0.1705$, $a = 50$, $T = 10$, $n = 10$, and $m = 10$. Then $h = 5$, $k = 1$, and $\lambda = 3/440 \approx 0.0068$.

TIME	X = 5	X = 10	X = 15	X = 20	X = 25	X = 30	X = 35	X = 40	X = 45
0	30.0000	30.0000	30.0000	30.0000	30.0000	30.0000	30.0000	30.0000	30.0000
1	29.7955	30.0000	30.0000	30.0000	30.0000	30.0000	30.0000	30.0000	29.7955
2	29.5937	29.9986	30.0000	30.0000	30.0000	30.0000	30.0000	29.9986	29.5937
3	29.3947	29.9959	30.0000	30.0000	30.0000	30.0000	30.0000	29.9959	29.3947
4	29.1984	29.9918	30.0000	30.0000	30.0000	30.0000	30.0000	29.9918	29.1984
5	29.0047	29.9864	29.9999	30.0000	30.0000	30.0000	29.9999	29.9864	29.0047
6	28.8136	29.9798	29.9998	30.0000	30.0000	30.0000	29.9998	29.9798	28.8136
7	28.6251	29.9720	29.9997	30.0000	30.0000	30.0000	29.9997	29.9720	28.6251
8	28.4391	29.9630	29.9995	30.0000	30.0000	30.0000	29.9995	29.9630	28.4391
9	28.2556	29.9529	29.9992	30.0000	30.0000	30.0000	29.9992	29.9529	28.2556
10	28.0745	29.9416	29.9989	30.0000	30.0000	30.0000	29.9989	29.9416	28.0745

(c) We identify $c = 50/27 \approx 1.8519$, $a = 20$, $T = 10$, $n = 10$, and $m = 10$. Then $h = 2$, $k = 1$, and $\lambda = 25/54 \approx 0.4630$.

TIME	X = 2	X = 4	X = 6	X = 8	X = 10	X = 12	X = 14	X = 16	X = 18
0	18.0000	32.0000	42.0000	48.0000	50.0000	48.0000	42.0000	32.0000	18.0000
1	16.1481	30.1481	40.1481	46.1481	48.1481	46.1481	40.1481	30.1481	16.1481
2	15.1536	28.2963	38.2963	44.2963	46.2963	44.2963	38.2963	28.2963	15.1536
3	14.2226	26.8414	36.4444	42.4444	44.4444	42.4444	36.4444	26.8414	14.2226
4	13.4801	25.4452	34.7764	40.5926	42.5926	40.5926	34.7764	25.4452	13.4801
5	12.7787	24.2258	33.1491	38.8258	40.7407	38.8258	33.1491	24.2258	12.7787
6	12.1622	23.0574	31.6460	37.0842	38.9677	37.0842	31.6460	23.0574	12.1622
7	11.5756	21.9895	30.1875	35.4385	37.2238	35.4385	30.1875	21.9895	11.5756
8	11.0378	20.9636	28.8232	33.8340	35.5707	33.8340	28.8232	20.9636	11.0378
9	10.5230	20.0070	27.5043	32.3182	33.9626	32.3182	27.5043	20.0070	10.5230
10	10.0420	19.0872	26.2620	30.8509	32.4400	30.8509	26.2620	19.0872	10.0420

(d) We identify $c = 260/159 \approx 1.6352$, $a = 100$, $T = 10$, $n = 10$, and $m = 10$. Then $h = 10$, $k = 1$, and $\lambda = 13/795 \approx 00164$.

TIME	X = 10	X = 20	X = 30	X = 40	X = 50	X = 60	X = 70	X = 80	X = 90
0	8.0000	16.0000	24.0000	32.0000	40.0000	32.0000	24.0000	16.0000	8.0000
1	8.0000	16.0000	23.6075	31.3459	39.2151	31.6075	23.7384	15.8692	8.0000
2	8.0000	15.9936	23.2279	30.7068	38.4452	31.2151	23.4789	15.7384	7.9979
3	7.9999	15.9812	22.8606	30.0824	37.6900	30.8229	23.2214	15.6076	7.9937
4	7.9996	15.9631	22.5050	29.4724	36.9492	30.4312	22.9660	15.4769	7.9874
5	7.9990	15.9399	22.1606	28.8765	36.2228	30.0401	22.7125	15.3463	7.9793
6	7.9981	15.9118	21.8270	28.2945	35.5103	29.6500	22.4610	15.2158	7.9693
7	7.9967	15.8791	21.5037	27.7261	34.8117	29.2610	22.2112	15.0854	7.9575
8	7.9948	15.8422	21.1902	27.1709	34.1266	28.8733	21.9633	14.9553	7.9439
9	7.9924	15.8013	20.8861	26.6288	33.4548	28.4870	21.7172	14.8253	7.9287
10	7.9894	15.7568	20.5911	26.0995	32.7961	28.1024	21.4727	14.6956	7.9118

9. (a) We identify $c = 15/88 \approx 0.1705$, $a = 20$, $T = 10$, $n = 10$, and $m = 10$. Then $h = 2$, $k = 1$, and $\lambda = 15/352 \approx 0.0426$.

TIME	X = 2.00	X = 4.00	X= 6.00	X = 8.00	X = 10.00	X = 12.00	X = 14.00	X = 16.00	X = 18.00
0.00	30.0000	30.0000	30.0000	30.0000	30.0000	30.0000	30.0000	30.0000	30.0000
1.00	28.7733	29.9749	29.9995	30.0000	30.0000	30.0000	29.9998	29.9916	29.5911
2.00	27.6450	29.9037	29.9970	29.9999	30.0000	30.0000	29.9990	29.9679	29.2150
3.00	26.6051	29.7938	29.9911	29.9997	30.0000	29.9999	29.9970	29.9313	28.8684
4.00	25.6452	29.6517	29.9805	29.9991	30.0000	29.9997	29.9935	29.8839	28.5484
5.00	24.7573	29.4829	29.9643	29.9981	29.9999	29.9994	29.9881	29.8276	28.2524
6.00	23.9347	29.2922	29.9421	29.9963	29.9997	29.9988	29.9807	29.7641	27.9782
7.00	23.1711	29.0836	29.9134	29.9936	29.9995	29.9979	29.9711	29.6945	27.7237
8.00	22.4612	28.8606	29.8782	29.9899	29.9991	29.9966	29.9594	29.6202	27.4870
9.00	27.7999	28.6263	29.8362	29.9848	29.9985	29.9949	29.9454	29.5421	27.2666
10.00	21.1829	28.3831	29.7878	29.9783	29.9976	29.9927	29.9293	29.4610	27.0610

(b) We identify $c = 15/88 \approx 0.1705$, $a = 50$, $T = 10$, $n = 10$, and $m = 10$. Then $h = 5$, $k = 1$, and $\lambda = 3/440 \approx 0.0068$.

TIME	X = 5.00	X = 10.00	X= 15.00	X = 20.00	X = 25.00	X = 30.00	X = 35.00	X = 40.00	X = 45.00
0.00	30.0000	30.0000	30.0000	30.0000	30.0000	30.0000	30.0000	30.0000	30.0000
1.00	29.7968	29.9993	30.0000	30.0000	30.0000	30.0000	30.0000	29.9998	29.9323
2.00	29.5964	29.9973	30.0000	30.0000	30.0000	30.0000	30.0000	29.9991	29.8655
3.00	29.3987	29.9939	30.0000	30.0000	30.0000	30.0000	30.0000	29.9980	29.7996
4.00	29.2036	29.9893	29.9999	30.0000	30.0000	30.0000	30.0000	29.9964	29.7345
5.00	29.0112	29.9834	29.9998	30.0000	30.0000	30.0000	29.9999	29.9945	29.6704
6.00	28.8212	29.9762	29.9997	30.0000	30.0000	30.0000	29.9999	29.9921	29.6071
7.00	28.6339	29.9679	29.9995	30.0000	30.0000	30.0000	29.9998	29.9893	29.5446
8.00	28.4490	29.9585	29.9992	30.0000	30.0000	30.0000	29.9997	29.9862	29.4830
9.00	28.2665	29.9479	29.9989	30.0000	30.0000	30.0000	29.9996	29.9827	29.4222
10.00	28.0864	29.9363	29.9986	30.0000	30.0000	30.0000	29.9995	29.9788	29.3621

(c) We identify $c = 50/27 \approx 1.8519$, $a = 20$, $T = 10$, $n = 10$, and $m = 10$. Then $h = 2$, $k = 1$, and $\lambda = 25/54 \approx 0.4630$.

TIME	X = 2.00	X = 4.00	X = 6.00	X = 8.00	X = 10.00	X = 12.00	X = 14.00	X = 16.00	X = 18.00
0.00	18.0000	32.0000	42.0000	48.0000	50.0000	48.0000	42.0000	32.0000	18.0000
1.00	16.4489	30.1970	40.1562	46.1502	48.1531	46.1773	40.3274	31.2520	22.9449
2.00	15.3312	28.5350	38.3477	44.3130	46.3327	44.4671	39.0872	31.5755	24.6930
3.00	14.4219	27.0429	36.6090	42.5113	44.5759	42.9362	38.1976	31.7478	25.4131
4.00	13.6381	25.6913	34.9606	40.7728	42.9127	41.5716	37.4340	31.7086	25.6986
5.00	12.9409	24.4545	33.4091	39.1182	41.3519	40.3240	36.7033	31.5136	25.7663
6.00	12.3088	23.3146	31.9546	37.5566	39.8880	39.1565	35.9745	31.2134	25.7128
7.00	11.7294	22.2589	30.5939	36.0884	38.5109	38.0470	35.2407	30.8434	25.5871
8.00	11.1946	21.2785	29.3217	34.7092	37.2109	36.9834	34.5032	30.4279	25.4167
9.00	10.6987	20.3660	28.1318	33.4130	35.9801	35.9591	33.7660	29.9836	25.2181
10.00	10.2377	19.5150	27.0178	32.1929	34.8117	34.9710	33.0338	29.5224	25.0019

(d) We identify $c = 260/159 \approx 1.6352$, $a = 100$, $T = 10$, $n = 10$, and $m = 10$. Then $h = 10$, $k = 1$, and $\lambda = 13/795 \approx 00164$.

TIME	X = 10.00	X = 20.00	X = 30.00	X = 40.00	X = 50.00	X = 60.00	X = 70.00	X = 80.00	X = 90.00
0.00	8.0000	16.0000	24.0000	32.0000	40.0000	32.0000	24.0000	16.0000	8.0000
1.00	8.0000	16.0000	24.0000	31.9979	39.7425	31.9979	24.0000	16.0026	8.3218
2.00	8.0000	16.0000	23.9999	31.9918	39.4932	31.9918	24.0000	16.0102	8.6333
3.00	8.0000	16.0000	23.9997	31.9820	39.2517	31.9820	24.0001	16.0225	8.9350
4.00	8.0000	16.0000	23.9993	31.9687	39.0176	31.9687	24.0002	16.0392	9.2272
5.00	8.0000	16.0000	23.9987	31.9520	38.7905	31.9521	24.0003	16.0599	9.5103
6.00	8.0000	15.9999	23.9978	31.9323	38.5701	31.9324	24.0005	16.0845	9.7846
7.00	8.0000	15.9999	23.9966	31.9097	38.3561	31.9098	24.0008	16.1126	10.0506
8.00	8.0000	15.9998	23.9951	31.8844	38.1483	31.8846	24.0012	16.1441	10.3084
9.00	8.0000	15.9997	23.9931	31.8566	37.9463	31.8569	24.0017	16.1786	10.5585
10.00	8.0000	15.9996	23.9908	31.8265	37.7499	31.8270	24.0023	16.2160	10.8012

12. We identify $c = 1$, $a = 1$, $T = 1$, $n = 5$, and $m = 20$. Then $h = 0.2$, $k = 0.04$, and $\lambda = 1$. The values below were obtained using *Excel*, which carries more than 12 significant digits. In order to see evidence of instability use $0 \leq t \leq 2$.

TIME	X = 0.2	X = 0.4	X = 0.6	X = 0.8
0.00	0.5878	0.9511	0.9511	0.5878
0.04	0.3633	0.5878	0.5878	0.3633
0.08	0.2245	0.3633	0.3633	0.2245
0.12	0.1388	0.2245	0.2245	0.1388
0.16	0.0858	0.1388	0.1388	0.0858
0.20	0.0530	0.0858	0.0858	0.0530
0.24	0.0328	0.0530	0.0530	0.0328
0.28	0.0202	0.0328	0.0328	0.0202
0.32	0.0125	0.0202	0.0202	0.0125
0.36	0.0077	0.0125	0.0125	0.0077
0.40	0.0048	0.0077	0.0077	0.0048
0.44	0.0030	0.0048	0.0048	0.0030
0.48	0.0018	0.0030	0.0030	0.0018
0.52	0.0011	0.0018	0.0018	0.0011
0.56	0.0007	0.0011	0.0011	0.0007
0.60	0.0004	0.0007	0.0007	0.0004
0.64	0.0003	0.0004	0.0004	0.0003
0.68	0.0002	0.0003	0.0003	0.0002
0.72	0.0001	0.0002	0.0002	0.0001
0.76	0.0001	0.0001	0.0001	0.0001
0.80	0.0000	0.0001	0.0001	0.0000
0.84	0.0000	0.0000	0.0000	0.0000
0.88	0.0000	0.0000	0.0000	0.0000
0.92	0.0000	0.0000	0.0000	0.0000
0.96	0.0000	0.0000	0.0000	0.0000
1.00	0.0000	0.0000	0.0000	0.0000

TIME	X = 0.2	X = 0.4	X = 0.6	X = 0.8
1.04	0.0000	0.0000	0.0000	0.0000
1.08	0.0000	0.0000	0.0000	0.0000
1.12	0.0000	0.0000	0.0000	0.0000
1.16	0.0000	0.0000	0.0000	0.0000
1.20	−0.0001	0.0001	−0.0001	0.0001
1.24	0.0001	−0.0002	0.0002	−0.0001
1.28	−0.0004	0.0006	−0.0006	0.0004
1.32	0.0010	−0.0015	0.0015	−0.0010
1.36	−0.0025	0.0040	−0.0040	0.0025
1.40	0.0065	−0.0106	0.0106	−0.0065
1.44	−0.0171	0.0277	−0.0277	0.0171
1.48	0.0448	−0.0724	0.0724	−0.0448
1.52	−0.1172	0.1897	−0.1897	0.1172
1.56	0.3069	−0.4965	0.4965	−0.3069
1.60	−0.8034	1.2999	−1.2999	0.8034
1.64	2.1033	−3.4032	3.4032	−2.1033
1.68	−5.5064	8.9096	−8.9096	5.5064
1.72	14.416	−23.326	23.326	−14.416
1.76	−37.742	61.067	−61.067	37.742
1.80	98.809	−159.88	159.88	−98.809
1.84	−258.68	418.56	−418.56	258.685
1.88	677.24	−1095.8	1095.8	−677.245
1.92	−1773.1	2868.9	−2868.9	1773.1
1.96	4641.9	−7510.8	7510.8	−4641.9
2.00	−12153	19663	−19663	12153

EXERCISES 16.3

The Wave Equation

3. (a) Identifying $h = 1/5$ and $k = 0.5/10 = 0.05$ we see that $\lambda = 0.25$.

TIME	X = 0.2	X = 0.4	X = 0.6	X = 0.8
0.00	0.5878	0.9511	0.9511	0.5878
0.05	0.5808	0.9397	0.9397	0.5808
0.10	0.5599	0.9059	0.9059	0.5599
0.15	0.5256	0.8505	0.8505	0.5256
0.20	0.4788	0.7748	0.7748	0.4788
0.25	0.4206	0.6806	0.6806	0.4206
0.30	0.3524	0.5701	0.5701	0.3524
0.35	0.2757	0.4460	0.4460	0.2757
0.40	0.1924	0.3113	0.3113	0.1924
0.45	0.1046	0.1692	0.1692	0.1046
0.50	0.0142	0.0230	0.0230	0.0142

(b) Identifying $h = 1/5$ and $k = 0.5/20 = 0.025$ we see that $\lambda = 0.125$.

TIME	X = 0.2	X = 0.4	X = 0.6	X = 0.8
0.00	0.5878	0.9511	0.9511	0.5878
0.03	0.5860	0.9482	0.9482	0.5860
0.05	0.5808	0.9397	0.9397	0.5808
0.08	0.5721	0.9256	0.9256	0.5721
0.10	0.5599	0.9060	0.9060	0.5599
0.13	0.5445	0.8809	0.8809	0.5445
0.15	0.5257	0.8507	0.8507	0.5257
0.18	0.5039	0.8153	0.8153	0.5039
0.20	0.4790	0.7750	0.7750	0.4790
0.23	0.4513	0.7302	0.7302	0.4513
0.25	0.4209	0.6810	0.6810	0.4209
0.28	0.3879	0.6277	0.6277	0.3879
0.30	0.3527	0.5706	0.5706	0.3527
0.33	0.3153	0.5102	0.5102	0.3153
0.35	0.2761	0.4467	0.4467	0.2761
0.38	0.2352	0.3806	0.3806	0.2352
0.40	0.1929	0.3122	0.3122	0.1929
0.43	0.1495	0.2419	0.2419	0.1495
0.45	0.1052	0.1701	0.1701	0.1052
0.48	0.0602	0.0974	0.0974	0.0602
0.50	0.0149	0.0241	0.0241	0.0149

6. We identify $c = 24944.4$, $k = 0.00010022$ seconds $= 0.10022$ milliseconds, and $\lambda = 0.25$. Time in the table is expressed in milliseconds.

TIME	X = 10	X = 20	X = 30	X = 40	X = 50
0.00000	0.2000	0.2667	0.2000	0.1333	0.0667
0.10022	0.1958	0.2625	0.2000	0.1333	0.0667
0.20045	0.1836	0.2503	0.1997	0.1333	0.0667
0.30067	0.1640	0.2307	0.1985	0.1333	0.0667
0.40089	0.1384	0.2050	0.1952	0.1332	0.0667
0.50111	0.1083	0.1744	0.1886	0.1328	0.0667
0.60134	0.0755	0.1407	0.1777	0.1318	0.0666
0.70156	0.0421	0.1052	0.1615	0.1295	0.0665
0.80178	0.0100	0.0692	0.1399	0.1253	0.0661
0.90201	−0.0190	0.0340	0.1129	0.1184	0.0654
1.00223	−0.0435	0.0004	0.0813	0.1077	0.0638
1.10245	−0.0626	−0.0309	0.0464	0.0927	0.0610
1.20268	−0.0758	−0.0593	0.0095	0.0728	0.0564
1.30290	−0.0832	−0.0845	−0.0278	0.0479	0.0493
1.40312	−0.0855	−0.1060	−0.0639	0.0184	0.0390
1.50334	−0.0837	−0.1237	−0.0974	−0.0150	0.0250
1.60357	−0.0792	−0.1371	−0.1275	−0.0511	0.0069
1.70379	−0.0734	−0.1464	−0.1533	−0.0882	−0.0152
1.80401	−0.0675	−0.1515	−0.1747	−0.1249	−0.0410
1.90424	−0.0627	−0.1528	−0.1915	−0.1595	−0.0694
2.00446	−0.0596	−0.1509	−0.2039	−0.1904	−0.0991
2.10468	−0.0585	−0.1467	−0.2122	−0.2165	−0.1283
2.20491	−0.0592	−0.1410	−0.2166	−0.2368	−0.1551
2.30513	−0.0614	−0.1349	−0.2175	−0.2507	−0.1772
2.40535	−0.0643	−0.1294	−0.2154	−0.2579	−0.1929
2.50557	−0.0672	−0.1251	−0.2105	−0.2585	−0.2005
2.60580	−0.0696	−0.1227	−0.2033	−0.2524	−0.1993
2.70602	−0.0709	−0.1219	−0.1942	−0.2399	−0.1889
2.80624	−0.0710	−0.1225	−0.1833	−0.2214	−0.1699
2.90647	−0.0699	−0.1236	−0.1711	−0.1972	−0.1435
3.00669	−0.0678	−0.1244	−0.1575	−0.1681	−0.1115
3.10691	−0.0649	−0.1237	−0.1425	−0.1348	−0.0761
3.20713	−0.0617	−0.1205	−0.1258	−0.0983	−0.0395
3.30736	−0.0583	−0.1139	−0.1071	−0.0598	−0.0042
3.40758	−0.0547	−0.1035	−0.0859	−0.0209	0.0279
3.50780	−0.0508	−0.0889	−0.0617	0.0171	0.0552
3.60803	−0.0460	−0.0702	−0.0343	0.0525	0.0767
3.70825	−0.0399	−0.0478	−0.0037	0.0840	0.0919
3.80847	−0.0318	−0.0221	0.0297	0.1106	0.1008
3.90870	−0.0211	0.0062	0.0648	0.1314	0.1041
4.00892	−0.0074	0.0365	0.1005	0.1464	0.1025
4.10914	0.0095	0.0680	0.1350	0.1558	0.0973
4.20936	0.0295	0.1000	0.1666	0.1602	0.0897
4.30959	0.0521	0.1318	0.1937	0.1606	0.0808
4.40981	0.0764	0.1625	0.2148	0.1581	0.0719
4.51003	0.1013	0.1911	0.2291	0.1538	0.0639
4.61026	0.1254	0.2164	0.2364	0.1485	0.0575
4.71048	0.1475	0.2373	0.2369	0.1431	0.0532
4.81070	0.1659	0.2526	0.2315	0.1379	0.0512
4.91093	0.1794	0.2611	0.2217	0.1331	0.0514
5.01115	0.1867	0.2620	0.2087	0.1288	0.0535

CHAPTER 16 REVIEW EXERCISES

3. (a)

TIME	X = 0.0	X = 0.2	X = 0.4	X = 0.6	X = 0.8	X = 1.0
0.00	0.0000	0.2000	0.4000	0.6000	0.8000	0.0000
0.01	0.0000	0.2000	0.4000	0.6000	0.5500	0.0000
0.02	0.0000	0.2000	0.4000	0.5375	0.4250	0.0000
0.03	0.0000	0.2000	0.3844	0.4750	0.3469	0.0000
0.04	0.0000	0.1961	0.3609	0.4203	0.2922	0.0000
0.05	0.0000	0.1883	0.3346	0.3734	0.2512	0.0000

(b)

TIME	X = 0.0	X = 0.2	X = 0.4	X = 0.6	X = 0.8	X = 1.0
0.00	0.0000	0.2000	0.4000	0.6000	0.8000	0.0000
0.01	0.0000	0.2000	0.4000	0.6000	0.8000	0.0000
0.02	0.0000	0.2000	0.4000	0.6000	0.5500	0.0000
0.03	0.0000	0.2000	0.4000	0.5375	0.4250	0.0000
0.04	0.0000	0.2000	0.3844	0.4750	0.3469	0.0000
0.05	0.0000	0.1961	0.3609	0.4203	0.2922	0.0000

(c) The table in part (**b**) is the same as the table in part (**a**) shifted downward one row.

17   Functions of a Complex Variable

EXERCISES 17.1
Complex Numbers

3. $i^8 = (i^2)^4 = (-1)^4 = 1$

6. $-3 - 9i$

9. $11 - 10i$

12. $-2 - 2i$

15. $\dfrac{2-4i}{3+5i} \cdot \dfrac{3-5i}{3-5i} = \dfrac{-14-22i}{34} = -\dfrac{7}{17} - \dfrac{11}{17}\, i$

18. $\dfrac{3-i}{11-2i} \cdot \dfrac{11+2i}{11+2i} = \dfrac{35-5i}{125} = \dfrac{7}{25} - \dfrac{1}{25}\, i$

21. $(1+i)(10+10i) = 10(1+i)^2 = 20i$

24. $(2+3i)(-i)^2 = -2 - 3i$

27. $\dfrac{x}{x^2 + y^2}$

30. 0

33. $2x + 2yi = -9 + 2i$ implies $2x = -9$ and $2y = 2$. Hence $z = -\frac{9}{2} + i$.

36. $x^2 - y^2 - 4x + (-2xy - 4y)i = 0 + 0i$ implies $x^2 - y^2 - 4x = 0$ and $y(-2x - 4) = 0$. If $y = 0$ then $x(x - 4) = 0$ and so $z = 0$ and $z = 4$. If $-2x - 4 = 0$ or $x = -2$ then $12 - y^2 = 0$ or $y = \pm 2\sqrt{3}$. This gives $z = -2 + 2\sqrt{3}\, i$ and $z = -2 - 2\sqrt{3}\, i$.

39. $|z_1 - z_2| = |(x_1 - x_2) + i(y_1 - y_2)| = \sqrt{(x_1 - x_2)^2 + (y_1 - y_2)^2}$ which is the distance formula in the plane.

EXERCISES 17.2

Powers and Roots

3. $3\left(\cos\dfrac{3\pi}{2} + i\sin\dfrac{3\pi}{2}\right)$

6. $5\sqrt{2}\left(\cos\dfrac{7\pi}{4} + i\sin\dfrac{7\pi}{4}\right)$

9. $\dfrac{3\sqrt{2}}{2}\left(\cos\dfrac{5\pi}{4} + i\sin\dfrac{5\pi}{4}\right)$

12. $z = -8 + 8i$

15. $z_1 z_2 = 8\left[\cos\left(\dfrac{\pi}{8} + \dfrac{3\pi}{8}\right) + i\sin\left(\dfrac{\pi}{8} + \dfrac{3\pi}{8}\right)\right] = 8i;$ $\dfrac{z_1}{z_2} = \dfrac{1}{2}\left[\cos\left(\dfrac{\pi}{8} - \dfrac{3\pi}{8}\right) + i\sin\left(\dfrac{\pi}{8} - \dfrac{3\pi}{8}\right)\right] = \dfrac{\sqrt{2}}{4} - \dfrac{\sqrt{2}}{4}i$

18. $\left[4\sqrt{2}\left(\cos\dfrac{\pi}{4} + i\sin\dfrac{\pi}{4}\right)\right]\left[\sqrt{2}\left(\cos\dfrac{3\pi}{4} + i\sin\dfrac{3\pi}{4}\right)\right] = 8\left[\cos\left(\dfrac{\pi}{4} + \dfrac{3\pi}{4}\right) + i\sin\left(\dfrac{\pi}{4} + \dfrac{3\pi}{4}\right)\right] = -8$

21. $2^9\left[\cos\dfrac{9\pi}{3} + i\sin\dfrac{9\pi}{3}\right] = -512$

24. $(2\sqrt{2})^4\left[\cos\dfrac{8\pi}{3} + i\sin\dfrac{8\pi}{3}\right] = -32 + 32\sqrt{3}\,i$

27. $8^{1/3} = 2\left[\cos\dfrac{2k\pi}{3} + i\sin\dfrac{2k\pi}{3}\right],$ $k = 0, 1, 2$

$w_0 = 2[\cos 0 + i\sin 0] = 2;$ $w_1 = 2\left[\cos\dfrac{2\pi}{3} + i\sin\dfrac{2\pi}{3}\right] = -1 + \sqrt{3}\,i$

$w_2 = 2\left[\cos\dfrac{4\pi}{3} + i\sin\dfrac{4\pi}{3}\right] = -1 - \sqrt{3}\,i$

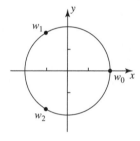

30. $(-1+i)^{1/3} = 2^{1/6}\left[\cos\left(\dfrac{\pi}{4} + \dfrac{2k\pi}{3}\right) + i\sin\left(\dfrac{\pi}{4} + \dfrac{2k\pi}{3}\right)\right],$ $k = 0, 1, 2$

$w_0 = 2^{1/6}\left[\cos\dfrac{\pi}{4} + i\sin\dfrac{\pi}{4}\right] = \dfrac{1}{\sqrt[3]{2}} + \dfrac{1}{\sqrt[3]{2}}i = 0.7937 + 0.7937i$

$w_1 = 2^{1/6}\left[\cos\dfrac{11\pi}{12} + i\sin\dfrac{11\pi}{12}\right] = -1.0842 + 0.2905i$

$w_2 = 2^{1/6}\left[\cos\dfrac{19\pi}{12} + i\sin\dfrac{19\pi}{12}\right] = 0.2905 - 1.0842i$

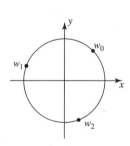

33. The solutions are the four fourth roots of -1;

$$w_k = \cos\dfrac{\pi + 2k\pi}{4} + i\sin\dfrac{\pi + 2k\pi}{4},\quad k = 0, 1, 2, 3.$$

We have

$w_1 = \cos\dfrac{\pi}{4} + i\sin\dfrac{\pi}{4} = \dfrac{\sqrt{2}}{2} + \dfrac{\sqrt{2}}{2}i$ $w_3 = \cos\dfrac{5\pi}{4} + i\sin\dfrac{5\pi}{4} = -\dfrac{\sqrt{2}}{2} - \dfrac{\sqrt{2}}{2}i$

$w_2 = \cos\dfrac{3\pi}{4} + i\sin\dfrac{3\pi}{4} = -\dfrac{\sqrt{2}}{2} + \dfrac{\sqrt{2}}{2}i$ $\qquad w_4 = \cos\dfrac{7\pi}{4} + i\sin\dfrac{7\pi}{4} = \dfrac{\sqrt{2}}{2} - \dfrac{\sqrt{2}}{2}i.$

36. $\dfrac{\left[8\left(\cos\dfrac{3\pi}{8} + i\sin\dfrac{3\pi}{8}\right)\right]^3}{\left[2\left(\cos\dfrac{\pi}{16} + i\sin\dfrac{\pi}{16}\right)\right]^{10}} = \dfrac{2^9}{2^{10}}\left[\cos\left(\dfrac{9\pi}{8} - \dfrac{10\pi}{16}\right) + i\left(\dfrac{9\pi}{8} - \dfrac{10\pi}{16}\right)\right] = \dfrac{1}{2}\left(\cos\dfrac{\pi}{2} + i\sin\dfrac{\pi}{2}\right) = \dfrac{1}{2}i$

39. (a) $\text{Arg}(z_1) = \pi$, $\text{Arg}(z_2) = \dfrac{\pi}{2}$, $\text{Arg}(z_1 z_2) = -\dfrac{\pi}{2}$, $\text{Arg}(z_1) + \text{Arg}(z_2) = \dfrac{3\pi}{2} \neq \text{Arg}(z_1 z_2)$

(b) $\text{Arg}(z_1/z_2) = -\dfrac{\pi}{2}$, $\text{Arg}(z_1) - \text{Arg}(z_2) = \pi - \dfrac{\pi}{2} = \dfrac{\pi}{2} \neq \text{Arg}(z_1/z_2)$

EXERCISES 17.3
Sets in the Complex Plane

3.

6.

9.

12.

15.

18.

21.

24. $|\text{Re}(z)| = |x|$ is the same as $\sqrt{x^2}$ and $|z| = \sqrt{x^2 + y^2}$. Since $y^2 \geq 0$ the inequality $\sqrt{x^2} \leq \sqrt{x^2 + y^2}$ is true for all complex numbers.

EXERCISES 17.4

Functions of a Complex Variable

3. $x = 0$ gives $u = -y^2$, $v = 0$. Since $-y^2 \leq 0$ for all real values of y, the image is the origin and the negative u-axis.

6. $y = -x$ gives $u = 0$, $v = -2x^2$. Since $-x^2 \leq 0$ for all real values of x, the image is the origin and the negative v-axis.

9. $f(z) = (x^2 - y^2 - 3x) + i(2xy - 3y + 4)$

12. $f(z) = (x^4 - 6x^2y^2 + y^4) + i(4x^3y - 4xy^3)$

15. (a) $f(0 + 2i) = -4 + i$

    (b) $f(2 - i) = 3 - 9i$

    (c) $f(5 + 3i) = 1 + 86i$

18. (a) $f\left(0 + \dfrac{\pi}{4}i\right) = \dfrac{\sqrt{2}}{2} + \dfrac{\sqrt{2}}{2}i$

    (b) $f(-1 - \pi i) = -e^{-1}$

    (c) $f\left(3 + \dfrac{\pi}{3}i\right) = \dfrac{1}{2}e^3 + \dfrac{\sqrt{3}}{2}e^3 i$

21. $\displaystyle\lim_{z \to i} \frac{z^4 - 1}{z - i} = \lim_{z \to i} \frac{(z^2 - 1)(z - i)(z + i)}{z - i} = -4i$

24. Along the line $x = 1$, $\displaystyle\lim_{z \to 1} \frac{x + y - 1}{z - 1} = \lim_{y \to 0} \frac{y}{iy} = \frac{1}{i} = -i$, whereas along the x-axis, $\displaystyle\lim_{z \to 1} \frac{x + y - 1}{z - 1} = \lim_{x \to 1} \frac{x - 1}{x - 1} = 1$.

27. $f'(z) = 12z^2 - (6 + 2i)z - 5$

30. $f'(z) = (z^5 + 3iz^3)(4z^2 + 3iz^2 + 4z - 6i) + (z^4 + iz^3 + 2z^2 - 6iz)(5z^4 + 9iz^2)$

33. $f'(z) = \dfrac{(2z + i)3 - (3z - 4 + 8i)2}{(2z + 1)^2} = \dfrac{8 - 13i}{(2z + i)^2}$

36. $0, 2 - 5i$

39. We have
$$\lim_{\Delta z \to 0} \frac{\overline{z + \Delta z} - \overline{z}}{\Delta z} = \lim_{\Delta z \to 0} \frac{\overline{\Delta z}}{\Delta z}.$$

If we let $\Delta z \to 0$ along a horizontal line then $\Delta z = \Delta x$, $\overline{\Delta z} = \Delta x$, and
$$\lim_{\Delta z \to 0} \frac{\overline{\Delta z}}{\Delta z} = \lim_{\Delta x \to 0} \frac{\Delta x}{\Delta x} = 1.$$

If we let $\Delta z \to 0$ along a vertical line then $\Delta z = i\Delta y$, $\overline{\Delta z} = -i\Delta y$, and
$$\lim_{\Delta z \to 0} \frac{\overline{\Delta z}}{\Delta z} = \lim_{\Delta y \to 0} \frac{-i\Delta y}{i\Delta y} = -1.$$

Since these two limits are not equal, $f(z) = \bar{z}$ cannot be differentiable at any z.

42. The system $\dfrac{dx}{dt} = -y$, $\dfrac{dy}{dt} = x$ can be solved as in Section 3.11. We obtain $x(t) = c_1 \cos t + c_2 \sin t$, $y(t) = c_1 \sin t - c_2 \cos t$.

45. If $y = \frac{1}{2}x^2$ the equations $u = x^2 - y^2$, $v = 2xy$ give $u = x^2 - \frac{1}{4}x^4$, $v = x^3$. With the aid of a computer, the graph of these parametric equations is shown.

EXERCISES 17.5
Cauchy-Riemann Equations

3. $u = x$, $v = 0$; $\dfrac{\partial u}{\partial x} = 1$, $\dfrac{\partial v}{\partial y} = 0$. Since $1 \neq 0$, f is not analytic at any point.

6. $u = x^2 - y^2$, $v = -2xy$; $\dfrac{\partial u}{\partial x} = 2x$, $\dfrac{\partial v}{\partial y} = -2x$; $\dfrac{\partial u}{\partial y} = -2y$, $-\dfrac{\partial v}{\partial x} = 2y$

The Cauchy-Riemann equations hold only at $(0,0)$. Since there is no neighborhood about $z = 0$ within which f is differentiable we conclude f is nowhere analytic.

9. $u = e^x \cos y$, $v = e^x \sin y$; $\dfrac{\partial u}{\partial x} = e^x \cos y = \dfrac{\partial v}{\partial y}$; $\dfrac{\partial u}{\partial y} = -e^x \sin y = -\dfrac{\partial v}{\partial x}$. f is analytic for all z.

12. $u = 4x^2 + 5x - 4y^2 + 9$, $v = 8xy + 5y - 1$; $\dfrac{\partial u}{\partial x} = 8x + 5 = \dfrac{\partial v}{\partial y}$, $\dfrac{\partial u}{\partial y} = -8y = -\dfrac{\partial v}{\partial x}$. f is analytic for all z.

15. $\dfrac{\partial u}{\partial x} = 3 = b = \dfrac{\partial v}{\partial y}$; $\dfrac{\partial u}{\partial y} = -1 = -a = -\dfrac{\partial v}{\partial x}$. f is analytic for all z when $b = 3$, $a = 1$.

18. $u = 3x^2 y^2$, $v = -6x^2 y^2$; $\dfrac{\partial u}{\partial x} = 6xy^2$, $\dfrac{\partial v}{\partial y} = -12x^2 y$; $\dfrac{\partial u}{\partial y} = 6x^2 y$, $-\dfrac{\partial v}{\partial x} = 12xy^2$

u and v are continuous and have continuous first partial derivatives. The Cauchy-Riemann equations are satisfied whenever $6xy(y + 2x) = 0$ and $6xy(x - 2y) = 0$. The point satisfying $y + 2x = 0$ and $x - 2y = 0$ is $z = 0$. The points that satisfy $6xy = 0$ are the points along the y-axis ($x = 0$) or along the x-axis ($y = 0$). The function f is differentiable but not analytic on either axis; there is no neighborhood about any point $z = x$ or $z = iy$ within which f is differentiable.

21. Since f is entire,
$$f'(z) = \frac{\partial u}{\partial x} + i\frac{\partial v}{\partial x} = e^x \cos y + ie^x \sin y = f(z).$$

24. $\dfrac{\partial^2 u}{\partial x^2} = 0$, $\dfrac{\partial^2 u}{\partial y^2} = 0$ gives $\dfrac{\partial^2 u}{\partial x^2} + \dfrac{\partial^2 u}{\partial y^2} = 0$. Thus u is harmonic. Now $\dfrac{\partial u}{\partial x} = 2 - 2y = \dfrac{\partial v}{\partial y}$ implies $v = 2y - y^2 + h(x)$, $\dfrac{\partial u}{\partial y} = -2x = -\dfrac{\partial v}{\partial x} = -h'(x)$ implies $h'(x) = 2x$ or $h(x) = x^2 + C$. Therefore $f(z) = 2x - 2xy + i(2y - y^2 + x^2 + C)$.

27. $\dfrac{\partial^2 u}{\partial x^2} = \dfrac{2y^2 - 2x^2}{(x^2 + y^2)^2}$, $\dfrac{\partial^2 u}{\partial y^2} = \dfrac{2x^2 - 2y^2}{(x^2 + y^2)^2}$ gives $\dfrac{\partial^2 u}{\partial x^2} + \dfrac{\partial^2 u}{\partial y^2} = 0$. Thus u is harmonic. Now $\dfrac{\partial u}{\partial x} = \dfrac{2x}{x^2 + y^2} = \dfrac{\partial v}{\partial y}$ implies $v = 2\tan^{-1}\dfrac{y}{x} + h(x)$, $\dfrac{\partial u}{\partial y} = \dfrac{2y}{x^2 + y^2} = -\dfrac{\partial v}{\partial x} = \dfrac{2y}{x^2 + y^2} - h'(x)$ implies $h'(x) = 0$ or $h(x) = C$.

Therefore $f(z) = \log_e(x^2 + y^2) + i\left(\tan^{-1}\dfrac{y}{x} + C\right), \quad z \neq 0.$

30. $f(x) = \dfrac{x}{x^2 + y^2} - i\dfrac{y}{x^2 + y^2}$. The level curves $u(x,y) = c_1$ and $v(x,y) = c_2$ are the family of circles $x = c_1(x^2 + y^2)$

and $-y = c_2(x^2 + y^2)$, with the exception that $(0,0)$ is not on the circumference of any circle.

3. $e^{-1+\frac{\pi}{4}i} = e^{-1}\cos\dfrac{\pi}{4} + ie^{-1}\sin\dfrac{\pi}{4} = e^{-1}\left(\dfrac{\sqrt{2}}{2} + \dfrac{\sqrt{2}}{2}i\right)$

6. $e^{-\pi+\frac{3\pi}{2}i} = e^{-\pi}\cos\dfrac{3\pi}{2} + ie^{-\pi}\sin\dfrac{3\pi}{2} = -e^{-\pi}i$

9. $e^{5i} = \cos 5 + i\sin 5 = 0.2837 - 0.9589i$

12. $e^{5+\frac{5\pi}{2}i} = e^5\cos\dfrac{5\pi}{2} + ie^5\sin\dfrac{5\pi}{2} = e^5i$

15. $e^{z^2} = e^{x^2-y^2+2xyi} = e^{x^2-y^2}\cos 2xy + ie^{x^2-y^2}\sin 2xy$

18. $\dfrac{e^{z_1}}{e^{z_2}} = \dfrac{e^{x_1}\cos y_1 + ie^{x_1}\sin y_1}{e^{x_2}\cos y_2 + ie^{x_2}\sin y_2} = \dfrac{(e^{x_1}\cos y_1 + ie^{x_1}\sin y_1)(e^{x_2}\cos y_2 - ie^{x_2}\sin y_2)}{e^{2x_2}}$

$= e^{x_1-x_2}[(\cos y_1\cos y_2 + \sin y_1\sin y_2) + i(\sin y_1\cos y_2 - \cos y_1\sin y_2)]$

$= e^{x_1-x_2}[\cos(y_1-y_2) + i\sin(y_1-y_2)] = e^{x_1-x_2+i(y_1-y_2)} = e^{(x_1+iy_1)-(x_2+iy_2)} = e^{z_1-z_2}$

21. $u = e^x\cos y, \quad v = -e^x\sin y; \quad \dfrac{\partial u}{\partial x} = e^x\cos y, \quad \dfrac{\partial v}{\partial y} = -e^x\cos y; \quad \dfrac{\partial u}{\partial y} = -e^x\sin y, \quad -\dfrac{\partial v}{\partial x} = e^x\sin y$

Since the Cauchy-Riemann equations are not satisfied at any point, f is nowhere analytic.

24. $\ln(-ei) = \log_e e + i\left(-\dfrac{\pi}{2} + 2n\pi\right) = 1 + \left(-\dfrac{\pi}{2} + 2n\pi\right)i$

27. $\ln(\sqrt{2} + \sqrt{6}i) = \log_e 2\sqrt{2} + i\left(\dfrac{\pi}{3} + 2n\pi\right) = 1.0397 + \left(\dfrac{\pi}{3} + 2n\pi\right)i$

30. $\text{Ln}(-e^3) = \log_e e^3 + \pi i = 3 + \pi i$

33. $\text{Ln}(1 + \sqrt{3}i)^5 = \text{Ln}(16 - 16\sqrt{3}i) = \log_e 32 - \dfrac{\pi}{3}i = 3.4657 - \dfrac{\pi}{3}i$

36. $\dfrac{1}{z} = \ln(-1) = \log_e 1 + i(\pi + 2n\pi) = (2n+1)\pi i$ and so $z = -\dfrac{i}{(2n+1)\pi}$.

39. $(-i)^{4i} = e^{4i\ln(-i)} = e^{4i[\log_e 1 + i(-\frac{\pi}{2}+2n\pi)]} = e^{(2-8n)\pi}$

42. $(1-i)^{2i} = e^{2i\ln(1-i)} = e^{2i[\log_e\sqrt{2}+i(-\frac{\pi}{4}+2n\pi)]} = e^{\frac{\pi}{2}-4n\pi}[\cos(\log_e 2) + i\sin(\log_e 2)] = e^{-4n\pi}[3.7004 + 3.0737i]$

45. If $z_1 = i$ and $z_2 = -1 + i$ then

$$\text{Ln}(z_1 z_2) = \text{Ln}(-1 - i) = \log_e\sqrt{2} - \dfrac{3\pi}{4}i,$$

whereas

$$\text{Ln}z_1 + \text{Ln}z_2 = \dfrac{\pi}{2}i + \left(\log_e\sqrt{2} + \dfrac{3\pi}{4}i\right) = \log_e\sqrt{2} + \dfrac{5\pi}{4}i.$$

48. (a) $(i^i)^2 = (e^{i\ln i})^2 = [e^{-(\frac{\pi}{2}+2n\pi)}]^2 = e^{-(\pi+4n\pi)}$ and $i^{2i} = e^{2i\ln i} = e^{-(\pi+4n\pi)}$

333

(b) $(i^2)^i = (-1)^i = e^{i\ln(-1)} = e^{-(\pi+2n\pi)}$, whereas $i^{2i} = e^{-(\pi+4n\pi)}$

EXERCISES 17.7
Trigonometric and Hyperbolic Functions

3. $\sin\left(\dfrac{\pi}{4} + i\right) = \sin\dfrac{\pi}{4}\cosh(1) + i\cos\dfrac{\pi}{4}\sinh(1) = 1.0911 + 0.8310i$

6. $\cot\left(\dfrac{\pi}{2} + 3i\right) = \dfrac{\cos(\frac{\pi}{2} + 3i)}{\sin(\frac{\pi}{2} + 3i)} = \dfrac{-i\sinh(3)}{\cosh(3)} = -0.9951i$

9. $\cosh(\pi i) = \cos(i(\pi i)) = \cos(-\pi) = \cos\pi = -1$

12. $\cosh(2 + 3i) = \cosh(2)\cos(3) + i\sinh(2)\sin(3) = -3.7245 + 0.5118i$

15. $\dfrac{e^{iz} - e^{-iz}}{2i} = 2$ gives $e^{2(iz)} - 4ie^{iz} - 1 = 0$. By the quadratic formula, $e^{iz} = 2i \pm \sqrt{3}\,i$ and so

$$iz = \ln[(2 \pm \sqrt{3})i]$$
$$z = -i\left[\log_e(2 \pm \sqrt{3}) + \left(\dfrac{\pi}{2} + 2n\pi\right)i\right] = \dfrac{\pi}{2} + 2n\pi - i\log_e(2 \pm \sqrt{3}), \quad n = 0, \pm1., \pm2, \ldots.$$

18. $\dfrac{e^z - e^{-z}}{2} = -1$ gives $e^{2z} + 2e^z - 1 = 0$. By the quadratic formula, $e^z = -1 \pm \sqrt{2}$, and so

$$z = \ln(-1 \pm \sqrt{2})$$
$$z = \log_e(\sqrt{2} - 1) + 2n\pi i \quad \text{or} \quad z = \log_e(\sqrt{2} + 1) + (\pi + 2n\pi)i,$$

$n = 0, \pm1, \pm2, \ldots.$

21. $\cos z = \cosh 2$ implies $\cos x\cosh y - i\sin x\sinh y = \cosh 2 + 0i$ and so we must have $\cos x\cosh y = \cosh 2$ and $\sin x\sinh y = 0$. The last equation has solutions $x = n\pi$, $n = 0, \pm1, \pm2, \ldots$, or $y = 0$. For $y = 0$ the first equation becomes $\cos x = \cosh 2$. Since $\cosh 2 > 1$ this equation has no solutions. For $x = n\pi$ the first equation becomes $(-1)^n\cosh y = \cosh 2$. Since $\cosh y > 0$ we see n must be even, say, $n = 2k$, $k = 0, \pm1, \pm2, \ldots$. Now $\cosh y = \cosh 2$ implies $y = \pm2$. Solutions of the original equation are then

$$z = 2k\pi \pm 2i, \quad k = 0, \pm1, \pm2, \ldots.$$

24. $\sinh z = \dfrac{e^{x+iy} - e^{-x-iy}}{2} = \dfrac{1}{2}(e^x e^{iy} - e^{-x}e^{-iy}) = \dfrac{1}{2}[e^x(\cos y + i\sin y) - e^{-x}(\cos y - i\sin y)]$

$\qquad = \left(\dfrac{e^x - e^{-x}}{2}\right)\cos y + i\left(\dfrac{e^x + e^{-x}}{2}\right)\sin y = \sinh x\cos y + i\cosh x\sin y$

27. $|\cosh z|^2 = \cosh^2 x\cos^2 y + \sinh^2 x\sin^2 y = (1 + \sinh^2 x)\cos^2 y + \sinh^2 x\sin^2 y$

$\qquad = \cos^2 y + \sinh^2 x(\cos^2 y + \sin^2 y) = \cos^2 y + \sinh^2 x$

30. $\tan z = \dfrac{\sin z}{\cos z} = \dfrac{\sin z \cos z}{|\cos z|^2} = \dfrac{[\sin x \cosh y + i \cos x \sinh y][\cos x \cosh y + i \sin x \sinh y]}{\cos^2 x + \sinh^2 y}$

$= \dfrac{(\sin x \cos x \cosh^2 y - \sin x \cos x \sinh^2 y)}{\cos^2 x + \sinh^2 y} + i\,\dfrac{\cos^2 x \sinh y \cosh y + \sin^2 x \sinh y \cosh y}{\cos^2 x + \sinh^2 y}$

$= \dfrac{\sin x \cos x (\cosh^2 y - \sinh^2 y)}{\cos^2 x + \sinh^2 y} + i\,\dfrac{\sin y \cosh y(\cos^2 x + \sin^2 x)}{\cos^2 x + \sinh^2 y}$

$= \dfrac{\sin x \cos x}{\cos^2 x + \sinh^2 y} + i\,\dfrac{\sinh y \cosh y}{\cos^2 x + \sinh^2 y} = \dfrac{\sin 2x}{2(\cos^2 x + \sinh^2 y)} + i\,\dfrac{\sinh 2y}{2(\cos^2 x + \sinh^2 y)}$

But

$$2\cos^2 x + 2\sinh^2 y = (2\cos^2 x - 1) + (2\sinh^2 y + 1) = \cos 2x + \cosh 2y.$$

Therefore $\tan z = u + iz$ where

$$u = \frac{\sin 2x}{\cos 2x + \cosh 2y}, \qquad v = \frac{\sinh 2y}{\cos 2x + \cosh 2y}.$$

EXERCISES 17.8
Inverse Trigonometric and Hyperbolic Functions

3. $\sin^{-1} 0 = -i \ln(\pm 1) = \begin{cases} 2n\pi + i \log_e 1 \\ (2n+1)\pi + i \log_e 1 \end{cases} = \begin{cases} 2n\pi \\ (2n+1)\pi \end{cases} = n\pi, \quad n = 0, \pm 1, \pm 2, \ldots$

6. $\cos^{-1} 2i = -i \ln[(2 \pm \sqrt{5})i] = \begin{cases} 2n\pi - \frac{\pi}{2} + i \log_e(2 + \sqrt{5}) \\ 2n\pi + \frac{\pi}{2} - i \log_e(2 + \sqrt{5}) \end{cases}, \quad n = 0, \pm 1, \pm 2, \ldots$

9. $\tan^{-1} 1 = \dfrac{i}{2} \ln \dfrac{i+1}{i-1} = \dfrac{i}{2} \ln(-i) = -n\pi + \dfrac{\pi}{4} + \dfrac{i}{2} \log_e 1 = \dfrac{\pi}{4} - n\pi, \quad n = 0, \pm 1, \pm 2, \ldots$

Note that this can also be written as $\tan^{-1} 1 = \frac{\pi}{4} + n\pi, \quad n = 0, \pm 1, \pm 2, \ldots$.

12. $\cosh^{-1} i = \ln[(1 + \pm\sqrt{2})i] = \begin{cases} \log_e(1 + \sqrt{2}) + (\frac{\pi}{2} + 2n\pi)i \\ \log_e(\sqrt{2} - 1) + (-\frac{\pi}{2} + 2n\pi)i \end{cases}, \quad n = 0, \pm 1, \pm 2, \ldots$

CHAPTER 17 REVIEW EXERCISES

3. $-7/25$

6. The closed annular region between the circles $|z + 2| = 1$ and $|z + 2| = 3$. These circles have center at $z = -2$.

9. $z = \ln(2i) = \log_e 2 + i\left(\dfrac{\pi}{2} + 2n\pi\right), \quad n = 0, \pm1, \pm2, \ldots$

12. $f(-1+i) = -33 + 26i$

15. $\operatorname{Ln}(-ie^3) = \log_e e^3 + \left(-\dfrac{\pi}{2}\right)i = 3 - \dfrac{\pi}{2}i$

18. $-\dfrac{1}{13} - \dfrac{17}{13}i$

21. The region satisfying $xy \le 1$ is shown in the figure.

24. The region satisfying $y < x$ is shown in the figure.

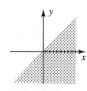

27. The four fourth roots of $1 - i$ are given by

$$w_R = 2^{1/8}\left[\cos\left(-\dfrac{\pi}{16} + \dfrac{k\pi}{2}\right) + i\sin\left(-\dfrac{\pi}{16} + \dfrac{k\pi}{2}\right)\right], \quad n = 0, 1, 2, 3$$

$$w_0 = 2^{1/8}\left[\cos\left(-\dfrac{\pi}{16}\right) + i\sin\left(-\dfrac{\pi}{16}\right)\right] = 1.0696 - 0.2127i$$

$$w_1 = 2^{1/8}\left[\cos\dfrac{7\pi}{16} + i\sin\dfrac{7\pi}{16}\right] = 0.2127 + 1.0696i$$

$$w_2 = 2^{1/8}\left[\cos\dfrac{15\pi}{16} + i\sin\dfrac{15\pi}{16}\right] = -1.0696 + 0.2127i$$

$$w_3 = 2^{1/8}\left[\cos\dfrac{23\pi}{16} + i\sin\dfrac{23\pi}{16}\right] = -0.2127 - 1.0696i$$

30. $\operatorname{Im}(z - 3\bar{z}) = 4y, \quad z\operatorname{Re}(z^2) = (x^3 - xy^2) + i(x^2y - y^3)$. Thus,

$$f(z) = (4y + x^3 - xy^2 - 5x) + i(x^2y - y^3 - 5y).$$

33. $z = z^{-1}$ gives $z^2 = 1$ or $(z-1)(z+1) = 0$. Thus $z = \pm 1$.

36. $z^2 = \bar{z}^2$ gives $xy = -xy$ or $xy = 0$. This implies $x = 0$ or $y = 0$. All real numbers $(y = 0)$ and all pure imaginary numbers $(x = 0)$ satisfy the equation.

39. $\operatorname{Ln}(1+i)(1-i) = \operatorname{Ln}(2) = \log_e 2; \quad \operatorname{Ln}(1+i) = \log_e \sqrt{2} + \dfrac{\pi}{4}i; \quad \operatorname{Ln}(1-i) = \log_e \sqrt{2} - \dfrac{\pi}{4}i.$
Therefore,

$$\operatorname{Ln}(1+i) + \operatorname{Ln}(1-i) = 2\log_e \sqrt{2} = \log_e 2 = \operatorname{Ln}(1+i)(1-i).$$

18 Integration in the Complex Plane

EXERCISES 18.1
Contour Integrals

3. $\displaystyle \int_C z^2 \, dz = (3+2i)^3 \int_{-2}^{2} t^2 \, dt = \frac{16}{3}(3+2i)^3 = -48 + \frac{736}{3}i$

6. $\displaystyle \int_C |z|^2 \, dz = \int_1^2 \left(2t^5 + \frac{2}{t}\right) dt - i \int_1^2 \left(t^2 + \frac{1}{t^4}\right) dt = 21 + \ln 4 - \frac{21}{8}i$

9. Using $y = -x+1$, $0 \le x \le 1$, $z = x + (-x+1)i$, $dz = (1-i)\,dx$,

$$\int_C (x^2 + iy^3) \, dz = (1-i) \int_1^0 [x^2 + (1-x)^3 i]\, dx = -\frac{7}{12} + \frac{1}{12}i.$$

12. $\displaystyle \int_C \sin z \, dz = \int_{C_1} \sin z \, dz + \int_{C_2} \sin z \, dz$ where C_1 and C_2 are the line segments $y = 0$, $0 \le x \le 1$, and $x = 1$, $0 \le y \le 1$, respectively. Now

$$\int_{C_1} \sin z \, dz = \int_0^1 \sin x \, dx = 1 - \cos 1$$

$$\int_{C_2} \sin z \, dz = i \int_0^1 \sin(1 + iy)\, dy = \cos 1 - \cos(1+i).$$

Thus

$$\int_C \sin z \, dz = (1 - \cos 1) + (\cos 1 - \cos(1+i)) = 1 - \cos(1+i) = (1 - \cos 1 \cosh 1) + i \sin 1 \sinh 1 = 0.1663 + 0.9889i.$$

15. We have $\displaystyle \oint_C z e^z \, dz = \int_{C_1} z e^z \, dz + \int_{C_2} z e^z \, dz + \int_{C_3} z e^z \, dz + \int_{C_4} z e^z \, dz$

On C_1, $y = 0$, $0 \le x \le 1$, $z = x$, $dz = dx$,

$$\int_{C_1} z e^z \, dz = \int_0^1 x e^x \, dx = x e^x - e^x \Big|_0^1 = 1.$$

On C_2, $x = 1$, $0 \le y \le 1$, $z = 1 + iy$, $dz = i\, dy$,

$$\int_{C_2} z e^z \, dz = i \int_0^1 (1 + iy) e^{1+iy} \, dy = i e^{i+1}.$$

On C_3, $y = 1$, $0 \le x \le 1$, $z = x + i$, $dz = dx$,

$$\int_{C_3} z e^z \, dz = \int_1^0 (x + i) e^{x+i} dx = (i - 1) e^i - i e^{1+i}.$$

337

On C_4, $x = 0$, $0 \leq y \leq 1$, $z = iy$, $dz = i\,dy$,

$$\int_{C_4} ze^z\,dz = -\int_1^0 ye^{iy}\,dy = (1-i)e^i - 1.$$

Thus

$$\oint_C ze^z\,dz = 1 + ie^{i+1} + (i-1)e^i - ie^{1+i} + (1-i)e^i - 1 = 0.$$

18. We have

$$\oint_C (2z-1)\,dz = \int_{C_1}(2z-1)\,dz + \int_{C_2}(2z-1)\,dz + \int_{C_3}(2z-1)\,dz$$

On C_1, $y = 0$, $0 \leq x \leq 1$, $z = x$, $dz = dx$,

$$\int_{C_1}(2z-1)\,dz = \int_0^1 (2x-1)\,dx = 0.$$

On C_2, $x = 1$, $0 \leq y \leq 1$, $z = 1 + iy$, $dz = i\,dy$,

$$\int_{C_2}(2z-1)\,dz = -2\int_0^1 y\,dy + i\int_0^1 dy = -1 + i.$$

On C_3, $y = x$, $z = x + ix$, $dz = (1+i)\,dx$,

$$\int_{C_3}(2z-1)\,dz = (1+i)\int_1^0 (2x - 1 + 2ix)\,dx = 1 - i.$$

Thus

$$\oint_C (2z-1)\,dz = 0 - 1 + i + 1 - i = 0.$$

21. On C, $y = -x + 1$, $0 \leq x \leq 1$, $z = x + (-x+1)i$, $dz = (1-i)\,dx$,

$$\int_C (z^2 - z + 2)\,dz = (1-i)\int_0^1 [x^2 - (1-x)^2 - x + 2 + (3x - 2x^2 - 1)i]\,dx = \frac{4}{3} - \frac{5}{3}i.$$

24. On C, $x = \sin t$, $y = \cos t$, $0 \leq t \leq \pi/2$ or $z = ie^{-it}$, $dz = e^{-it}\,dt$,

$$\int_C (z^2 - z + 2)\,dz = \int_0^{\pi/2}(-e^{-2it} - ie^{-it} + 2)e^{-it}\,dt = \int_0^{\pi/2}(-e^{-3it} - ie^{-2it} + 2e^{-it})\,dt$$

$$= -\frac{1}{3}ie^{-3\pi i/2} + \frac{1}{2}e^{-\pi i} + 2ie^{-\pi i/2} + \frac{1}{3}i - \frac{1}{2} - 2i = \frac{4}{3} - \frac{5}{3}i.$$

27. The length of the line segment from $z = 0$ to $z = 1 + i$ is $\sqrt{2}$. In addition, on this line segment

$$|z^2 + 4| \leq |z|^2 + 4 \leq |1 + i|^2 + 4 = 6.$$

Thus $\left| \int_C (z^2 + 4)\,dz \right| \leq 6\sqrt{2}$.

30. With $z_k^* = z_k$,

$$\int_C z\,dz = \lim_{\|P\|\to 0} \sum_{k=1}^n z_k(z_k - z_{k-1})$$

$$= \lim_{\|P\|\to 0} [(z_1^2 - z_1 z_0) + (z_2^2 - z_2 z_1) + \cdots + (z_n^2 - z_n z_{n-1})]. \tag{1}$$

With $z_k^* = z_{k-1}$,

$$\int_C z\,dz = \lim_{\|P\|\to 0} \sum_{k=1}^n z_{k-1}(z_k - z_{k-1})$$

$$= \lim_{\|P\|\to 0}[(z_0 z_1 - z_0^2) + (z_1 z_2 - z_1^2) + \cdots + (z_{n-1}z_n - z_{n-1}^2)]. \tag{2}$$

Adding **(1)** and **(2)** gives

$$2 \int_C z \, dz = \lim_{\|P\| \to 0} (z_n^2 - z_0^2) \quad \text{or} \quad \int_C z \, dz = \frac{1}{2}(z_n^2 - z_0^2).$$

33. For $f(z) = 2z$, $\overline{f(z)} = 2\bar{z}$, so on $z = e^{it}$, $\bar{z} = e^{-it}$, $dz = ie^{it} \, dt$, and

$$\oint_C \overline{f(z)} \, dz = \int_0^{2\pi} (e^{-it})(ie^{it} \, dt) = 2i \int_0^{2\pi} dt = 4\pi i.$$

Thus circulation $= \text{Re}\left(\oint_C \overline{f(z)} \, dz \right) = 0$, and net flux $= \text{Im}\left(\oint_C \overline{f(z)} \, dz \right) = 4\pi.$

EXERCISES 18.2
Cauchy-Goursat Theorem

3. $f(z) = \dfrac{z}{2z + 3}$ is discontinuous at $z = -3/2$ but is analytic within and on the circle $|z| = 1$.

6. $f(z) = \dfrac{e^z}{2z^2 + 11z + 15}$ is discontinuous at $z = -5/2$ and at $z = -3$ but is analytic within and on the circle $|z| = 1$.

9. By the principle of deformation of contours we can choose the more convenient circular contour C_1 defined by $|z| = 1$. Thus

$$\oint_C \frac{1}{z} \, dz = \oint_{C_1} \frac{1}{z} \, dz = 2\pi i$$

by (4) of Section 18.2.

12. By Theorem 18.4 and (4) of Section 18.2,

$$\oint_C \left(z + \frac{1}{z^2} \right) dz = \oint_C \frac{1}{z} \, dz + \oint_C \frac{1}{z^2} \, dz = 0 + 0 = 0.$$

15. By partial fractions, $\oint_C \dfrac{2z + 1}{z(z + 1)} \, dz = \oint_C \dfrac{1}{z} \, dz + \oint_C \dfrac{1}{z + 1} \, dz.$

(a) By Theorem 18.4 and (4) of Section 18.2,

$$\oint_C \frac{1}{z} \, dz + \oint_C \frac{1}{z + 1} \, dz = 2\pi i + 0 = 2\pi i.$$

(b) By writing $\oint_C = \oint_{C_1} + \oint_{C_2}$ where C_1 and C_2 are the circles $|z| = 1/2$ and $|z + 1| = 1/2$, respectively, we have by Theorem 18.4 and (4) of Section 18.2,

$$\oint_C \frac{1}{z} \, dz + \oint_C \frac{1}{z + 1} \, dz = \oint_{C_1} \frac{1}{z} \, dz + \oint_{C_1} \frac{1}{z + 1} \, dz + \oint_{C_2} \frac{1}{z} \, dz + \oint_{C_2} \frac{1}{z + 1} \, dz$$

$$= 2\pi i + 0 + 0 + 2\pi i = 4\pi i.$$

(c) Since $f(z) = \dfrac{2z+1}{z(z+1)}$ is analytic within and on C it follows from Theorem 18.4 that

$$\oint_C \frac{2z+1}{z^2+z}\, dz = 0.$$

18. (a) By writing $\displaystyle\oint_C = \oint_{C_1} + \oint_{C_2}$ where C_1 and C_2 are the circles $|z+2| = 1$ and $|z-2i| = 1$, respectively, we have by Theorem 18.4 and (4) of Section 18.2,

$$\oint_C \left(\frac{3}{z+2} - \frac{1}{z-2i} \right) dz = \oint_{C_1} \frac{3}{z+2}\, dz - \oint_{C_1} \frac{1}{z-2i}\, dz + \oint_{C_2} \frac{3}{z+2}\, dz - \oint_{C_2} \frac{1}{z-2i}\, dz$$

$$= 3(2\pi i) - 0 + 0 - 2\pi i = 4\pi i.$$

21. We have
$$\oint_C \frac{8z-3}{z^2-z}\, dz = \oint_{C_1} \frac{8z-3}{z^2-z}\, dz - \oint_{C_2} \frac{8z-3}{z^2-z}\, dz$$

where C_1 and C_2 are the closed portions of the curve C enclosing $z = 0$ and $z = 1$, respectively. By partial fractions, Theorem 18.4, and (4) of Section 18.2,

$$\oint_{C_1} \frac{8z-3}{z^2-z}\, dz = 5\oint_{C_1} \frac{1}{z-1}\, dz + 3\oint_{C_1} \frac{1}{z}\, dz = 5(0) + 3(2\pi i) = 6\pi i$$

$$\oint_{C_1} \frac{8z-3}{z^2-z}\, dz = 5\oint_{C_2} \frac{1}{z-1}\, dz + 3\oint_{C_2} \frac{1}{z}\, dz = 5(2\pi i) + 3(0) = 10\pi i.$$

Thus
$$\oint_C \frac{8z-3}{z^2-z}\, dz = 6\pi i - 10\pi i = -4\pi i.$$

24. Write
$$\oint_C (z^2 + z + \mathrm{Re}(z))\, dz = \oint_C (z^2 + z)\, dz + \oint_C \mathrm{Re}(z)\, dz.$$

By Theorem 18.4, $\displaystyle\oint_C (z^2 + z)\, dz = 0$. However, since $\mathrm{Re}(z) = x$ is not analytic,

$$\oint_C x\, dz = \oint_{C_1} x\, dz + \oint_{C_2} x\, dz + \oint_{C_3} x\, dz$$

where C_1 is $y = 0$, $0 \leq x \leq 1$, C_2 is $x = 1$, $0 \leq y \leq 2$, and C_3 is $y = 2x$, $0 \leq x \leq 1$. Thus,

$$\oint_C x\, dz = \int_0^1 x\, dx + i\int_0^2 dy + (1+2i)\int_1^0 x\, dx = \frac{1}{2} + 2i - \frac{1}{2}(1+2i) = i.$$

EXERCISES 18.3

Independence of Path

3. The given integral is independent of the path. Thus

$$\int_C 2z\,dz = \int_{-2+7i}^{2-i} 2z\,dz = z^2 \Big|_{-2+7i}^{2-i} = 48 + 24i.$$

6. $\int_{-2i}^{1}(3z^2 - 4z + 5i)\,dz = z^3 - 2z^2 + 5iz \Big|_{-2i}^{1} = -19 - 3i$

9. $\int_{-i/2}^{1-i}(2z+1)^2\,dz = \frac{1}{6}(2z+1)^3 \Big|_{-i/2}^{1-i} = -\frac{7}{6} - \frac{22}{3}i$

12. $\int_{1-i}^{1+2i} ze^{z^2}\,dz = \frac{1}{2}e^{z^2}\Big|_{1-i}^{1+2i} = \frac{1}{2}[e^{-3+4i} - e^{-2i}] = \frac{1}{2}(e^{-3}\cos 4 - \cos 2) + \frac{1}{2}(e^{-3}\sin 4 + \sin 2)i = 0.1918 + 0.4358i$

15. $\int_{\pi i}^{2\pi i}\cosh z\,dz = \sinh z\Big|_{\pi i}^{2\pi i} = \sinh 2\pi i - \sinh \pi i = i\sin 2\pi - i\sin \pi = 0$

18. $\int_{1+i}^{4+4i}\frac{1}{z}\,dz = \mathrm{Ln}\,z\Big|_{1+i}^{4+4i} = \mathrm{Ln}(4+4i) - \mathrm{Ln}(1+i) = \log_e 4\sqrt{2} + \frac{\pi}{4}i - \left(\log_e \sqrt{2} + \frac{\pi}{4}i\right) = \log_e 4 = 1.3863$

21. Integration by parts gives

$$\int e^z \cos z\,dz = \frac{1}{2}e^z(\cos z + \sin z) + C$$

and so

$$\int_{\pi}^{i} e^z \cos z\,dz = \frac{1}{2}e^z(\cos z + \sin z)\Big|_{\pi}^{i} = \frac{1}{2}[e^i(\cos i + \sin i) - e^{\pi}(\cos \pi + \sin \pi)]$$

$$= \frac{1}{2}[(\cos 1\cosh 1 - \sin 1\sinh 1 + e^{\pi}) + i(\cos 1\sinh 1 + \sin 1\cosh 1)] = 11.4928 + 0.9667i.$$

24. Integration by parts gives

$$\int z^2 e^z\,dz = z^2 e^z - 2ze^z + 2e^z + C$$

and so

$$\int_{0}^{\pi i} z^2 e^z\,dz = e^z(z^2 - 2z + 2)\Big|_{0}^{\pi i} = e^{\pi i}(-\pi^2 - 2\pi i + 2) - 2 = \pi^2 - 4 + 2\pi i.$$

EXERCISES 18.4

Cauchy's Integral Formulas

3. By Theorem 18.9 with $f(z) = e^z$,

$$\oint_C \frac{e^z}{z - \pi i}\, dz = 2\pi i e^{\pi i} = -2\pi i.$$

6. By Theorem 18.9 with $f(z) = \frac{1}{3}\cos z$,

$$\oint_C \frac{\frac{1}{3}\cos z}{z - \frac{\pi}{3}}\, dz = 2\pi i \left(\frac{1}{3}\cos\frac{\pi}{3}\right) = \frac{\pi}{3} i.$$

9. By Theorem 18.9 with $f(z) = \frac{z^2 + 4}{z - i}$,

$$\oint_C \frac{\frac{z^2 + 4}{z - i}}{z - 4i}\, dz = 2\pi i \left(-\frac{12}{3i}\right) = -8\pi.$$

12. By Theorem 18.10 with $f(z) = z$, $f'(z) = 1$, $f''(z) = 0$, and $f'''(z) = 0$,

$$\oint_C \frac{z}{(z - (-i))^4}\, dz = \frac{2\pi i}{3!}(0) = 0.$$

15. **(a)** By Theorem 18.9 with $f(z) = \frac{2z + 5}{z - 2}$,

$$\oint_C \frac{\frac{2z + 5}{z - 2}}{z}\, dz = 2\pi i \left(-\frac{5}{2}\right) = -5\pi i.$$

(b) Since the circle $|z - (-1)| = 2$ encloses only $z = 0$, the value of the integral is the same as in part **(a)**.

(c) From Theorem 18.9 with $f(z) = \frac{2z + 5}{z}$,

$$\oint_C \frac{\frac{2z + 5}{z}}{z - 2}\, dz = 2\pi i \left(\frac{9}{2}\right) = 9\pi i.$$

(d) Since the circle $|z - (-2i)| = 1$ encloses neither $z = 0$ nor $z = 2$ it follows from the Cauchy-Goursat Theorem, Theorem 18.4, that

$$\oint_C \frac{2z + 5}{z(z - 2)}\, dz = 0.$$

18. **(a)** By Theorem 18.10 with $f(z) = \frac{1}{z - 4}$, $f'(z) = -\frac{1}{(z - 4)^2}$, and $f''(z) = \frac{2}{(z - 4)^3}$,

$$\oint_C \frac{\frac{1}{z - 4}}{z^3}\, dz = \frac{2\pi i}{2!}\left(\frac{2}{-64}\right) = -\frac{\pi}{32} i.$$

(b) By the Cauchy-Goursat Theorem, Theorem 18.4,

$$\oint_C \frac{1}{z^3(z-4)}\, dz = 0.$$

21. We have

$$\oint_C \frac{1}{z^3(z-1)^2}\, dz = \oint_{C_1} \frac{\dfrac{1}{(z-1)^2}}{z^3}\, dz + \oint_{C_2} \frac{\dfrac{1}{z^3}}{(z-1)^2}\, dz$$

where C_1 and C_2 are the circles $|z| = 1/3$ and $|z-1| = 1/3$, respectively. By Theorem 18.10,

$$\oint_{C_1} \frac{\dfrac{1}{(z-1)^2}}{z^3}\, dz = \frac{2\pi i}{2!}(6) = 6\pi i, \qquad \oint_{C_2} \frac{\dfrac{1}{z^3}}{(z-1)^2}\, dz = \frac{2\pi i}{1!}(-3) = -6\pi i.$$

Thus

$$\oint_C \frac{1}{z^3(z-1)^2}\, dz = 6\pi i - 6\pi i = 0.$$

24. We have

$$\oint_C \frac{e^{iz}}{(z^2+1)^2}\, dz = \oint_{C_1} \frac{\dfrac{e^{iz}}{(z+i)^2}}{(z-i)^2}\, dz - \oint_{C_2} \frac{\dfrac{e^{iz}}{(z-i)^2}}{(z-(-i))^2}\, dz$$

where C_1 and C_2 are the closed portions of the curve C enclosing $z = i$ and $z = -i$, respectively. By Theorem 18.10,

$$\oint_{C_1} \frac{\dfrac{e^{iz}}{(z+i)^2}}{(z-i)^2}\, dz = \frac{2\pi i}{1!}\left(\frac{-4e^{-1}}{-8i}\right) = \pi e^{-1}, \qquad \oint_{C_2} \frac{\dfrac{e^{iz}}{(z-i)^2}}{(z-(-i))^2}\, dz = \frac{2\pi i}{1!}\left(\frac{0}{8i}\right) = 0.$$

Thus

$$\oint_C \frac{e^{iz}}{(z^2+1)^2}\, dz = \pi e^{-1}.$$

CHAPTER 18 REVIEW EXERCISES

3. True

6. $\pi(-16+8i)$

9. True (Use partial fractions and write the given integral as two integrals.)

12. 12π

15. $\displaystyle\int_C |z^2|\, dz = \int_0^2 (t^4 + t^2)\, dt + 2i\int_0^2 (t^5 + t^3)\, dt = \frac{136}{15} + \frac{88}{3}i$

18. $\displaystyle\int_{3i}^{1-i} (4z - 6)\, dz = 2z^2 - 6z \,\Big|_{3i}^{1-i} = 12 + 20i$

21. On $|z| = 1$, let $z = e^{it}$, $dz = ie^{it}\, dt$, so that

$$\oint_C (z^{-2} + z^{-1} + z + z^2)\, dz = i\int_0^{2\pi} (e^{-2it} + e^{-it} + e^{it} + e^{2it})e^{it}\, dt = -e^{-it} + it + \frac{1}{2}e^{2it} + \frac{1}{3}e^{3it}\Big|_0^{2\pi} = 2\pi i.$$

24. By Theorem 18.10 with $f(z) = \dfrac{\cos z}{z-1}$ and $f'(z) = \dfrac{\sin z - \cos z - z\sin z}{(z-1)^2}$,

$$\oint_C \frac{\dfrac{\cos z}{z-1}}{z^2}\, dz = \frac{2\pi i}{1!}\left(\frac{-1}{1}\right) = -2\pi i.$$

27. Using the principle of deformation of contours we choose C to be the more convenient circular contour $|z+i| = \frac{1}{4}$. On this circle $z = -i + \frac{1}{4}e^{it}$ and $dz = \frac{1}{4}ie^{it}\, dt$. Thus

$$\oint_C \frac{z}{z+i}\, dz = i\int_0^{2\pi} \left(\frac{1}{4}e^{it} - i\right) dt = 2\pi.$$

30. We have

$$\left|\int_C \text{Ln}(z+1)\, dz\right| \le |\text{max of Ln}(z+1) \text{ on } C| \cdot 2,$$

where 2 is the length of the line segment. Now

$$|\text{Ln}(z+1)| \le |\log_e(z+1)| + |\text{Arg}(z+1)|.$$

But $\max \text{Arg}(z+1) = \pi/4$ when $z = i$ and $\max |z+1| = \sqrt{10}$ when $z = 2+i$. Thus,

$$\left|\int_C \text{Ln}(z+1)\, dz\right| \le \left(\frac{1}{2}\log_e 10 + \frac{\pi}{4}\right)2 = \log_e 10 + \frac{\pi}{2}.$$

344

19 Series and Residues

3. 0, 2, 0, 2, 0

6. Converges. To see this write the general term as $\left(\dfrac{2}{5}\right)^n \dfrac{1 + n2^{-n}i}{1 + 3n5^{-n}i}$.

9. Diverges. To see this write the general term as $\sqrt{n}\left(1 + \dfrac{1}{\sqrt{n}}i^n\right)$.

12. Write $z_n = \left(\dfrac{1}{4} + \dfrac{1}{4}i\right)^n$ in polar form as $z_n = \left(\dfrac{\sqrt{2}}{4}\right)^n \cos n\theta + i\left(\dfrac{\sqrt{2}}{4}\right)^n \sin n\theta$. Now

$$\operatorname{Re}(z_n) = \left(\frac{\sqrt{2}}{4}\right)^n \cos n\theta \to 0 \text{ as } n \to \infty \quad \text{and} \quad \operatorname{Im}(z_n) = \left(\frac{\sqrt{2}}{4}\right)^n \sin n\theta \to 0 \text{ as } n \to \infty$$

since $\sqrt{2}/4 < 1$.

15. We identify $a = 1$ and $z = 1 - i$. Since $|z| = \sqrt{2} > 1$ the series is divergent.

18. We identify $a = 1/2$ and $z = i$. Since $|z| - 1$ the series is divergent.

21. From

$$\lim_{n \to \infty} \left| \frac{\dfrac{1}{(1 - 2i)^{n+2}}}{\dfrac{1}{(1 - 2i)^{n+1}}} \right| = \frac{1}{|1 - 2i|} = \frac{1}{\sqrt{5}}$$

we see that the radius of convergence is $R = \sqrt{5}$. The circle of convergence is $|z - 2i| = \sqrt{5}$.

24. From

$$\lim_{n \to \infty} \left| \frac{\dfrac{1}{(n+1)^2(3 + 4i)^{n+1}}}{\dfrac{1}{n^2(3 + 4i)^n}} \right| = \lim_{n \to \infty} \left(\frac{n}{n+1}\right)^2 \frac{1}{|3 + 4i|} = \frac{1}{5}$$

we see that the radius of convergence is $R = 5$. The circle of convergence is $|z + 3i| = 5$.

27. From

$$\lim_{n \to \infty} \sqrt[n]{\left| \frac{1}{5^{2n}} \right|} = \lim_{n \to \infty} \frac{1}{25} = \frac{1}{25}$$

we see that the radius of convergence is $R = 25$. The circle of convergence is $|z - 4 - 3i| = 25$.

30. (a) The circle of convergence is $|z| = 1$. Since the series of absolute values

$$\sum_{k=1}^{\infty} \left| \frac{z^k}{k^2} \right| = \sum_{k=1}^{\infty} \frac{|z|^k}{k^2} = \sum_{k=1}^{\infty} \frac{1}{k^2}$$

converges, the given series is absolutely convergent for every z on $|z| = 1$. Since absolute convergence implies convergence, the given series converges for all z on $|z| = 1$.

(b) The circle of convergence is $|z| = 1$. On the circle, $n|z|^n \to \infty$ as $n \to \infty$. This implies $nz^n \nrightarrow 0$ as $n \to \infty$. Thus by Theorem 19.3 the series is divergent for every z on the circle $|z| = 1$.

EXERCISES 19.2
Taylor Series

3. Differentiating $\dfrac{1}{1+2z} = 1 - 2z + 2^2 z^2 - 2^3 z^3 + \cdots$ gives $\dfrac{-2}{(1+2z)^2} = -2 + 2 \cdot 2^2 z - 3 \cdot 2^3 z^2 + \cdots$. Thus

$$\frac{1}{(1+2z)} = 1 - 2 \cdot (2z) + 3 \cdot (2z)^2 - \cdots = \sum_{k=1}^{\infty} (-1)^{k-1} k (2z)^{k-1} \text{ where } R = \frac{1}{2}.$$

6. Replacing z in $e^z = \displaystyle\sum_{k=0}^{\infty} \frac{z^k}{k!}$ by $-z^2$ and multiplying the result by z gives $ze^{-z^2} = \displaystyle\sum_{k=0}^{\infty} \frac{(-1)^k}{k!} z^{2k+1}$ where $R = \infty$.

9. Replacing z in $\cos z = \displaystyle\sum_{k=0}^{\infty} (-1)^k \frac{z^{2k}}{(2k)!}$ by $z/2$ gives $\cos \dfrac{z}{2} = \displaystyle\sum_{k=0}^{\infty} \frac{(-1)^k}{(2k)!} \left(\frac{z}{2}\right)^{2k}$ where $R = \infty$.

12. Using the identity $\cos z = \dfrac{1}{2}(1 + \cos 2z)$ and the series $\cos z = \displaystyle\sum_{k=0}^{\infty} (-1)^k \frac{z^{2k}}{(2k)!}$ gives

$$\cos^2 z = \frac{1}{2} + \frac{1}{2} \sum_{k=0}^{\infty} (-1)^k \frac{(2z)^{2k}}{(2k)!} = 1 + \sum_{k=1}^{\infty} (-1)^k \frac{2^{2k-1}}{(2k)!} z^{2k} \text{ where } R = \infty.$$

15. Using (5) of Section 19.1,

$$\frac{1}{3-z} = \frac{1}{3 - 2i - (z - 2i)} = \frac{1}{3 - 2i} \cdot \frac{1}{1 - \dfrac{z - 2i}{3 - 2i}}$$

$$= \frac{1}{3 - 2i} \left[1 + \frac{z - 2i}{3 - 2i} + \frac{(z - 2i)^2}{(3 - 2i)^2} + \frac{(z - 2i)^3}{(3 - 2i)^3} + \cdots \right]$$

$$= \frac{1}{3 - 2i} + \frac{z - 2i}{(3 - 2i)^2} + \frac{(z - 2i)^2}{(3 - 2i)^3} + \frac{(z - 2i)^3}{(3 - 2i)^4} + \cdots = \sum_{k=0}^{\infty} \frac{(z - 2i)^k}{(3 - 2i)^{k+1}} \text{ where } R = \sqrt{13}.$$

18. Using (5) of Section 19.1,

$$\frac{1+z}{1-z} = -1 + \frac{2}{1-z} = -1 + \frac{2}{1 - i - (z - i)} = -1 + \frac{2}{1 - i} \cdot \frac{1}{1 - \dfrac{z - i}{1 - i}}$$

$$= -1 + \frac{2}{1 - i} \left[1 + \frac{z - i}{1 - i} + \frac{(z - i)^2}{(1 - i)^2} + \frac{(z - i)^3}{(1 - i)^3} + \cdots \right]$$

$$= -1 + \frac{2}{1 - i} + \frac{2(z - i)}{(1 - i)^2} + \frac{2(z - i)^2}{(1 - i)^3} + \frac{2(z - i)^3}{(1 - i)^4} + \cdots = -1 + \sum_{k=0}^{\infty} \frac{2(z - i)^k}{(1 - i)^{k+1}} \text{ where } R = \sqrt{2}.$$

21. Using $e^z = e^{3i} \cdot e^{z-3i}$ and (12) of Section 19.2, $e^z = e^{3i} \sum_{k=0}^{\infty} \dfrac{(z-3i)^k}{k!}$ where $R = \infty$.

24. Using (8) of Section 19.2, $e^{1/(1+z)} = e - ez + \dfrac{3e}{2} z^2 - \cdots$.

27. The distance from $2 + 5i$ to i is $|2 + 5i - i| = |2 + 4i| = 2\sqrt{5}$.

30. The series are

$$f(z) = \sum_{k=0}^{\infty} (-1)^k \frac{(z-3)^k}{3^{k+1}} \text{ where } R = 3$$

and

$$f(z) = \sum_{k=0}^{\infty} (-1)^k \frac{(z-1-i)^k}{(1+i)^{k+1}} \text{ where } R = \sqrt{2}.$$

33. From $e^z \approx 1 + z + \dfrac{z^2}{2}$ we obtain

$$e^{(1+i)/10} \approx 1 + \frac{1+i}{10} + \frac{(1+i)^2}{100} = 1.1 + 0.12i.$$

36. $e^{iz} = \sum_{k=0}^{\infty} \dfrac{(iz)^k}{k!} = 1 + i\dfrac{z}{1!} - \dfrac{z^2}{2!} - i\dfrac{z^3}{3!} + \dfrac{z^4}{4!} + i\dfrac{z^5}{5!} - \dfrac{z^6}{6!} - i\dfrac{z^7}{7!} + \cdots$

$$= \left(1 - \frac{z^2}{2!} + \frac{z^4}{4!} - \frac{z^6}{6!} + \cdots\right) + i\left(\frac{z}{1!} - \frac{z^3}{3!} + \frac{z^5}{5!} - \frac{z^7}{7!} + \cdots\right) = \cos z + i \sin z$$

EXERCISES 19.3

Laurent Series

3. $f(z) = 1 - \dfrac{1}{1! z^2} + \dfrac{1}{2! z^4} - \dfrac{1}{3! z^6} + \cdots$

6. $f(z) = z\left(1 - \dfrac{1}{2! z^2} + \dfrac{1}{4! z^4} - \dfrac{1}{6! z^6} + \cdots\right) = z - \dfrac{1}{2! z} + \dfrac{1}{4! z^3} - \dfrac{1}{6! z^5} + \cdots$

9. $f(z) = \dfrac{1}{z-3} \cdot \dfrac{1}{3+z-3} = \dfrac{1}{3(z-3)} \cdot \dfrac{1}{1 + \dfrac{z-3}{3}} = \dfrac{1}{3(z-3)} \left[1 - \dfrac{z-3}{3} + \dfrac{(z-3)^2}{3^2} - \dfrac{(z-3)^3}{3^3} + \cdots\right]$

$$= \frac{1}{3(z-3)} - \frac{1}{3^2} + \frac{z-3}{3^3} - \frac{(z-3)^2}{3^4} + \cdots$$

12. $f(z) = \dfrac{1}{3}\left[\dfrac{1}{z-3} - \dfrac{1}{z}\right] = \dfrac{1}{3}\left[\dfrac{1}{-4+z+1} - \dfrac{1}{z+1-1}\right] = \dfrac{1}{3}\left[-\dfrac{1}{4} \cdot \dfrac{1}{1 - \dfrac{z+1}{4}} - \dfrac{1}{z+1} \cdot \dfrac{1}{1 - \dfrac{1}{z+1}}\right]$

$$= \frac{1}{3}\left[-\frac{1}{4}\left(1 + \frac{z+1}{4} + \frac{(z+1)^2}{4^2} + \frac{(z+1)^3}{4^3} + \cdots\right) - \frac{1}{z+1}\left(1 + \frac{1}{z+1} + \frac{1}{(z+1)^2} + \frac{1}{(z+1)^3} + \cdots\right)\right]$$

$$= \cdots - \frac{1}{(z+1)^2} - \frac{1}{z+1} - \frac{1}{12} - \frac{z+1}{3 \cdot 4^2} - \frac{(z+1)^2}{3 \cdot 4^3} - \cdots$$

15. $f(z) = \dfrac{1}{z-1} \cdot \dfrac{-1}{1-(z-1)} = \dfrac{-1}{z-1}[1+(z-1)+(z-1)^2+(z-1)^3+\cdots] = -\dfrac{1}{z-1}-1-(z-1)-(z-1)^2-\cdots$

18. $f(z) = \dfrac{1}{3(z+1)} + \dfrac{2}{3} \cdot \dfrac{1}{(z+1)-3} = \dfrac{1}{3(z+1)} + \dfrac{2}{3(z+1)} \cdot \dfrac{1}{1-\dfrac{3}{z+1}}$

$ = \dfrac{1}{3(z+1)} + \dfrac{2}{3(z+1)} \left(1 + \dfrac{3}{z+1} + \dfrac{3^2}{(z+1)^2} + \dfrac{3^3}{(z+1)^3} + \cdots \right)$

$ = \dfrac{1}{z+1} + \dfrac{2}{(z+1)^2} + \dfrac{2 \cdot 3}{(z+1)^3} + \dfrac{2 \cdot 3^2}{(z+1)^4} + \cdots$

21. $f(z) = \dfrac{1}{z}(1-z)^{-2} = \dfrac{1}{z}\left(1 + (-2)(-z) + \dfrac{(-2)(-3)}{z!}(-z)^2 + \dfrac{(-2)(-3)(-4)}{3!}(-z)^3 + \cdots \right) = \dfrac{1}{z} + 2 + 3z + 4z^2 + \cdots$

24. $f(z) = \dfrac{1}{(z-3)^3} \cdot \dfrac{-1}{1-(z-1)} = \dfrac{-1}{(z-1)^3}[1+(z-1)+(z-1)^2+(z-1)^3+\cdots]$

$ = -\dfrac{1}{(z-1)^3} - \dfrac{1}{(z-1)^2} - \dfrac{1}{z-1} - 1 - (z-1) - \cdots$

27. $f(z) = z + \dfrac{2}{z-2} = 1 + (z-1) + \dfrac{2}{-1+z-1} = 1 + (z-1) + \dfrac{2}{z-1} \cdot \dfrac{1}{1-\dfrac{1}{z-1}}$

$ = 1 + (z-1) + \dfrac{2}{z-1}\left(1 + \dfrac{1}{z-1} + \dfrac{1}{(z-1)^2} + \dfrac{1}{(z-1)^3} + \cdots \right) = \cdots + \dfrac{2}{(z-1)^2} + \dfrac{2}{z-1} + 1 + (z-1)$

EXERCISES 19.4
Zeros and Poles

3. Since $f(-2+i) = f'(-2+i) = 0$ and $f''(z) = 2$ for all z, $z = -2+i$ is a zero of order two.

6. Write $f(z) = (z^2+9)/z = (z-3i)(z+3i)/z$ to see that $3i$ and $-3i$ are zeros of f. Now $f'(z) = 1 - 9/z^2$ and $f'(3i) = f'(-3i) = 2 \neq 0$. This indicates that each zero is of order one.

9. From $$f(z) = z(1 - \cos z^2) = z\left(-\dfrac{z^4}{2!} + \dfrac{z^8}{4!} - \cdots \right) = z^5\left(-\dfrac{1}{2!} + \dfrac{z^4}{4!} - \cdots \right)$$

we see that $z = 0$ is a zero of order five.

12. From the series $e^z = -\displaystyle\sum_{k=0}^{\infty} \dfrac{(z-\pi i)^k}{k!}$ centered at πi and

$$f(z) = 1 - \pi i + z + e^z = 1 - \pi i + z + \left(-1 - \dfrac{z-\pi i}{1!} - \dfrac{(z-\pi i)^2}{2!} - \dfrac{(z-\pi i)^3}{3!} - \cdots \right)$$

$$= -\dfrac{(z-\pi i)^2}{2!} - \dfrac{(z-\pi i)^3}{3!} - \cdots = (z-\pi i)^2\left(-\dfrac{1}{2!} - \dfrac{z-\pi i}{3!} - \cdots \right)$$

we see that $z = \pi i$ is a zero of order two.

15. From $f(z) = \dfrac{1+4i}{(z+2)(z+i)^4}$ and Theorem 19.11 we see that -2 is a simple pole and $-i$ is a pole of order four.

18. From $z^2 \sin \pi z = z^3\left(\pi - \dfrac{\pi^3 z^2}{3!} + \cdots \right)$ we see $z = 0$ is a zero of order three. From $f(z) = \dfrac{\cos \pi z}{z^2 \sin \pi z}$ and Theorem 19.11 we see 0 is a pole of order three. The numbers n, $n \pm 1$, ± 2, ... are simple poles.

21. From $\;1 - e^z = 1 - \left(1 + \dfrac{z}{1!} + \dfrac{z^2}{2!} + \cdots\right) = z\left(-1 - \dfrac{z}{2!} - \cdots\right)\;$ we see that $z = 0$ is a zero of order one. By

periodicity of e^z it follows that $z = 2n\pi i$, $n = 0, \pm 1, \pm 2, \ldots$ are zeros of order one. From $\;f(z) = \dfrac{1}{1 - e^z}\;$ and

Theorem 19.11 we see that the numbers $2n\pi i$, $n = 0, \pm 1, \pm 2, \ldots$ are simple poles.

24. From the Laurent series

$$f(z) = z^3\left[\frac{1}{z} - \frac{1}{3!}\left(\frac{1}{z}\right)^3 + \frac{1}{5!}\left(\frac{1}{z}\right)^5 - \frac{1}{7!}\left(\frac{1}{z}\right)^7 + \cdots\right] = z^2 - \frac{1}{3!} + \frac{1}{5!z^2} - \frac{1}{7!z^4} + \cdots, \quad 0 < |z|,$$

we see that the principal part contains an infinite number of nonzero terms. Hence $z = 0$ is an essential singularity.

EXERCISES 19.5

Residues and Residue Theorem

3. $f(z) = -\dfrac{3}{z} - \dfrac{1}{z - 2} = -\dfrac{3}{z} + \dfrac{1}{2} \cdot \dfrac{1}{1 - \dfrac{z}{2}} = -\dfrac{3}{z} + \dfrac{1}{2}\left(1 + \dfrac{z}{2} + \dfrac{z^2}{2^2} + \dfrac{z^3}{2^3} + \cdots\right) = -\dfrac{3}{z} + \dfrac{1}{2} + \dfrac{z}{2^2} + \dfrac{z^2}{2^3} + \cdots$

$\operatorname{Res}(f(z), 0) = -3$

6. $f(z) = \dfrac{e^{-2}}{(z - 2)^2}\, e^{-(z-2)} = \dfrac{e^{-2}}{(z - 2)^2}\left(1 - \dfrac{z - 2}{1!} + \dfrac{(z - 2)^2}{2!} - \dfrac{(z - 2)^3}{3!} + \cdots\right)$

$\quad = \dfrac{e^{-2}}{(z - 2)^2} - \dfrac{e^{-2}}{z - 2} + \dfrac{e^{-2}}{2} - \dfrac{e^{-2}(z - 2)}{3!} + \cdots$

$\operatorname{Res}(f(z), 2) = -e^{-2}$

9. $\operatorname{Res}(f(z), 1) = \lim\limits_{z \to 1}(z - 1)\dfrac{1}{z^2(z + 2)(z - 1)} = \lim\limits_{z \to 1}\dfrac{1}{z^2(z + 2)} = \dfrac{1}{3}$

$\operatorname{Res}(f(z), -2) = \lim\limits_{z \to -2}(z + 2)\dfrac{1}{z^2(z + 2)(z - 1)} = \lim\limits_{z \to -2}\dfrac{1}{z^2(z - 1)} = -\dfrac{1}{12}$

$\operatorname{Res}(F(z), 0) = \dfrac{1}{1!}\lim\limits_{z \to 0}\dfrac{d}{dz}\left[z^2 \cdot \dfrac{1}{z^2(z + 2)(z - 1)}\right] = \lim\limits_{z \to 0}\dfrac{-2z - 1}{(z + 2)^2(z - 1)^2} = -\dfrac{1}{4}$

12. $\operatorname{Res}(f(z), -3) = \lim\limits_{z \to -3}(z + 3) \cdot \dfrac{2z - 1}{(z - 1)^4(z + 3)} = \lim\limits_{z \to -3}\dfrac{2z - 1}{(z - 1)^4} = -\dfrac{7}{256}$

$\operatorname{Res}(f(z), 1) = \dfrac{1}{3!}\lim\limits_{z \to 1}\dfrac{d^3}{dz^3}\left[(z - 1)^4 \cdot \dfrac{2z - 1}{(z - 1)^4(z + 3)}\right] = \dfrac{1}{6}\lim\limits_{z \to 1}\dfrac{-42}{(z + 3)^4} = -\dfrac{7}{256}$

15. Using $\dfrac{d}{dz}\cos z = -\sin z$ and the result in (4) in the text,

$$\operatorname{Res}\left(f(z), (2n + 1)\frac{\pi}{2}\right) = \frac{1}{-\sin z}\bigg|_{z = (2n+1)\frac{\pi}{2}} = \frac{1}{-\sin(2n + 1)\frac{\pi}{2}} = (-1)^{n+1}.$$

18. (a) $\displaystyle\oint_C \dfrac{z + 1}{z^2(z - 2i)}\, dz = 2\pi i\, \operatorname{Res}(f(z), 0) = \pi\left(-1 + \dfrac{1}{2}i\right)$

(b) $\displaystyle\oint_C \dfrac{z + 1}{z^2(z - 2i)}\, dz = 2\pi i\, \operatorname{Res}(f(z), 2i) = \pi\left(1 - \dfrac{1}{2}i\right)$

349

(c) $\displaystyle\oint_C \frac{z+1}{z^2(z-2i)}\,dz = 2\pi i[\operatorname{Res}(f(z),0) + \operatorname{Res}(f(z),2i)] = 2\pi i\left[\frac{1}{4} + \frac{1}{2}i + \left(-\frac{1}{4} - \frac{1}{2}i\right)\right] = 0$

21. $\displaystyle\oint_C \frac{1}{z^2 + 4z + 13}\,dz = 2\pi i\operatorname{Res}(f(z),-2+3i) = \frac{\pi}{3}$

24. $\displaystyle\oint_C \frac{z}{(z+1)(z^2+1)}\,dz = 2\pi i[\operatorname{Res}(f(z),i) + \operatorname{Res}(f(z),-i)] = 2\pi i\left[\frac{1}{4} - \frac{1}{4}i + \frac{1}{4} + \frac{1}{4}i\right] = \pi i$

27. $\displaystyle\oint_C \frac{\tan z}{z}\,dz = 2\pi i\operatorname{Res}\left(f(z),\frac{\pi}{2}\right) = -4i$. Note: $z=0$ is not a pole. See Example 1, Section 19.4.

30. $\displaystyle\oint_C \frac{2z-1}{z^2(z^3+1)}\,dz = 2\pi i\left[\operatorname{Res}(f(z),0) + \operatorname{Res}(f(z),-1) + \operatorname{Res}\left(f(z),\frac{1}{2} + \frac{\sqrt{3}}{2}i\right)\right]$

$$= 2\pi i\left[2 + (-1) + \left(-\frac{1}{2} - \frac{1}{6}\sqrt{3}\,i\right)\right] = \pi\left(\frac{\sqrt{3}}{3} + i\right)$$

EXERCISES 19.6

Evaluation of Real Integrals

3. $\displaystyle\int_0^{2\pi} \frac{\cos\theta}{3 + \sin\theta}\,d\theta = \oint_C \frac{z^2+1}{z(z^2 + 6iz - 1)}\,dz = 2\pi i[\operatorname{Res}(f(z),0) + \operatorname{Res}(f(z),-3+2\sqrt{2}\,i)] = 0$

6. $\displaystyle\int_0^{\pi} \frac{d\theta}{1 + \sin^2\theta} = \frac{1}{2}\int_0^{2\pi} \frac{d\theta}{1 + \sin^2\theta} = -\frac{2}{i}\oint_C \frac{z}{z^4 - 6z^2 + 1}\,dz$

$$= \left(-\frac{2}{i}\right)2\pi i[\operatorname{Res}(f(z),\sqrt{3-2\sqrt{2}}) + \operatorname{Res}(f(z),-\sqrt{3-2\sqrt{2}})] = \frac{\pi}{\sqrt{2}}$$

9. We use $\cos 2\theta = (z^2 + z^{-2})/2$.

$$\int_0^{2\pi} \frac{\cos 2\theta}{5 - 4\cos\theta}\,d\theta = \frac{i}{2}\oint_C \frac{z^4+1}{z^2(2z^2 - 5z + 2)}\,dz = \left(\frac{i}{2}\right)2\pi i\left[\operatorname{Res}(f(z),0) + \operatorname{Res}\left(f(z),\frac{1}{2}\right)\right] = \frac{\pi}{6}$$

12. $\displaystyle\int_{-\infty}^{\infty} \frac{1}{x^2 - 2x + 25}\,dx = 2\pi i\operatorname{Res}(f(z),1+2\sqrt{6}\,i) = \frac{\pi}{2\sqrt{6}}$

15. $\displaystyle\int_{-\infty}^{\infty} \frac{1}{(x^2+1)^3}\,dx = 2\pi i\operatorname{Res}(f(z),i) = \frac{3\pi}{8}$

18. $\displaystyle\int_{-\infty}^{\infty} \frac{dx}{(x^2+1)^2(x^2+9)} = 2\pi i[\operatorname{Res}(f(z),i) + \operatorname{Res}(f(z),3i)] = \frac{5\pi}{96}$

21. $\displaystyle\int_{-\infty}^{\infty} \frac{e^{ix}}{x^2+1}\,dx = 2\pi i\operatorname{Res}(f(z),i) = \pi e^{-1}$. Therefore, $\displaystyle\int_{-\infty}^{\infty} \frac{\cos x}{x^2+1}\,dx = \operatorname{Re}\left(\int_{-\infty}^{\infty} \frac{e^{ix}}{x^2+1}\,dx\right) = \pi e^{-1}$.

24. $\displaystyle\int_{-\infty}^{\infty} \frac{e^{ix}}{(x^2+4)^2}\,dx = 2\pi i\operatorname{Res}(f(z),2i) = \frac{3e^{-2}}{16}\pi;\quad \int_{-\infty}^{\infty} \frac{\cos x}{(x^2+4)^2}\,dx = \operatorname{Re}\left(\int_{-\infty}^{\infty} \frac{e^{ix}}{(x^2+4)^2}\,dx\right) = \frac{3e^{-2}}{16}\pi$.

Therefore,

$$\int_0^{\infty} \frac{\cos x}{(x^2+4)^2}\,dx = \frac{1}{2}\left(\frac{3e^{-2}}{16}\pi\right) = \frac{3e^{-2}}{32}\pi.$$

27. $\displaystyle\int_{-\infty}^{\infty} \frac{e^{2ix}}{x^4+1}\,dx = 2\pi i\left[\text{Res}\left(f(z), \frac{1}{\sqrt{2}} + \frac{1}{\sqrt{2}}\,i\right) + \text{Res}\left(f(z), -\frac{1}{\sqrt{2}} + \frac{1}{\sqrt{2}}\,i\right)\right]$

$\displaystyle \qquad\qquad = 2\pi i\left[\left(-\frac{\sqrt{2}}{8} - \frac{\sqrt{2}}{8}\,i\right)e^{(-\sqrt{2}+\sqrt{2}\,i)} + \left(\frac{\sqrt{2}}{8} - \frac{\sqrt{2}}{8}\,i\right)e^{(-\sqrt{2}-\sqrt{2}\,i)}\right]$

$\displaystyle \qquad\qquad = \pi e^{-\sqrt{2}}\left[\frac{\sqrt{2}}{2}\cos\sqrt{2} + \frac{\sqrt{2}}{2}\sin\sqrt{2}\right]$

$\displaystyle\int_{-\infty}^{\infty} \frac{\cos 2x}{x^4+1}\,dx = \text{Re}\left(\int_{-\infty}^{\infty} \frac{e^{2ix}}{x^4+1}\,dx\right) = \pi e^{-\sqrt{2}}\left[\frac{\sqrt{2}}{2}\cos\sqrt{2} + \frac{\sqrt{2}}{2}\sin\sqrt{2}\right]$

Therefore

$$\int_0^{\infty} \frac{\cos 2x}{x^4+1}\,dx = \pi e^{-\sqrt{2}}\frac{\sqrt{2}}{4}(\cos\sqrt{2} + \sin\sqrt{2}).$$

30. $\displaystyle\int_{-\infty}^{\infty} \frac{xe^{ix}}{(x^2+1)(x^2+4)}\,dx = 2\pi i[\text{Res}(f(z), i) + \text{Res}(f(z), 2i)] = 2\pi i\left[\frac{1}{6}e^{-1} - \frac{1}{6}e^{-2}\right] = \frac{\pi}{3}(e^{-1} - e^{-2})i;$

$\displaystyle\int_{-\infty}^{\infty} \frac{x\sin x}{(x^2+1)(x^2+4)}\,dx = \text{Im}\left(\int_{-\infty}^{\infty} \frac{xe^{ix}}{(x^2+1)(x^2+4)}\,dx\right) = \frac{\pi}{3}(e^{-1} - e^{-2}).$

Therefore,

$$\int_0^{\infty} \frac{x\sin x}{(x^2+1)(x^2+4)}\,dx = \frac{1}{2}\left[\frac{\pi}{3}(e^{-1} - e^{-2})\right] = \frac{\pi}{6}(e^{-1} - e^{-2}).$$

33. $\displaystyle\int_0^{\pi} \frac{d\theta}{(a+\cos\theta)^2} = \frac{1}{2}\int_0^{2\pi} \frac{d\theta}{(a+\cos\theta)^2} = \frac{2}{i}\oint_C \frac{z}{(z^2+2az+1)^2}\,dz$ $(C$ is $|z| = 1)$ $= \frac{2}{i}\oint_C \frac{z}{(z-r_1)^2(z-r_2)^2}\,dz$

where $r_1 = -a + \sqrt{a^2-1}$, $r_2 = -a - \sqrt{a^2-1}$. Now

$$\oint_C \frac{z}{(z-r_1)^2(z-r_2)^2}\,dz = 2\pi i\,\text{Res}(f(z), r_1) = 2\pi i\frac{a}{4(\sqrt{a^2-1})^3} = \frac{a\pi}{2(\sqrt{a^2-1})^3}\,i.$$

Thus,

$$\int_0^{\pi} \frac{d\theta}{(a+\cos\theta)^2} = \frac{2}{i}\cdot\frac{a\pi}{2(\sqrt{a^2-1})^3}\,i = \frac{a\pi}{(\sqrt{a^2-1})^3}.$$

When $a = 2$ we obtain

$$\int_0^{\pi} \frac{d\theta}{(2+\cos\theta)^2} = \frac{2\pi}{(\sqrt{3})^3} \qquad\text{and so}\qquad \int_0^{2\pi} \frac{d\theta}{(2+\cos\theta)^2} = \frac{4\pi}{3\sqrt{3}}.$$

36. Using the Fourier sine transform with respect to y the partial differential equation becomes $\dfrac{d^2 U}{dx^2} - \alpha^2 U = 0$

and so

$$U(x, \alpha) = c_1\cosh\alpha x + c_2\sinh\alpha x.$$

The boundary condition $u(0, y)$ becomes $U(0, \alpha) = 0$ and so $c_1 = 0$. Thus $U(x, \alpha) = c_2\sinh\alpha x$. Now to evaluate

$$U(\pi, \alpha) = \int_0^{\infty} \frac{2y}{y^4+4}\sin\alpha y\,dy = \int_{-\infty}^{\infty} \frac{y}{y^4+4}\sin\alpha y\,dy$$

we use the contour integral $\displaystyle\int_C \frac{ze^{i\alpha z}}{z^4+4}\,dz$ and

$$\int_{-\infty}^{\infty} \frac{xe^{i\alpha x}}{x^4+4}\,dx = 2\pi i[\text{Res}(f(z), 1+i) + \text{Res}(f(z), -1+i)] = 2\pi i\left[-\frac{1}{8}ie^{(-1+i)\alpha} + \frac{1}{8}ie^{(-1-i)\alpha}\right] = \frac{\pi}{2}(e^{-\alpha}\sin\alpha)i$$

$$\int_{-\infty}^{\infty} \frac{x \sin \alpha x}{x^4 + 4}\, dx = \text{Im}\left(\int_{-\infty}^{\infty} \frac{x e^{i\alpha x}}{x^4 + 4}\, dx\right) = \frac{\pi}{2} e^{-\alpha} \sin \alpha.$$

Finally, $U(\pi, \alpha) = \dfrac{\pi}{2} e^{-\alpha} \sin \alpha = c_2 \sinh \alpha \pi$ gives $c_2 = \dfrac{\pi}{2} \dfrac{e^{-\alpha} \sin \alpha}{\sinh \alpha \pi}$. Hence $U(x, \alpha) = \dfrac{\pi}{2} \dfrac{e^{-\alpha} \sin \alpha}{\sinh \alpha \pi} \sinh \alpha x$ and

$$u(x, y) = \int_0^{\infty} \frac{e^{-\alpha} \sin \alpha}{\sinh \alpha \pi} \sinh \alpha x \sin \alpha y\, d\alpha.$$

CHAPTER 19 REVIEW EXERCISES

3. False

6. True

9. $1/\pi$

12. False

15. $f(z) = \dfrac{1}{z^4}\left[1 - \left(1 + \dfrac{iz}{1!} + \dfrac{i^2 z^2}{2!} + \dfrac{i^3 z^3}{3!} + \dfrac{i^4 z^4}{4!} + \cdots\right)\right] = -\dfrac{i}{z^3} + \dfrac{1}{2! z^2} + \dfrac{i}{3! z} - \dfrac{1}{4!} - \dfrac{iz}{5!} + \cdots$

18. $\dfrac{1 - \cos z^2}{z^5} = \dfrac{1}{z^5}\left[1 - \left(1 - \dfrac{z^4}{2!} + \dfrac{z^8}{4!} - \dfrac{z^{12}}{6!} + \dfrac{z^{16}}{8!} - \cdots\right)\right] = \dfrac{1}{2! z} - \dfrac{z^3}{4!} + \dfrac{z^7}{6!} - \dfrac{z^{11}}{8!} + \cdots$

21. $\displaystyle\oint_C \dfrac{2z + 5}{z(z + 2)(z - 1)^4}\, dz = 2\pi i[\text{Res}(f(z), 0) + \text{Res}(f(z), -2)] = \dfrac{404}{81}\pi i$

24. $\displaystyle\oint_C \dfrac{z + 1}{\sinh z}\, dz = 2\pi i[\text{Res}(f(z), 0) + \text{Res}(f(z), \pi i)] = 2\pi i[1 + (-\pi i - 1)] = 2\pi^2$

27. $\displaystyle\oint_C \dfrac{1}{z(e^z - 1)}\, dz = 2\pi i\, \text{Res}(f(z), 0) = -\pi i.$ Note: $z = 0$ is a pole of order two, and so

$$\text{Res}(f(z), 0) = \lim_{z \to 0} \frac{d}{dz}\, z^2 \cdot \frac{1}{z^2\left(1 + \dfrac{z}{2!} + \dfrac{z^2}{3!} + \cdots\right)} = \lim_{z \to 0} -\frac{\left(\dfrac{1}{2!} + \dfrac{2z}{3!} + \cdots\right)}{\left(1 + \dfrac{z}{2!} + \dfrac{z^2}{3!} + \cdots\right)^2} = -\frac{1}{2}.$$

30. $\displaystyle\oint_C \csc \pi z\, dz = 2\pi i[\text{Res}(f(z), 0) + \text{Res}(f(z), 1) + \text{Res}(f(z), 2)] = 2\pi i\left[\dfrac{1}{\pi} + \left(-\dfrac{1}{\pi}\right) + \dfrac{1}{\pi}\right] = 2i$

33. $\displaystyle\int_0^{2\pi} \dfrac{\cos^2 \theta}{2 + \sin \theta}\, d\theta = \dfrac{1}{2}\oint_C \dfrac{z^4 + 2z^2 + 1}{z^2(z^2 + 4iz - 1)}\, dz$ (C is $|z| = 1$) $= \pi i[\text{Res}(f(z), 0) + \text{Res}(f(z), (-2 + \sqrt{3}\,)i)]$

$$= \pi i[-4i + 2\sqrt{3}\, i] = (4 - 2\sqrt{3}\,)\pi$$

[Note: The answer in the text is correct but not simplified.]

36. We have

$$Ce^{-a^2 z^2} e^{ibz}\, dz = \int_{-r}^{r} + \int_{C_1} + \int_{C_2} + \int_{C_3} = 0$$

352

by the Cauchy-Goursat Theorem. Therefore,

$$\int_{-r}^{r} = -\int_{C_1} - \int_{C_2} - \int_{C_3}.$$

Let C_1 and C_3 denote the vertical sides of the rectangle. By the ML-inequality, $\int_{C_1} \to 0$ and $\int_{C_3} \to 0$ as $r \to \infty$. On C_2, $z = x + \dfrac{b}{2a^2}\, i$, $-r \le x \le r$, $dz = dx$,

$$\int_{-\infty}^{\infty} e^{-ax^2} e^{ibx}\, dx = -\int_{\infty}^{-\infty} e^{-a^2\left(x+\frac{b}{2a^2}i\right)^2} e^{ib\left(x+\frac{b}{2a^2}i\right)}\, dx = \int_{-\infty}^{\infty} e^{-a^2 x^2} e^{-b^2/4a^2}\, dx$$

$$\int_{-\infty}^{\infty} e^{-ax^2} \left(\cos bx + i \sin bx\right) dx = e^{-b^2/4a^2} \int_{-\infty}^{\infty} e^{-a^2 x^2}\, dx.$$

Using the given value of $\displaystyle\int_{-\infty}^{\infty} e^{-a^2 x^2}\, dx$ and equating real and imaginary parts gives

$$\int_{-\infty}^{\infty} e^{-ax^2} \cos bx\, dx = \frac{\sqrt{\pi}}{a}\, e^{-b^2/4a^2} \qquad \text{and so} \qquad \int_{0}^{\infty} e^{-ax^2} \cos bx\, dx = \frac{\sqrt{\pi}}{2a}\, e^{-b^2/4a^2}.$$

20 Conformal Mappings

EXERCISES 20.1
Complex Functions as Mappings

3. For $w = z^2$, $u = x^2 - y^2$ and $v = 2xy$. If $xy = 1$, $v = 2$ and so the hyperbola $xy = 1$ is mapped onto the line $v = 2$.

6. If $\theta = \pi/4$, then $v = \theta = \pi/4$. In addition $u = \log_e r$ will vary from $-\infty$ to ∞. The image is therefore the horizontal line $v = \pi/4$.

9. For $w = e^z$, $u = e^x \cos y$ and $v = e^x \sin y$. Therefore if $e^x \cos y = 1$, $u = 1$. The curve $e^x \cos y = 1$ is mapped into the line $u = 1$. Since $v = \dfrac{\sin y}{\cos y} = \tan y$, v varies from $-\infty$ to ∞ and the image is the line $u = 1$.

12. For $w = \dfrac{1}{z}$, $u = \dfrac{x}{x^2 + y^2}$ and $v = \dfrac{-y}{x^2 + y^2}$. The line $y = 0$ is mapped to the line $v = 0$, and, from Problem 2, the line $y = 1$ is mapped onto the circle $|w + \frac{1}{2} i| = \frac{1}{2}$. Since $f(\frac{1}{2} i) = -2i$, the region $0 \le y \le 1$ is mapped onto the points in the half-plane $v \le 0$ which are on or outside the circle $|w + \frac{1}{2} i| = \frac{1}{2}$. (The image does not include the point $w = 0$.)

15. The mapping $w = z + 4i$ is a translation which maps the circle $|z| = 1$ to a circle of radius $r = 1$ and with center $w = 4i$. This circle may be described by $|w - 4i| = 1$.

18. Since $w = (1 + i)z = \sqrt{2} e^{i\pi/4} z$, the mapping is the composite of a rotation through $45°$ and a magnification by $\alpha = \sqrt{2}$. The image of the first quadrant is therefore the angular wedge $\pi/4 \le \operatorname{Arg} w \le 3\pi/4$.

21. We first let $z_1 = z - i$ to map the region $1 \le y \le 4$ to the region $0 \le y_1 \le 3$. We then let $w = e^{-i\pi/2} z_1$ to rotate this strip through $-90°$. Therefore $w = -i(z - i) = -iz - 1$ maps $1 \le y \le 4$ to the strip $0 \le u \le 3$.

24. The mapping $w = iz$ will rotate the strip $-1 \le x \le 1$ through $90°$ so that the strip $-1 \le v \le 1$ results.

27. By Example 1, Section 20.1, $z_1 = e^z$ maps the strip $0 \le y \le \pi$ onto the upper half-plane $y_1 \ge 0$, or $0 \le \operatorname{Arg} z_1 \le \pi$. The power function $w = z_1^{3/2}$ changes the opening of this wedge by a factor of $3/2$ so the wedge $0 \le \operatorname{Arg} w \le 3\pi/2$ results. The composite of these two mappings is $w = (e^z)^{3/2} = e^{3z/2}$.

30. The mapping $z_1 = -(z - \pi i)$ lowers R by π units in the vertical direction and then rotates the resulting region through $180°$. The image region R_1 is upper half-plane $y_1 \ge 0$. By Example 1, Section 20.1, $w = \operatorname{Ln} z_1$ maps R_1 onto the strip $0 \le v \le \pi$. The composite of these two mappings is $w = \operatorname{Ln}(\pi i - z)$.

EXERCISES 20.2

Conformal Mappings

3. $f'(z) = 1 + e^z$ and $1 + e^z = 0$ for $z = \pm i \pm 2n\pi i$. Therefore f is conformal except for $z = \pi i \pm 2n\pi i$.

6. The function $f(z) = \pi i - \frac{1}{2}[\mathrm{Ln}(z+1) + \mathrm{Ln}(z-1)]$ is analytic except on the branch cut $x - 1 \leq 0$ or $x \leq 1$, and

$$f'(z) = -\frac{1}{2}\left(\frac{1}{z+1} + \frac{1}{z-1}\right) = -\frac{z}{z^2 - 1}$$

is non-zero for $z \neq 0, \pm 1$. Therefore f is conformal except for $z = x$, $x \leq 1$.

9. $f(z) = (\sin z)^{1/4}$ is the composite of $z_1 = \sin z$ and $w = z_1^{1/4}$. The region $-\pi/2 \leq x \leq \pi/2$, $y \geq 0$ is mapped to the upper half-plane $y_2 \geq 0$ by $z_1 = \sin z$ (See Example 2) and the power function $w = z_1^{1/4}$ maps this upper half-plane to the angular wedge $0 \leq \mathrm{Arg}\, w \leq \pi/4$. The real interval $[-\pi/2, \pi/2]$ is first mapped to $[-1, 1]$ and then to the union of the line segments from $e^{i\frac{\pi}{4}}$ to 0 and 0 to 1. See the figures below.

12. Using C-3 $w = e^z$ maps R onto the target region R'. The image of AB is shown in the figure.

15. Using H-6, $z_1 = \dfrac{e^{\pi/z} + e^{-\pi/z}}{e^{\pi/z} - e^{-\pi/z}}$ maps R onto the upper half-plane $y_1 \geq 0$, and $w = z_1^{1/2}$

maps this half-plane onto the target region R'. Therefore

$$w = \left(\frac{e^{\pi/z} + e^{-\pi/z}}{e^{\pi/z} - e^{-\pi/z}}\right)^{1/2}$$

and the image of AB is shown in the figure.

18. Using E-9, $z_1 = \cosh z$ maps R onto the upper half-plane $y_1 \geq 0$. Using M-7, $w = z_1 + \mathrm{Ln}\, z_1 + 1$ maps this half-plane onto the target region R'. Therefore $w = \cosh z + \mathrm{Ln}(\cosh z) + 1$ and the image of AB is shown in the figure.

21. $f(z) = i\dfrac{1-z}{1+z}$, using H-1, and so $u = U(f(z)) = \dfrac{1}{\pi}\mathrm{Arg}\left(\dfrac{1-z}{1+z}\right)$. The solution may also be written as

$$u(x, y) = \frac{1}{\pi}\tan^{-1}\left(\frac{1 - x^2 - y^2}{2y}\right).$$

24. In this exercise, we find a conformal mapping $w = f(z)$ that maps the given region R onto the upper half-plane $v \geq 0$ and transfers the boundary conditions so that the resulting Dirichlet problem is as shown in the figure.

The mapping $z_1 = z^2$ maps R onto the region R_1 defined by $y_1 \geq 0$, $|z_1| \geq 1$ and shown in H-3, and $w = \dfrac{1}{2}\left(z_1 + \dfrac{1}{z_1}\right)$ maps R_1 onto the upper half-plane $v \geq 0$. Letting $c_0 = 5$,

$$u = \frac{5}{\pi}\left[\mathrm{Arg}\left(\frac{1}{2}\left[z^2 + \frac{1}{z^2}\right] - 1\right) - \mathrm{Arg}\left(\frac{1}{2}\left[z^2 + \frac{1}{z^2}\right] + 1\right)\right].$$

27. (a) If $u = \dfrac{\partial^2 \phi}{\partial x^2} + \dfrac{\partial^2 \phi}{\partial y^2}$,

$$\frac{\partial^2 u}{\partial x^2} + \frac{\partial^2 u}{\partial y^2} = \frac{\partial^4 \phi}{\partial x^4} + 2\frac{\partial^4 \phi}{\partial x^2 \partial y^2} + \frac{\partial^4 \phi}{\partial y^4} = 0$$

since ϕ is assumed to be biharmonic.

(b) If $g = u + iv$, then $\phi = \mathrm{Re}(\bar{z}g(z)) = xu + yv$.

$$\frac{\partial^2 \phi}{\partial x^2} = 3\frac{\partial u}{\partial x} + x\frac{\partial^2 u}{\partial x^2} + y\frac{\partial^2 v}{\partial x^2}$$

$$\frac{\partial^2 \phi}{\partial y^2} = 2\frac{\partial v}{\partial y} + x\frac{\partial^2 u}{\partial y^2} + y\frac{\partial^2 v}{\partial y^2}.$$

Since u and v are harmonic and $\dfrac{\partial u}{\partial x} = \dfrac{\partial v}{\partial y}$,

$$\frac{\partial^2 \phi}{\partial x^2} + \frac{\partial^2 \phi}{\partial y^2} = 2\frac{\partial u}{\partial x} + 2\frac{\partial v}{\partial y} = 4\frac{\partial u}{\partial x}.$$

Now $u_1 = \dfrac{\partial u}{\partial x}$ is also harmonic and so $\dfrac{\partial^2 u_1}{\partial x^2} + \dfrac{\partial^2 u_1}{\partial y^2} = 0$. But

$$\frac{\partial^2 u_1}{\partial x^2} + \frac{\partial^2 u_1}{\partial y^2} = \frac{1}{4}\left[\frac{\partial^4 \phi}{\partial x^4} + 2\frac{\partial^4 \phi}{\partial x^2 \partial y^2} + \frac{\partial^4 \phi}{\partial y^4}\right]$$

and so ϕ is biharmonic.

EXERCISES 20.3
Linear Fractional Transformations

3. (a) For $T(z) = \dfrac{z+1}{z-1}$, $T(0) = -1$, $T(1) = \infty$, and $T(\infty) = 1$.

(b) The circle $|z| = 1$ passes through the pole at $z = 1$ and so the image is a line. Since $T(-1) = 0$ and $T(i) = -i$, the image is the line $u = 0$. If $|z - 1| = 1$,

$$|w - 1| = \left|\frac{z+1}{z-1} - 1\right| = \frac{2}{|z-1|} = 2$$

and so the image is the circle $|w - 1| = 2$ in the w-plane.

(c) Since $T(0) = -1$, the image of the disk $|z| \leq 1$ is the half-plane $u \leq 0$.

6. $S^{-1}(T(z)) = \dfrac{az+b}{cz+d}$ where

$$\begin{bmatrix} a & b \\ c & d \end{bmatrix} = \mathrm{adj}\left(\begin{bmatrix} 2 & 1 \\ 1 & 1 \end{bmatrix}\right)\begin{bmatrix} i & 0 \\ 1 & -2i \end{bmatrix} = \begin{bmatrix} -1+i & 2i \\ 2-i & -4i \end{bmatrix}.$$

Therefore,

$$S^{-1}(T(z)) = \frac{(-1+i)z+2i}{(2-i)z-4i} \quad \text{and} \quad S^{-1}(w) = \frac{w-1}{-w+2}.$$

9. $T(z) = \dfrac{(z-z_1)(z_2-z_3)}{(z-z_3)(z_2-z_1)}$ maps z_1, z_2, z_3 to 0, 1, ∞. Therefore, $T(z) = \dfrac{(z+1)(-2)}{(z-2)(1)} = -2\dfrac{z+1}{z-2}$ maps -1, 0, 2 to 0, 1, ∞.

12. As in Exercise 11, $z = \dfrac{(w-1-i)(-1+i)}{(w-1+i)(-1-i)}$ and, solving for w, $w = \dfrac{2z-2}{(1+i)z-1+i}$ maps 0, 1, ∞ to $1+i$, 0, $1-i$.

15. Using the cross-ratio formula (7),

$$S(w) = \frac{(w+1)(-3)}{(w-3)(1)} = \frac{(z-1)(2i)}{(z+i)(i-1)} = T(z).$$

We can solve for w to obtain

$$w = 3\frac{(1+i)z+(1-i)}{(-3+5i)z-3-5i}.$$

Alternatively we can apply the matrix method to compute $w = S^{-1}(T(z))$.

18. The mapping $T(z) = \dfrac{1}{2}\dfrac{z+1}{z}$ maps -1, 1, 0 to 0, 1, ∞ and maps each of the two circles in R to lines since both circles pass through the pole at $z = 0$. Since $T(\frac{1}{2} + \frac{1}{2}i) = 1 - i$ and $T(1) = 1$, the circle $|z - \frac{1}{2}| = \frac{1}{2}$ is mapped onto the line $u = 1$. Likewise, the circle $|z + \frac{1}{2}| = \frac{1}{2}$ is mapped onto the line $u = 0$. The transferred boundary conditions are shown in the figure and $U(u, v) = u$ is the solution. The solution to the Dirichlet problem in Figure 20.38 is

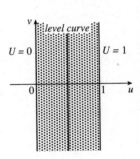

$$u = U(T(z)) = \mathrm{Re}\left(\frac{1}{2}\frac{z+1}{z}\right) = \frac{1}{2} + \frac{1}{2}\frac{x}{x^2+y^2}.$$

The level curves $u = c$ are the circles with centers on the x-axis which pass through the origin. The level curve $u = \frac{1}{2}$, however, is the vertical line $x = 0$.

21. $T_2(T_1(z)) = \dfrac{a_2 T_1(z) + b_2}{c_2 T_1(z) + d_2} = \dfrac{a_2\dfrac{a_1 z + b_1}{c_1 z + d_1} + b_2}{c_2\dfrac{a_1 z + b_1}{c_1 z + d_1} + d_2} = \dfrac{a_1 a_2 z + a_2 b_1 + b_2 c_1 z + b_2 d_1}{a_1 c_2 z + b_1 c_2 + c_1 d_2 z + d_1 d_2} = \dfrac{(a_1 a_2 + c_1 b_2)z + (b_1 a_2 + d_1 b_2)}{(a_1 c_2 + c_1 d_2)z + (b_1 c_2 + d_1 d_2)}$

EXERCISES 20.4

Schwarz-Christoffel Transformations

3. $\arg f'(t) = -\dfrac{1}{2}\mathrm{Arg}(t+1) + \dfrac{1}{2}\mathrm{Arg}(t-1) = \begin{cases} 0, & t < -1 \\ \pi/2, & -1 < t < 1 \\ 0, & t > 1 \end{cases}$

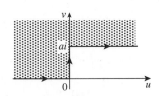

and $\alpha_1 = \pi/2$ and $\alpha_2 = 3\pi/2$. Since $f(-1) = 0$, the image of the upper half-plane is the region shown in the figure.

6. Since $\alpha_1 = \pi/3$ and $\alpha_2 = \pi/2$, $\alpha_1/\pi - 1 = -2/3$ and $\alpha_2/\pi - 1 = -1/2$ and so $f'(z) = A(z+1)^{-2/3}z^{-1/2}$ for some constant A.

9. Since $\alpha_1 = \alpha_2 = \pi/2$, $f'(z) = A(z+1)^{-1/2}(z-1)^{-1/2} = A/(z^2-1)^{1/2}$. Therefore, $f(z) = A\cosh^{-1} z + B$. But $f(-1) = \pi i$ and $f(1) = 0$. Since $\cosh^{-1} 1 = 0$, $B = 0$. Since $\cosh^{-1}(-1) = \pi i$, $\pi i = A(\pi i)$ and so $A = 1$. Hence $f(z) = \cosh^{-1} z$.

12. From (3), $f'(z) = Az^{-3/4}(z-1)^{(\alpha_2/\pi)-1}$. But $\alpha_2 \to \pi$ as $\theta \to 0$. This suggests that we examine $f'(z) = Az^{-3/4}$. Therefore, $f(z) = A_1 z^{1/4} + B_1$. But $f(0) = 0$ and $f(1) = 1$ so that $B_1 = 0$ and $A_1 = 1$. Hence $f(z) = z^{1/4}$ and we recognize that this power function maps the upper half-plane onto the wedge $0 \le \mathrm{Arg}\, w \le \pi/4$.

EXERCISES 20.5

Poisson Integral Formulas

3. The harmonic function

$$u_1 = \frac{5}{\pi}[\pi - \mathrm{Arg}(z-1)] = \begin{cases} 5, & x > 1 \\ 0, & x < 1 \end{cases}, \quad \text{and} \quad u_2 = -\frac{1}{\pi}\mathrm{Arg}\left(\frac{z+1}{z+2}\right) + \frac{1}{\pi}\mathrm{Arg}\left(\frac{z}{z+1}\right)$$

from (3) satisfies all boundary conditions except that $u_2 = 0$ for $x > 1$. Therefore $u = u_1 + u_2$ is the solution to the given Dirichlet problem.

6. From Theorem 20.5,

$$u(x,y) = \frac{y}{\pi}\int_{-\infty}^{\infty} \frac{\cos t}{(x-t)^2 + y^2}\, dt = \frac{y}{\pi}\int_{-\infty}^{\infty} \frac{\cos(x-s)}{s^2 + y^2}\, ds$$

letting $s = x - t$. But $\cos(x-s) = \cos x \cos x + \sin x \sin s$. It follows that

$$u(x,y) = \frac{y\cos x}{\pi}\int_{-\infty}^{\infty} \frac{\cos s}{s^2 + y^2}\, ds + \frac{y\sin x}{\pi}\int_{-\infty}^{\infty} \frac{\sin s}{s^2 + y^2}\, ds = \frac{y\cos x}{\pi}\left(\frac{\pi e^{-y}}{y}\right) = e^{-y}\cos x, \quad y > 0.$$

9. Using H-1, $f(z) = i\dfrac{1-z}{1+z}$ maps R onto the upper half-plane R'. The corresponding Dirichlet problem in R' is shown in the figure. From (3),

$$U = -\frac{1}{\pi}\operatorname{Arg}\left(\frac{w}{w+1}\right) + \frac{1}{\pi}\operatorname{Arg}\left(\frac{w-1}{w}\right) = \frac{1}{\pi}\left[\operatorname{Arg}\left(\frac{w-1}{w}\right) - \operatorname{Arg}\left(\frac{w}{w+1}\right)\right].$$

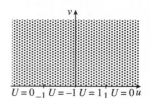

$U=0_{-1}\,U=-1\,U=1_1\,U=0\,u$

The harmonic function $u = U(f(z))$ may be simplified to

$$u = \frac{1}{\pi}\left[\operatorname{Arg}\left(\frac{(1-i)z - (1+i)}{1-z}\right) - \operatorname{Arg}\left(\frac{1-z}{-(1+i)z + 1 - i}\right)\right]$$

and is the solution to the original Dirichlet problem in R.

12. From Theorem 20.6, $u(x,y) = \dfrac{1}{2\pi}\displaystyle\int_{-\pi}^{\pi} e^{-|t|}\dfrac{1-|z|^2}{|e^{it}-z|^2}\,dt$. Therefore,

$$u(0,0) = \frac{1}{2\pi}\int_{-\pi}^{\pi} e^{-|t|}\,dt = \frac{1}{\pi}\int_0^{\pi} e^{-t}\,dt = \frac{1}{\pi}(1 - e^{-\pi}).$$

With the aid of Simpson's Rule, $u(0.5, 0) = 0.5128$ and $u(-0.5, 0) = 0.1623$.

15. For $u(e^{i\theta}) = \sin\theta + \cos\theta$, the Fourier series solution (6) reduces to
$$u(r,\theta) = r\sin\theta + r\cos\theta \quad \text{or} \quad u(x,y) = y + x.$$
The corresponding system of level curves is shown in the figure.

EXERCISES 20.6

Applications

3. $g(z) = \dfrac{x}{x^2+y^2} - \dfrac{y}{x^2+y^2}\,i = \dfrac{1}{z}$ is analytic for $z \neq 0$ and so div $\mathbf{F} = 0$ and curl $\mathbf{F} = 0$ by Theorem 20.7. A complex potential is $G(z) = \operatorname{Ln}z$ and

$$\phi(x,y) = \operatorname{Re}(G(z)) = \frac{1}{2}\log_e(x^2+y^2).$$

The equipotential lines $\phi(x,y) = c$ are circles $x^2 + y^2 = e^{2c}$ and are shown in the figure.

6. The function $f(z) = \dfrac{1}{z}$ maps the original region R to the strip $-\frac{1}{2} \le v \le 0$ (see Example 2, Section 20.1). The boundary conditions transfer as shown in the figure. $U = -2v$ is the solution in the horizontal strip and so

$$\phi(x, y) = -2\mathrm{Im}\left(\frac{1}{z}\right) = \frac{2y}{x^2 + y^2}$$

is the potential in the original region R. The equipotential lines $\dfrac{2y}{x^2 + y^2} = c$ may be written as

$$x^2 + \left(y + \frac{1}{c}\right)^2 = \left(\frac{1}{c}\right)^2 \quad \text{for } c \ne 0 \text{ and are circles. If } c = 0, \text{ we obtain the line } y = 0. \text{ Note that}$$

$\phi(x, y) = \mathrm{Re}\left(\dfrac{2i}{z}\right)$ and so $G(z) = \dfrac{2i}{z}$ is a complex potential. The corresponding vector field is

$$\mathbf{F} = \overline{G'(z)} = \frac{2i}{\bar{z}^2} = \left(\frac{-4xy}{(x^2 + y^2)^2}, \frac{2(x^2 - y^2)}{(x^2 + y^2)^2}\right).$$

9. (a) $\psi(x, y) = \mathrm{Im}(z^4) = 4xy(x^2 - y^2)$ and so $\psi(x, y) = 0$ when $y = x$ and $y = 0$.

(b) $\mathbf{V} = \overline{G'(z)} = \overline{4z^3} = 4(x^3 - 3xy^2, y^3 - 3x^2 y)$

(c) In polar coordinates $r^4 \sin 4\theta = c$ or $r = (c \csc 4\theta)^{1/4}$, for $0 < \theta < \pi/4$, are the streamlines. See the figure.

12. (a) The image of R under $w = i\sin^{-1} z$ is the horizontal strip (see E-6) $-\pi/2 \le v \le \pi/2$ and

$$\psi(x, y) = \mathrm{Im}(i\sin^{-1} z) = \begin{cases} \pi/2, & x \ge 1 \\ -\pi/2, & x \le -1 \end{cases}.$$

Each piece of boundary is therefore a streamline.

(b) $\mathbf{V} = \overline{G'(z)} = \overline{\dfrac{i}{(1 - z^2)^{1/2}}} = \dfrac{-i}{(1 - \bar{z}^2)^{1/2}}$

(c) The streamlines are the images of the lines $v = b$, $-\pi/2 < b < \pi/2$ under $z = -i\sin w$ and are therefore hyperbolas. See Example 2, Section 20.2, and the figure. Note that at $z = 0$, $v = -i$ and the flow is downward.

15. (a) For $f(z) = \pi i - \frac{1}{2}[\mathrm{Ln}(z + 1) + \mathrm{Ln}(z - 1)]$

$$f(t) = \pi i - \frac{1}{2}[\log_e |t + 1| + \log_e |t - 1| + i\mathrm{Arg}(t + 1) + i\mathrm{Arg}(t - 1)]$$

and so $\mathrm{Im}(f(t)) = \begin{cases} 0, & t < -1 \\ \pi/2, & -1 < t < 1 \\ \pi, & t > 1 \end{cases}$. Hence $\mathrm{Im}(G(z)) = \psi(x, y) = 0$ on the boundary of R.

(b) $x = -\dfrac{1}{2}[\log_e |t + 1 + ic| + \log_e |t - 1 + ic|], \quad y = \pi - \dfrac{1}{2}[\mathrm{Arg}(t + 1 + ic) + \mathrm{Arg}(t - 1 + ic)]$ for $c > 0$

(c)

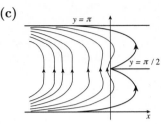

18. (a) For $f(z) = 2(z+1)^{1/2} + \text{Ln}\left(\dfrac{(z+1)^{1/2} - 1}{(z+1)^{1/2} + 1}\right),$

$$f(t) = 2(t+1)^{1/2} + \text{Ln}\frac{(t+1)^{1/2} - 1}{(t+1)^{1/2} + 1}.$$

If we write $(t+1)^{1/2} = |t+1|^{1/2}e^{(i/2)\text{Arg}(t+1)}$, we may conclude that

$$\text{Im}(f(t)) = \begin{cases} 0, & t > 0 \\ \pi, & -1 < t < 0 \end{cases} \quad \text{and} \quad \text{Re}(f(t)) = 0 \text{ for } t < -1.$$

Therefore $\text{Im}(G(z)) = \psi(x, y) = 0$ on the boundary of R.

(b) $x = \text{Re}\left[2(t + ic + 1)^{1/2} + \text{Ln}\dfrac{(t+ic+1)^{1/2} - 1}{(t+ic+1)^{1/2} + 1}\right]$

$y = \text{Im}\left[2(t + ic + 1)^{1/2} + \text{Ln}\dfrac{(t+ic+1)^{1/2} - 1}{(t+ic+1)^{1/2} + 1}\right]$ for $c > 0$

(c)

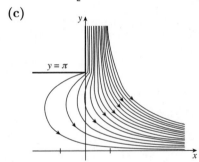

21. $f(z) = z^2$ maps the first quadrant onto the upper half-plane and $f(\xi_0) = f(1) = 1$. Therefore $G(z) = \text{Ln}(z^2 - 1)$ is the complex potential, and so

$$\psi(x, y) = \text{Arg}(z^2 - 1) = \tan^{-1}\left(\frac{2xy}{x^2 - y^2 - 1}\right)$$

is the streamline function where $\tan^{-1}$ is chosen to be between 0 and π. If $\psi(x, y) = c$, then $x^2 + Bxy - y^2 - 1 = 0$ where $B = -2\cot c$. Each hyperbola in the family passes through $(1, 0)$ and the boundary of the first quadrant satisfies $\psi(x, y) = 0$.

24. (a) $\mathbf{V} = \dfrac{a + ib}{\bar{z}} = \left(\dfrac{ax - by}{x^2 + y^2}, \dfrac{bx + ay}{x^2 + y^2}\right)$ and since $(x'(t), y'(t)) = \mathbf{V}$, the path of the particle satisfies

$$\frac{dx}{dt} = \frac{ax - by}{x^2 + y^2}, \qquad \frac{dy}{dt} = \frac{bx + ay}{x^2 + y^2}.$$

(b) Switching to polar coordinates,

$$\frac{dr}{dt} = \frac{1}{r}\left(x\frac{dx}{dt} + y\frac{dy}{dt}\right) = \frac{1}{r}\left(\frac{ax^2 - bxy}{r^2} + \frac{bxy + ay^2}{r^2}\right) = \frac{a}{r}$$

$$\frac{d\theta}{dt} = \frac{1}{r^2}\left(-y\frac{dx}{dt} + x\frac{dy}{dt}\right) = \frac{1}{r^2}\left(\frac{-axy + by^2}{r^2} + \frac{bx^2 + axy}{r^2}\right) = \frac{b}{r^2}.$$

Therefore $\dfrac{dr}{d\theta} = \dfrac{a}{b} r$ and so $r = ce^{a\theta/b}$.

(c) $\dfrac{dr}{dt} = \dfrac{a}{r} < 0$ if and only if $a < 0$, and in this case r is decreasing and the curve spirals inward. $\dfrac{d\theta}{dt} = \dfrac{b}{r^2} < 0$

if and only if $b < 0$, and in this case θ is decreasing and the curve is traversed clockwise.

CHAPTER 20 REVIEW EXERCISES

3. The wedge $0 \le \operatorname{Arg} w \le 2\pi/3$. See figure 20.6 in the text.

6. A line, since $|z - 1| = 1$ passes through the pole at $z = 2$.

9. False. $\overline{g(z)} = P - iQ$ is analytic. See Theorem 20.7.

12. First use $z_1 = z^2$ to map the first quadrant onto the upper half-plane $y_1 \ge 0$, and segment AB to the negative real axis. We then use $w = \frac{1}{\pi}(z_1 + \operatorname{Ln} z_1 + 1)$ to map this half-plane onto the target region R'. The composite transformation is

$$w = \frac{1}{\pi}[z^2 + \operatorname{Ln}(z^2) + 1]$$

and the image of AB is the ray extending to the left from $w = i$ along the line $v = 1$.

15. The inversion $w = 1/z$ maps R onto the horizontal strip $-1 \le v \le -1/2$ and the transferred boundary conditions are shown in the figure. The solution in the strip is $U = 2v + 2$ and so

$$u = U\left(\frac{1}{z}\right) = 2\operatorname{Im}\left(\frac{1}{z}\right) + 2 = \frac{-2y}{x^2 + y^2} + 2$$

is the solution to the original Dirichlet problem in R.

18. (a) From Theorem 20.5,

$$u(x, y) = \frac{y}{\pi} \int_{-\infty}^{\infty} \frac{\sin t}{(x - t)^2 + y^2}\, dt = \frac{y}{\pi} \int_{-\infty}^{\infty} \frac{\sin(x - s)}{s^2 + y^2}]\, ds \quad (\text{letting } s = x - t).$$

But $\sin(x - s) = \sin x \cos s - \cos x \sin s$. We now proceed as in the solution to Problem 6, Section 20.5 to show that $u(x, y) = e^{-y} \sin x$.

(b) For $u(e^{i\theta}) = \sin \theta$, the Fourier Series solution (6) in Section 20.5 reduces to $u(r, \theta) = r \sin \theta$.